How Humankind Created Science

Falin Chen · Fang-Tzu Hsu

How Humankind Created Science

From Early Astronomy to Our Modern Scientific Worldview

 Springer

Falin Chen
Institute of Applied Mechanics
National Taiwan University
Taipei, Taiwan

Fang-Tzu Hsu
School of Medicine
National Taiwan University
Taipei, Taiwan

ISBN 978-3-030-43134-1   ISBN 978-3-030-43135-8   (eBook)
https://doi.org/10.1007/978-3-030-43135-8

This Springer imprint is published by the registered company Springer Nature Switzerland AG
The registered company address is: Gewerbestrasse 11, 6330 Cham, Switzerland

Preface

Have you ever found yourself looking up at the night sky in search of your favorite stars? For those living in smog-covered cities drenched in light pollution, this search is much like happening upon a school of rare fish roaming the open ocean—there is very little chance of you coming across your target in such vastness. But if you were to head to the wilderness and gaze into the clear night sky, you would see a galaxy of stars strewn across the heavens; a grand expanse stretched over thousands of miles. Such majesty is a visceral reminder of just how small our world is. It was under these star-filled skies that ancient people living in vast open spaces could not help but explore the mysteries of the cosmos. Their efforts resulted in remarkable achievements that we can still admire today. Our ancestors did not just study the heavens, but put their cosmic knowledge to use as the basis for agricultural cycles, thus formulating calendar systems to gauge nature's rhythmical yearly course. This was humanity's earliest definition of time—a concept completely abstract, but key in determining the nature of our universe. And so, astronomy is considered the very first science advanced by the human race.

This dazzling star-studded sky appears like an abstract painting elucidating four-dimensional information across space and time, bringing generations of scientists, philosophers and poets to show their admiration for its inexplicable beauty. To this day, many of the theories in physics we know of are an accumulation of knowledge stemming from changes observed in the stars above. But what kind of changes did the scientists responsible for this wisdom see? What questions did they think of? What methods did they use? What difficulties did they come across? And what kind of persecution might they have faced on the road to discovering these wonderful, sometimes even mystical, ideas? This book's very purpose is to investigate these questions. It will take you through the stories behind scientific advancements in order to explain their theories, and derive associated examples and propositions. By observing how these theories, propositions and examples appear one after the other, we can get a closer look at what scientists experienced when exploring the natural world, and understand both the substance of these scientific theories and the influence they had. These propositions and examples can also show us real-life instances of scientific theories applied to natural phenomena or commonplace technology.

Whenever we delve into stories about scientific progress or the birth of scientific theories, a closer look at how these achievements were made often leaves us in a state of astonishment. Take, for example, the case of a young graduate, who after offending one of his professors, was unable to find work and had to rely on tutoring to support his wife and daughter. What set the wheels in motion for this young graduate, Albert Einstein, to later embark on his theory of relativity? He had originally intended to go in a different direction—to explore why electrons in an electric field develop a different effective mass when moving parallel or perpendicular to the field.

Einstein took on a unique way of thinking, abandoning typical views held about the study of electricity and magnetism prevailed at the time. He instead began making deductions based on the most basic definition of time and space. This made it possible for Einstein to eventually propose his formula for mass-energy equivalence, unraveling the mystery behind the enormity of energy stored in an object's mass, and giving the world his theory of special relativity.

This, however, was not Einstein's ultimate goal. He could not stop thinking of just where gravity comes from—an issue that Newton had never managed to explain clearly in his law of universal gravitation. When tackling this issue, Einstein's line of thinking regarding time and space was the same as it had been for his mass-energy equivalence formula but for one thing—this time he needed to find four-dimensional mathematical operations that could deal with distortions in spacetime. It took nearly ten years of searching, but Einstein was eventually able to find an equation that could propose rules explaining how celestial bodies move. This led him to the discovery that gravity is not a force exerted by recoiling springs or wrenching muscles, nor is it a result of air or liquid pressure, but is caused by distortions in time and space, with the extent of its force connected to the curvature of spacetime. This was the guiding idea behind Einstein's general theory of relativity.

The equation derived from this theory opened up a number of fields to unprecedented scientific inquiry, such as gravitational waves, gravitational lenses and black holes among others. The hundred years following the general theory of relativity have seen physicists clambering to explore the possibilities Einstein presented them. Each time one manages to prove anything theorized in Einstein's equations, they might as well start preparing an acceptance speech for the Nobel Prize in Physics.

We next turn our attention to the eternal bachelor, the lonely academic: Isaac Newton. How is it that the survivor of premature birth, who had never set his eyes on the sea but could explain how its tides worked, would go on to develop the law of universal gravitation?

It all started with classes at Cambridge being suspended after an outbreak of bubonic plague and its scholars leaving the campus to avoid infection. Newton was on his way back to his countryside home when he turned his head skywards and noticed that the way the moon hangs in the night sky seems quite different to the way an apple falls from a tree. In order to find answers to this enigma that only a mind like Newton's would gravitate toward, he had to stand on the shoulders of a scientific giant—Johannes Kepler. With Kepler's laws of planetary motion, Newton could prove that the snail-like movement of this heavenly body along Kepler's elliptical orbit is precisely the result of the Sun's gravitational force. The principle behind the Moon's orbit was, in fact, the very same as that governing how an apple falls to the ground.

After conceiving of his law in its embryonic form, Newton stashed his findings away in a drawer for twenty years. It was only when Edmond Halley, filled with academic zeal, approached him with questions regarding how comets move in orbit that Newton awoke to the significance of his discovery. He thus dedicated a further two years of his life to the laws of gravity, working eighteen painstaking hours a day and using every conceivable geometric tool at his disposal. Finally, by incorporating the infinitesimal calculus that he himself had conceived of, Newton put forward the law of universal gravitation. This law, in all its elaborate detail, was later inscribed in Newton's most emblematic work: the *Mathematical Principles of Natural Philosophy*.

The emergence of his gravitational law meant that disputes over astronomy that had been raging for millennia were promptly flung out from the halls of academia, replaced with a mathematical equation based on the masses of and the distance between two objects. Newton's discovery showed the world that the universe creator, if there is one, must be a bit of a joker. He (or She) needs a good sense of humor to create a universe in which matter and motion are governed by such a simple equation and one that can be applied to both on earth and in space to boot.

Before Newton and Einstein, there were, of course, numerous other pioneers these two considered scientific giants, who's intellectual odysseys gave rise to vital contributions to our understanding of nature. Giants like Copernicus, the man who opened the door for the astronomical revolution, put his academic integrity on the line in placing the Sun at the center of the universe. He could not bear the cosmic chaos brought on by the ancient Greek scholar Ptolemy's system with more than 80 epicycles jammed into the heavens. Supported by the focus of Neo-Platonism placed on using mathematics to interpret the natural world, Copernicus combined geometry and astronomical data to create a model of the universe with the Sun at its center. With this heliocentric system, he could explain a variety of planetary movements without the need to rely on any of Ptolemy's constructs of epicycles, deferents or equants. From this point on the Earth was removed from its central position, placed behind Mercury and Venus in the pecking order, third on the planetary podium.

And then there was Kepler whose Protestant faith made him a target of the Catholic Church, forcing him into a nomadic lifestyle. He maintained that theories defining how the universe works must be both simple and harmonious—an opinion brought on by his steadfast reverence of God's wisdom. And so, when inheriting Tycho's treasured collection of astronomic measurements and data, Kepler used the variations in the relative positions of the Sun, Earth and Mars to make a model of their orbits. He used over 1000 pages, repeatedly drafting calculations of 19 different orbits in his simulation. After five years of continuous calculations and comparisons, Kepler finally confirmed that Mars travels around the Sun, its focal point, on an elliptical orbit. When the red planet gets closer to the Sun, it moves at a faster pace; when further away, it slows down. He even managed to propose the first quantitative relational equation in all physics: the square of Mars's orbital period is proportional to the cube of its orbit's semi-major axis.

There was also Galilei, who had before earned additional income using the constellations to divine fitting times for medical students to begin bloodletting operations, and invented a refracting telescope to aid naval officers in sighting distant enemy ships. This telescope brought Galilei both wealth and fame, and also gave him a tool with which to observe the night sky and its wandering stars. This led him to the discovery that Jupiter is orbited by four satellites, giving the universe yet another potential center and gob-smacking those in the two opposing schools of thought still arguing over the geocentric and heliocentric models. It was Galilei's immense talent and the obstinate nature that came with it that brought him to offend the Roman Catholic Church with his defense of heliocentrism. In the end, he was sentenced to a life of imprisonment by the Roman Inquisition and forced to publicly declare his research untrue, as well as write daily confessions.

The above-mentioned scientists were all organized and critical thinkers, who, when faced with any challenges would investigate them to their logical conclusions. Their stories often entail wisdom worthy of our admiration along with mind-blowing discoveries, while their achievements have had a far-reaching impact on scientific progress. Inevitably, a desire to understand the laws of nature they uncovered and their associated theories often wells up inside us. If we were to apply these theories to natural phenomena that have consistently attracted humanity's attention, there would surely be endless curiosity over what might be uncovered next.

Attempting to satisfy this craving for knowledge is a key objective of this book. And so it is structured as follows: stories, theories, examples and propositions are combined in a timeline covering 3,000 years of the scientific laws discovered and passed down by these critical thinkers. Cases from the past will be reexamined using modern mathematics, while questions arising from the confirmation of past theories will be proposed, and a reflection-heavy approach will be taken to expose the very essence of these scientific laws. Today the scientists who gave us these discoveries are long gone, looking up at the heavens from their resting places. Whenever we gaze up at the sky, it is almost impossible not to apply their laws to the ocean of stars above and imagine the magnificent four-dimensional image of the universe they painted with such methodical mindsets.

This group of scientists' pursuit of the truth brings to mind a parable written over 1,000 years ago by one of the eight giants of Tang and Song prose—Su Shi. In his piece "The Blind Do Not Know the Sun" Su writes: "People who speak of the Way today, whether they describe what they have seen, or speculate about something they have never seen, all pursue it in the wrong way." We can interpret Su's wisdom as follows: those who seek the truth are either carefully explaining what they have seen or inferring conclusions from what they cannot see, but regardless of how they go about this their methods are all incomplete. Su Shi's writing rings true—history's stargazers were indeed using both what was visible and invisible to them to investigate universal truths. But their results were only just a piece of this puzzle, as later generations of astronomers were handed the responsibility of filling in the missing pieces. It is no wonder that this is the parable Einstein himself quoted when describing his inner feelings regarding the theory of relativity.[1]

Here we suggest you, the reader, look at the stories and facts in this book from your viewpoint. We hope you can rearrange this book's narrative in your way and have a go at inferring information that was not mentioned within. In this way, once you have read this book, you will have embedded 3,000 years encompassing the history of scientific development deep in your mind, able to pull out interesting tidbits to reflect on at your convenience. Reaching this level of knowledge is surely a cause for celebration!

From our point of view, this book is well suited to university students taking subjects like physics, engineering, medicine or agricultural studies as a textbook on general education about the physical sciences. From the lives of these scientists, we can learn how unique environments and opportunities gave them the motivation they needed to explore specific subjects, as well as the problems they faced and solutions they found along their journeys. At selected points, we have included explanations of scientific laws and their related background information, and also use examples and propositions to illustrate the practical applications and scientific value of certain theories. Those who have an interest in scientific development but don't have a background in the above-mentioned fields can choose to skip over some theories and examples as they read, and instead view them as extra information embellishing these stories. These readers still ought to be able to fully experience the emotional voyages this group of scientists went through.

Finally, we must explain the use of mathematics used in this book. Chapters 1–4, none of the mathematical constructs we use are any more complex than Pythagoras' theorem, nor are they any more difficult than first-year university calculus. Chapter 5 when introducing special relativity, we occasionally use some basic tensor symbols. The highest volume of and most complex mathematics appears in Chap. 6 which discusses general relativity. But when deriving field equations we place more emphasis on the deducing and illustration of logical thinking. For example, we will progressively explain questions such as why Minkowski space is not applicable in a gravitational field, and must instead be replaced with curved spacetime and the Riemannian manifold; how the result of this is interchangeable with Newton's gravitational field; and at what point we can introduce the concepts of covariance and invariance in order to comply with the consistent nature of coordinate transformation. Our whole narrative is

---

[1] Lin Yutang, The Gay Genius: The Life and Times of Su Tung-Po.

focused on explaining the meaning and intention of these mathematical functions, enabling you to have a clearer understanding of the fundamentals of every formula and what role they play in the theory of relativity. We hope our efforts will bring readers closer to these scientific laws as well as allow them to appreciate the immense force of Einstein's intellect. We are not aiming to force readers into learning a load of meddlesome mathematics, as we would not want to extinguish any burgeoning passion they may have for physics.

This book uses several images to explain relevant theories. Many of these images were drafted by Tzu-Chin Lin, Chien-Long Chiang and Chi-Cheng Lin, students of the Institute of Applied Mechanics of National Taiwan University. We want to express our gratitude to them for their assistance.

Taipei, Taiwan

Falin Chen
Fang-Tzu Hsu

Prologue: Our Universe

At 5:51 AM, September 14, 2015, the Laser Interferometer Gravitational-Wave Observatory (LIGO), a four-kilometer long large-scale physics experiment and observatory located on the outskirts of Livingston, Louisiana, in the southeastern US, detected a very weak signal from deep space that lasted only one-fifth of a second. About 6.9 microseconds later at the LIGO Hanford Observatory, located in Washington State, northwestern US, the same signal was also intercepted. The signal detected by these two massive $2-billion[2] observatories was generated and began to transmit about one billion years ago, the result of two 36- and 29-solar-mass black holes on the other side of the distant Universe colliding and merging. The change in gravity due to the conversion of the matter of about 3 times that of the Sun into energy caused a series of distortions of spacetime. These distortions were propagated outward, from the location where the black holes merged, in the form of gravitational waves traveling at the speed of light, and it took one billion years for these waves to reach Earth.

From the amplitude of the gravitational waves measured by the two LIGO sites, estimated to be the diameter of several protons, the gravitational force generated by the merger of the two black holes is about fifty times the sum total of the gravitational force of all celestial objects in the entire Universe. LIGO's detection of gravitational waves confirmed the general

---

[2] A Special Report, The Discovery of Gravitational Waves—All you need to know about the ripples in spacetime detected by LIGO, Scientific American, February 14, 2016.

theory of relativity proposed by Einstein in 1915 (exactly a century before): the Universe is a spacetime entity formed by gravity, and it is subject to distortion by gravitational force. This distortion is then transmitted in spacetime at the speed of light in the form of gravitational waves. As of the beginning of 2018, three additional instances of such gravitational waves had been detected by LIGO, confirming the predictive power of Einstein's general theory of relativity.

The LIGO observatories in the US are not the only experimental facilities of its kind designed to explore the mysteries of the Universe. Comparable facilities constructed for the purposes of detecting gravitational waves have also been installed in Japan, Germany, Italy and other countries, starting in 2000. Many advanced countries around the world have set up similar expensive instruments or funded experiments of staggering costs since the twentieth century with the aim of confirming the theory of relativity, believed to describe the laws that govern our Universe. For example, HIgh Precision PARallax COllecting Satellite (*Hipparcos*), launched in 1989, was employed to measure the parallaxes of stars caused by gravitational lensing. As of 1996, the parallax data for a million stars had been collected. A gravity probe, equipped with four fused quartz spheres the size of table tennis balls and coated with superconducting materials, was launched in 2004 and has been used to observe the geodetic effect in spacetime and the axial precession effect of stars closely orbiting a supermassive black hole. Another satellite was launched in 2012 to continue with the same experiment; it is known as the Laser Relativity Satellite.

The above examples are just a few of the many costly experiments carried out for the purpose of verifying Einstein's relativistic Universe. The real reason behind these endeavors, of course, is to satisfy our curiosity and ambitions for the Universe. If one day scientists are able to completely reveal the secrets of the Universe, human beings will be able to traverse the vast spacetime with time machines! For now, however, our knowledge of the Universe remains quite limited.

The experiment conducted by LIGO confirmed the existence of gravitational waves, as predicted by Einstein's theory of relativity, and this provides evidence that Universe was formed after the Big Bang that occurred about 13.8 billion years ago. Two hundred million years after the Big Bang, the Universe was still shrouded in darkness and filled with gases and dust clouds (most of which was helium). As the Universe began to cool, these clouds condensed under gravitation and eventually celestial objects were formed. Within these celestial objects, more and more chemical elements assembled and gravitational forces increased rapidly, thus allowing gas molecules to fuse under high pressure. This, in turn, caused the release of massive amounts of energy that continued to fuel the fusion of even larger molecules. The fusion process released huge quantities of energy and light, and the resulting celestial objects became the stars that see today.

Our Sun, for example, was born in this manner 6 billion years ago. Many celestial objects gathered instead to form clusters or nebulae under gravitation. Our home galaxy, the Milky Way, for instance, was born 11 billion years ago. Stars continued to attract the residual substances in the Universe, and these masses would aggregate to form planets, which would in turn orbit the stars in elliptical orbits. In our Solar System, for example, the Sun is located at the center and eight planets revolve around it in elliptical orbits. The Earth's orbital speed around the Sun is about 107,000 km per hour. The Solar System itself also revolves around the center of the Milky Way in an orbit of about 250 quadrillions (10^{15}) km in radius, with an orbital speed of 778,000 km per hour, and it takes about 226 million years for the Solar System to complete one revolution. There are at least 200 billion stars in our galaxy, and each star is revolving around its center in a similar way.

So just how large is the Universe? We know that the light emanating from Earth will reach the Sun in about eight minutes. The same light will leave the Milky Way in about 100,000 years and reach our nearest major galactic neighbor, the Andromeda Galaxy, in about 2.5 million years. If the Universe is regarded as a sphere, then its diameter will be about 100 billion light-years. The Solar System is located about 100,000 light-years from the center of the Milky Way galaxy. When we look up at the night sky, the hazy band of light is actually the side projection of the galaxy on Earth, which indicates that our galaxy has a flat disk-shaped structure but with an incomplete or even fragmented disk. Our galaxy may be located near the edge of the Universe, and it is said to contain hundreds of billions of (sun-like) stars. There are perhaps up to a hundred billion galaxies in the entire Universe. In August 2009, the Fermi Gamma-ray Space Telescope (FGST), launched and operated by NASA, transmitted to Earth the data converted from signals in the gamma-ray bands that it had detected. The data was analyzed by scientists and a hypothesis regarding the existence of a black hole at least four million solar mass at the center of the Milky Way galaxy, now known as the "Fermi Bubbles," which was formed only millions of years ago.[3] Over the past 100 years, the gamma rays emitted by the high-energy jets from the black hole have not been able to penetrate Earth's

---

[3]The suggestion of the existence of a massive bubble structure at the center of the Galaxy was first published in *The Astrophysical Journal* (U.S.) in May 2010. The article was written by Meng Su, Tracy Slatyer and Douglas Finkbeiner. They named this structure "Fermi Bubbles", a tribute to Fermi Gamma-ray Space Telescope. For this discovery, the authors were awarded the Bruno Rossi Prize, the highest honor for high-energy astrophysics, in 2014.

atmosphere, so they have not been detected by ground-based observatories. With FGST deployed in Earth's low orbit, the discovery was finally made in 2010.

Although scientists have some idea about the shape of the Universe, they remain uncertain about the constituent elements of the Universe. Nowadays, scientists employ different bands of the electromagnetic spectrum to detect the substances present in the Universe, and they have discovered that all visible celestial bodies constitute only 5% of the total mass and energy of the Universe. Two other substances also exist in the Universe: dark matter and dark energy, which account for about 27% and 68%, respectively, of the total mass-energy of the Universe. At present, scientists have been successful in exploring the ordinary (visible) matter, and have developed the so-called Standard Model of particle physics, which states that ordinary matter (5% of the entire Universe), including Earth, the Moon, the Sun, the planets and the galaxies, as well as all creatures that exist on Earth, are made up of twelve elementary particles. The forces that bring together these particles to form matter and those that govern the interaction between matter are the four fundamental interactions (or forces): the electromagnetic force controls the interaction between the electric field and the magnetic field; the strong interaction confines nuclear particles within the atomic nucleus; the weak interaction is responsible for the radioactive decay of atoms. Finally, the universal gravitational force governs the motions of celestial bodies.

On the other hand, although scientists are beginning to explore the nature of the invisible dark matter and dark energy, which constitute 95% of the Universe, what is known is still extremely limited, not unlike the knowledge of prehistoric humans about the sky. What is reasonably certain is that dark matter interacts with ordinary matter through gravity, or else it would not have been discovered in the first place. Dark matter can also collide with itself due to gravity, from which electrons or positrons may be generated, and therefore it is possible that dark matter is also be composed of particles. These collisions, however, do not happen routinely. Nor are the particles produced abundant enough to form anything close to a black hole, so dark matter may also exist in an alternative form at the same time. The scant evidence and conjectures show that our understanding of dark matter is extremely inadequate at the moment, and even the most basic definitions are still lacking. As for dark energy, scientists remain completely baffled as to its very nature. Anything from kinetic energy, momentum and heat in the Universe, or a combination thereof, is a possibility. It will likely also appear in the term in Einstein's field equations that represents the source of gravity, the existence and the variation of which will alter the spacetime curvature of the Universe. In any event, the generation, existence, changes and any other properties of dark energy currently lie in the realm of unknown as far as we are concerned. We have at best barely scratched the surface.

As we explore these unknown areas of science, we are keenly made aware of how ignorant we still are. Apart from such ignorance, there is also one aspect that is difficult to imagine. It is the distortion of spacetime defined by general relativity, which also depicts the depths of space and time. The "depth of space" clearly points to a three-dimensional image of celestial bodies at different distances from Earth. The "depth of time", on the other hand, represents collages of the images of heavenly bodies all created at different points in time. For example, standing on Earth, the sunlight we see constitutes photons emitted from the Sun more than eight minutes ago; an image of Jupiter we see in a telescope is 50 minutes old; and an image of the Alpha Centauri system, our closest stellar neighbor about four light-years away, is actually four years old. Similarly, the starlight from Vega in the constellation of Lyra is about 25 years old, and the light from other stars farther away can be decades, tens of thousands, or even hundreds of millions of years old. This Universe, created with the finite speed of light time and possessing depth in time, constitutes an entity with a distinctive contrast between space and time. Someday if we could take a ride on a spaceship and travel close to the speed of light in this distorted spacetime continuum, how would the Universe appear on our viewscreen is anybody's guess!

Despite the Universe being full of unknowns and hard to imagine, we are going to explore the process by which the development of science has been brought to its current form. In other words, how did science begin as a curiosity of Bronze Age humans about the starry sky and end up with the principles of the theory of relativity that enable us to understand how the Universe works? In the following chapters, we will explore the history of scientific development over a period of four millennia, starting with the Babylonians in 1000 BC and concluding with Albert Einstein's theory of relativity in the early twentieth century. During this time many scientists were inspired by various natural phenomena and proposed theories and hypotheses, many of which evolved into important theorems and scientific knowledge. The occurrences of a series of scientific events allowed us to catch a glimpse of the intelligence, strength and endurance exhibited by previous generations of scientists in successive waves of scientific discoveries, and we will also examine whether human beings have the ability to go forward and proposer in the unknown future.

Contents

1

Ancient Wisdom and Natural Philosophers

人法地.
地法天.
天法道.
道法自然.

He models himself on Earth,
Earth on Heaven,
Heaven on the Tao,
The Tao models itself.
On Nature,
On the So-of-Itself.

From Tao Te Ching, *Chapter 25, by Lao Tzu (571–471 BC), an ancient Chinese philosopher.*
Translated by John Minford (b. 1946).

In 1881, the Assyrian-British archaeologist Hormuzd Rassam (1826–1910) discovered a clay tablet covered with cuneiform inscriptions in the ruins of the ancient city Sippar, some 30 km southwest of Baghdad, the capital of Iraq (then part of the Ottoman Empire). The clay tablet is said to be the oldest map ever discovered in history. On this tablet, now known as the Babylonian "Map of the World" (or *Imago Mundi*), two concentric circles enclose several haphazardly arranged ellipses and arcs, and the circumference of the outer circle touches the bases of eight triangles, each of which is almost equidistant

© The Editor(s) (if applicable) and The Author(s), under exclusive license to Springer Nature Switzerland AG 2020
F. Chen and F.-T. Hsu, *How Humankind Created Science*,
https://doi.org/10.1007/978-3-030-43135-8_1

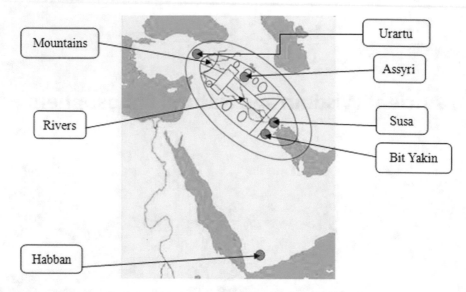

Fig. 1.1 The world of Babylon superimposed on a modern map. Inspired by https://en.wikipedia.org/wiki/Babylonian_Map_of_the_World

from its two triangular neighbors on the circle. The wide belt region between the two concentric circles is labeled as the Aral Sea, which signifies that the world inhabited by humans is surrounded by an ocean. The two arcs in the interior of the inner circle represent the Euphrates, which flows from north to south and passes through Babylon, represented by a rectangle. The mouth of the Euphrates is another rectangle marked "swamp". Placed around Babylon are a number of small ovals, which represent cities, such as Susa, Bit Yakin, Habban, Urartu, Der and Assyria. The map is also marked with animals such as cows, monkeys, lions, wolves, and sheep.

This is an abstract and stylized map. The mapmaker employed triangles, circles and rectangles to signify, respectively, mountains, bodies of water and cities on Earth's surface. The map is centered on the Babylonian world, with text and graphics incorporated into the clay tablet. Based on the names of the cities depicted on the map, we have displayed their locations on a modern map to produce an elliptically shaped region shown in Fig. 1.1. The major axis of the ellipse happens to be superimposed over the Tigris–Euphrates river system, and the high elevations on the outer rim represent mountain ranges in modern-day Turkey and the Caucasus, as well as the Arabian Desert (the triangles on the clay tablet indicating mountains are not shown here). The surrounding bodies of water include the Mediterranean, the Black Sea, the Caspian Sea, the Persian Gulf, and the Red Sea. This area, known as Mesopotamia, was settled by the Babylonians and is a vast floodplain of the

Euphrates and Tigris. It is also known as the Fertile Crescent due to its shape and its fertile soil. This was the world known to the Babylonians, a self-sufficient community surrounded by mountains and seas, where its residents lived and worked in peace.

1.1   The Universe in Mythology: Enlightenment by the Laws of Nature

The multitude of cities marked on the clay tablet was essentially agricultural in nature, developed over a long period of time beginning around 4000 BC. Early settlers discovered the fertile land between the rivers Tigris and Euphrates and decided that the nomadic lifestyle, which they had led for generations, was no longer the only option available to them. They began to build irrigation works and cultivate the land, thus laying down the foundation of an agricultural society. With a more reliable and abundant food supply, raising families became the norm. The higher birth rates and reduced mortality enabled the population to grow rapidly, transforming nomadic existence into settlements, and from tribal societies into cities. Apart from tilling the fields, these settlers also raised cattle, sheep, hog and horses, which provided them with not only meat and dairy products but also woven garments from animal hair. They also learned to produce beer and wine. All the ingredients required for people to live as a cohesive community were at their disposal.

Nevertheless, a prosperous and fast-growing population also necessitated effective administration, and eventually, a loosely structured social hierarchy came into existence. At the very top of the social classes sat the religious authorities who served as an intermediary between ordinary citizens and the gods. In an agrarian society before the emergence of powerful military leaders, the most effective way of governing the populace was to introduce awe-inspiring deities that would instill fear and respect among the people, in response to their anxieties toward the unpredictability of nature. In the beginning, there would be a tutelary deity in each city responsible for all matters great and small, from public administration to household affairs. Citizens would be able to seek guidance and comfort from the deity via their religious authorities, who would intercede on their behalf. The responses communicated by these religious leaders would often become the principles by which ordinary people lived their lives. Over time, religion would transform into an effective means of managing a city, where a majestic temple would be built at its center. The priests of the temple would become what people considered mediatory agents between themselves and the deities, as well as their common leaders. At the height of this theocratic regime, the deities being worshipped in the temple would number in the dozens, ranging from those in charge of agriculture, commerce, craft, wine production and animal husbandry to those who administered justice and even death.

1.1.1 Marduk, Patron Deity of Babylon

To ensure harmony and to maintain social order in the community, relying merely on the admonishments by religious leaders was far from adequate. It was necessary to lay down written rules and regulations for the citizens to follow. The responsibilities of administrators were limited to adjudicating disputes and enforcing judgments. From a modern point of view, this is equivalent to making a set of laws that regulate citizens' behavior and to mete out justice whenever the laws are violated. The establishment and enforcement of these ancient laws depended on religious norms and constraints. In the case of Babylon, the rulers proclaimed that it was the great god Marduk who created the heavens, the earth, human beings and animals out of the chaotic primeval ocean to form a world centered at Babylonia, thus giving the rulers supreme authority to govern the world and to pass judgment on infringements of the law according to Marduk's will. In 1772 BC, Hammurabi (c. 1792–1750 BC), the sixth king of the First Babylonian dynasty, even claimed to have received divine instructions from the great god Marduk to administer justice and to eradicate all evil in the world. He duly promulgated the Code of Hammurabi, one of the first legal codes in human history. King Hammurabi ordered the text of the code carved onto a black basalt stele measuring 2.4 m high, the purpose of which was to make it clear to the subjects of his kingdom that the law was to be obeyed without question.

Currently, on display at the Louvre Museum in Paris, this black stele was uncovered in 1901 by archaeologists at the ruins of the ancient city of Susa (or Shushan, in modern-day Iran). It is said that Marduk not only instructed Hammurabi to enact laws for the purpose of keeping human behavior in check but also lay down a set of principles (or laws) of nature that describe how the natural world actually functions. These laws of nature were quite distinct from the legal code that applied to people because one cannot "order" nature to do or not to do something. Nor could nature be "punished" if it ever violates these laws. What is more unsatisfactory is that these laws, designed to describe how nature works, were not clearly enunciated like its human law counterpart but were instead phrased in very vague terms. From a modern perspective, these were merely a set of self-serving, subjectively determined hypotheses that could never be proven. Below we will elaborate on this point with examples of the "laws" of nature in the Code of Hammurabi.

After the Babylonians founded their kingdom in Mesopotamia in the 18th century BC, they inherited the cosmological view of the Sumerians but replaced the Sumerian gods with their own new deity, Marduk. According to the laws of nature in the Code of Hammurabi, Marduk was born out of the primordial ocean before time, who then proceeded to create the world, took control of power and maintained order in his new realm. In the process of creation, Marduk breathed into the primordial ocean, and after inflating it, he pierced it with an arrow. Half of the resulting fragments became the water of the sky and the ground, and the other half became Earth. The Earth is like a disc surrounded by rivers, and beyond the rivers are impassable mountains, much like the depiction on the clay tablet discovered by Rassam. A cluster of surrounding lofty peaks supports the six-level heavens, the uppermost level of which being composed of special stones. The second level is the throne of Marduk made of lapis lazuli; the third level contains the sacred jasper stars; the fourth level is the land inhabited by humans. The fifth and sixth levels are the middle and lower terrestrial levels where the gods reside. Marduk's abode in Babylon is the center of the Universe, with the second level of the heavens reachable via the high tower atop it. The tower's foundation passes through the lower terrestrial level. This explains why the temples found in West Asia are often tall towers, as the residence of the gods is supposed to span both the heavens and the earth in order to bring people and the gods together.

The temple is also the center of the Universe, and the gods residing in it can therefore maintain the order of the Universe from a divine vantage point. Having been endowed with such superior status and sacred power, this center-of-the-Universe narrative began to be emulated by other peoples. The Hebrews, for example, designated Jerusalem as the Navel of the World, and the ancient Greeks believed Delphi to be the center of the Universe. We have covered in the preceding paragraphs the laws of nature of the Babylonians. The stories, however, contrived they may seem, vividly portray Marduk and the universe he created. Yet they never took the effort to demonstrate or question the veracity of these assertions. This type of culture, where the rulers' words were taken at face value and never challenged, would continue until the inauguration of Greek natural philosophy, at which time the "true" laws of nature were also entirely up to the natural philosophers.

This kind of law of nature indeed carried the ruler's subjective views and self-serving motives, but Hammurabi was far from the first person to have proposed such theocentric laws. The earliest laws of nature may have been established before 3000 BC by the Sumerians, who built large-scale irrigation systems and developed urban cultures in Mesopotamia. In addition, despite having lands that were fertile, irregular flooding, as well as dry seasons lasting six months or longer, the Sumerians were entirely at the mercy of nature and had to learn to tame the lands into submission in order to be productive. For this reason, the Sumerians fashioned a deity that was harsh, short-tempered and mighty, and always strict but fair in meting out rewards and punishments, when they constructed their cosmological view. They believed that when the Universe first came into existence, it was composed of a mixture of primordial waters and solid matter. Then the gods emerged, separated the heavens from the earth, and provided the space on land needed for the growth of seeds (i.e., all forms of life, including humans).

1.1.2   Atum, the Great Egyptian God

Also around 3000 BC, another great civilization was born out of the alluvial plains of the Nile River: ancient Egypt. On these plains, the prosperous inhabitants also established their own set of laws of nature centered on the Nile Delta. The ancient Egyptians lived in an area bordered by the Mediterranean to the north, the Red Sea to the east, and a desert to the southwest. The only external passage was through the Sinai Peninsula to the northeast, which allowed interactions and cultural exchanges with the peoples living in Mesopotamia. The Egyptian laws of nature, therefore, were influenced by the Sumerians, but what distinguished their system from their counterparts in Mesopotamia was that the Nile had regular flooding periods and the river basins received an abundant supply of water, which allowed ample crops to be harvested throughout the year. Also, because of the surrounding natural barriers, the Egyptians encountered few invasions, and the result was a self-sufficient, cyclical view of cosmology. From their observations of natural phenomena, the Egyptians discovered that the Sun, the Moon and the stars follow a regular, routine cycle of movements across the sky. For example, the Sun is "born" from one end of the horizon at dawn and "dies" at the other end at dusk. To them, these cyclical patterns appeared to stay constant and never changed. They subsequently connected human life with the cycle of nature and concluded that the end of human life in this world signifies a rebirth in another, and the ruler is responsible for maintaining order and harmony in the Universe.

With respect to the origins and composition of the Universe, the Egyptians believed that the Universe was at the beginning a chaotic and invisible ocean, from which a mound covered with papyrus, a plant endemic to the Nile Valley, slowly emerged. Then the god Atum appeared. He created light, air and water vapor, and out of these three elements he fashioned the heavens and the earth. This newly created heavens and earth, however, remained a merged entity initially, until air filled its space forcefully and separated the two. The night sky was similar to a solid sheet of metal. The stars were perforations resulting from falling meteorites that pierced through the sheet, and their light came from the Sun shining through these holes.

1.1.3   Jehovah, God of the Hebrews

Hebrew civilization was born in 2000 BC, a millennium later. Also influenced and enlightened by the Mesopotamian culture, the Hebrews developed their own set of laws of nature. The Book of Genesis of the Hebrew Bible described the creation of the world at the hands of the Hebrew God Jehovah: "In the beginning, God created the heaven and the earth, the earth was without form and void, the darkness was upon the face of the deep, and the Spirit of God moved upon the face of the waters."[1] Then God created all things in six days and imposed order on them. The process of creation is divided into two stages: from the first to the third days, the Universe was formed in its intended sequence, including light, darkness, the sky, land and oceans, as well as plants on the ground; from the fourth to the sixth days, creatures came into being in the order of the first three days: on the fourth day, light in the sky was created to distinguish it from that created on the first day; on the fifth day, birds in the sky and the aquatic life were made to complete the sky and the oceans created on the second day; on the sixth day, animals and the first person Adam was created, and they would take the plants and fruits created on the third day as food. Finally, God emphasized that humans were created in his own image, and he gave them dominion over everything on the earth.

1.1.4   Greek Pantheon

The laws of nature introduced by the Hebrews clearly define the relationship between human beings and God, and when the human view of nature gradually shifted from a theocentric perspective to a humanistic one, these concepts began to spread among the tribal peoples of Mesopotamia. After having evolved for more than a thousand years, the anthropomorphic deities and the human-centered ideas and philosophy began to take hold on mainland

---

[1]Genesis 1:1–2, King James Bible.

Greece, in southern Balkans, around 800 BC. The people who took a leading role in propagating these thoughts were the prehistoric Indo-Europeans who lived in the areas around the Black Sea and the Caspian Sea, who for centuries lived a nomadic existence. To escape wars and famines, they gradually scattered and migrated to other parts of the world and formed different branches and tribes. Some of them moved toward the southeast and settled in present-day Iran and India. Others migrated in the southwestern direction to modern-day Greece, Italy and Spain. Still others, a smaller group, made their way to the west of Europe. What is known as Greek civilization was established by the Indo-European tribe who entered the Balkans and settled in Greece. Unlike the monotheism of the Sumerians, the Egyptians and the Hebrews, the Greeks advocated polytheism, providing vivid accounts of the relationships between their anthropomorphized deities and all levels of human societies. All these epic stories were told in their entirety in Homer's *Iliad* and *Odyssey*, as well as in Hesiod's *Theogony*. The Greeks believed that the beginning of the Universe was empty, chaotic, filled with darkness, and in a state of serenity and isolation. Then out of the void (or Chaos) emerged three gods: Gaia (the personification of Earth), Nyx (goddess of the night) and Erebus (god of darkness). Then came Eros (god of love), who established order. With Erebus, Nyx gave birth to Aether, the god of the brightness of the upper sky; and then to Hemera, the god of the day and light of the earth. Many other gods were also born to symbolize different natural phenomena, such as fate, death, sleep and dreams.

The theological view of the Greek culture brought not only polytheism but also their personifications. A personified god is characterized by his or her duty to carry out what is required of causality. The causality model has been the mode of thought adopted by humans when changes in certain things or phenomena cannot be explained, i.e., a certain cause (or antecedent) resulting in a certain outcome (or consequent), and either the cause or effect may be liable to intervention by the gods. For example, if a person is ill, he is being punished by the gods, and if he becomes rich, it is because of the gods' blessings, and so on. Though fraught with subjectivity and without basis, this type of logical thinking gradually led to the emergence of science, which was born out of practical applications. Science flourished along with the rise of Greek civilization. The resulting vast empire established by the Greeks, which spanned Europe and Asia, brought about the further development of science as well as the conquests by military force. The Greek empire helped to promote the exchange of goods and cultural integration between originally independent city-states. The resulting large-scale ethnic integration and consolidation of territories, as well as the obliteration of state boundaries, contributed to a unified Mediterranean culture and an excellent opportunity for the evolution of European civilization. To explore the origins of scientific development, one must go back in history and try to understand how Greek culture came into being.

1.2 Rise of Empires[2]

Like many other civilizations in history, the rise of ancient Greek civilization was preceded by a series of wars and plunders, and the origins of these upheavals can be traced back to 800 BC. During this period, many city-states (or *poleis*; singular *polis*) along the Aegean coast in the eastern Mediterranean Sea emerged. For each of these city-states, the *acropolis* was its nucleus at the center, surrounded by tracts of farmland assigned by the ruler to peasants who were citizens. The ruler would then collect harvested crops from the farmers as taxes. The city itself was home to the principal citizens of the city-state, and the aristocrats had a monopoly on political and military matters. When city-states later grew into stable and prosperous polities, they began to institute a preliminary democratic system that allowed citizens to participate in public administration and decision-making in specific areas. In general, early city-states were moderate in size, but in order to maintain its stability and integrity, its population would require at least several thousand men to maintain economic self-reliance and the ability to resist foreign aggression. In an effort to achieve a self-sustaining scale, many city-states aggressively solicited dispersed towns and villages in their vicinity to join their alliance, and very large city-states were born. Examples include Athens, located at the southern tip of the Balkans, and Sparta, in the south of the Peloponnese, both of which being populated by tens of thousands of citizens in their heydays. The population of these city-states would exceed 100,000 if the non-citizen farmers and fishermen in the surrounding areas were included. The proliferation of these city-states, which numbered as high as a thousand, thus made the Aegean Sea a community bustling with commercial activities.

1.2.1 Assyrian Empire

While the Greek city-states were flourishing, the Assyrian Empire emerged on the east coast of the Aegean Sea. This ancient empire had been built by the Assyrians in Mesopotamia, and at its height, its territories included parts

---

[2]Referenced from Abulafia [1]

of modern-day Turkey and Egypt. To participate in maritime trade, Assyria joined their neighboring city-states and became part of the commercial sphere in the Aegean. Trade with the Assyrians was welcomed and in fact sought after by the Greek city-states, as Assyrian technology and culture were exotic to them. These city-states, regardless of their size, were eager to explore trading opportunities with Assyria because the Assyrians possessed goods and technologies that they desired, including sculpture, architecture, metallurgy, and processing techniques, all of which were valuable commodities and knowledge previously unknown to the Greeks. Having access to them would ensure a competitive advantage over other city-states. Indeed, the success of Greece in later centuries in transforming into the great empire that it became, spanning across the Mediterranean, owed much to what they gained from Assyria's advanced civilization. The Assyrian Empire only traded directly with economically and militarily superior city-states, such as Athens and Sparta. Smaller city-states were only able to sell their goods to Assyria through their much mightier neighbors. Under such a trading arrangement, therefore, Athens and Sparta were the principal beneficiaries, as they dealt directly with Assyria and had access to the best technologies and precious goods and resources in the struggle for supremacy among the city-states. Indeed, the leaders of these two dominant powers were exemplary.

In Sparta, profits from trade further strengthened the city-state, and it continued to expand in size. The powerful king also demonstrated supreme confidence and allowed the more senior members of its citizenry to participate fully in political and military activities. He also forged an alliance with all the city-states in Peloponnese and formed a confederation known as the Peloponnesian League, the strength of which approached that of an empire similar to Assyria. Although Athens emerged as a military power at about the same time as Sparta, a rudimentary form of democracy would not be introduced until Solon became a chief magistrate. Following nearly a century of evolution and expansion, Athens also entered into an alliance with nearby city-states, thus creating a prosperous commercial, industrial and agricultural community where citizens could live a harmonious life. However, all of these completely changed when the Assyrian Empire was wiped out by the Persians.

1.2.2 Persian Empire

The Persians, whose ancestors had lived in Central Asia, became a regional power following a period of more than two hundred years of growth, and they soon began to expand their territory and were invincible in their military campaigns. In 546 BC the Persian Empire subjugated the kingdom of Lydia (in today's Turkey). It conquered the Babylonian Empire, in Mesopotamia, seven years later and then Egypt in 525 BC. At its peak, the Persian Empire's territory extended as far as India, and it is the first empire in history to have spanned both Europe and Asia. Darius I was the Persian king who ushered in the rise of this great empire. This gifted king not only excelled at his military leadership but also implemented policies benevolent to his conquered subjects. He built roadways and irrigation ditches in the newly established administrative regions to keep the people fed and safe, hoping that they would accept the new reality of being ruled by a foreign power. The Greek city-states led by Athens and Sparta, however, did not appreciate the Persians' good deeds or intentions.

In 492 BC, Darius I was outraged by Athens' support of the Ionian Revolt in Asia Minor (in modern-day Turkey), and he sent troops to Athens to punish the Athenians, thus igniting the Greco-Persian Wars that lasted more than a decade. The outcomes of several of these battles, however, were a great disappointment to Darius I. The Persians were first routed by the Athenian army at the Battle of Marathon, and then their fleet was annihilated at the Battle of Salamis by Greek naval forces. Finally, the Persian army suffered heavy losses at the Battle of Plataea. After this unprecedented victory in the Greco-Persian Wars, Athens' leadership position among the Greek city-states was decisively established, and the Delian League, led by Athens, was born. Nevertheless, Sparta, leader of the Peloponnesian league, was unwilling to recognize Athens' supremacy. Sparta launched the Peloponnesian War in 431 BC that lasted nearly three decades. Shortly after the war began, Sparta decimated the Athenian fleet and soon replaced Athens as the paramount power of the Greek city-states. Spartan hegemony, however, did not last long. Supported financially by the Persians, Athens formed an alliance with other Greek city-states that were dissatisfied with the Spartan rule, and they jointly declared war on Sparta. This conflict is known as the Corinthian War. After nearly a decade of fighting, with both sides of the war weakened militarily and resources severely depleted, Sparta and Athens entered into a peace treaty known as the King's Peace, which applied to all Greek city-states. The agreement, however, failed to bring lasting peace to Greece. Over the following thirty years, wars continued to be fought on the peninsula between city-states. The only party who profited was the Persian Empire, which funded the conflicts among the Greeks. This decades-long civil war was devastating to the citizens, whose discontent and frustration were widespread. It was under this backdrop of turmoil, chaos, violence and fragmentation that the kingdom of Macedon, located in the north of the Balkans, rapidly emerged as a regional power.

1.2.3   Macedonian Empire

The ancestors of the ancient Macedonians were nomadic tribes who inhabited in the mountainous regions of the Aegean's northern shore, in the areas around today's Thessaloniki. They lived in the mountains during the summer and relocated to the plains during the winter. With the rise of Greek city-states to the south, organized military forces became a threat to the Macedonian nomads who did not have well-defined territories of their own. These threats forced the Macedonians to relocate to other pastures, and their relatively unbridled and free lifestyle was gradually replaced by hunting and brutal conflicts. Although threatened territorially by the Greek city-states, the Macedonians became drawn to Greek civilization and took every opportunity to involve themselves with the mainstream of Greek culture. They even covertly participated in the affairs of the warring city-states to gain interests or advantages. This parasitical existence of the Macedonians continued with a powerless monarch acting as the figurehead until the 23-year-old Philip II (382–336 BC) succeeded to the Macedonian throne and embarked on a nation-building journey.

King Philip II's reign began at a time when Macedon was ill-treated by the Greek city-states. The young and rash king refused to bow down to the humiliation or to swallow his country's pride. He introduced Macedon's phalanx infantry corps, armed with the famous sarissa (a very long spear) based on the military skills and experiences acquired in Thebes while he was held as a hostage there. Combining this infantry corps with traditional Macedonian light cavalry, he created a unique and well-equipped combat force in the history of the Balkans. By marrying a total of seven noblewomen from city-states around Macedon during his lifetime, Philip practiced the nuances of diplomacy learned from his time in Lydia, namely to ensure the security of his kingdom through royal marriages. Philip understood the conflicts and disputes among the Greek city-states and, being a master of political maneuvering, he exploited the discords among members of the Delphic Amphictyonic League, which consisted of twelve Greek city-states, and managed to defeat the kingdoms of Phocis and Locris. Philip won a conclusive victory at the Battle of Chaeronea in 338 BC over the forces of the Balkans-alliance led by Athens and Thebes. The outcome of this battle, described as the most decisive of ancient Europe by historians, was Macedonian hegemony over the Balkans. Philip II himself became the *hegemon* of the newly formed League of Corinth, which consisted of the majority of Greek city-states, with the notable exception of Sparta.

Having secured his power in the Balkans, Philip turned his attention to the land across the Aegean, the Persian Empire. In 337 BC, the Macedonian army, led by two generals, crossed the Dardanelles into Asia Minor and captured the Greek city-states under Persian control one after another. When the news of victory from the battlefield reached Macedon, Philip was at the ancient capital of Vergina to preside over the wedding of his daughter. This supposedly joyous event turned out to be more perilous than the battlefield. The king, who was not armed for the occasion, was assassinated by his own bodyguard Pausanias of Orestis, who plunged a concealed blade into his chest. Philip died at the age of 46.

1.2.4   Conquests of Alexandria the Great[3]

In the wake of King Phillip II's death, the 20-year-old prince Alexander III (356–323 BC), commonly known as Alexander the Great, ascended to the throne of Macedon. This was a tumultuous time. Alexander was immediately confronted by the rebellion of the Macedonian nobleman and general Attalus, as well as the revolts of Athens and Thebes. Despite being besieged on all sides, Alexander demonstrated his superb political wisdom and military talent by dispatching troops, in the name of reinforcements, to Asia Minor to assassinate Attalus, who was stationed with the Macedonian advance army on the front line against Persia. After the death of Attalus, Alexander persuaded the Delphic Amphictyonic League to surrender and pledge allegiance to Macedon. He then expedited troops to subjugate Thebes, traveling at a staggering 30 km a day to reach that rebellious city. When the Thebans saw Alexander's heavily armed troops approaching their city gates, they immediately surrendered. After the Athenians learned of the fate of Thebes, they sent an emissary to Alexander at once to convince the king of their undivided loyalty to Macedon. Finally, all Greek city-states surrendered to Macedon in the name of the League of Corinth, and the political dominance of Macedon over the Balkans prior to the late King Philip II's assassination was restored.

After his emphatic victory in Greece, Alexander immediately crossed the Balkan Mountains to the north and reached the territories of the Thracians and the Celts along the Danube. The coalition forces assembled by the Thracians and Celts consisted of 10,000 infantry and 4,000 cavalries. Stationed on the northern bank of the Danube, they awaited the arrival of the Macedonian army, which was only half their military strength. Alexander's troops employed canoes acquired locally, as well as leather buoys they constructed

---

[3]Referenced from Moritani [25].

themselves, and forced their way across the river overnight, and they began their frenzied attacks at dawn. Sensing that the battle had been lost, the Thracian-Celtic coalition forces dispersed in disgrace. Alexander therefore annexed the territories along the Danube into Macedonia, fulfilling the dying wishes of his father, Philip II. Meanwhile, news of the rebellion of Thebes, which had recently surrendered to Macedon, reached Alexander, who was enraged and decided to March his army yet again toward the unrepentant city-state. Upon arrival at Thebes, Alexander burned the prosperous city to the ground. Six thousand Theban troops perished, and 30,000 women and children were sold into slavery. At this point, the Balkans were fully under the dominance of Macedon, and this laid a solid foundation for the future empire of Alexander the Great. Now Alexander turned his attention to his country's sworn enemy located to the east of Macedon: the great Persian Empire.

In the spring of 334 BC, Alexander led an expeditionary force of 47,000 troops and marched across the Dardanelles into Asia Minor. The army consisted of three combat forces. The first was his cavalry consisting of 14,000 war horses, on which Alexander relied heavily. Next was his infantry known as *pezhetairoi* (singular *pezhetairos*, literally "foot companion"), who were armed with the sarissa, a 5.5 m long spear or spike weighing 7 kg. The third was the *peltasts*, the "shield-bearing" troops specializing in guerrilla campaigns and raids, as well as desert and mountain warfare. Apart from soldiers and horses, the expeditionary army also brought with them many pieces of heavy siege equipment, such as 40 m-high mobile siege towers that housed archers and spearmen, battering rams for breaching castle gates, and catapults capable of projecting rocks and rockets from 50 m away. Alexander's elite troops continued their unstoppable campaigns along the Aegean coast, steamrolling through one Persian-controlled Greek city-state after another. In the autumn of 333 BC, after more than a year of advancing deeper into Asia Minor, the Macedonian army arrived at the town of Issos, located just outside the Cilician Gates and close to today's Turkish-Syrian border. There they were confronted by the 60,000-strong Persian army led by King Darius III. Following a brief standoff, the two armies met near the mouth of the Pinarus River and were determined to battle to the death. From the outset, Alexander's peltasts organized wedge-shaped formations and quickly breached the Persian cavalry's defense, seriously disrupting the troop formations of Persia's principal fighting forces and weakening their infantry's position. The Persian army subsequently collapsed and the soldiers scattered. This battle, fought and won by Alexander in just one day, marked the beginning of the end of Persian dominance.

A month later, having rapidly reorganized his troops, Alexander advanced southward and marched along the Mediterranean coast. His destination was Egypt, located in North Africa. The resistance to Alexander's army along the way was minimal. In January 332 BC, Alexander arrived at the outskirts of Tyre, the capital of Carthage in Palestine. This prosperous ancient city of commerce was located on an island in the Mediterranean Sea only 800 m off the Palestinian coast. The island city had fortifications right up to the sea on all sides, and the towering walls formed a strong fortress easy to defend and difficult to attack. What is even more intimidating was that Tyre also possessed a powerful naval fleet of hundreds of warships and no doubt constituted a strong resistance to Alexander's assault from the east. With no advantage whatsoever in naval strength to speak of, Alexander's only recourse was to mount a terrestrial battle. The tactics he adopted were very simple: dismantling the buildings captured on land and building a causeway that allowed his troops to approach and breach Tyre's fortifications. After five months of construction, the kilometer-long and 8-meter wide causeway was completed ahead of schedule. Armed with siege equipment, the Macedonian army approached the walls of Tyre along the narrow causeway. To their credit, the Tyrians were well prepared to defend themselves against the invaders: archers were not only deployed on the city walls to attack the Macedonian army but also on Tyrian warships to mount bow-and-arrow assaults from long distances. Isolated on the causeway, the Macedonians were unable to capture the city with such strong resistance. The two sides of the battle were deadlocked for nearly two months, when the Phoenicians and Cypriots, who possessed a keen sense of who the victors would be in this battle, dispatched 200 war galleys and joined Alexander against the Tyrians. With the support of his new powerful fleet and the ability to mount counterattacks, the fortunes of Alexander's attacking forces were rapidly reversed. The Macedonian troops finally forced their way into the city and slaughtered 8,000 Tyrian men in one fell swoop. In addition, over 30,000 women and children were later sold into slavery.

Soon after the destruction of Tyre, the Macedonian army entered Egypt. Egypt had been under Persian rule for over two centuries via the viceroy appointed by the king of Persia, and the country was experiencing a period of national decline. Therefore, when the Macedonian army reached the perimeters of the city's walls, the Persian army was in no shape to defend themselves. In the winter of 332 BC, Alexander's army entered Egypt virtually with no resistance. The army marched upstream along the Nile and entered the capital city of Memphis. The Egyptians welcomed Alexander as their king, placed him on the throne of the Pharaohs at the temple of Apis, and declared him the son of the Egyptian deity Zeus Ammon. On this glorious day, Alexander decided to build a great city on the west bank of the Nile Delta, the future Alexandria, as the trading center for the countries around the Mediterranean. On April 7, 331 BC the groundbreaking ceremony of the new city was held. Like all the newly built cities in the Mediterranean, Alexandria was laid out as a grid of parallel streets. The city had a rectangular shape, measuring 5 km from east to west and 3 km from north to south, and it was also a large port city facing the Mediterranean. Eight years later, Alexander died of contracting malaria during the Indian campaign, and the empire he had built was divided between three of his generals, who gained control of Macedon, Greece and Egypt. In Egypt, Ptolemy I Soter (367–283 BC) continued to build and shape the great city of Alexandria after becoming the Pharaoh of Egypt.

1.3   The Great Library of Alexandria

In order to expand Alexandria into a flourishing city, Ptolemy I sent a large army to Syria and seized the sarcophagus of Alexander the Great, whose funeral procession was on its way to Macedon. His intention was to boost the status of the Ptolemaic dynasty by burying the king's remains in Alexandria, while at the same time announcing to the world that he was the legitimate successor to Alexander. This legitimacy would naturally have to be built upon the fulfillment of Alexander's dying wishes, namely to transform Alexandria into the greatest city in the world. The key to making Alexander's dream come true was to construct two majestic buildings in the city: the Mouseion, which was dedicated to the Muses, and the famous Library of Alexandria.

With tremendous resources provided by the pharaohs, the Mouseion at Alexandria was not merely a magnificent piece of architecture; it was also equipped with a number of rooms and creative spaces for music, philosophy, the arts and science. Many thinkers, philosophers and artists from around the Mediterranean would travel long distances to reach Alexandria. Those who were selected as scholars to work there would no longer need to worry about their livelihoods and be able to discuss philosophy, create new and ingenious ideas, and explore different theories. The Library of Alexandria was even more impressive than the Mouseion. Although initially, the library collected only books in scrolls made of papyrus, which was the dominant material used as writing surface in Egypt at the time, under the Ptolemaic kings' aggressive order and well-funded policies for procuring texts from all over the known world, the scope of the library's collection expanded to those from various cultures around the entire Mediterranean world and even as far as Persia and India. Texts collected from other cultures on behalf of the library were translated into Greek and copied onto papyrus scrolls. The library's most famous work of translation, commissioned by Ptolemy II, was completed by 72 Jewish scholars, six from each of the Twelve Tribes of Israel. They were summoned to Alexandria from Jerusalem and were each assigned a separate chamber within the library to translate the Hebrew Pentateuch into Greek independently. This translation of the Hebrew Bible is known as the Septuagint. According to tradition, scholars in later generations compared these 72 translations and discovered that they were almost identical word for word, which suggests that the translation skills of the Jewish scholars were indeed miraculous or, indeed, divinely inspired.

Under the deliberate and effective policies pursued by the Ptolemaic kings, the Library of Alexandria achieved an extensive collection of over 400,000 scrolls at its height. The collection can be divided into two categories: mixed books (multiple works within a papyrus scroll) and unmixed books (each scroll of papyrus containing at most one transcribed work). Ptolemy I spared no expense to acquire the master copies of precious texts, from which manual transcriptions were made, by manipulating their owners into offering the books to the library on loan. Later the original texts were confiscated and kept in the library, and the copies delivered to their original owners, who were then properly compensated for their losses. This type of unapologetically forceful manner in appropriating cultural assets from other parts of the world made the Library of Alexandria not only a massive repository of books of every kind; it also made Alexandria a prime destination for literati and scientists from elsewhere in the Hellenistic world. Prominent scholars who made their way to Alexandria include Euclid[4] (325–265 BC), a geometer from Athens; Eratosthenes (276–194 BC), the astronomer from Libya who accurately deduced the diameter of the Earth and who became the director of the Library of Alexandria; Aristarchus of Samos (310–230 BC), the first Greek mathematician to conclude that the Earth in fact revolves around the Sun; and Archimedes (287–212 BC), the philosopher from Sicily who designed many intricate machines that fascinated Ptolemy I.

Supporting cultural activities of such enormous magnitude required tremendous financial resources. The financial might that supported the Library of Alexandria was quickly and assiduously amassed under the reigns of the first two pharaohs of the Ptolemaic dynasty, after they had successfully

---

[4]Very little is known about the life of Euclid. Some scholars have suggested that *Euclid* might actually have been the name of not one person but a school of thought used by a group of scholars with substantial contributions to the study of geometry.

transformed Alexandria into the busiest trading port of the Mediterranean. In order to make Alexandria an effective trading port, in 297 BC Ptolemy I ordered the construction of a 135-meter lighthouse on the island of Pharos, located on the western edge of the Nile Delta. Fourteen years later, in 283 BC, the Lighthouse of Alexandria was completed. The lighthouse's tower consisted of four stories. Located at the lowest level was a base with a square cross-section, and further up the shape of the structure became slightly trapezoidal to reduce the cross-sectional area. The middle story was a platform with an octagonal observation deck, and on this level one would find several large reflective mirrors installed along the interior walls, which would allow the illumination generated from the burning oil to be projected onto the sea 60 km away. Here a lookout tower designed by Thales (see Fig. 1.2 in Sect. 1.6) several centuries earlier would have been installed to provide surveillance of incoming ships. Finally, an enormous statue of Zeus stood atop the lighthouse structure, a symbol of Hellenistic defense and conquest.

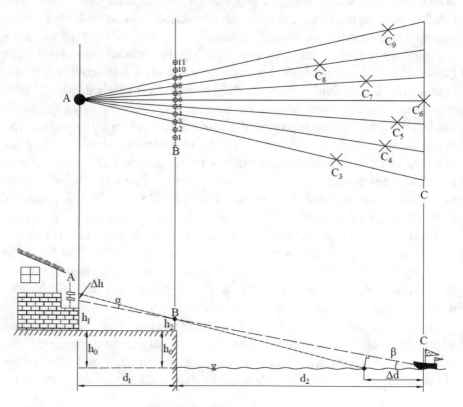

Fig. 1.2 The lookout tower of Thales

With the navigational assistance of the lighthouse, vessels were able to identify safe waterways and be guided into port securely at night. In short, the Lighthouse of Alexandria allowed the port city to prosper as a center of commercial activities.

Apart from the lighthouse, Ptolemy I also built a fleet of naval vessels, then the largest in the known world, to maintain its hegemony in the Mediterranean and to establish effective control over maritime trade. At its height, Ptolemy I's fleet is said to have consisted of over 300 warships of various sizes. The smallest of these vessels had three rows of oars and were manned by dozens of sailors and soldiers, whereas the largest ones had as many as forty rows of oars and were capable of carrying a thousand troops. Merchandise being traded included bulk raw materials such as lumber, gold, silver, copper, iron, tin, as well as various types of processed products such as textile fabrics and earthenware. Extensively traded agricultural products included nuts and cheese from the Black Sea region; olive oil, figs, honey and wool from the northern shores of the Mediterranean; and cured and preserved meat products such as beef, mutton and ham. The most sought-after and profitable product, however, was wine, which was popular among members of the upper classes. Commodities that possessed the highest values may have been gold, frankincense and myrrh, said to have been the three most valuable and treasured items of the Mediterranean world. With such rich varieties of goods and tremendous volumes of trade, merchants in Alexandria became immensely wealthy. The government also benefited greatly from the high taxes that it levied on the traded goods to boost the country's treasury. Despite this dual exploitation, Alexandria's maritime trade continued to thrive for hundreds of years, helping to foster the development of Hellenistic and Roman cultures across the Mediterranean. It was in the great city of Alexandria that the scientific civilization of Greece took shape and flourished.

1.4 The Hellenistic Civilization

After being crowned Pharaoh of Egypt in 332 BC, Alexander the Great continued his military campaigns and eventually founded a vast empire, at an unprecedented pace, that spanned an area exceeding five million square kilometers across Europe, Asia and Africa. His empire, built over the period of his very short lifetime, extended as far as the eastern part of Central Asia and even to modern-day India and Pakistan. The result of this political consolidation enabled the first wave of great advancement in ideas in the history of civilization. According to the German historian Johann G. Droysen

(1808–1884), this period of Greek civilization, also known as the Hellenistic period, began with Alexander's establishment of the Greek empire in 334 BC to the annexation of the Ptolemaic Egypt by the Roman Republic in 30 BC[5], which lasted about three centuries. During this period, the Greeks excelled in areas such as geometry, architecture, sculpture, painting, literature, drama and music, leaving a rich cultural legacy for posterity. Scientific research, in particular, thrived in Alexandria[6], the capital of the Ptolemaic dynasty, and is considered the greatest contribution to the world among all great achievements of the Greek civilization. During a period of about two centuries, the natural philosophers working at the Library of Alexandria made great strides in research that elevated Western science to an unprecedented level. These accomplishments would continue to be the guiding force of Western thought over the next two millennia, and it was not until the 16th century that Greek science was supplanted by the so-called Scientific Revolution, which continued for a century.

Alexander's army brought to the conquered lands not only a new ruler but also the magnificence of Greek culture and thought. The political and cultural transformations were quite successful in the vast plains of Mesopotamia. The culture built around the Tigris–Euphrates river system had existed for over a thousand years, and scattered across the plains lay many centuries-old ancient cities with majestic palaces and gardens interspersed with irrigation ditches. Despite the rich legacy left behind by their ancestors, the inhabitants of these ancient realms began to read and recite Greek poetry, practice Greek wrestling, imitate Greek sculpture and architecture, and so on, following the Macedonian conquest. They even adopted Greek tax laws and replaced their official documents recorded on clay tablets with the more lightweight papyrus. In this part of Fertile Crescent known as the cradle of civilization, the inhabitants abandoned their cultural baggage overnight and completed the Hellenization of everything from art to public administration. Why was Greek culture so aggressive yet so attractive? The answer lies in the region of Ionia on the east coast of the Aegean Sea, in present-day western Turkey.

In the ancient Greek era, several centuries before Alexander the Great created his vast empire, the belief system of Greeks remained centered around myths and legends. They believed that natural phenomena, such as lightning,

---

[5]The Egyptian pharaoh at the time of the Roman conquest was Ptolemy XV, the son of Roman dictator and general Julius Caesar (100–44 BC) and Queen Cleopatra VII Philopator (69–30 BC).

[6]Alexandria remained the largest city of the Hellenic world. It later became one of the largest city of the Roman Empire, second only to Rome.

earthquakes and severe storms, had to do with the gods that ruled the universe, so they would seek divine guidance whenever they engaged in anything from harvesting crops to warfare. Whenever these ancient Greeks looked up at the sky at night, they would admire the world that the mysterious gods created out of chaos. These inspirations and thoughts would eventually turn into oral history and passed down through the generations. At around 700 BC, the legendary poet Homer recorded these oral histories in words, and his texts would become the main guiding doctrines in the education of the Greeks, to be followed by generations of thinkers and philosophers. All of this changed, however, after 500 BC, and the changes began in Miletus, the small ancient Greek city with a warm and sunny climate located in the prosperous and fertile region of Ionia. Overlooking the Aegean Sea, this coastal city had been home to the Carians since 1000 BC. They had built the city into one with a population of a hundred thousand over several centuries. Miletus had an endless supply of fish and riches from the sea, abundant fruits and vegetables as well as olive trees, and it developed into a significant trading center in the Aegean and across the Mediterranean. The city exported fresh produce and foodstuffs and imported ores, timber as well as pottery and furniture from Egypt. This important trading center thus began to accumulate wealth and soon became the wealthiest city in the Greek world.

Wealth fosters the birth of culture, and wealth leads to the creation of leisure. Miletus soon became a living paradise as well as a center of learning, providing a superior environment for cultivating new ideas and technological innovation. In this atmosphere, a group of natural philosophers emerged in the Ionian region, with Miletus at its very center, and one of the most magnificent waves of advancement in philosophy and science in the history of civilization had begun. Among these great thinkers were the trio of master and disciples, Thales (624–546 BC), Anaximander (610–546 BC) and Anaximenes (570–526 BC). The scientific community hailed Thales as the first of the Seven Sages of Greece and the pioneer of science and philosophy, and he was the first Western thinker to have left a legacy of scientific works.

1.5 Natural Philosopher 1: Thales

Thales, who pioneered the field of natural philosophy, did not live a life of poverty as many of his peers did. He was a shrewd businessman who became wealthy by monopolizing the olive oil trade and raising prices to maximize his profits. However, rather than indulging in his wealth and luxuries, Thales made many significant contributions in the field of science and technology with his natural-born abilities, great interest in natural phenomena, and profound insights. In addition to making pioneering scientific discoveries, Thales also abandoned the ideas prevalent at the time that the universe was under the absolute control of the gods, and he embraced the belief that natural changes can be attributed to their own causes and will lead to inevitable consequences. He thus created the theory of causality that would become a mainstream philosophical doctrine centuries later during Aristotle's time. Curious about nature and brave in pursuing the truth of the universe, Thales was also the first person who attempted to explain the cause of solar eclipses, and he accurately predicted the solar eclipse that occurred in Ionia in 585 BC. Whether or not he made this prediction by consulting Babylonian astronomical data available to him, being able to accurately predict the occurrences of such highly local phenomena required extraordinary abilities. What sort of extraordinary abilities could this have been? As suggested by Professor Leonard Mlodinow of the California Institute of Technology, the beginning of science starts with asking the right questions, and Thales indeed did possess this particular talent. He raised many questions in his lifetime, but the tools to answer them were not available. What is the size and shape of the Earth? How do the four seasons come about? How are the Earth, the Sun and the Moon interrelated? All these questions that are closely associated with nature but were hidden, completely out of reach and extremely challenging to tackle, would become the puzzles that scientists of future generations would do everything in their power to piece together.

Although these questions were difficult to resolve, Thales attempted to find satisfactory answers to them. He adopted an approach that would be familiar to modern scientists: formulate a hypothesis and try to find evidence for it. For example, Thales advocated that the Moon's illumination is in fact reflected sunlight. He also once boldly argued that earthquakes are not caused by the wrath of Poseidon, the god of the sea, but instead the Earth is rocked by waves caused by the movement of water on which the world floats. Thales used the approaches taken by the Babylonians to observe lunar and solar eclipses to calculate how far ships are from the shore. He also utilized the shadows caused by the Sun to measure the height of the pyramid. Thales made attempts to estimate the size of the Earth and the Moon, and he was the first astronomer to propose a 365-day year. To appreciate the level of scientific accomplishments in Thales' time, we will reconstruct the lookout tower that Thales built on the coast of Miletus in 500 BC. As mentioned earlier, the design of Thales' lookout tower was inspired by the Babylonians' observations of lunar and solar eclipses, so Thales was likely to have utilized the similarity of right triangles, already familiar to scholars at the time, to translate the triangular relationships of observing the Sun and the Moon in the sky into a similar geometric shape on Earth, which includes ships in a distance, as seen from the tower. The calculations relevant to similar triangles were then carried out to obtain the distance between any two points.

Based on the above discussion, we have designed a lookout tower, as illustrated in Fig. 1.2, for the purposes of conducting surveillance of coastal waters.[7] In this design, we have chosen to set up a piece of lookout equipment at point A, at an altitude of h_1, where a sighting tube is located on the balcony of a building on the shore, which is in turn located h_0 above sea level. At point B, located at a horizontal distance of d_1 from the lookout tower, a row of bamboo poles are erected, each with a height of h_2 and numbered 1, 2, 3,..., up to 11. The distance between the lookout equipment at A and the 11 bamboo poles are, respectively, $d_{11}, d_{12}, d_{13}, \ldots, d_{111}$.

Suppose now an unidentified vessel appears on the sea at a distance. The guards on the lookout tower immediately adjust the height of the sighting tube and at the same time select a bamboo pole as a reference point, from which a straight line can be drawn to connect the unidentified ship, the top of the bamboo pole and the sighting tube. Once this alignment is made, the ship's distance d_2 can be computed from the following relationship:

$$d_2 = \frac{H}{1 - H} d_{1n}, \quad H = \frac{h_0 + h_2}{h_0 + h_1} \tag{1.1}$$

Here d_{1n} represents one of $d_{11}, d_{12}, \ldots, d_{111}$. Combining the two equalities in (1.1), we have

$$d_2 = \frac{h_2 + h_0}{h_1 - h_2} d_{1n} \tag{1.2}$$

From (1.2) we know that the closer h_1 and h_2 are, the farther away from the observed distance of the ship d_2 is, and the opposite is true if h_1 and h_2 are farther apart. Therefore, being able to adjust the height of the sighting tube h_1 is an essential function. When the tube is raised by Δh, d_2 is reduced, so the observed distance of the object on the sea can change by Δd. Their relationship is as follows:

$$\Delta d = d_2(h_1) - d_2(h_1 + \Delta h) = \frac{h_2 + h_0}{h_1 - h_2} d_{1n} - \frac{h_2 + h_0}{h_1 + \Delta h - h_2} d_{1n} \tag{1.3}$$

---

[7] See also Pickover [27].

Simplifying the above, we obtain:

$$\Delta d = \frac{\Delta h(h_2 + h_0)}{(h_1 - h_2)(h_1 + \Delta h - h_2)} d_{1n} \tag{1.4}$$

The above equation can help the designer to determine the maximum distance of a ship d_2 that can be observed and the maximum difference in height Δh that can be adjusted, given the selected bamboo pole's distance d_{1n} and the height h_2. Suppose $\Delta h = 1$ m, $h_0 = 4$ m, $h_1 = 10$ m, $h_2 = 8$ m, and $d_{16} = 100$ m. Then it can be shown that $d_2 = 6d_{16} = 600$ m. Substituting these values into (1.4), we obtain $\Delta d = 200$ m. From these results, it can also be seen that the effect of adjusting the height of the sighting tube is quite sensitive because adjusting it by 1 m will expand the distance of surveillance on the sea by 200 m, which truly leverages the lookout tower's design objectives. Suppose many ships appear on the sea simultaneously, such as the scenario illustrated in Fig. 1.2 with seven ships, C_3 through C_9, each being located at a different distance from the shore. The guards on the tower will then choose the appropriate bamboo poles while at the same time skillfully adjust the height of the sighting tube and use the above equation to compute the required data. Legend has it that the design of the Thales' lookout tower was implemented on the second level of the three-story lighthouse tower built at the port of Alexandria a few centuries later, and it was used for the surveillance of distant ships several kilometers away.

In addition to the lookout tower, Thales also designed a number of cleverly conceived and practical apparatuses, such as the catapult for besieging a city. Alexander the Great used catapults two hundred years later in his conquests of countries across Europe and Asia. Thales' catapult then continued to evolve over the next thousand years into different forms at the hands of numerous designers. Even as late as the 16th century during Galilei's time, the essence of Thales' design remained. Below we will present a design of a catapult using the knowledge and technology in Thales' time. In this design, illustrated in Fig. 1.3, the principal working component is a robust lever measuring several meters long. A rock weighing several tons is attached to its left end, and an iron cannonball to be hurled toward castles is placed on its right end. When preparing for an attack, the right end of the lever will be pulled down to the lowest possible position and locked, and when an attack command is issued, the locking mechanism will be released immediately. Reacting to the torque generated by the massive rock on its left end, the lever rotates in the counterclockwise direction and the iron cannonball is accelerated and projected forward along the tangential direction of its initial circular trajectory.

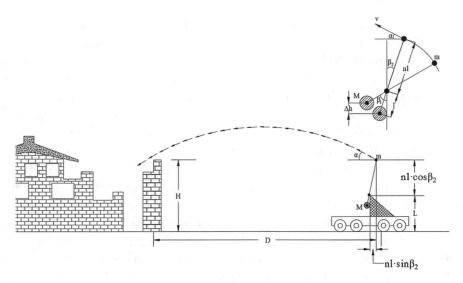

Fig. 1.3 Structural diagram and projection mechanism of Thales' catapult

Based on the design in Fig. 1.3, we analyze the operation of the device using the law of conservation of energy below. For the sake of brevity, we present only the final results here. Refer to "Appendix 1.1 The Catapult of Thales" for their derivations.

$$D = \xi \tan \beta_2 + \sqrt{\xi^2 \tan^2 \beta_2 + 2(-H + L + nl \cos \beta_2)\xi} - nl \sin \beta_2$$
(1.5)

$$\xi = 2n^2 \cos^2 \beta_2 (\cos \beta_2 - \cos \beta_1) \frac{M - nm}{M + n^2 m} l$$
(1.6)

The mathematical notations used in the above two equations are the same as those in Fig. 1.3, and the units of measurement are meter (m) for length and kilogram (kg) for mass. The parameter values likely to have been used at the time are as follows: $M = 10$ metric tons (or 10,000 kg), $m = 100$ kg, $L = 10$ m, $l = 2$ m, $\beta_1 = 180°$ (lever's right end being at the lowest position so that M is at the top), $\beta_2 = 45°$, and $n = 6$. Substituting these numbers into Eq. (1.5), we can obtain the relationship between D and H. In order words, if a cannonball is to be projected successfully over a wall of height H, the distance from which the catapult must be located with respect to the wall must be D. If the wall is $H = 20$ m high, for example, the cannonball should be projected at $D = 230$ m from the wall. The requirement that the expression inside the square root sign of (1.5) must be a non-negative real number also implies that the maximum projection height of the catapult cannot exceed 61 m, and in which case the projection distance must be 76 m.

This design obviously offers much higher assault power than that of those used by Alexander's expeditionary forces. The maximum projection distance of Alexander's catapults was merely 50 m, and the projectiles used were rocks weighing only dozens of kilograms. Despite their limited power, these types of catapults were widely deployed in warfare over a period of several millennia. According to Wikipedia,[8] a certain catapult designed in the mid-14th century was capable of projecting iron cannonballs weighing 450 kg over a distance of 270 m. In modern times, experiments involving working replicas of ancient catapult designs have been carried out. The results show that if the length of the lever is increased to 15 m and the counterweight rock raised to 10 metric tons, then the catapult will be capable of projecting iron cannonballs each weighing 100 kg across a distance of over 270 m, which is similar to the result of the analysis above. Similar designs produce similar levels of performance, so it would seem.

From the ways his lookout tower and catapult were constructed, it is apparent that to be able to design such practical machines, Thales must have been quite familiar with the techniques of trigonometry, widely in use at the time, including actual numerical computations. Thales' reputation as the first of the Seven Sages of Greece is therefore well deserved. He was a doer with the technical know-how and was innovative in his engineering practices. He even proposed a visionary theory about the nature of the Universe, which was purportedly inspired by his observations of the Nile's seasonal flooding and the life created from these cyclical changes. Thales' theory suggests that because water plays a crucial role in maintaining life, possesses both solid and gaseous forms, and is the most abundant liquid on Earth, it is the source of all things and the ultimate origins of the Universe itself. This particular perspective regarding the fundamental constituent elements of the Universe, despite its pioneering nature, sparked many different viewpoints and debates. The first dissenter was none other than Thales' own student Anaximenes, who believed that air would have made a better candidate if the quantity was the only criterion. According to his observations, when air is thin, heat will be generated to form fire, and when air becomes dense, coldness is the result. Then come wind, vapor and water, in that order. Water, if further compressed, will turn into earth and rock. As air is continuously flowing, the changes described above are always present and continue to occur. For this reason, air deserves its place as the ultimate origins of the Universe. Another student of Thales,

---

[8]Wikipedia, https://en.wikipedia.org/wiki/Catapult.

Anaximander, also offered his own theory. He argued that everything in the Universe is made of an endless, boundless and ever-changing element. One day this element suddenly burst into flames and began to burn along the edges of what was to become Earth. Over time the fire subsided and was eventually extinguished, and rings of stars, planets, the Sun and the Moon gradually appeared from top to bottom and began to orbit the Earth, and the Universe as we know it came into existence.

The fact that Thales' students were able to confront their teacher with their own diametrically opposite theories without being accused of disloyalty or disrespect is a testament to the scholastic freedom of this era, which was not constrained by social class or institutional bias. Nor was whether the master-student bond was weakened or even broken over a question. Ultimately the trio even joined forces to propose the first-ever cosmological law in history: The celestial bodies are equidistant and their positions are fixed relative to one another. Located at the very center of the Universe, Earth is cylindrical in shape with a height one-third that of its diameter, and it is superior to all things in nature. To highlight the superior status of the Greeks in the Universe, the trio even designated the Temple of Apollo in Delphi (located in southern Balkans) as the center of the Universe, and they believed that human life originated from the moisture of the Universe. This metaphysical naturalism, which explains the origins of humankind and the Universe based on natural theories, was a significant progress compared with the deity-based mysticism that had existed for a long time, even though this improvement was the culmination of 2,000 years of slow evolution that had begun earlier around 3000 BC.

Great thinkers like Thales, who were active in Ionia and were used to thinking and seeking answers in accordance with the laws of nature, were described by Aristotle (Aristotélēs in Greek, 384–322 BC) as the first physicists in history. However, this group of physicists was dispersed in 550 BC after Miletus was conquered by the Persians, and subsequently the general atmosphere of formulating hypotheses from observing nature gradually declined. Worse still, the Miletians were almost completely wiped out by the Persians fifty years later in an unsuccessful coup, which ended the prosperity of Miletus for good. During this same period, however, another group of thinkers emerged on the island of Samos, right across the Aegean Sea from Miletus. These philosophers were skilled at applying mathematics to model the Universe, the brightest of them being Pythagoras of Samos (c. 570–495 BC). According to legend, one of the reasons for his success was the journey he made to Miletus at the age of 18, where he drew great inspiration from Thales himself. Pythagoras also traveled to Egypt, Babylonia, Phoenicia and other contemporary civilizations, and he developed a comprehensive knowledge about the different cultures around the Mediterranean. When he was 40, he left Samos alone for Croton, a city in the southern Italian Peninsula, to escape the incompetence and brutality of the ruler of Samos. He chose to leave the beautiful island where he had lived for nearly half a century and settled in the south of Italy, where he founded his mathematical kingdom.

1.6 Natural Philosopher 2: Pythagoras

Around 3000 BC, ancient Egyptians living in the Nile valley began to use the Egyptian numerals, a sexagesimal (base-60) numeral system written in hieroglyphs, to perform fairly complex arithmetic operations. They also developed simple algebraic techniques to solve quadratic equations using verbal descriptions. About a millennium later, this system spread to the Babylonian Empire in Mesopotamia, and the Sumerians modified the system into the following form: each numeral either counts "units" or "tens", e.g., 1 is represented by Y and 2 by YY, 3 by YYY, and so on, up to 9. The number 10 is written as <, 20 as ≪, and so on, up to 50 which is a complicated combination of five < (see Fig. 1.4[9]). The sexagesimal scheme is still in use today in specific systems of

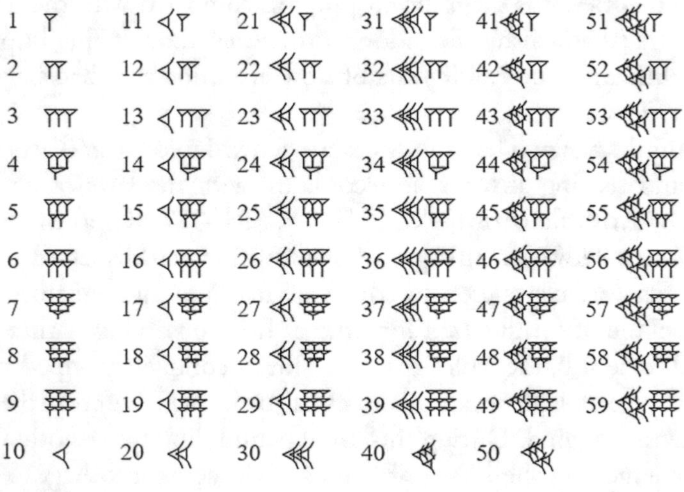

Fig. 1.4 The sexagesimal numeral system used by Sumerians, All icons have been recreated based on the picture of following website, https://commons.wikimedia.org/wiki/File:Babylonian_numerals.svg

measurement. For example, in angular measurement, a circle is divided into 360 degrees (°), each degree is divided into 60 arcminutes (arcmin, '), and each minute is divided into 60 arcseconds (arcsec, "). In time measurement, an hour is divided into 60 min, and a minute is divided into 60 s.

Using 60 symbols to represent a number created complicated arithmetic operations, and therefore errors were common and it was often even impossible to obtain answers. For this reason, humans later developed a decimal (base-10) numeral system, which requires only 10 symbols, and this system is still in use today everywhere. However, computer engineers prefer the binary system of Boolean algebra to represent numbers to simplify computational procedures. Each number in the binary system is represented by only two symbols. Below we illustrate their differences using the number 183. In its decimal representation, $183 = 1 \times 10^2 + 8 \times 10^1 + 3 \times 10^0$, three numerals, 1, 8, and 3, are required. The binary representation of the same number is $183 = 1 \times 2^7 + 0 \times 2^6 + 1 \times 2^5 + 1 \times 2^4 + 0 \times 2^3 + 1 \times 2^2 + 0 \times 2^1 + 1 \times 2^0$, or, 10110101_2. Although the expression is longer, the representation requires only 2 symbols, 1 and 0. This digital logic, with much simplified computational procedures due to its limited number of digits, is indeed the optimal choice for computers, which excel at computing and possess large storage space.

At first, the cumbersome numeral system invented by the Sumerians (as depicted in Fig. 1.4) was used only for everyday activities, such as the creation of running accounts to keep track of materials, labor and livestock. As human endeavors became more ambitious and demanded more complex quantitative tools, arithmetic based on addition and subtraction with five fingers evolved into multiplication and division. Algebraic techniques involving quadratic or even cubic equations with unknown variables were developed over time. Evolving from arithmetic to mathematics, the numeral system was not only used in commerce but also in engineering designs. For example, when designing irrigation ditches, the engineer must first calculate how many cubic units of earth should be hollowed out, and then divide it by the amount of earth a worker must remove per day in order to come up with the manpower and the number of days required to complete the construction project. From the information contained in the ancient tablets that have been excavated from

---

[9]In his book, Evans has explained the algorithm of the sexagesimal numeral system of Sumerians. Please refer to J. Evans, "The history & practice of ancient astronomy," Oxford University Press, Oxford, UK, 1998, pp. 13–14.

archaeological sites, this type of logic was repeatedly applied to solve problems, which shows that the basic ability to reason logically has existed since ancient times, and it is part of our natural-born abilities.

Mathematics is more than just numbers. There is also the mathematics of shapes or geometry. In Egypt, the regular flooding of the Nile caused land boundaries to be blurred and problematic. In an agrarian society where land was more important than life, correctly calculating land area was an urgent task. Out of this necessity, the Egyptians developed techniques to calculate the area of their lands and defined various plane geometric shapes, such as the square, the triangle, the circle, and the trapezoid as well as solid figures, such as the cube and the cylinder. All these shapes were used in everyday tasks such as measuring land areas and calculating the volume of harvested grain. The geometry of the Egyptians and the arithmetic and algebra developed by the Babylonians were essential tools for advancing civilization in ancient times. However, the purpose of developing these tools was practical, and the tools were rarely used for the pursuit of knowledge or truth per se. This practical orientation, nevertheless, gradually changed when Pythagoras burst onto the scene, as he pioneered a number of geometric principles and provided original proofs to raise the profile of geometry to the theoretical level from its mundane beginnings as a practical tool.

One of the most famous examples of this geometric theory is the familiar Pythagorean theorem[10]: The square of the hypotenuse of a right triangle is equal to the sum of the squares of the other two sides. This theorem may have been the masterpiece of Pythagoras, although no evidence of his actual proof exists. Below we will demonstrate the development of geometry as a discipline at the time of Pythagoras with the proof provided by Euclid (a contemporary of Pythagoras) using the diagram in Fig. 1.5. In this diagram the squares and the right triangle are connected in two ways: (1) Area of the square $BCDE = 2\times$ area of the triangle $\triangle ABE = 2\times$ area of triangle $\triangle BCM =$ area of $MBPQ$. Here the second equality holds because $\triangle ABE$ and $\triangle MBC$ are congruent. (2) Area of the square $ACFG =$ area of the rectangle $NAPQ$. Combining (1) and (2) we obtain the Pythagorean theorem. Historically there have been a number of proofs of the theorem. In addition to Euclid's proof above, there is also a proof published by James A. Garfield (1831–1881), the 20th President of the United States, when he was a member of the House of Representatives. A third proof uses the proportional relationships between the sides of similar triangles. See "Appendix 1.2 Proofs of the Pythagorean Theorem."

---

[10]According to Hsiang et al. [15], p. 3, a version of the Pythagorean theorem was used in ancient China.

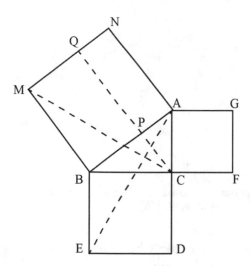

Fig. 1.5 Geometric configuration employed by Euclid to prove the Pythagorean theorem

Despite its apparent simplicity, the Pythagorean theorem is broadly applicable and has contributed extensively to the techniques of measuring lengths and distances. Some, however, utilized it as a weapon to eliminate enemies. According to legend, Hippasus (c. 500 BC), who was a Pythagorean from the south of the Italian Peninsula, once described the Pythagorean theorem as follows: the sum of the areas of the squares formed by a right triangle's perpendicular sides a and b, $a^2 + b^2$, should be equal to the area of the square constructed with the triangle's hypotenuse, c^2. However, if we take $a = b = 1, then c = \sqrt{2}$, which cannot be expressed as the ratio of two non-zero integers,[11] and this violates the first axiom of Greek geometry, commensurability, an axiom treasured by the Pythagoreans. According to Hsiang et al.[12] Hippasus disproved the axiom of commensurability using a regular pentagon $AEDCB$ with the vertex B at the top (see Fig. 1.6). Two lines CC_1 and DD_1 are extended from the pentagon's two sides BC and ED, respectively, and the lengths of these extensions are exactly the same as OC (or OD). Joining C_1 and D_1, we obtain another regular pentagon OCC_1D_1D, and repeating this process an infinite sequence of regular pentagons with decreasing areas can be generated indefinitely. Therefore, the lengths of the sides

---

[11]The first axiom of Greek geometry was known as the axiom of commensurability, which stated that all line segments are commensurable ("any two lengths have a third length which divides both an integer number of times"). From this axiom one can deduce the area of a rectangle or a triangle, and even the Pythagorean theorem.

[12]Refer to Hsiang et al. [15], p. 8.

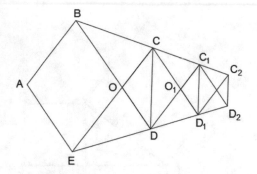

Fig. 1.6 Regular pentagons used by Hippasus to prove non-commensurability. Inspired by Fig. 1.5 of Hsiang et al. [15]

of these pentagons cannot possibly be related through integer multiples, and commensurability cannot be valid.

It is said that Hippasus was exiled and even murdered by the Pythagoreans for overturning the first axiom. Losing one's life over mathematical research is unimaginable. This could have just been a myth with no factual basis, or perhaps a very unusual case, so there was no negative impact on the development of mathematics at that time. Pythagoras therefore was able to explore musical rhythms and patterns at a leisurely pace. The process by which Pythagoras explored musical temperament can be seen from the following legend. One day young Pythagoras was taking a stroll in the city. When he passed by a blacksmith, he heard something interesting: there was an intriguing relationship between the sound and rhythm as the blacksmith struck the iron. Upon returning home, Pythagoras set up a five-stringed musical instrument and held a wooden stick in each of his two hands, one for striking the string and the other for adjusting the string's length. One end of each of the five strings is supported by the instrument's bridge, and the other end is attached to a stone used as a counterweight to set the string's tension. In modern terms, Pythagoras would be described as a mathematician with absolute pitch, since he used clear logical reasoning to define the musical scales audible to the human ear. According to legend, Pythagoras formulated the scales by dividing the sounds produced by two strings with a 2:1 ratio in length into seven notes (frequencies), thus forming an octave. He also divided the sounds produced by two strings with a 3:2 ratio in length into four notes to produce a "fifth". Finally, he put these two scales together by ordering the notes to get 12 semitones.

The musical chords that are familiar to us today are also based on the principle: two strings with the same diameter and made from the same material will produce notes that are an octave apart if their lengths have a ratio of 1/2, and the two notes will be represented by the same letter. A length ratio of 2/3 corresponds to an interval of a fifth, and a ratio of 3/4 to a fourth. Each of these three ratios will produce a pleasant harmony. On the other hand, if two strings have a length ratio that cannot be expressed as a ratio of two integers, e.g., 1.32/3.45, then together they will produce a harsh sound like that of an alarm or siren. Pythagoras' acoustic principles were widely influential on the development of music in subsequent generations, whose incremental revisions over time have evolved into modern-day music theory, as well as the theory of wave mechanics. Refer to "Appendix 1.3 Pythagoras' Vibrating String: Length and Harmony" for its history.

Whether or not Pythagoras actually proved the theorem named after him or discovered the laws of acoustics and harmonics can no longer be ascertained. But one thing that everyone can agree on is that Pythagoras transformed mathematics from its mundane role in everyday commercial transactions to the scientific language employed by scientists in their research, thus successfully providing a mathematical framework for describing how nature operates. Pythagoras' attempts were later abandoned by the Aristotelians, but his quantitative approach was re-established by great scientific minds such as Galilei, Kepler and Newton almost two thousand years later in the 17th century. From this point forward, science began to take shape as a quantitative discipline, based on facts and logic, and the approaches to studying natural phenomena began to evolve from statements made by natural philosophers to computations in the minds of natural scientists.

1.7   Natural Philosopher 3: Euclid

Apart from mathematics, another discipline that contributed substantially to the understanding of the laws of nature and their perfect order is geometry. This useful and sophisticated tool was utilized by the Egyptians and the Babylonians to create architectural designs as early as 3000 BC. Geometry was later expanded upon and improved further by the natural philosophers of Greece, the most iconic of whom was Euclid. Euclid wrote the *Elements*, a mathematical treatise that has been tremendously influential on the development of geometry for two millennia. During the time of Euclid (a contemporary of Plato), around 300 BC, Greek philosophers were already familiar with making calculations based on the geometric relationships of triangles and polyhedra. Born in Athens, Euclid was invited by Ptolemy I to conduct research on geometry in Alexandria at the age of 30. It was there that Euclid completed the *Elements*, the most important work of his life. The treatise consists of 15 books and is a compilation of propositions on a plane and solid geometry based on works by earlier Greek mathematicians, and Euclid provided many rigorous proofs to them. Euclid's *Elements*, with its comprehensive coverage and logical rigor, continued to be relevant and its status was unchallenged until the 18th century.

From the propositions found in the *Elements*, the reasoning of Euclid's geometric proof is considered quite rigorous. Taking the calculation of the circle's circumference as an example, Euclid divides the circle into numerous isosceles triangles, as shown in Fig. 1.7. The sum of the bases of all these triangles provides an approximation to the circumference S. If we divide this sum by r, we obtain a value. Now repeat this exercise with another circle of a different size (radius), we obtain the same value, which is 2π. Therefore we have $S = 2\pi r$. The same diagram can also be used to calculate the area of the circle by adding the total area of all the triangles, as follows. Each triangle has an area of approximately $\frac{1}{2}r\Delta S$, and summing over all triangles we have $\frac{1}{2}r \cdot \Sigma_{i=1}^{n}\Delta S = \frac{1}{2}rS = \pi r^2$, which is the familiar equation for the circle's area. This is what is known as the theory of approximation, a method often employed to estimate the area or volume of a geometric figure in Euclid's time.

In addition, the equation for the cone's volume is also proved in detail in the *Elements*. Legend has it that Democritus (c. 460–370 BC) was the first to observe that a cone has one-third the area of its base multiplied by its height, which is correct, although he did not provide its proof. It would have to wait until Eudoxus of Cnidus (390–337 BC) proved it using the theory

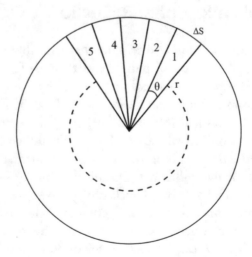

Fig. 1.7 Euclid's proof of the area of the circle via the theory of approximation

(a) (b)

Fig. 1.8 Illustration of Euclid's equation for the volume of a prism **a** volume of a cylinder **b** volume of a triangular prism

of approximation.[13] Euclid later provided a concrete proof, which we will explain with the following logic. First of all, we outline the conditions under which two triangular pyramids have the same volume. From the two stacks of coins illustrated in Fig. 1.8a, we know that both stacks contain the same number of coins and each coin has the same volume, so the two stacks have the same total volume, despite the different ways the coins are stacked. By analogy, we arrive at the following theorem: If two objects have the same cross-sectional area at the same height, they also have the same volume. This theorem can also be stated more succinctly as: Two pyramids having the same base area and the same height also have the same volume.

Now, let us examine the triangular prism in Fig. 1.8b. Firstly, we divide the triangular prism, with base area A, into three triangular pyramids. Here pyramids 1 and 2 have the same volume, as they have the same base area and the same height. Pyramids 3 and 2, in addition, also have identical base areas and identical heights, so they too have the same volume. Pyramids 1 and 3 have the same volume as well using the same argument. Since all pyramids have the same volume, the volume of each pyramid V must be one-third that of the prism, i.e., $V = \frac{1}{3}Ah$. This equation also applies to a prism with an arbitrary polygonal shape for its base, such as a quadrilateral pyramid and a hexagonal pyramid. The proof is simply to divide the base into triangles, and hence the prism into many triangular prisms with a common vertex and the

---

[13]The theory of approximation was used by Eudoxus to prove that area equations, the Pythagorean theorem, and the like, which are valid under the commensurability hypothesis, are also generally justified under incommensurability conditions. This elevated geometry to a scientific discipline that lends itself to rigorous treatment. Euclid's *Elements* has collected a large number of geometric principles proved by Eudoxus using the theory of approximation, the volume equation for the triangular pyramid (also a tetrahedron) being one of them.

same height. Adding the total volumes of these triangular prisms, we obtain $\frac{1}{3} \Sigma_{i=1}^{n} \Delta A_i h = \frac{1}{3} A h$.

It can be said that the geometric reasoning and proofs provided in the preceding paragraphs transformed geometry into a rigorous scientific discipline. At the same time, geometry has been hailed as the very first scientific discipline ever created during the course of development of the Civilization of Rational Mind. On the whole, geometers have exhausted all the principles and techniques in order to comprehend the three aspects that represent the essence of space[14]: (1) symmetry: antisymmetry exists with respect to a plane in space, it can be used for a plane. (2) straightness: The sum of the measures of the interior angles of a triangle is always 180°, a straight angle, which is equivalent to the parallel postulate; (3) continuity: the straight line is continuous, and it retains its continuity after losing a point (e.g., having been cut at that point) and becoming two straight lines. Because of its uniqueness, to be described in Sect. 1.8, geometry was propelled into a genuine branch of science based on rigor and rational argument, and indeed a great achievement in the history of civilization, after it underwent the following two developments: first, Hippasus' refutation of the axiom of commensurability; and secondly, the derivation of the equation for cones and pyramids using the theory of approximation pioneered by Eudoxus.

Geometry is not only a kind of science with a foundation in logical reasoning. It is also an extremely powerful tool in a large number of practical applications. The person who took geometry and applied its principles to the next level of practicality is Archimedes (c. 287–212 BC), who was hailed as the greatest scientist and inventor in ancient Greece.

1.8　Natural Philosopher 4: Archimedes

Archimedes was born in an era of peace after the empire of Alexander the Great had disintegrated and science began to flourish under the rule of the Ptolemaic dynasty in Egypt. His birthplace was the ancient city of Syracuse, Sicily, an island deeply influenced by Greek culture. Archimedes traveled to the Egyptian city of Alexandria to study, and he conducted extensive research on Euclid's *Elements*, especially on problems of the infinite using the method of approximation, which was tremendously helpful in deriving the area enclosed by a circle and the volume inside a sphere. In a scroll of parchment rediscovered in the 20th century, known as the *Archimedes Palimpsest*,

---

[14]Refer to Hsiang et al. [15], p. 3.

his research on mathematics and mechanics was described in detail, and he has been lauded by modern historians of science as the most influential scholar in the development of mathematics.[15] Apart from his contributions to geometry, Archimedes also invented various levers and pulleys capable of lifting heavy objects and also the screw pump (or Archimedes' screw) for pumping water,[16] iron claws used in battles to overturn and sink enemy ships, and even fortifications equipped with light and heat-focusing concave mirrors to set invading Roman warships on fire.[17] What is now most familiar to us about him is Archimedes' principle: the upward buoyant force exerted on a body immersed in water, whether fully submerged or partially submerged (floating), is equal to the weight of the water displaced by that body.

From his principle of buoyancy, described above, Archimedes also derived a technique that he used to measure the density of an object, which would have discouraged goldsmiths from mixing lesser metals such as brass and silver into pure gold when making crowns. Archimedes' most remarkable accomplishments, however, lie in geometry. An impressive body of knowledge of geometry had already been accumulated at the time, including the valuable contributions by Eudoxus of Cnidus, Euclid and Apollonius of Perga (262–190 BC). Archimedes' efforts were by no means any less impressive. For example, he was the first person to offer an approximation of π as $3\frac{1}{7}$ using nothing more than purely geometric methods. He also calculated the areas of figures in plane geometry as well as the surface areas and volumes of various solid objects. These geometric arguments and reasoning would be tremendously helpful to Sir Isaac Newton two thousand years later in his development of the theory of calculus, which perfectly blended physics and mathematics together. According to one account, when the Roman general Marcus C. Marcellus (268–208 BC) captured the city of Syracuse in 212 BC, he had Archimedes, who was at the time contemplating a geometric problem, summarily executed. The tombstone of Archimedes was reported to be engraved with a diagram illustrating a sphere and a circumscribed cylinder of the same height and diameter. It is speculated that the geometers who was charged with burying Archimedes intended to showcase his proof that "the volume

---

[15] See Cooter [7].

[16] In an Archimedes' screw, water from low-lying places is pumped to an elevated level by manually turning a screw-shaped surface inside a wooden or bamboo pipe. Examples include transferring water from rivers to farmland located on higher altitudes, and pumping groundwater to irrigation ditches.

[17] In 212 BC, Archimedes assembled a concave mirror using multiple plane mirrors. The combined mirror was reportedly used to focus reflected sunlight at noon on a distant warship, causing it to burn. While theoretically possible, the criteria under which this could be carried out would be quite stringent, since it would require a long time for the focused sunbeam to produce enough heat to set the ship's hull on fire. As a result this instrument would more likely be used for surprise attacks rather than in a regular combat situation.

and surface area of the sphere is two thirds that of the cylinder including its bases" (from his work *On the Sphere and Cylinder*) as one of the most important scientific discoveries of Archimedes' life.

Archimedes arrived at the approximation of π, $3\frac{1}{7}$, using purely geometric reasoning. How was he able to produce such precision? One account is that he employed the method of approximation and a 96-sided regular polygon. To estimate the upper bound of π, he computed the perimeter of the circumscribed 96-gon, as illustrated in Fig. 1.9. Here we attempt to derive it using a modern trigonometric function (the tangent). Suppose the radius of the circle in Fig. 1.9 is 1, and its circumscribed regular polygon with n sides has a perimeter of $2n\tan(\pi/n)$. The n-gon's perimeter is therefore greater than the circle's circumference, and we have

$$2\pi < 2n \tan \frac{\pi}{n} \Rightarrow \pi < n \tan \frac{\pi}{n} \tag{1.7}$$

Since we know that $\tan(\pi/6) = \sqrt{3}/3$, the identity $\tan 2\theta = 2\tan\theta/(1 - \tan^2\theta)$ can be used to compute the values $\tan(\pi/12)$, $\tan(\pi/24)$, $\tan(\pi/48)$, etc. Finally $\tan(\pi/96)$ is approximated as

$$96 \tan \frac{\pi}{96} \approx 3.1427146 \approx \frac{22}{7} \tag{1.8}$$

In the above derivation, we employ the method of approximation together with a trigonometric function to reconstruct Archimedes' computational procedure. Note that trigonometry did not yet exist at Archimedes' time, so

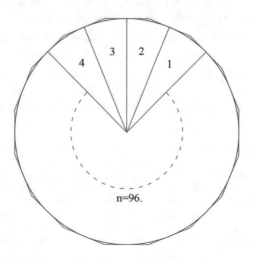

Fig. 1.9 Diagram illustrating Archimedes' calculation of π

here we use trigonometry purely as a device of convenience and also because trigonometric functions can be represented with the ratios of sides in a right triangle. The upper and lower bounds of the value of π can also be derived using the methods of calculus (i.e., non-geometric approaches). Refer to "Appendix 1.4: Deriving Archimedes' π Using Integration." It can be seen from the above trigonometry-based method that natural philosophers over two millennia ago, including Archimedes, already possessed a great deal of knowledge about geometry, which is truly astounding.

However, Archimedes' proof of the volume of the sphere using methods of mechanics (i.e., the principles behind levers together with the calculations of the volumes of cylinders and cones) can be regarded as the pinnacle of the study of geometry up to that point in history. The logical reasoning in Archimedes' derivational approach, which combines mechanics and geometry, is intriguing and thought-provoking. Let us now reconstruct this process.[18] First of all, the geometric algorithm for deriving the volume of the sphere requires the following results, which are already available from Euclid's *Elements*: a circle has a circumference of $S = 2\pi r$ and an area of $A = \pi r^2$, and a cylinder has a volume of $V = Ah$, where r is the radius of the circle and h is the height of the cylinder; and the triangular (or trigonal) pyramid has a volume of $V = \frac{1}{3}Ah$, where A is the area of its triangular base and h is its height. Armed with these equations, we can now begin to derive the volume of the sphere $V = \frac{4}{3}\pi r^3$, where r is the radius.

We begin with the rectangle $IFGH$ in Fig. 1.10a, which has a width of $2r$ and a height of $4r$. A circle of radius r is placed at the exact center of the rectangle, and the two longer sides of the rectangle become tangent to the circle at points B and C (so that BC forms a diameter of the circle). Now, take $\overline{BC}$ as an axis of rotation and form a cylinder by revolving the rectangle on this axis, which will produce a volume of $V_2 = \pi(2r)^2 \cdot 2r = 8\pi r^3$. Meanwhile, the same revolution applying to the triangle $\triangle BGF$ will form a cone with a volume of $V_1 = \frac{1}{3}\pi(2r)^2 \cdot 2r = \frac{8}{3}\pi r^3$, and likewise, the circle at the center will form a sphere. Next, we consider a straight line $L\left(\overline{JK}\right)$ between and parallel to the rectangle's two longer sides. L intersects with the sphere's diameter $\overline{BC}$, the triangle $\triangle BGF$ and the circle at D, E and A, respectively. As $\triangle ABC$ and $\triangle DBA$ are similar, we have $\overline{BD} : \overline{AB} = \overline{AB} :$

[18]The geometric configurations in Fig. 1.10 are based on the lecture notes provided by Professor Yon-Ping Chen, Department of Electrical and Computer Engineering, National Chiao Tung University, Hsinchu, Taiwan. The notes provide detailed descriptions of Archimedes' derivations of the sphere's volume and surface area. Download the file here (in Chinese):
http://jsjk.cn.nctu.edu.tw/JSJK/Pre_U\%20School/Archimedes\%20and\%20spherical\%20area.
pdf. For further information on the principles behind these derivations, the reader is referred to Polya [28], Chapter 9, pp 166–170.

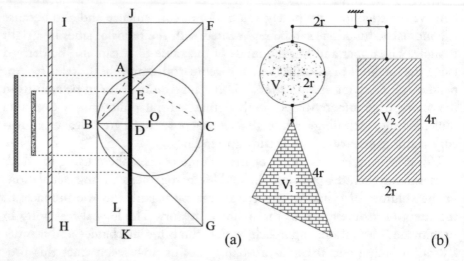

Fig. 1.10 Geometric figures used by Archimedes to compute the sphere's volume: **a** geometric relationships between the sphere, the cylinder, and the cone; **b** how the Law of the Lever governs the sphere, the cylinder, and the cone Inspired by Fig. 9.13 of Polya [28]

$\overline{BC}$, or $\overline{AB}^2 = \overline{BD} \times \overline{BC}$. From the Pythagorean theorem, we have $\overline{AB}^2 = \overline{DA}^2 + \overline{BD}^2$, and combining these two expressions results in $\overline{DA}^2 + \overline{BD}^2 = \overline{BD} \times \overline{BC}$. Since $\overline{BC} = 2r$ and $\overline{DB} = \overline{DE}$, we then have $\overline{DA}^2 + \overline{DE}^2 = \overline{BD} \times 2r$ (see Fig. 1.10a). Multiplying both sides by $2\pi r$, we obtain the following:

$$2r\left(\pi \overline{DA}^2 + \pi \overline{DE}^2\right) = \overline{BD} \times \pi(2r)^2 = \overline{BD} \times \pi \overline{DJ}^2 \qquad (1.9)$$

Here $\pi\overline{DA}^2$, $\pi\overline{DE}^2$, and $\pi\overline{DJ}^2$ represent, respectively, the areas of the cross-sections (or slices) of the sphere (centered at O), the cone (revolution of ΔBGF) and the cylinder formed by $IFGH$. Suppose the sphere, the cone and the cylinder are made from homogeneous materials and also have the same density, we multiply both sides of (1.9) with the density ρ, the disc's thickness w and the acceleration due to gravity g, we obtain

$$2r \times \left[\rho\left(w\,\pi\,\overline{DA}^2\right)g + \rho\left(w\,\pi\,\overline{DE}^2\right)g\right] = \overline{BD} \times \rho\left(w\,\pi\,\overline{DJ}^2\right)g \qquad (1.10)$$

In the above equation, $\rho\left(w\pi\overline{DA}^2\right)g$, $\rho\left(w\pi\overline{DE}^2\right)g$ and $\rho\left(w\pi\overline{DJ}^2\right)g$ are the weights of the cross-sectional slices of the sphere, the cone and the cylinder, respectively. If we multiply a weight by a distance, then the moment of force (or torque) comes to mind. Now, we rearrange these three objects into the beam-balance configuration illustrated in Fig. 1.10b, where the sphere and the cone are placed on the left-hand side and the cylinder on the right. If we align the cylinder's left base with the vertical plane containing the fulcrum (as shown in Fig. 1.10b), then the right-hand side of (1.10) is the moment of force created by of the cylinder's cross-sectional slice relative to the fulcrum. Next, since (1.10) is equally valid for all parallel cross-sectional slices in the cylinder, if we sum across all these slices, the right-hand side of (1.10) will represent the moment of the entire cylinder about the fulcrum. On the left-hand side of (1.10), since the quantity inside the square brackets represents the weight of the cross-sectional slices of the cone and the sphere combined, summing all the slices up will give us the total weight of the two objects. Therefore, (1.10) has the following interpretation:

$$2r \times (weight\ of\ cone\ and\ sphere) = moment\ of\ cylinder's\ weight\ about\ a\ fulcrum \tag{1.11}$$

What follows is where the clever part comes in. Given that the moment generated by a rigid body's weight originates from the rigid body's center of mass, so if the cone and the sphere's point of suspension has a horizontal distance of $2r$ from the fulcrum, then the entire equation represents an equation of moment equilibrium, and the left-hand side of (1.11) can also be expressed as $r \times (\text{weight of cylinder})$.

Next, we compute the moments produced on each side of the fulcrum. Suppose the volumes of the sphere O, the cone $\triangle BGF$ and the cylinder $IFGH$ are V, V_1 and V_2, respectively. The moment equilibrium equation can be expressed as:

$$2r \times \rho(V + V_1)g = r \times \rho V_2 g \tag{1.12}$$

The volumes of the cone and the cylinder are $V_1 = \frac{8}{3}\pi r^3$ and $V_2 = 8\pi r^3$, as have been derived previously. Substituting these two expressions into

(1.12), we arrive at the sphere's volume:[19]

$$V = \frac{4}{3} \pi r^3 \qquad (1.13)$$

Equation (1.13) gives us the volume of the sphere. Once this volume is available, we can partition the sphere into many small hexagonal pyramids with a common vertex at the center of the sphere, each having a volume of $\frac{1}{3}\Delta Ar$ (ΔA being the area of its hexagonal base). Aggregating the volumes of all these pyramids, we get[20]

$$\frac{1}{3}\Sigma_{i=1}^{n} \Delta A_i r = \frac{1}{3}Sr = \frac{4}{3} \pi r^3 \qquad (1.14)$$

and the sphere's surface area can be deduced from (1.14) as $S = 4\pi r^2$.

Also worth noting is Apollonius of Perga, a distinguished Greek geometer and astronomer whose scientific career began a little later than that of Archimedes. In his work *Conic Sections*, Apollonius mentions that intersecting a plane with a cone at different angles will produce different types of curves. As illustrated in Fig. 1.11, an ellipse is formed by the intersection of the cone and a plane inclined to the cone's circular base, but without the plane cutting through the cone's base. A parabola is formed with a cutting plane parallel to a side in the conic surface of the axial triangle. The hyperbola is formed by a cutting plane perpendicular to the base (or parallel to the cone's axis of rotation). These curves were later instrumental to the development of Kepler's laws of planetary motion.

---

[19]There is yet another simple method to calculate the volume of a sphere. Suppose a sphere of radius r is inscribed in a cylinder of height $2r$. Assuming that the two bases of the cylinder are coplanar with the two horizontal planes $z = -r$ and $z = r$, respectively, then for any horizontal plane of height $z \in [-r, r]$ will intersect with the sphere to form a circle of radius $\sqrt{r^2 - z^2}$ and an area of $B = \pi(r^2 - z^2)$. The same horizontal plane also intersects with the cylinder to form a circle of radius r and an area of $A = \pi r^2$. The difference in area between these two circles is $C(z) = A - B = \pi z^2$. Note that this area is exactly the same as the cross-sectional area of a cone of radius is r and height r at a height of z from its base. This is because if two objects have the same cross-sectional area at any arbitrary height, their volumes will also be identical (as illustrated by two stacks of coins in Fig. 1.8). Therefore, subtracting the sphere's volume from the that of the cylinder results in exactly twice the volume of the cone. Since we know that the cone's volume is $\frac{1}{3}\pi r^3$ and the cylinder's is $2\pi r^3$, the sphere will have a volume of $\frac{4}{3}\pi r^3$.

$V = \frac{4}{3}\pi r^3$ 1.13

[20]Referenced from Chu [6].

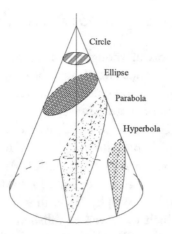

Fig. 1.11 Relationships between various conic sections and their respective cutting planes within on a conical surface

From the discoveries made by Archimedes and Apollonius in geometry, we can see that the early development of Greek science relied heavily on geometric reasoning rather than manipulating algebraic equations. Thus geometry exerted a considerable influence on science that persisted well into the 16th century, at the dawn of the Scientific Revolution. For example, Galilei's *The Assayer*, published in 1623, is representative of the type of works that employed geometry to explain physical phenomena. Even in *Philosophiæ Naturalis Principia Mathematica* (usually shortened to *Principia*) Newton's unprecedented work on the law of universal gravitation, which influenced later generations of scientists, geometric tools were employed to the fullest extent to describe the principles of calculus before the law of universal gravitation was finally derived.[21] The transformation of geometry into algebra, however, would have to wait until the 17th century, when René Descartes (1596–1650) developed analytic geometry. This classical branch of mathematics describes geometric properties with mathematical equations, which are then manipulated using algebraic techniques to get the desired results. By then, geometry had begun to take on the true characteristics of mathematics.

---

[21] Newton specified the law only in words at the time. The modern, well-known equation that describes the universal law of gravitation was first proposed by Henry Cavendish (1731–1810) in his paper.

1.9   Natural Philosopher 5: Aristotle

Ideas and different schools of thought of natural philosophy flourished in Classical Greece. Great advances were made in geometry and practical technology, even by modern-day standards. Aristotle, the great natural philosopher, was at the pinnacle of these achievements. Born in the small town of Stagira in northeastern Greece in 384 BC, Aristotle had a father who was a physician at the court of King Amyntas II of Macedon (?–370 BC). Both of Aristotle's parents died when he was a child. At the age of 17, Aristotle traveled to Athens by himself and joined the Platonic Academy to study various disciplines in natural philosophy, including botany, zoology, astronomy and critical thinking, which were well-established branches of knowledge in Greece at the time. During his 20-year career at the Academy, Aristotle established a unique style of research and earned the title "Heart of the Academy". However, all of this came to an end in 347 BC when Speusippus, Plato's nephew, took over as head of the Academy after the death of Plato. Aristotle decided to leave Athens to steer clear of provocations from Speusippus and also the anti-Macedonian sentiments in the city. He lived in solitude in a city-state across the Aegean Sea for four years. At the age of 41 in 343 BC, Aristotle was invited by King Philip II of Macedon to become the tutor to his 14-year-old son, the future Alexander the Great. Aristotle taught the talented young prince about morals, politics, natural philosophy and other disciplines. As a result, Alexander developed an interest in and came to rely on scientific knowledge throughout his life. Later, Aristotle founded the Lyceum in Athens with the support of Alexander and recruited Greek orators and scholars to promote free thinking and debate on the basis of sufficient data and information. The Lyceum served as the breeding ground for the adherents of the Peripatetic school of philosophy and also what would become Aristotelianism.

Aristotelianism was an all-encompassing school of thought. Here we will present aspects of the Aristotelian doctrines that are concerned only with natural philosophy. Aristotle advocated that changes to the world are divided into two types, the natural and the unnatural, both of which can be attributed to the interaction among the four basic elements that make up the Universe: earth, air, fire and water. For example, changes in an object's position (i.e., its motion) is natural; a rock rolling down the hill or rain falling on the surface of the sea are natural manifestations of the elements earth and water returning to their original states. However, if one wishes to make a rock fly in the air, external intervention, which is unnatural, is required. Similarly, a piece of meat rotting is a natural change, but if it is to be preserved and prevented from going bad, unnatural methods such as salting are required. On the other hand, when something is undergoing a natural change and is disturbed, an unnatural change will occur. Suppose a stone is thrown into the air and it hits a flying eagle. The stone will stop ascending and begin to fall. Another example is that a piece of meat must be cooked before it rots in order for it to become edible. Aristotle named such external interventions "force." So when there is an external force, the object's motion changes. Aristotle attempted to understand changes in the lives of people and in the natural world by observing them closely and attentively. He then identified and analyzed the commonalities of these changes, which ranged from volcanic eruptions and the rising and receding of tides to others like sunrise and sunset, the waxing and waning of the Moon, water evaporating into the air, and rain falling. He referred to the hidden logic or truth behind these changes as "physics."

Aristotle not only founded a school of philosophy to explore nature but also built a well-funded museum of natural history that housed an extensive collection of manuscripts and hand-painted maps, along with carefully compiled results of research done by members of the Peripatetic school. The museum's collection was so extensive that nothing else could compare to it. During his 12 years in Athens, Aristotle fashioned himself as a pioneer in reinvigorating human civilization and knowledge exploration, creating over 170 works during his lifetime, about a third of which have survived to the present day. His works encompassed a wide range of disciplines, including different branches of philosophy (*Metaphysics, Ethics, On Youth, Old Age, Life and Death, and Respiration;* and *On the Soul*); natural sciences (*Meteorology, On Sleep and Sleeplessness, On Dreams, On the Heavens,* and *On Motions*); and literature (*Rhetoric* and *Poetics*). Aristotle's prolific career created the first encyclopedia of knowledge in the history of civilization, and his *Physics* has been hailed as the greatest scientific achievement in ancient Greece. The cosmological view, in particular, was influential to the development of Western civilization for nearly two millennia.

This corpus of works, which has been referred to as a single-author encyclopedia by later generations, is based almost entirely on teleology, the core theoretical foundation of Aristotelianism. The so-called teleology associates all changes in nature with some specific purpose or goal. For example, the purpose of rainfall is to provide moisture to the living world, plants absorb water to grow, animals run to hunt or avoid being hunted, and they mate to propagate their species. All these observed phenomena support the rational thinking behind teleology, and teleology has since become an intrinsic principle for exploring nature. It has even become the core value of today's scientific research in modern society where utilitarianism prevails. Precisely because teleology is easy to understand, Aristotelianism further emphasized that natural phenomena are always rational. For example, pushing a cart makes it go forward; when the sky is full of storm clouds, it is bound to rain heavily; and one will get sick if one becomes too tired. On the basis of common sense, Aristotle believed that everything in the world is composed of two factors: domination and subordination, and for this reason he endorsed slavery. Those with more dominating factors become masters, and those who possess more subordinate factors become slaves.

The most distinctive feature of Aristotelianism is its qualitative doctrine, because its adherents were adamantly opposed to the idea of imposing mathematics on nature. When Aristotelians described changes in speed, for example, they would depict them as certain objects run faster than others within the same time span, which is a type of qualitative rather than quantitative concept. As a result, it would be fruitless to offer a further explanation of the physical phenomenon, e.g., what acceleration is, based solely on such qualitative information. However, Aristotelians were not to blame entirely for this undesirable outcome, because the concept of mathematics or numbers was still in its infancy, and the basic arithmetic operations were not yet widely studied. Neither did fundamental units of physical quantities exist. It would therefore be unreasonable to employ mathematics to make sense of natural phenomena with such a shaky foundation, and any attempts to produce meaningful results would be full of errors at best. A quantitative approach to science would simply be a myth to most. Some scholars, such as the Pythagoreans, might have achieved some success in a few special cases, but attempting to generalize the quantitative methodology further to accommodate a wider collection of natural phenomena would undoubtedly be a futile exercise.

However, the qualitative teleological theory of Aristotelianism was consistent with what people encountered in everyday life. Examples include illness leading to death, eating contaminated food causing illness, and not improperly cleaning food leading to contamination. These are a series of causes and effects that give rise to changes in people's lives. This seemingly vague yet sufficiently concrete theory of causality has been prevalent in the philosophy of the world's major religions, including Judaism, Christianity, Islam and Buddhism. Almost any faith-based activities have inherited this idea of cause and effect. It is no coincidence that Aristotelianism stood firmly and was unchallenged in the history of civilization for more than two millennia until Newton proposed the equations of universal gravitation in the 17th century. His theory confirmed that the motion of an object thrown into the air obeys the law of universal gravitation with no specific goals or purposes. Aristotle's doctrines then began to fall apart. The study of physics, led by astronomical investigations, was the most fruitful and fascinating aspect in Aristotelianism, as they proposed a cosmological model that appeared to be comprehensive.

1.10 Aristotle's Universe

The profound influence of Aristotelianism stems from its foundation on a robust system of logical deduction. Aristotle's followers believed that scientific knowledge is the result of rectifying incorrect thinking under universal principles. Therefore, all knowledge must be based on a broad range of observational data and adequate data collection. The universal principles can then be generalized from a series of interpretations and reasonings. Based on these principles, we can perform logical deduction on our understanding of natural phenomena to construct independent theories based on the specific circumstances under study. In other words, scientific research involves repeatedly exploring and observing the physical world, categorizing and summarizing the characteristics of the things observed, and finally determining the nature of these things. This allows us to understand the operating principles of the natural world. The Aristotelian cosmological view was precisely founded on such rigorous logical reasoning. For example, one of the arguments that the Aristotelians made about their belief of a spherical Earth is the fact that when a lunar eclipse occurs, the shadow cast on the surface of the Moon is curved. The second argument for a round Earth is the different celestial objects, e.g. constellations, that can be observed in the night sky as we travel to locations of different latitudes. Stars that used to be high up in the sky may disappear below the curved horizon as we move from one location to another. Similarly, stars that used to be obscured will reappear above the horizon. The third argument for Earth's spherical shape is that the Universe, as the Aristotelians saw it, is made up of multiple layers of concentric spheres, with the stationary Earth at the center of the Universe. Going outward from Earth, the spheres that contain the Moon, Mercury, Venus, the Sun, Mars, Jupiter and Saturn can be found, in that order. The outermost sphere is home to a large number of fixed stars. This cosmological view centered on concentric spheres was obviously the conclusion drawn from a long period of observation from the ground: Earth is motionless, while the Moon, the Sun and five visible planets move in a cyclical fashion, and all the other stars do not have relative motion.

Aristotle himself also attempted to expound on the formation of this view about the cosmos. He believed that the Universe is divided between two regions, with the lunar spherical shell acting as the boundary between them.

The celestial, or *superlunar*, region is the zone in the Universe beyond the Moon, whereas the terrestrial, or *sublunar*, region is the division below the Moon. In particular, the sublunar region contains the Earth and the atmosphere interior to the lunar spherical shell, and it is composed of the four classical elements: fire (dry and hot), air (wet and hot), earth (dry and cold), and water (wet and cold). The motion of matter is either up or down. Matter moving upward is driven by air and fire, whereas matter moving downward is pulled by earth and water. For this reason, both earth and water are concentrated at the center of the Universe (i.e., Earth), and air and fire exist in the vicinity of the sublunar region. This region is characterized by birth, death, and the changes in all species. On the other hand, the superlunar region encompasses all stars and planets, all of which are composed of the eternal fifth element, the aether, the purity of which increases with its distance from the sublunar region. An object entering this region will either be in a perpetually circular motion or becomes stationary. This cosmological view with celestial bodies that are eternal and an Earth with a finite lifespan is far from what we know as facts today, but the theory is nonetheless based on the results of countless careful observations and logical deduction. It covers a vast area of knowledge in detail, and consequently its impact continued for a long time until the advent of modern physics in the 17th century.

From the perspective of natural philosophy, Aristotle's cosmological view was deeply influenced by the number "four". From about 400 BC, the ancient Greeks started to interpret natural phenomena through the number "four". "Four" was even regarded as a sacred number by the Pythagoreans. Aristotle also proposed four axioms in his *Physics*, attempting to direct the world's attention away from the heavenly gods and to focus on the physical matter on Earth. These four axioms are:

1. Physical elements: Aristotle inherited the narratives of philosophers of previous generations regarding the composition of matter and proposed the four-element theory, which advocates that everything on Earth (or the sublunar region) is composed of fire (dry and hot), air (wet and hot), water (wet and cold), and earth (dry and cold). These four elements can synthesize new physical matter via their own changes or motions.

2. Motion of objects: Motion is a process, and the motion of an object is carried out according to a special purpose. When its purpose has been fulfilled, the motion terminates. For example, water and earth, two of the four classical elements, are heavier than the other two by nature. For the purpose of gravitating toward its destination, which is Earth and also the center of the Universe, earth and water will continue their downward

movement. On the other hand, being lighter in by nature, fire and air will float upward to their destination, the celestial bodies. From an alternative perspective, all four elements have their proper positions determined by their nature, and their movements will stop after reaching their appropriate positions.

3. Earth is at the center of the Universe: Since earth and water are heavy elements, they fall toward Earth and congregate to form the center of the Universe. Air and fire are light elements, so by nature they move upward toward the edge of the sublunar region before stopping. Aristotle believed that these descriptions were verifiable from his observations of the celestial phenomena, such as the sky appears to be an Earth-centered spherical surface with all celestial objects revolving around the Earth on this sphere.

4. Composition of celestial bodies: He believed that celestial bodies (i.e., objects beyond Earth, including the Sun, the Moon, the planets and the stars) are made up of a solid substance that is weightless and transparent. This substance aether, commonly known as the fifth element, is also the constituent element of the celestial spheres and the only substance that carries the planets. Aether, however, is different from the substances that exist on Earth. It is immutable and perfect for eternity. Therefore, things that undergo changes in the sky, such as meteors, comets and novas are all below the sky or in Earth's atmosphere, rather than among the stars or the planets. The natural movement of aether is not a linear motion toward Earth but a constant-speed circular motion around the earth. The shape of a celestial body formed from aether is also a perfectly smooth sphere, like the sphere or the circle is the most perfect geometric shape.

Using modern scientific terminology, the motion of a body in Earth's vicinity (in the sublunar region) is subject to terrestrial mechanics, whereas and the motion of a body beyond the Moon (superlunar region) is governed by celestial mechanics. In particular, there are two categories of motions for objects in the sublunar region: natural movement and forced movement. Natural movement is a linear motion, and its orientation is either upward or downward but is in no way a perpetual motion. The reason the speed of an object's downward motion increases is that acceleration is a natural tendency as the object approaches its target (Earth's surface). When an object rises above, its speed becomes slower and slower, because it is gradually moving away from its target. The forced movement comes into existence when the object is exposed to an external force, which is later referred to as the Causal Mechanical model. Aristotle further argued that because the Earth that we inhabit is too massive to be moved by any force, it has to be completely stationary, and the sky is a dome that moves in a regular manner.

The principal framework of the above theory was based on the ideas left behind by the Babylonians a thousand years before Aristotle's time. However, following further interpretations made on circular motion by Greek natural philosophers on faith, celestial movements that had been observed in the night sky were defined as the natural philosophy that would withstand the test of time for two millennia: Earth is the center of the Universe, and the Sun, the Moon and the other celestial bodies are all revolving around Earth on their own concentric spheres of different diameters. This perfect Greek philosophy about the cosmos later evolved into the metaphysical Stoicism, the name of which derives from the Stoa Poikile, the colonnade of the Ancient Agora of Athens where teachers and students gathered to discuss philosophy. Stoicism, a school of Hellenistic philosophy, has its origins in Plato's *Sophist*, which advocated that only physical entities can exist, but certain intangibles such as time and space are also part of the Universe. They also argued that the two divine powers of the Universe advocated by Empedocles, Love and Strife, cause elements in the Universe to come together or disperse, respectively. They stressed that the Universe will eventually release a massive amount of energy through ekpyrosis, thus ushering in a new cycle of rebirth: When the great fire begins to recede, fire and air (two of the four elements) begin to contract and condense, gradually transforming into water and earth. This is how the Universe is reborn. This theory of cosmological cycle has survived to this day. In 2001, several contemporary physicists incorporated it into their Ekpyrotic universe model to explain the formation of the early Universe. This model states that the Universe was born out of two colliding 3D branes in a four-dimensional space. Despite its divergence from the ideas of the cosmic inflation model, the Ekpyrotic universe is compatible with the ΛCDM model, a parameterization of the Big Bang cosmological model. Aristotle himself probably never foresaw that the cosmological view advocated by his followers would be revitalized more than two thousand years later.

However, there are many things that Aristotle failed to anticipate, especially matters that had to do with his own future or ultimate fate. In 321 BC, Alexander the Great infected with malaria and died, after which his vast Macedonian Empire was divided among his four generals. Aristotle was all too familiar with the political climate of the time. Anyone associated with Alexander would likely be dealt with ruthlessly or purged, as he recalled the trial and execution of Socrates some years before when the philosopher was forced to drink a mixture containing poison hemlock. When news of Alexander's untimely demise reached Aristotle, he at once decided to flee Athens and go to Chalcis to avoid persecution, given the likelihood that the Athenians might choose to commit yet another crime against philosophy. Although he managed to escape the aftermath of Alexander's death, he became afflicted with a stomach ailment, from which he died the following year at the age of 62. From then on, scholars at the Lyceum and the followers of the Aristotelian doctrines continued to spread his teachings. Having been passed down from generation to generation on the European continent, Aristotelianism spread to the Arab world and the works of Aristotle and his followers were translated into local languages; it subsequently became part of the Roman Catholic Church's doctrines and a force that dominated the scientific thoughts of the Western world for nearly two millennia.

Appendix 1.1: The Catapult of Thales

Here we have recreated a catapult design based on the knowledge in geometry and technical craftsmanship in Thales' time, as depicted in Fig. 1.3. The main purpose of the catapult (diagram's right-hand side) is to throw a projectile (e.g. an iron cannonball) on the platform over a castle's rampart of height h (diagram's left-hand side) along a parabolic trajectory. The principal mechanism of the catapult is a lever installed on the platform.

On the left of the lever's fulcrum, there is a rock of mass M, whose center of mass is located l from the fulcrum. To the right of the fulcrum is an iron cannonball of mass m used to bombard the wall, with its center of mass at a distance of nl from the fulcrum. When preparing for an attack, the projectile end of the lever is pulled down to a position where it forms an angle of β_1 with the vertical line. When an attack command is issued, a soldier will immediately let go of the tight drawstring and allow the lever to rotate quickly via the moment of force generated by the rock's weight. The projectile at the other end of the lever will accelerate along with a circular motion of radius nl, until the rock at the other end hits the stopping plate on the platform. The projectile will then be hurled into the air upon the sudden cessation of the lever's rotational movement, at which time the lever forms an angle of β_2 with the vertical line and the speed at which the iron cannonball is projected along the circumferential tangent line is v.

$$U(\beta) = mg \cdot nl \cos \beta - Mg \cdot l \cos \beta \qquad (1.13)$$

Therefore, when the lever changes from its attack position, i.e., being at an angle of β_1 with the vertical line, to the instant when the iron cannonball is released, i.e., being at an angle of β_2 with the vertical line, the gravitational potential energy of the lever will experience the following changes:

$$U(\beta_1) - U(\beta_2) = (M - nm)(\cos \beta_2 - \cos \beta_1)gl \cos \beta \qquad (1.14)$$

From the above equations, we can see that the gravitational potential is reduced in the process. Ignoring the loss in potential energy due to the friction of the overall lever mechanism, the change in quantity of the potential energy in (1.14) is converted entirely into kinetic energy. From this energy conservation relationship, we can deduce the catapult's attack position (location from the rampart) and other required conditions. The derivations are as follows.

$$E_K = \frac{1}{2}M(l\omega)^2 + \frac{1}{2}m(nl\omega)^2 = \frac{1}{2}\left(M + n^2m\right)l^2\omega^2 \qquad (1.15)$$

$$\omega = \sqrt{\frac{2(M - nm)(\cos \beta_2 - \cos \beta_1)g}{(M + n^2m)l}} \qquad (1.16)$$

$$v = n\sqrt{\frac{2(M - nm)(\cos \beta_2 - \cos \beta_1)gl}{M + n^2m}} \qquad (1.17)$$

$$\alpha = \beta_2 \qquad (1.18)$$

$$\Delta x = v_x t \qquad (1.19)$$

$$\Delta y = v_y t - \frac{1}{2}gt^2 \qquad (1.20)$$

Substituting the relevant physical quantities into (1.19) and (1.20) and eliminating the time variable t:

$$(D + nl \sin \beta_2) \tan \beta_2 - \frac{1}{2}g\left(\frac{D + nl \sin \beta_2}{v \cos \beta_2}\right)^2 = H - L - nl \cos \beta_2 \qquad (1.21)$$

Substituting (1.16) into (1.21), we can derive the relationship between β_1 and β_2:

$$\cos \beta_1 = \cos \beta_2 - \frac{1}{4n^2 \cos^2 \beta_2} \frac{M + n^2 m}{M - nm} \frac{(D + nl \sin \beta_2)^2}{l[(D + nl \sin \beta_2) \tan \beta_2 - H + L + nl \cos \beta_2]}$$

$$(1.22)$$

Now, let us consider a siege that would have likely occurred at Thales' time. The relevant parameters are as follows: $M = 10$ metric tons, $m = 100$ kg, $L = 10$ m, $l = 2$ m, $\beta_1 = 180°$ (the projectile end of the lever lowered to the fullest extent possible; M at the highest level), $\beta_2 = 45°$, $n = 6$. We substitute these numbers into (1.22) to obtain the relationship between D and H, which specifies how far the catapult should be positioned to be able to throw projectiles over a wall of a specific height. Simplifying (1.23) further we have

$$D = \xi \tan \beta_2 + \sqrt{\xi^2 \tan^2 \beta_2 + 2(-H + L + nl \cos \beta_2)\xi} - nl \sin \beta_2$$
$$= 76.46 + \sqrt{10357 - 169.9H} \qquad (1.23)$$

The units of the quantities D and H above are in meter (m), and the parameter ξ satisfies the following:

$$\xi = 2n^2 \cos^2 \beta_2 (\cos \beta_2 - \cos \beta_1) \frac{M - nm}{M + n^2 m} l \approx 84.95 \text{m} \qquad (1.24)$$

If the height of the rampart is $H = 20$ m, then the catapult should be placed at a distance of $D = 160$ m. Since the expression inside the square root sign of (1.24) is a non-negative real number, we have $10357 - 169.9H \geq 0$. This means that the maximum height of the rampart the catapult is capable of besieging cannot exceed $\frac{10357}{169.9} \approx 61$ m and the corresponding distance of projection is 76 m.

Note that we set $\beta_1 = 180°$ above to make (1.24) more tractable, although this would create issues in real-world designs. The interested reader may assume an arbitrary value for β_1 and repeat the derivation process. The equations also reveal a design limitation: the catapult's position from the rampart cannot exceed 160 m.

Appendix 1.2: Proofs of the Pythagorean Theorem

There are a number of different ways to prove the Pythagorean theorem. Apart from the proof provided by Euclid in his *Elements*, the 12th U.S. President James A. Garfield also published a proof when he was a U.S. Representative (1876). There is also one that uses the principle of similar right-angled triangles and the ratio of their sides. Both are outlined below.

1. President Garfield's proof
   As depicted in Fig. 1.12a of the diagram below, the trapezoid is composed of two congruent right-angled triangles and an isosceles triangle. The area of the trapezoid is easily calculated as

   $$\frac{1}{2}(a+b)(a+b) = \frac{1}{2}ab + \frac{1}{2}ab + \frac{1}{2}c^2$$

   The left-hand side is simply (based + height)/2, and the right-hand side is the sum of the areas of three triangles. After simplifying the expression, we obtain the Pythagorean theorem $a^2 + b^2 = c^2$.

2. A proof based on similar right triangles
   As shown in Fig. 1.12b. From the similarity of the triangles $\triangle ABC$ and $\triangle DBA$, we have $\overline{AB}^2 = \overline{BD} \times \overline{BC}$;. Likewise, the similarity of $\triangle ABC$ and $\triangle DAC$ implies $\overline{AC}^2 = \overline{CD} \times \overline{CB}$. Adding these two equations together, we have $\overline{AB}^2 + \overline{AC}^2 = (\overline{BD} + \overline{DC}) \times \overline{BC} = \overline{BC}^2$, which is the Pythagorean theorem.

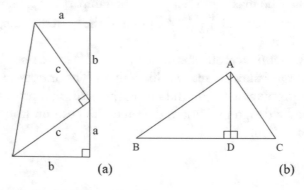

Fig. 1.12 The proofs of Pythagorean theorem

Appendix 1.3: Pythagoras' Vibrating String: Length and Harmony

In a block print dating back to the Renaissance period, Pythagoras is vividly portrayed to be next to a five-stringed instrument, tuning it with two wooden sticks in his hands. The protagonist is striking a string with one of the sticks and using the other to adjust the length of the vibrating portion of the string. One end of each of the five strings is supported by the instrument's bridge, and the other end is attached to a stone used as a counterweight to set the string's tension. The artist's inspiration must have come from the desire to appreciate the music theory legacy of Pythagoras, which contributed to the development of Western music for over a thousand years.

In modern terms, Pythagoras would be described as a mathematician with absolute pitch (often called "perfect pitch"), since he used clear logical reasoning to define the musical scales audible to the human ear. According to legend, Pythagoras formulated the scales by dividing the sounds produced by two strings with a 2:1 ratio in length into seven notes (frequencies), thus forming an octave. He also divided the sounds produced by two strings with a 3:2 ratio in length into four notes to produce a "fifth". Finally, he put these two scales together by ordering the notes to get 12 semitones.

However, to produce the beauty of harmony, someone later made gradual adjustments to the octave to produce the so-called "just intonation" on the basis of these 12 semitones. The purpose of tuning was then to make the harmonies of musical chords more natural and pleasing. However, the frequency difference between each pair of notes is not the same, which means that the chords of musical pieces of different keys are not interchangeable, causing many problems in composing and performing music. A good example is that when a musical piece changes keys midway, the performer may make a mistake without realizing it.[22] To solve this problem, the German musicians known for their insistence on uniformity, introduced the 12-tone equal temperament toward the end of the 17th century. The person who advocated this system of tuning was Andreas Wreckmeister (1645–1706). He divided the octave into 12 parts (semitones), all of which are equal on a logarithmic scale. See the keyboard shown in Fig. 1.13.

Given a note, the note one octave above it has a frequency twice that of the first note, so the frequency of the note one semitone higher would be $\sqrt[12]{2} = 1.059463094$ times that of the first note. If the frequency of the C

---

[22]In the article *Just Intonation and Equal Temperament* (in Chinese), written by Ai Xia, the problems associated with just intonation due to changed keys are discussed in detailed. Refer to http://dejavu.city/node/85

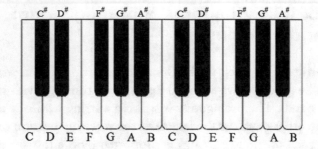

Fig. 1.13 The notes of the modern keyboard

note is taken to be 1, then the frequency of C# (one semitone above C) is the product of 1 and $\sqrt[12]{2}$, the frequency of D is $(\sqrt[12]{2})^2$, and D# is $(\sqrt[12]{2})^3$, and so on. The keyboard of the modern piano, depicted in Fig. 1.13, is constructed using the 12-tone equal temperament scale.

This common multiple of $\sqrt[12]{2}$, between two adjacent semitones, can be explained by the following mathematical relationship. Modern wave theory, familiar to many, points out that the frequency f of the sound emitted by a vibrating string is related to the length of the string L, the density ρ of the string, and the string's tension T. Therefore, the velocity of propagation of a wave in the string v can be expressed as $v = \sqrt{\frac{T}{\rho}}$. When the string is vibrating at the fundamental frequency or first harmonic ($n = 1$), the following relationship holds:

$$f = \frac{1}{2L}\sqrt{\frac{T}{\rho}} \tag{1.25}$$

This equation tells us that if we reduce the string's density, or shorten the string's length, or increase the string's tension, we will produce a higher vibrating frequency. Now suppose a string is fixed at both ends. Provided that the density and force of tension are both constant, plucking the string will produce a standing (or stationary) wave of wavelength $\lambda/2 = L/n$ and frequency $f = nv/2L$, where n is a positive integer and v is the velocity of wave propagation on the string. The equation $\lambda/2 = L/n$ can then be interpreted as follows: when the string's length is an integer multiple of half of the wavelength, the frequency of vibration is inversely proportional to the string's length.

Now suppose we have two strings with a length ratio of 1:2. This implies the ratio of their frequencies is 2:1. In other words, the notes they produce are one octave, or 12 semitones, apart. This can be expressed mathematically as $\log_{\sqrt[12]{2}}\left(\frac{2}{1}\right) = 12$. Since a note of a semitone higher has a frequency being $\sqrt[12]{2}$ times of the original note, if we want to double the frequency, we need to multiply the original frequency by 12th power of $\sqrt[12]{2}$. Similarly, if two strings have a length ratio of 2:3, their frequency ratio will be 3:2, and the difference in pitch will be $\log_{\sqrt[12]{2}}\left(\frac{3}{2}\right) = 7.02 \approx 7$ semitones, so it is necessary to multiply the original frequency by the 7th power of $\sqrt[12]{2}$, about five scale degrees apart. If the two strings have a length ratio of is 3:4, their frequency ratio is 4:3, and the difference in pitch will be $\log_{\sqrt[12]{2}}\left(\frac{4}{3}\right) = 4.98 \approx 5$ semitones, so it is necessary to multiply the original frequency by the 5th power of $\sqrt[12]{2}$, about four scale degrees apart. These notes with integer multiples of frequencies are important components of musical chords. If the frequencies of the notes do not have integer ratios, the notes produced will not sound harmonious at all and will even be harsh to the ear. Examples are alarms, sirens, noisy machines, screams, and the fluttering sounds made by insects.

The principles behind musical chords were applied by the clergymen in monasteries in the 12th century to design Gothic-style cathedrals and churches. For example, the ratio of the lateral width of a church's main hall to the distance between lengthwise stone pillars is 2:1; the ratio of the main hall's length to the altar's width is 8:5; and the ratio of main hall's height and width is 3:2. A church built with musical chord frequency ratios in mind not only exhibits visual beauty but also a serene auditory sensation. The reason is that the choral harmonies and musical chords performed in the church are amplified, while discordant music is suppressed. With both visual and auditory harmony, the church attendants naturally enjoy peace of mind.

Appendix 1.4: Computing Archimedes' π Using Integration

Archimedes' achievements in geometry are even more brilliant. He was the first to employ a purely geometric method to estimate π to be $3\frac{1}{7}$, which we already introduced in Sect. 1.4. Here we will take advantage of the techniques in modern calculus to derive the upper and lower bounds of π. We first consider the integral[23]

$$I = \int_0^1 \frac{x^4(1-x)^4}{1+x^2}dx = \int_0^1 \frac{x^4 - 4x^5 + 6x^6 - 4x^7 + x^8}{1+x^2}dx$$

$$= \int_0^1 \left(x^6 - 4x^5 + 5x^4 - 4x^2 + 4 - \frac{4}{1+x^2} \right)dx$$

$$= \left[\frac{x^7}{7} - \frac{2x^6}{3} + x^5 - \frac{4x^3}{3} + 4x - 4\tan^{-1}x \right]\Big|_0^1 = \frac{22}{7} - \pi \quad (1.26)$$

From the result of this integral, we can estimate its upper and lower bounds. First, we substitute $x = 1$ into the right-hand side of the first equality in (1.26) above to get $I = \int\limits_0^1 \frac{x^4(1-x)^4}{2} \geq \frac{1}{1260}$. Next, we substitute $x = 0$ to get $I = \int\limits_0^1 \frac{x^4(1-x)^4}{1} \leq \frac{1}{630}$. Combining these two results we have $\frac{1}{1260} < \frac{22}{7} - \pi \leq \frac{1}{630}$, or $\frac{22}{7} - \frac{1}{630} \leq \pi \leq \frac{22}{7} - \frac{1}{1260}$, which gives the lower and upper bounds for π.

In arriving at the same result achieved by Archimedes over two millennia ago, both of the approaches above utilize modern trigonometry and the machinery of integral calculus, which shows that ancient natural philosophers already possessed an impressive grounding in geometric reasoning. Today almost everybody knows that quantities related to the circle (e.g., its area and circumference) and the sphere (e.g., its volume and surface area) are all associated with the number π. The circle and the sphere are the most abundant shapes in the Universe, as well as on Earth's surface. Even in physical or mathematical equations that are ostensibly unrelated to the circle, π can be found everywhere, making this irrational number an important universal constant. Examples include: the period of a simple pendulum, $T = 2\pi\sqrt{\frac{L}{g}}$; the Stefan–Boltzmann constant in the calculation of black-body radiation, $\sigma = \frac{2\pi^5 k^4}{15c^2 h^3}$; and the probability density function of the normal (or Gaussian) distribution, $f(x) = \frac{1}{\sqrt{2\pi}}e^{-\frac{x^2}{2}}$, used in computational statistics. If we look at the decimal representation of π, we can feel its profound nature: it is irrational, with an infinite, non-repeating number of digits after the decimal point. However, any short sequences of digits anyone considers important, such as birthdays, license plates and phone numbers, can all be found

---

[23] *Source* https://en.wikipedia.org/wiki/Proof_that_22/7_exceeds_\%CF\%80.

somewhere in this infinite sequence, and we only need a precision of only 39 decimal places to calculate the circumference of the Universe and the size of an atom.

2

The Blooming Greek Astronomy

明月幾時有?把酒問青天。
不知天上宮闕,今夕是何年?

How rare the moon, so round and bright
With cup in hand, I ask the blue sky,
"I do not know in the celestial sphere
What name this festive night goes by?"

From Shuidiao Getou *by Song dynasty poet Su Shi (1037–1101)*
Translated by Lin Yutang (1895–1976)

In an atmosphere in which mathematics and geometry were burgeoning, ancient Greek astronomers naturally took every opportunity to put forth brand new theories once they believed they had unraveled the mystery of nature from their nightly observation of the heavens. These theories were progressive and all-encompassing. Some, for instance, proposed theories about what they believed to be the fundamental elements of the Universe. Others formulated models to describe the structure of the Universe. Still, others simply picked up simple astronomical instruments, aimed at the stars and carefully but passionately performed observations on the sizes of celestial objects and their distances from Earth. The results of these endeavors brought astronomy to a new level and spurred explosive growth in the field. The development of Greek astronomy was an evolutionary process in the history of science and was replete with novel and far-reaching ideas. It is no less significant or exciting than the applied technologies or knowledge in geometry described in Chap. 1. In the following sections we will review the development of Greek astronomy from several different perspectives. We will first present theories about the constituent elements of the Universe, the idea of which was as controversial as it was creative.

2.1   Elements that Make up Our Universe

Following the introduction of a number of theories put forth by Thales and two of his students regarding the fundamental elements of the Universe, around 500 BC Pythagoras of Samos (570–495 BC), known for his mathematical discoveries, also proposed the theory that the Universe was composed of four basic elements: fire (hot), air (cold), water (wet), and earth (dry). Pythagoras' theory provided a concrete, widely accepted and satisfying conclusion to the multitudes of competing theories proposed in the era of Thales. According to Pythagoras' own description, he reached this conclusion after having made careful observations of various phenomena in the heavens. He discovered that order, harmony and balance could be found throughout the natural world, and from this observation, he offered three insights: First, all matter is made up of the four fundamental elements: earth, air, water and fire. Second, Earth is a sphere similar to the Moon and the Sun, and it is located at the center of the Universe. Finally, all of the stars are also spherical objects and revolve around Earth in elliptical orbits. Pythagoras provided further details on these spheres: Earth and the stars are all perfect spheres (naturally the ideal perception by a mathematician). Each of these spheres travels along its own orbit and is powered by a mysterious and invisible "fire." This fire, however, is distinct from the Sun, as the Sun itself is one of the stellar objects that revolve around Earth.

Later, the Greek philosopher Empedocles (495–435 BC) from Sicily elaborated on the theory of Pythagoras: all things are made up of the aggregation and segregation of the four elements of Pythagoras, which are brought into union and separated from one another by two divine powers, Love and Strife (or Hatred). According to his hypothesis, moonlight comes from the Sun, and the Sun and the Moon revolve around Earth. A solar eclipse is the result of the Moon passing between the Sun and Earth. It is rather astonishing that from this Love/Strife hypothesis Empedocles was able to derive these conclusions that are not too far from the truth. Based on this Sun-Moon-Earth relationship, he further conjectured that the sky is a transparent crystalline sphere that drives the revolutions of celestial bodies around Earth, which sits at the center of the sphere. Even though this hypothesis was merely speculation with no real scientific basis, the Love/Strife doctrine was accepted for several decades before it was refuted by Philolaus (470–385 BC), an astronomer of the new generation. Philolaus believed that Earth is not at the center of the Universe but instead, like all other celestial objects, revolves around a hypothetical astronomical object he referred to as the Central Fire. Philolaus, however, agreed with Pythagoras and did not identify the Central Fire with the Sun. This could well have been the very first non-geocentric view of the Universe in human history.

Apart from the four-element theory, two other natural philosophers, Anaxagoras (500–428 BC) and Democritus (c. 460–370 BC) proposed an atomic hypothesis of the Universe. They formulated a theory where the Universe is composed of countlessly many atoms and something called the "void." There are an infinite number of atoms and the void has an infinite volume. The atoms have identical composition but differ in shape and size. All atoms travel through the void freely in a diffuse but perpetual motion. They collide to form celestial objects or decompose slowly to form other atoms. This was the first atomic theory introduced in history. Granted that this ancient atomic theory had no direct bearing on the development of modern science, the concepts that it embraced were nevertheless resurrected and amplified two millennia later by its modern incarnation, nuclear physics, a discipline that emerged toward the end of the 19th century.

Plato (427–347 BC), who appreciated the beauty of geometry, was eager to make his ideas known as well. He offered a rather unique cosmological theory based on aesthetics. Plato combined several existing theories with geometry and advocated the hypothesis that the four elements of Pythagoras and the Universe itself are composed of five distinct regular polyhedra, as shown in Fig. 2.1. A regular polyhedron is a three-dimensional solid with identical faces (same shape and area) as well as identical dihedral angle between each pair of intersecting faces. Examples are the regular tetrahedron (Fig. 2.1a) which is composed of four identical equilateral triangles and the regular hexahedron (Fig. 2.1b), or the cube, which consists of six identical regular quadrilaterals (or squares), and so on, up to the regular icosahedron (Fig. 2.1e), which has 20 faces.

How do these regular polyhedra correspond to the elements of the Universe? Plato associated earth with the regular hexahedron (the cube), possibly because it tessellates three-dimensional Euclidean space (i.e., being packed

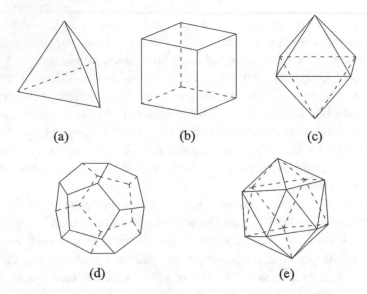

Fig. 2.1 The Platonic solids. **a** tetrahedron; **b** hexahedron (cube); **c** octahedron; **d** dodecahedron; **e** icosahedron (all regular polyhedra)

without leaving gaps) and reflects the hardness and rigidity of earth. He associated air with the regular octahedron, as objects of this shape have sharp vertices and bounce around freely, which demonstrates the amorphous nature of air. Fire was associated with regular tetrahedrons, perhaps due to its robust and unbreakable nature, reflecting the fact that proximity to fire is undesirable. Water was associated with the icosahedron, a shape closest to the sphere, possibly due to its capability to change into any form and flow like little spheres. Finally, Plato associated the last regular polyhedron, the dodecahedron, with the substance that binds the Universe together to reflect the Universe's extraordinary diversity. Plato's use of aesthetics and geometry was rather innovative thinking.[1] He also emphasized that, apart from aesthetics, these are the only regular polyhedra that exist in the Universe; no other regular polyhedra are possible. A theory with both aestheticism and uniqueness was entirely in line with Plato's philosophy of pursuing perfection. A design of the inhabited Universe incorporating such flawless geometric shapes is nothing short of perfection. In his mind, only his beloved geometry could manifest the truth of the Universe. If Plato had such a high expectation for this theory, it wouldn't be a stretch to say that he must have proved it rigorously.

---

[1]In modifying Plato's theory, Aristotle added a fifth element, the "ether", and postulated that the heavens beyond the Moon are made of this element. He chose to ignore his teacher's regular dodecahedron completely, perhaps because he was biased or had doubts about Plato's teachings.

Despite the absence of direct evidence for this proof by Plato, in his *Elements* the Greek mathematician Euclid did provide such proof, the logic of which will be explained below in the language of modern mathematics.

Suppose a symmetric regular polyhedron (or alternatively referred to as a polyhedron whose surface is homeomorphic to a sphere) has A vertices, B edges and C faces. The Euler-Descartes equation states that $A - B + C = 2$. When we calculate the number of edges based on the number of faces or vertices, each edge is counted twice, so we have $2B = nC = mA$, where n is the number of edges on each given face, and m is the number of edges incident on each given vertex, with $n, m > 3$. Solving the above three equations simultaneously, we have $\frac{1}{n} + \frac{1}{m} = \frac{1}{2} + \frac{1}{B} > \frac{1}{2}$. If n and m are both integers, then (n, m) can only take on the following possible combinations: $(3, 3), (4, 3), (3, 4), (5, 3), (3, 5)$, which correspond to the tetrahedron, hexahedron, octahedron, dodecahedron, and icosahedron, respectively.

Granted that these five regular polyhedra are unique and indeed aesthetically pleasing, regarding them as the fundamental elements of the Universe would be self-serving and a bit far-fetched. However, some theories do stand up to the test of time, as over the past thousand years of human history, many substances on planet Earth have been identified to possess the shapes of the regular polyhedra. Examples include the enormous cubic salt crystals found in some salt mines and believed to have formed under tremendously high temperatures and pressure over a long period of time. These crystals are a compound formed from sodium (Na^+) and chloride (Cl^-) ions with a cubic lattice structure. Recent medical research has also discovered that some viral proteins possess the shape of the regular dodecahedron. The molecules or ionic lattices of these natural substances directly reflect the fact that their crystalline structures are none other than the Platonic solids. Why does nature tend to shape certain substances in the forms of regular polyhedra? Today scientists have proposed the so-called Principle of Minimum Energy. Refer to "Appendix 2.1 Principle of Minimum Energy: Cases and Proofs" at the end of this chapter for further details.

Theories abound as to what the constituent elements of the Universe really could be, and thus far these theories were nothing but speculations of natural philosophers based on their subjective views. No one ever dared to ask the question today's schoolchildren would ask with minimal hesitation: Why (these elements)? Well, it was probably a good thing that this question was never asked, as one would be hard-pressed to find anyone with the ability to offer a satisfactory answer. Despite these inexplicable mysteries of nature, the progress of astronomy never stalled. Nor was the discipline in danger of fading into obscurity. In fact, a different group of scholars rose to the occasion and focused their attention on astronomical observations as they realized that those astronomical phenomena that appeared to be completely random yet exhibiting some degree of regularity were intimately tied to many important human endeavors, including marine navigation, farming, and even divination.

2.2   Astronomical Observations: Navigation, Divination and Agriculture

Various branches of natural philosophy emerged in ancient Greece, but astronomy stood out as the dominant field of study. This was mainly because many highly regular astronomical phenomena, such as sunrise, sunset, and the waxing and waning of the Moon, were of great interest to the public. On the other hand, unlike some often kaleidoscopic natural events (such as weather patterns and seismic events) on the ground, astronomical phenomena seemed to lend themselves to more systematic exploration. Stars that appeared to be stationary in the night sky over a long period of time provided a natural setting as well as simplicity and uniformity to the study of astronomy. Apart from the pure pursuit of knowledge, the results of the astronomical study were also applicable to people's daily lives, as well as to important and highly relevant economic and military activities, such as farming, navigation and divination. Around 3000 BC, for instance, the Egyptians were already aware that every year on a particular day in June the star Sirius would rise together with the Sun in the east. On the previous day, Sirius could not be found anywhere in the dark sky just before sunrise. On the day after, however, Sirius would be up in the sky at dawn. The Egyptians also determined that the day Sirius and the Sun rose in tandem in the eastern sky would mark the beginning of Nile's flooding. It would, therefore, be necessary to harvest the crops and store them away as quickly as possible.

The discovery of the link between an astronomical phenomenon and the timing of farming practices thus gave rise to this uniquely important event on the Egyptian agricultural calendar. Similarly, the Greeks also designated the day the star Arcturus would rise together with the Sun the day for harvesting grapes. The end of the growing season would be marked by the day when the Pleiades, an open star cluster in the constellation Taurus, would set in the west. The Pleiades occupies a special place in human history. Made up of nine blue stars, the cluster is also known by the name the Seven Sisters (corresponding to the seven brightest stars) and is a nebula closest to Earth. Due to its growing brightness in the northern hemisphere's night sky during summer, there have been many written records about the Pleiades in both Asia and Europe since ancient times. During the Spring and Autumn period (771–476 or 403 BC) in Chinese history, sages supposedly from the Pleiades communicated divine revelation to Confucius. In the Book of Job of the Hebrew Bible, it mentioned that God's creation of Heaven and Earth also included the Pleiades. An even more interesting example is the following. Ancient navigators already possessed the knowledge that the relative positions of the stars are constant, and they appear to rotate westward around the north celestial pole (where Earth's axis of rotation meets the northern celestial hemisphere). When Odysseus (Ulysses in Latin), the protagonist of Homer's epic poem *Odyssey*, was sailing back home after seven years in captivity, the nymph Calypso urged him to keep the Great Bear[2] visible in the night sky above the port bow at all times while he crossed the sea, so that the correct eastward sailing direction always followed.

This technique of pinpointing one's location on Earth's surface according to the positions of the stars gradually gave rise to Western astrology. It was based on what is now known as the twelve constellations of the zodiac. References to these constellations can be found in Babylonian records going back to 2000 BC. The Babylonians were already aware of the fact that the Sun not only rises in the east and sets in the west but its positions in the sky during the year, relative to the fixed stars in the background, would slowly move eastward along the zodiac (a belt of space extending 9° either side of the ecliptic). Stars within the zodiac are divided among twelve regions on the celestial sphere, namely the constellations Aries, Taurus, Gemini, Cancer, Leo, Virgo, Libra, Scorpio, Sagittarius, Capricorn, Aquarius and Pisces. These twelve constellations then divide the ecliptic into twelve equally spaced longitudinal arcs, each of which is assigned to one of the twelve months. The first celestial coordinate system thus came into existence. The beginning of this longitudinal system is set at Aries, the first house of astrological signs, with the Sun located at the vernal equinox.

We now know that Earth revolves around the Sun, and its orbital plane is coplanar with the ecliptic plane, where the twelve constellations are located. Earth's equatorial plane (the plane of the celestial equator) is inclined to the ecliptic plane at a certain angle. These phenomena were known to the ancients for a long time and their observations were recorded as follows. The plane of the Sun's daily westward motion (the ecliptic plane) is observed tilting at an angle with respect to the equatorial plane, which is precisely the angle between the axis of Earth's rotation and the vertical axis of Earth's

---

[2]This is the constellation Ursa Major, of which Big Dipper or the Plough is a part.

orbital plane. The zodiac, therefore, contains the specific path of the Sun's annual eastward motion through the twelve constellations. The positions of the Sun fall onto different constellations from one season to another. During the day, however, the Sun illuminates the entire sky, so in the absence of modern-day instruments, how were the ancients[3] able to superimpose the Sun over the constellations, which are visible only at night? It turned out that they took advantage of the fact that twinkling stars are visible to the naked eye at both dawn and dusk.

It was in such an exciting atmosphere within the world of astronomical research that Hipparchus of Rhodes (190–120 BC) called on other astronomers to stop focusing their efforts disproportionately on creating models of the Universe. Instead, he advocated that emphasis should be placed on the systematic collection and processing of data. To practice what he preached, Hipparchus spent much time compiling a star catalog that contained 850 stars and incorporated existing astronomical observations. This was the first document of its kind ever created in human history. Compared with modern star catalogs, which may contain as many as 118,000 stars, the one created by Hipparchus might seem rather insignificant, but the fact remains that at a time when astronomical observations were made with nothing but the naked eye, this was no small feat at all. Compiling star catalogs was an essential step in making astronomical observations, such as the movements of the Sun, the Moon and the "wandering stars." Using the armillary sphere that he developed with the ecliptic and equatorial coordinates,[4] Hipparchus was able to annotate the positions and brightness of the 850 stars in his catalog. This catalog was also used by Claudius Ptolemy (Ptolemaeus in Latin; 90–168) in the first century AD as the basis for his model of the Universe, which remained an essential foundation for astronomical observations until the 16th century.

In addition to the armillary sphere, Hipparchus also invented the astrolabe to measure the latitude (altitude above the horizon) of a star and created the dioptra to estimate the diameters of the Sun and the Moon. Additionally, he pioneered the equatorial ring to observe the positions of the Sun with high

---

[3]From the wreckage retrieved off the coast of the Greek island Antikythera, archaeologist Valerios Stais (1857–1923) discovered a mechanical device, believed to have been manufactured between 150 and 100 BC, that was used to predict astronomical positions. It is a complex clockwork mechanism composed of over 30 meshing bronze gears, with a dial assembly equipped with fine gradations and a pointer to indicate the measurement.

[4]To create a star catalog, one must select a reference coordinate system. Judging from records going back to Roman Empire (around third century AD), the ecliptic and equatorial reference coordinates selected by Hipparchus were fixed on an invisible sphere, the so-called celestial sphere, where all the fixed stars are located. In other words, this positioning method did not take distances into account as it was assumed that all stars were equidistant from Earth.

accuracy. When the Sun is seen to set on the equatorial ring (i.e., above the equator), one can be certain that the vernal or autumnal equinox has arrived. Apart from creating star catalogs and producing instruments, Hipparchus also utilized these tools to observe the changes in the lengths of the four seasons on Earth. He discovered that the Sun's observed constant speed of revolving motion around Earth has a circular orbit whose center is something other than the circle's center (i.e., Earth). This was the first time in history that the theory of eccentricity was introduced. Later Hipparchus would apply this theory to study the Moon's orbit around Earth.

Besides the observation of distant stars, ancient Greek astronomers also noticed that there were five "wandering stars" (or planets) that behaved quite differently from the other "fixed" stars in the night sky. These wandering celestial objects changed their positions slightly on a nightly basis. Awestruck over such magnificent natural wonders, Greek astronomers proceeded to name these planets after their gods. That is how the five planets closest to Earth, i.e., Mercury, Venus, Mars, Jupiter and Saturn, got their names. Following the Babylonian tradition, the Romans later added the Sun and the Moon to the list of five planets. The naming of the seven days of the week was also associated with these seven "planets."[5] These astronomers were not only avid stargazers but also serious scientists. This is because they also made observations of the planets and recorded them for an extended period of time. Like the Sun and the Moon, the planets orbited around the plane of the ecliptic. From the records made by these astronomers, we learn that the lengths of the cycles of the planets returning to their respective original positions are: Mercury–88 days; Venus–225 days; Mars–1 year and 322 days; Jupiter–11 years and 315 days; and Saturn–29 years and 166 days.[6] This indicates that astronomical research was serious business, and many results were based on observational data obtained in a rigorous manner.

2.3   Observing the Sun: Gnomon

As much as they were interested in observing constellations in the night sky, the ancients in fact put most of their astronomical research efforts on what appeared to them the largest and closest celestial objects: the Sun and the

---

[5]The names used by the Greeks were Hermes (Mercury), Aphrodite (Venus), Ares (Mars), Zeus (Jupiter) and Cronos (Saturn). As for the English names for the days of the week, the correspondences are: Saturday–Saturn, Sunday–Sun, Monday–Moon; the rest are derived from the equivalents of Roman gods in Germanic mythology: Tuesday–Týr (Mars), Wednesday–Woden/Odin (Mercury), Thursday–Thor (Jupiter), Friday–Frigga (Venus).

[6]The year and day referred to here are Earth year and Earth day.

Moon. Moreover, the changes in the positions of the Sun and the Moon were intimately connected with all manners of human endeavors, and for this reason, studying these two objects was essential. The study of the relationships between the Sun's cyclical behavior and people's day-to-day lives relied on a structurally simple but versatile instrument known as a gnomon. The structure of the gnomon, colloquially called the Babylonian rod, is quite simple.[7] A stick is vertically positioned on a horizontal surface, and the shadow cast by the stick under the sunlight is measured for its length and angle, which in turn are used to determine the time of day and the season of the year. It is believed that the gnomon existed in ancient Babylon, but other accounts have suggested that the device was invented by Anaximander, a Greek philosopher who lived in Miletus and was a student of Thales. With the unaided eyes, we can use a gnomon to measure the angle between the horizon and the line of sight of a celestial body, just as Eratosthenes (276–194 BC) performed measurements using a gnomon to estimate the radius of Earth (see Sect. 2.5 for more details). With the advent of the gnomon, astronomy gradually evolved into a quantitative scientific discipline.

The gnomon is a simple but highly functional device. Here is how it may be used to determine the hour of the day: at noon, the Sun is at its highest point in the sky, so the shadow cast by the gnomon is the shortest. In contrast, the shadow is the longest at dawn or dusk. The gnomon can also determine the four cardinal directions. Take the region north of latitude 23.5° in the northern hemisphere as an example. A shadow cast by a gnomon located here will point due north at noon, in any season. Similarly, a shadow cast by a gnomon in the southern hemisphere (south of latitude 23.5° S) will point due south. The gnomon can also be used to create a calendar system. In Taiwan, for example, the Tropic of Cancer (the circle of latitude at 23.5° N) passes through the central part of the island, so at noon on the day of summer solstice there, the Sun will appear to be directly overhead, and the gnomon will cast no visible shadow on the ground. This signifies the day with the longest period of daylight of the year.[8] At noon on the day of the winter solstice, the Sun and the gnomon form the largest angle, so the shadow is the longest of the year, which corresponds to the shortest period of daylight of

---

[7]The simplest gnomon is in fact a stick vertically erected on the ground. Time can be determined by observing the shadow it casts. In the northern hemisphere the shadow revolves in a clockwise direction around the stick, whereas in the southern hemisphere it revolves in a counterclockwise direction. The speed of the shadow's revolving motion, however, is not uniform, and the angular span covered in each hour is different. If the stick is tilted toward the north celestial pole (currently Polaris), its shadow will not be subject to seasonal changes and the precision can be improved.

[8]The Chinese solar term for the summer solstice is xiàzhì: xià "summer", zhì "extreme" ("height of summer").

the year. In addition, as the Sun rises from the east every day, the gnomon's shadow will point due west at dawn and due east at dusk on the day of the vernal or autumnal equinox, signifying the middle of spring or autumn.

The gnomon also alludes to the fact that Earth's orbit around the Sun is not perfectly circular. As the gnomon is capable of indicating the change of seasons, the number of days each season spans can also be calculated by observing how the shadow changes over a period of time. As early as the time of Socrates (around 400 BC), it was already known from the use of the gnomon that the numbers of days of the four seasons are not identical. The reasons behind this, of course, would have to wait until the 17th century, when Johannes Kepler discovered that Earth's orbit is in fact elliptical, that Earth travels at uneven speeds in its orbit, and that the Sun is located in one of the two foci of Earth's elliptical orbit.

2.4   Observing the Moon: Calendars

Apart from the Sun, observations of the Moon by astronomers produced results that were even more relevant to many human activities. The ancients recognized that the cycle of lunar phases has to do with the positions of Earth relative to the Sun and the Moon, despite their incorrect assumption that both of these celestial bodies revolve around Earth. What's more curious is that early records also mentioned that solar eclipses would occur once every 18 years, as the plane of the Moon's orbit around Earth is tilted by an angle with respect to the plane of Earth's orbit around the Sun (the plane of the ecliptic). This angle varies between its minimum and maximum values in a cycle of about 18 years, 11 days and 8 h, also called the Saros. Because of this varying angle, the Moon does not block the sun (i.e., produce a solar eclipse) on a monthly basis. Viewed from Earth, the Moon behaves much like the stars and revolves around the north celestial pole (point of intersection between Earth's axis of rotation and the northern celestial hemisphere) nightly in a counterclockwise direction, assuming the observer is facing south. Unlike the Sun's movement, however, the Moon travels monthly along its orbital plane in an easterly direction. For the Sun this annual motion takes $365\frac{1}{4}$ days to complete a cycle, which is a solar year (also known as a tropical year). For the Moon, it takes $27\frac{1}{2}$ days to return to the original position, a period referred to as a lunar month.

Given these fixed cycles of phenomena observed in the sky, the Athenian astronomer Glaucon,[9] an opponent of the doctrines of Socrates, advocated the study of astronomy for its vital importance to farmers, sailors and soldiers. He believed that seasonal changes, the waxing and waning of the Moon, and other phenomena were very important because they affect agriculture, conditions on the seas and the weather in general. In fact, this dissenter's advocacy was unnecessary, as the same conclusion had already been reached prior to 1000 BC by the Egyptians and the Babylonians, whose calendars were indeed based on the lunar phases. So why did they base their calendars on the Moon rather than the Sun? It was because observing the cycles of the Moon's phases to determine the number of days since the last new moon would take less than 30 days, which is much shorter than the time required to observe the Sun's annual cyclical behavior. Thus was born the lunar calendar, which is a more convenient and practical system for organizing the days of the year and widely adopted in many Eastern and Western cultures.

However, a calendar that is truly useful and practical has to somehow incorporate seasonal changes. In other words, the Sun's behavior must be taken into account. As the Sun's cycle is year-long, creating a solar calendar would have been very time-consuming and exceedingly difficult. Astrologers at that time came up with a compromise: the lunisolar calendar, a system incorporates seasonal changes into an existing lunar calendar. At around 432 BC, Meton of Athens discovered from Babylonian star catalogs that 19 solar years and 235 lunar months are almost the same duration. This knowledge was nevertheless abandoned ultimately, as a calendar based on a 19-year cycle would be even harder to deal with. Later in Rome during the time of Julius Caesar, people noticed that not all months were of the same duration (e.g., 30 days), and neither was the length of the year a whole number (it is in fact 365.25 days). Therefore, the 366-day leap year, occurring once every four years, was added to create the new Julian calendar, which was, in fact, a revised version of an older calendar that had been in use for several centuries.

2.5   The Maps and Shape of the Earth

In addition to observing the night sky, Greek astronomy was also concerned with measuring the sizes of the Earth, the Sun and the Moon, as well as the distances between them. Despite the inaccuracies of these measurements, this bold attempt transformed purely qualitative statements about nature into

---

[9]Plato's older brother who also grew up in Socrates' time.

quantitative inquiries, and it represents a significant step forward in the history of science. To make meaningful measurements of these three bodies, however, it was first necessary to recognize the fact that the Earth, like the Sun and the Moon, is a spherical object. This was by no means a foregone conclusion. Many hypotheses about the shape of the Earth had been conceived, such as the cylindrical shape proposed by Anaximander, the Greek philosopher mentioned earlier credited with inventing the gnomon. In his description, the flat top of this cylinder forms the inhabited world. Anaxagoras, who once proposed a form of atomic theory, also believed a flat Earth. He managed to deduce that the illumination of the Moon originates from the sunlight that it reflects, and also that when moving behind the Earth, the Moon falls into its shadow, which is when a lunar eclipse occurs. When the Moon moves between the Sun and the Earth, on the other hand, the shadow it casts on Earth causes the solar eclipse. Anaxagoras is therefore said to have been the first person to give a correct explanation of eclipses.

Despite this achievement, Anaxagoras failed to recognize the fallacy of his flat-earth hypothesis given the fact that the shadow cast by Earth on the Moon during a lunar eclipse has a curved shaped. Perhaps he meant to advocate a flat Earth that is in the form of a disc? Whatever the circumstances, this went against the most widely accepted theory of the time, a spherical Earth. The earliest record that advocated a spherical Earth was *On the Heavens* by Aristotle, who, in his consistent narrative regarding the nature of matter, put forth the assumptions that heavier matter (i.e., earth and water) gravitates toward the center of the Universe, whereas lighter matter (i.e., fire and air) moves away from it. The conclusion was that the Earth became a spherical object with aggregated earth and water and has since existed at the center of the Universe.

Although the shape of the Earth remained controversial, some proceeded to conduct actual surveys of the Earth, that is, to explore its geography: distribution of mountains, rivers, land, oceans, ethnic groups, mineral resources and other assets on Earth's surface. The original purpose was to attempt to understand the topographic features on Earth's surface and also the subterranean resources. This knowledge would enable the rulers to manage their citizens and their countries more effectively. Such knowledge would also be tremendously helpful to those with imperialistic ambitions of expanding their territories by conquering others. During the one-thousand-year period of the Greek and later Roman Empires, these exploratory research endeavors were seamlessly incorporated into imperialism and led to the transformation of human civilizations. Measuring the Earth was widely pursued in the time of Alexander the Great. Pytheas of Massalia (350–285 BC), who lived on the

Italian Peninsula, was a pioneer in this field. At the age of 25, he began to explore the coasts of Western and Northern Europe. He identified the location known as *Thule* (possibly today's Iceland or Greenland) as the northernmost point of *oikumene* (the entire known inhabited world). He also pointed out the location of the North Pole. Moreover, he offered a detailed description of the relationship between the latitude and the duration of daytime, which proved quite helpful in understanding the Earth's structure. Dicaearchus of Messana (326–296 BC), a contemporary and student of Aristotle, advocated an inhabited world with a rectangular shape and an aspect ratio of 3:2, from which he produced a map describing the geographical image that spanned the land between the Iberian Peninsula and India.

Aristotle also commented on the structure of the terrestrial world. In his cosmological treatises *On the Heavens* (completed in 350 BC) and *Meteorology*, Aristotle wrote that from the curved shadow cast onto the Moon during a lunar eclipse, one could not conclude that Earth's horizon is a straight line. These intuitions point to the conclusion that the Earth has to be round. From information gathered from travelers and his awareness that climate and temperature vary from one region to another, Aristotle further deduced that the Earth can be divided into five climate zones: two polar regions, two temperate zones (on either side of the equator), and the "uninhabitable" tropical zone along the equator. He confidently pointed out that terrestrial temperatures vary with the latitude but not the longitude. He also estimated the areas of the temperate zones in both hemispheres and produced drawings of the terrains and surfaces, which essentially amount to what is known as maps today. These maps would allow Alexander the Great, whom Aristotle had tutored, to conquer all the lands between the Balkans and India from 335 to 332 BC along habitable temperate routes. Alexander learned from his mentor the importance of empirical observation, and during his military campaigns, he designated a group of scholars to keep records of everything from the animals, plants and culture to the history and geography that they encountered along the way. These records had far-reaching implications for ancient Greek culture, the cradle of Western civilization.

In the second century, Ptolemy, who lived in Alexandria, employed a grid made up of longitudes and latitudes to produce more accurate cartographic projections. Ptolemy was the author of *Geography*, written in Greek on eight scrolls of papyrus. The book recorded the coordinates of 8,000 geographic locations, including those in the Mediterranean, Europe, North Africa, Middle East and Western Asia. Only the Americas, Australia and East Asia were not included. The book also detailed the size, shape and features of the inhabited world as an illustration of the nature and position of the known physical world, including gulfs, rivers, cities, ethnic groups and other attributes that reflect civilization and geography. No other book up until that time in human history was as comprehensive in its scope of observation and description of the Earth. After its completion, *Geography* disappeared for a thousand years and did not resurface until the 13th century in the Byzantine Empire.

2.6   Estimating the Distances to the Sun and the Moon

Aristotle once wrote about the first human attempt at measuring the Universe, which was about the sizes of the Sun, the Moon and the Earth, as well as the distances between them. Aristarchus of Samos (310–230 BC) was the person who carried them out. In 270 BC, Aristarchus published *On the Sizes and Distances (of the Sun and Moon)*, considered by today's scientific community to be the first academic research paper published in history. In this article, Aristarchus took into account the relative distances between the three celestial spheres (including Earth) and the sizes of the shadows cast by them to arrive at the distance between the Earth and the Sun, which he estimated to be 19–20 times that of the Earth-Moon distance. The same data rendered the Sun's diameter as 361/60 to 215/27 times Earth's diameter.[10] According to the explanatory notes provided by physicist and Nobel laureate Steven Weinberg, Aristarchus based his calculations on the following three observations: (1) The angle between the line connecting Earth-Moon and the line connecting Earth-Sun during a half-moon is 87°; (2) During a total solar eclipse, the disc of the Sun is fully obscured by the Moon; and (3) The width of Earth's shadow is precisely twice that of the Moon. Below we will derive the preceding two estimates based on the data from these three observations.

Refer to Fig. 2.2. Suppose Earth has a radius of r_0, and the radii of the Moon and the Sun are r and R, respectively. Let the distance between Earth and the Sun be D and the distance between Earth and the Moon be d. During a half-moon, the angle between the Earth-Moon line and the Earth-Sun line is 87°. Therefore,

$$\frac{d}{D} = \cos 87° = \cos\left(87 \times \frac{\pi}{180}\right) = \frac{1}{19.1} \qquad (2.1)$$

The Earth-Sun distance D, according to this calculation, is approximately 19–20 times that of the Earth-Moon distance d. Besides, during a solar eclipse

---

[10]Refer to Weinberg [32], Chap. 7.

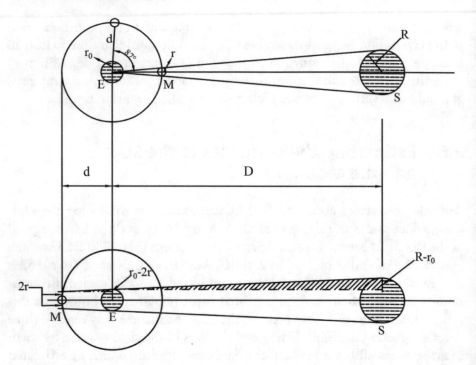

Fig. 2.2 The configurations of Aristarchus in his derivations of the relationships among Earth (E), the Moon (M), and the Sun (S)

(see the top half of Fig. 2.2), the Moon blocks the Sun completely, as viewed from Earth. Using the two similar right triangles, one with base r and the other R, together with the common vertex at the center of Earth, we arrive at the following:

$$\frac{d}{D} = \frac{r}{R} \tag{2.2}$$

Aristarchus later also discovered that the width of Earth's shadow is twice that of the Moon during a lunar eclipse, so from the two (shaded) similar right triangles at the bottom half of Fig. 2.2, which has a base of $r_0 - 2r$ and $R - r_0$, respectively, the following relationship holds:

$$\frac{r_0 - 2r}{d} = \frac{R - r_0}{D} \tag{2.3}$$

Substituting (2.1) and (2.2) into (2.3), we have:

$$\frac{R}{r_0} = \frac{\cos 87° + 1}{3 \cos 87°} = 6.70 \tag{2.4}$$

In addition, since

$$\frac{361}{60} \approx 6.01 < 6.70 < \frac{215}{27} \approx 7.96 \tag{2.5}$$

we deduce that the solar radius R is between 361/60 and 215/27 times Earth's radius r_0. Although quite far from their true values, the work that yielded these estimates was truly pioneering, considering this was done at a time when the most basic weights and measures had yet to be established. Despite the inaccuracies, the qualitative conclusion remains true: the Sun is much larger than the Earth. This result also led Aristarchus to conclude that the Earth, in fact, revolves around the Sun, rather than the other way around, as it would be inconceivable for an object as enormous as the Sun to orbit around a tiny Earth (refer to Sect. 2.8 for additional details).

In Aristarchus' time, measuring the Moon's distance from Earth was a much-researched topic. The technology that enabled such measurements, however, did not mature until Hipparchus improved it significantly in the period between 160 and 127 BC. Recall that Hipparchus had compiled star catalogs previously. Although his original work has now been lost to history, Ptolemy's *Almagest* contains detailed descriptions of his work on this topic. According to Ptolemy, the solar eclipse that occurred on March 14, 189 BC was the event that Hipparchus employed in his study. During the eclipse, the Sun in Alexandria was completely obscured by the Moon (a total eclipse), but the same event in Hellespont (a city on the shores of modern-day Dardanelles Strait, Turkey) was only a partial eclipse (with about 4/5 of the Sun being covered). The observations of the Sun and the Moon at Hellespont during this time yielded the visual angles of 0.55° and 0.11°, respectively. In addition, Hipparchus learned the latitudes of Alexandria and Hellespont from the gnomon measurements. Finally, the relative positions of the three bodies observed during the solar eclipse enabled him to calculate the Earth-Moon distance to be between 71 and 83 times Earth's radius, a factor that is not entirely accurate compared with the currently known value of 60.

The calculations of the distances between the Earth, the Sun and the Moon by Aristarchus and Hipparchus, as outlined above, were based on Earth's radius, still unknown at the time. Without a definitive value for Earth's radius, these calculations were merely abstract formulas without much practical utility. Measuring Earth's radius was therefore widely regarded as an absolutely essential task, the undertaking of which remained far from trivial for contemporary astronomers. This difficult issue was not resolved until several decades later when Eratosthenes employed a gnomon to measure the Earth's circumference successfully.

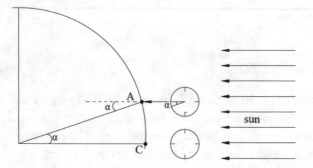

Fig. 2.3 Measurement of Earth's radius by Eratosthenes using a gnomon

Eratosthenes lived in Libya when he made the historical measurements. He combined his astronomical observations and known facts to calculate the Earth's circumference using a time-keeping device known as the "Babylonian rod" (i.e., gnomon) to perform a series of measurements of the shadow cast by the apparatus in Syene (modern-day Aswan, Egypt), a city located to the south of Alexandria.[11] He knew that at noon on the summer solstice in Syene, the Sun would be directly overhead and the gnomon would cast no shadow at all. At the same time in Alexandria, a gnomon located there would cast a shadow at an angle of about 1/50 of a circle (i.e., $2\pi/50$ radians, or 7.2°). He, therefore, deduced the circumference of Earth to be 50 times the distance between the two cities, a number rather close to the actual factor of 47.9 known today. The reasoning is in fact quite straightforward. Since the distance between the Sun and Earth is much larger than Earth's radius, we can treat rays coming from the Sun as being parallel when they reach Earth. As a result, the statement of "the angle of the Sun's shadow in Alexandria relative to the vertical gnomon being 7.2°" can be interpreted as "the distance between Alexandria and Syene corresponding to an arc that subtends an angle of 7.2° at the center of the Earth". Since a full circle corresponds to a total of 360°, the circumference of the Earth is $360/7.2 = 50$ times the distance between the two cities. Refer to Fig. 2.3 for details.

Having deduced this crucial factor of 50, in order to calculate the size of Earth the next task would be to measure the distance between Syene and Alexandria. Eratosthenes was reputed to have recorded the time required for a trained camel to travel between the two cities, which he used to convert to the total number of paces taken. Then he multiplied this number by the

---

[11]This story goes like this: Eratosthenes first discovered in Syene, in the south of Egypt, that once a year the sunlight would shine straight down a well. On that same day at the port of Alexandria (located in northern Egypt), a docked ship would cast a shadow. Eratosthenes then argued that Earth is a sphere and proceeded to measure the radius of the Earth with a gnomon.

distance covered by each pace to arrive at the desired result. After several experiments, Eratosthenes calculated this distance to be about 5,000 stadia (singular *stadium*, an ancient Greek unit of length). Research into ancient units of measurement indicates that one stadium is between 500 and 600 feet. Taking the midpoint of 550 feet, the conversion gives the circumference of the Earth as 550 × 5,000 × 50 = 137,500,000 feet, or about 42,000 km, an error of 5% compared with the actual value of 40,075 km. Despite such a simplistic approach as a camel's paces to measure Earth's circumference, the result is remarkably close to the true value. A thousand years later, Europe's natural philosophers, including Sir Isaac Newton, would continue to apply this result in their research.

2.7   Precession of the Equinoxes

Arriving at two sets of relatively accurate measurements of the Sun, the Earth and the Moon, as described in the previous sections, was no doubt a phenomenal achievement in the history of astronomy. What is considered by future generations to be the most amazing and impressive measurement, however, is Hipparchus' estimation of the cycle of Earth's axial precession (or the precession of the equinoxes). To understand the precession (wobbling) of the Earth's rotational axis, it is necessary to revisit the astronomical phenomena of the vernal and autumnal equinoxes, which we have mentioned several times before. The vernal (spring) equinox or the autumnal equinox is the result of the Earth's rotational axis being tilted at an angle of 23.5° relative to the ecliptic plane. As shown in Fig. 2.4, the Earth depicted on the left-hand side

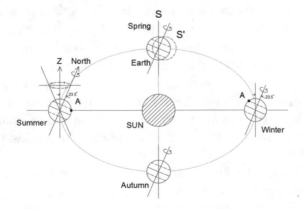

Fig. 2.4 Illustration of astronomical phenomena resulting from the precession of Earth's axis of rotation

("Summer") has a rotational axis tilted at 23.5°, and the Tropic of Cancer *A* (in the northern hemisphere) is facing the Sun directly. So this is summer in the northern hemisphere and the winter in the southern hemisphere. As point *A* gradually moves away from the Sun as the Earth revolves around it, *A* reaches the right-hand side position in Fig. 2.4, where the Tropic of Capricorn now faces the Sun directly. At this point, it is summer in the southern hemisphere and winter in the northern hemisphere. As the northern hemisphere transitions from summer (the Earth on the left) to winter (the Earth on the right), there is a certain day where Earth's axis of rotation will become perpendicular to the line connecting the Sun and the Earth, as shown in the Earth at the bottom of Fig. 2.4. Here the Sun is directly overhead on the equator. This day is defined as the autumnal equinox in the northern hemisphere. Similarly, during the transition from winter to summer in the northern hemisphere, the equator will once again be directly facing the Sun, and this day is defined as the vernal equinox in the northern hemisphere. Since ancient astrologers (that we know of) lived in the northern hemisphere, they tended to make observations of the sky during the vernal equinox as the weather was gradually warming up to make astrological predictions about uncertain things in the new year and how to deal with these uncertainties. Nevertheless, the vernal equinox did not always occur at the same time each year.

The vernal (or autumnal) equinox is determined by the relative positions of the Earth and the Sun as well as the orientation of Earth's rotational axis. If these conditions do not change, the positions of the distant stars observed at the same angle during the vernal equinox should also be the same. Yet actual observational results tell a different story. This is because the orientation of Earth's rotational axis is constantly shifting, albeit very slightly and slowly, as the planet revolves around the Sun. The Earth's position in its orbit at the vernal (or autumnal) equinox is also changing. To explain these changes, we have labeled the Earth on the left-hand side of Fig. 2.4 with the *Z* axis, which is perpendicular to the ecliptic plane. Here the *Z* axis has an angle of 23.5° relative to Earth's rotational axis. If one looks southward along the *Z* axis, Earth's rotational axis will be making a very slow rotation around the *Z* axis in a clockwise direction. This is what is known as the precession of Earth's axis of rotation. This axial precession also contributes to the difference between a tropical year and a sidereal year. What exactly are the tropical year and the sidereal year?

The tropical year (also known as the solar year) refers to the time interval for the Sun to go from one vernal (or autumnal) equinox to the next, as seen from Earth, or the interval between two consecutive vernal (or autumnal) equinoxes. The sidereal year is the time taken by Earth to orbit the Sun once, completing a 360° revolution. Now suppose the point S in Fig. 2.4 represents the current vernal equinox. The sidereal year refers to the time required for Earth to reach S in its next revolution around the Sun. However, this next point S no longer corresponds to the vernal equinox, since the precession of Earth's rotational axis occurs in a clockwise direction. Before Earth returns to the same point S of the previous year, the vernal equinox has already occurred (indicated by the point S' in the diagram. In other words, the vernal equinox point S will slowly move in a clockwise direction to point S' against the direction of Earth's revolution around the Sun. Despite this small-time discrepancy, the tropical year is measurably shorter than the sidereal year. This phenomenon is known as *suìchā*, or "annual difference," in Chinese and is the direct result of the precession of the equinoxes. According to modern-day observational data, which have much higher precision, the length of the sidereal year is 365.25636042 solar days and that of the tropical year is 365.24219907 solar days, a difference of 0.01416135 solar days (or about 20 min). From these numbers the period of Earth's axial precession can be deduced: it is approximately $365.25/0.01416135 \approx 25,800$ years.

Around 2000 BC, the Babylonians divided the stars scattered around the ecliptic into twelve zodiac constellations. They discovered that at the vernal equinox each year, one could always spot the same constellation in a given section of the night sky. After a long period of observation, however, it became known that the constellation at this position would shift over time. Today we know that these twelve constellations shift by one zodiac sign every 2,000 years or so. In 2000 BC, for example, the Babylonians would see Aries at the vernal equinox, but in the first years of the Common Era, Pisces would be spotted in that same location. Today at the start of the second millennium, the closest constellation at the vernal equinox is Aquarius. These changes are noticeable only on a thousand-year time scale, so a meticulous observation is required to spot the subtle shifts over a time scale as small as a decade. Nevertheless, these slight changes were observed by none other than the ancient astronomer Hipparchus, who not only compiled a star catalog consisting of 850 stars but also carried out the measurement of the Moon's distance from Earth, as described previously. Hipparchus is recognized to have discovered the precession of the equinoxes and estimated its cycle in 127 BC.

According to Weinberg's accounts,[12] the reason Hipparchus wanted to estimate Earth's precession cycle was that he had observed some spinning objects that exhibited precessional behavior, so it was natural for him to ask if the same phenomenon applies to Earth itself. In his work *Precession of the*

---

[12]Refer to Weinberg [32], which explains the possible computational approach taken by Hipparchus.

Equinoxes and Summer Solstices, Hipparchus pointed out that since the relative positions of the stars on the celestial sphere never change, if Earth's rotational axis does exhibit precession, then there will be visible changes in the positions of the stars on the celestial sphere as viewed from Earth's surface. To verify his claim, Hipparchus chose the star Spica in the constellation Virgo as his object of study. The reason was that observational records of Spica made 150 years before by the two astronomers Timocharis (320–260 BC) and Aristillus (fl. c. 261 BC) were available to him. By carefully comparing the observations, he had collected for Spica with the data from one and a half centuries earlier. Hipparchus was able to conclude that the right ascension[13] of Spica at the vernal equinox had shifted by 2°. He then made the following calculations using the 2° eastward movement of Spica over the past 150 years. Then he concluded that if the Earth's rotational axis has precessed one full revolution, the right ascension will have also rotated 360° and the precession takes a total of $360 \times (150/2) = 27,000$ years to complete one cycle, a number quite close to the modern value of 25,727 years. Hipparchus also calculated that the tropical year had been shortened by 1/300 years over the 150-year period, whereas the sidereal year had lengthened by 1/144 years.

Although Hipparchus' observational data has been lost to history, research on precession was revived in the second century by Ptolemy, who obtained similar results. However, as geocentrism was the prevailing cosmological model at the time, Ptolemy explained this phenomenon as the movement of the stars on the celestial sphere and did not attribute it to the precession of Earth's rotational axis. Therefore, although this phenomenon was discussed a number of times in the astronomical community, its cause was never investigated thoroughly. This scientific mystery remained unresolved until a thousand years later when Newton discovered the law of universal gravitation and applied this theory to calculate the precession cycle. We will provide a detailed discussion on its theoretical derivation in Chap. 4 (see Sect. 4.8).

2.8   The Elusive Cosmological Model

Although remarkable advances in the observations of the Sun, the Moon and Earth were achieved, the issue of how these three astronomical bodies interact with one another remained a matter of much debate. From the perspective of

---

[13] Right ascension is one of the two coordinate axes in the equatorial coordinate system for the celestial sphere, analogous to terrestrial longitude. Circles of constant right ascension run north-south from one celestial pole to the other and are perpendicular to the celestial equator. The other coordinate axis of the celestial sphere is declination.

everyday experience, the idea of the Sun, the Moon and the five "wandering stars" orbiting the Earth was not inconsistent with terrestrial observations and was quite reasonable and easy to comprehend. On the other hand, from the observed solar and lunar eclipses as well as the retrograde motions of planets, it was exceeding difficult to explain these complex phenomena based on a rather awkward geocentric model, i.e., a stationary Earth located at the center of the Universe. Distinctly different cosmological models coexisted for the next few centuries, each with its own merits and body of supporting evidence, yet none could produce sufficient compelling evidence to completely refute others' competing theories. Below we will review several of these models that found widespread support during this period.

The Pythagoreans were the first philosophers to propose an astronomical system that explains how celestial bodies interact. Adherents of this school of philosophy believed that Earth, the Moon, the Sun, the five planets and all the stars revolve in their own invisible celestial spheres around a "Central Fire" located at the center of the Universe. Earth was believed to be a disc, and human beings live on the surface of the disc that faces away from the Central Fire. More than 200 years later, however, this Central Fire was criticized by the Aristotelians as being absurd. They accused the Pythagoreans of fabricating the non-existent Central Fire to complete the number 10, in their views the perfect number. In other words, adding the Central Fire to the original eight celestial bodies (Earth, Sun, Moon, and the five planets) and the fixed stars (if treated as a single unit) produces the perfect number "10".

2.8.1 Geocentric Model with Concentric Celestial Spheres

Consequently, the Platonists proceeded to remove the Central Fire and constructed their own cosmological model from the nine real, observable astronomical bodies: The fixed stars are located on the invisible, outermost spherical shell, which rotates around Earth from east to west once a day at a constant speed. The Sun and the Moon, driven by their own distinct spherical shells, move around Earth from east to west daily (this explains the diurnal motions of the two celestial bodies). The inner spherical shell is attached to the outer spherical surface in its revolution, and it itself rotates in an easterly direction (which explains the annual motion of the Sun along the ecliptic and the Moon's monthly motion along its orbital plane). The five visible planets are each attached to four spherical shells surrounding the Earth, with the outer two layers associated with the Sun's and the Moon's shells and the inner two layers governing the planet's retrograde motions around the Earth.

Later, this concentric spherical cosmological model was modified by the Aristotelians, who incorporated the outermost spherical shell of Saturn into the fixed-stars shell and introduced three additional spherical shells between Saturn and Jupiter. After making several adjustments with specific purposes to the model, however, the results became utterly ridiculous. Jupiter would orbit Earth twice a day, Venus four times a day, Mercury five times a day, the Sun six times a day, and the Moon even seven times a day!

In defining the relationship between Earth and the celestial sphere, Greek mathematician and astronomer Eudoxus of Cnidus (c. 390–c. 337 BC), who was a student of Plato's, once constructed a cosmological system centered on Earth with celestial objects revolving around it on multi-layered spherical shells. Incorporating methods he himself had developed to compute the Earth-Sun and Earth-Moon distances as well as various observed movements of the Sun, the Moon and the five planets, his model consists of 27 concentric spheres: the fixed stars are on the outermost sphere, followed by Saturn, Jupiter, Mars, the Sun, Venus, Mercury, and the Moon. Except for the Sun and the Moon, each having three spheres, the five planets are assigned four spheres each. Each of these shells rotates at different angular speeds on various axes. The four-layer spherical model for each planet is specifically designed to explain its retrograde motions. Figure 2.5 illustrates how this concentric sphere model can explain planetary retrograde motions.

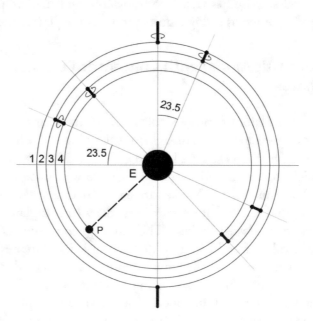

Fig. 2.5 The Eudoxan concentric spheres planetary models. Inspired by Fig. 2–7 of Hsiang et al. [15]

In this model, each planet is assigned four concentric spherical shells with different radii, and Earth is located at the center of all these spheres. These four spherical shells operate in accordance with the following rules: The outermost sphere (labeled "1" in Fig. 2.5) rotates westward around the north-south axis of rotation once in 24 h, which explains the planet's daily motion. The second sphere ("2") has a rotational axis tilted 23.5° relative to the north-south axis and completes one rotation per sidereal (orbital) period,[14] driving the planet's slow movement along the ecliptic. The third sphere ("3") has an axis of rotation tilted 23.5° relative to the equatorial plane of the outermost sphere. The innermost sphere ("4") has a rotational axis that is tilted at a small angle relative to the third sphere's rotational axis, which is meant to indicate that the five planets' orbital planes do not lie on the same plane (i.e., they are not coplanar), and the equator of the innermost sphere represents the planet's orbit. Finally, two adjoining spheres are linked by their rotational axes, so the outermost sphere is affected by the second sphere's rotation in addition to its own rotation, and the second is affected by the third, and so on. Eventually, the planet on the innermost sphere has a trajectory that represents the motions of all four spheres combined. With such an elaborate arrangement, the observer on Earth, which lies at the very center of all spheres, should be able to view the planets' retrograde motions quite clearly.[15]

From this ingenious and intricate model, it is undeniable that Eudoxus did indeed possess exceptional skills in geometry. He was also able to create an astronomical model capable of explaining complex phenomena, such as planetary retrograde motions, by adopting the "uniform circular motion" and other similar principles that convey simplicity and regularity, qualities sought after in geometry. To the natural philosophers who advocated that the Universe is the natural product of a Creator who produced the design in a rational manner, this was an encouraging accomplishment in both geometry and astronomy. Also because this system of concentric spheres successfully preserved the notion that the circle is a perfect geometric figure in Greek philosophy, Aristotle subsequently incorporated this cosmological model into his writings about the cosmos. The geocentric model of the Universe, therefore, became widely known. The most famous discovery of Eudoxus, a brilliant geometer and astronomer, however, is his calculation of the Earth's circumference, which he estimated to be between 60,000 and 70,000 km, an error between 50 and 80% compared with the true value of about 40,000 km, a

---

[14]It was already known at the time that the planets have the indicated sidereal periods: Mercury–88 (earth) days; Venus–225 days; Mars–1 (earth) year 322 days; Jupiter–11 years 315 days; Saturn–29 years 166 days.

[15]Refer to Fig. 2.8 of Hsiang et al. [15].

much less accurate result than that provided by Eratosthenes (which we discussed in Sect. 2.6).

2.8.2   Heliocentric Model with Concentric Orbits

Herakleides Ponticus (c. 350 BC), a contemporary of Aristotle, argued that the Earth actually spins around its rotational axis once a day, which explains the uniform daily motions of the stars, planets, the Sun and the Moon across the sky. To explain the perpetual proximity of both Mercury and Venus to the Sun, he suggested that these two planets are in fact orbiting the Sun, while the Sun, the Moon, Mars, Jupiter and Saturn are orbiting the Earth. On the other hand, however, in the article *On the Sizes and Distances (of the Sun and Moon)*, not only did Aristarchus derive the sizes of the Sun and the Earth, as well as the distance between them, but he also extended the estimation to the size of the entire Solar System, as described in Sect. 2.6. Aristarchus also argued that the Earth not only rotates but also revolves around the Sun, along with other planets. From his estimation of the sizes of the Sun and the Earth, Aristarchus concluded that the Sun's volume is 1.3 million times that of the Earth, and it would be unlikely for such a large object to revolve around a relatively tiny Earth. Consequently, he proposed that the Sun is the center of the Universe and that it is orbited by Earth and all the other planets. This could have been the first heliocentric theory introduced in history.

Although Aristarchus based his heliocentric theory on the large quantity of data that he had collected, such as the distances of the Sun and the Moon from Earth, it was clearly inconsistent with the daily sunrise and sunset experienced by common folks with their naked eyes. His theory was also seriously at odds with the Aristotelian system of concentric spheres. As a result, Aristarchus' ideas were often rejected by the astronomical community, which was dominated by supporters of the geocentric theories of Aristotle. Aristotle also refuted Aristarchus' model with his usual empiricism: If the Earth revolves around the Sun, why do the stars in the night sky always stay at the same locations? This refutation was, however, invalid, as careful observation shows that the positions of the fixed stars are not fixed at all. They gradually shift during the course of the year, and their trajectories form closed curves (ellipses). The sizes of these ellipses vary according to their distances from Earth. With respect to these subtle but widely known annual shifts, Aristarchus provided a rather satisfying explanation: Since the distance between Earth and the stars is much greater than that between Earth and the Sun, even though the positions of these stars do shift over time, the changes are too insignificant to be noticeable by observers on Earth.

Aristotle nevertheless remained unconvinced by Aristarchus' argument and deemed it utterly implausible, and he continued to refute heliocentrism with mundane questions that call into doubt the veracity of a moving Earth, with the most quoted being: "If the Earth is moving, why does an object thrown up vertically fall straight down?", "Why don't we feel the ground moving at all?" and other similar ones, which are admittedly difficult to answer. Even though opinions were widely divided in the astronomical community, geocentrism eventually won out. After all, Aristotelianism was the dominant school of thought at the time. A few centuries later, the Roman writer Marcus T. Cicero (106–43 BC) emphasized in his work *On the Republic* (*De re Publica*) that the cosmological model prevalent at the time was able to represent accurately the movements of the Sun, the Moon and the five planets around the Earth. No matter how far-fetched or inaccurate Cicero's premise was, it is apparent that the geocentric theory was now a well-established part of human knowledge, and this would continue for nearly two millennia, until it was eventually supplanted by the heliocentric model reintroduced during the Scientific Revolution begun by Nicolaus Copernicus (1473–1543) in the 16th century.

2.8.3 The Deferent-and-Epicycle Model

In addition to being involved in the controversy surrounding the rightful occupant of the center of the Universe, the astronomers were also baffled by the retrograde motion[16] of Mars. One theory to explain this phenomenon is the system of concentric spheres devised by Eudoxus, as discussed previously. Based on data provided by his 850-star catalog, however, Hipparchus considered the premise of the Eudoxan planetary model, where each celestial body moves at a uniform speed along with the spherical shell, to be erroneous. In his views, many celestial bodies (which we take to mean the five planets visible to the naked eye: Mercury, Venus, Mars, Jupiter, and Saturn) move in a certain direction, only to retrogress (turn back) at some point later. Their brightness would also change according to its position along this retrograde trajectory. This means that when a planet is in retrograde motion, its distance from Earth will change. Based on this reasoning, Apollonius of Perga

---

[16]Because of Earth's higher orbital speed compared with Mars, when Earth approaches Mars, its apparent motion changes from catching up with the red planet to moving away from it. Viewed from Earth's surface, the movement of Mars appears to change from a clockwise to a counterclockwise direction. This phenomenon of changing the direction of movement is known as retrograde motion, and it has long been recorded by astronomers since ancient times. See Appendix 2.2: Derivation of Formulas for Planetary Retrograde Motion.

(c. 240–c. 190 BC) first proposed a hypothesis known as the eccentric model to explain the apparently aberrant motion of the planets. A moving planet is carried on a circular path called an *eccentric*, the center of which moves in a circular orbit around the Earth. This hypothesis, however, was later modified by Hipparchus. In his *deferent-and-epicycle* model, the planet moves around a smaller circle called an epicycle, which itself moves around a larger circle called a deferent, with the Earth located at its center.

We illustrate these two models side by side in Fig. 2.6, as they are in theory equivalent. If we enlarge the circular path traced by the eccentric's center (top-left of Fig. 2.6) to match the circular path traced by the epicycle's center (top-right of Fig. 2.6), and then shrink an eccentric (top-left) down to the size of an epicycle (top-right), then the two diagrams become identical. In addition, if we also adjust the size of the epicycle and the angular speed of the planet it carries, we can duplicate the planet's retrograde trajectory, as shown in Fig. 2.6 (bottom). In this diagram, as the planet moves around the epicycle once, the epicycle's center advances by 40° on the deferent. What is also illustrated here is that when the epicycle has advanced by 120°, three

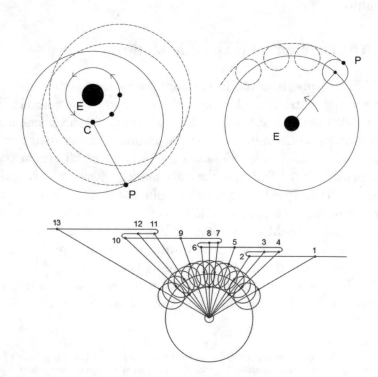

Fig. 2.6 Top: The eccentric (left) and epicycle (right) models developed by Hipparchus and Apollonius of Perga. Bottom: Three instances of retrograde motion resulting from parameter adjustments to the deferent-and-epicycle model

separate instances of retrograde motion have occurred. We know from actual observational data recently collected that there have been two occurrences of Jupiter's retrograde motion as viewed from Earth: once on June 4, 2007 and the other on July 9, 2008, with the two occasions separated by about 400 Earth days. As Jupiter's orbital period is 11.86 Earth years (or 4,329 Earth days), the two Jovian retrograde motions seen from Earth occurred over a span of only 33° in the gas giant's orbital movement around the Sun. We have also observed two separate occasions of Saturn's retrograde motion, once on February 10, 2007 and the other on February 24, 2008, which are 379 Earth days apart. Taking into account Saturn's orbital period of 29.46 Earth years, the planet moved only about 12.7° along its orbit before two events of its retrograde motion were observed from Earth. The conclusion is that the farther the planet is away from Earth, the shorter the time interval between its two successive retrograde motions. In Appendix 2.2: Derivation of Equation for Planetary Retrograde Motion, we will derive a set of equations that can compute the times a planet in the Solar System retrogresses, as view from Earth.

From the above comparison, if appropriate adjustments are made to the angular speeds of motion for both the epicycle and deferent, Hipparchus' epicycle model can indeed represent Jupiter or Saturn's retrograde trajectory. Over two hundred years later, this model was revised further by Claudius Ptolemy in Egypt based on data of higher precision. This revised model would influence the development of astronomy for the next thousand years. We will elaborate on this point in Sect. 2.10.

2.9   Alexandria as a Flourishing Center for Learning

Following the death of Alexander the Great, his vast Macedonian Empire was divided among four generals under this command. The Ptolemaic dynasty of Egypt was founded by General Ptolemy I Soter (367–282 BC) in 321 BC. The dynasty flourished for nearly three hundred years until the reign of Ptolemy XV, the last pharaoh of Egypt. He was the son of Cleopatra VII Philopator (69–30 BC) and G. Julius Caesar (100–44 BC), Roman dictator and general. The Ptolemaic dynasty ended in 31 BC when the combined forces of Roman general Mark Antony (83–30 BC) and Cleopatra were annihilated by the Octavian's fleet at the Battle of Actium. The demise of the 290-year-old Ptolemaic dynasty followed the murder of Pharaoh Ptolemy XV and the annexation of Egypt to the Roman Republic as a province.

During the reigns of the Ptolemaic pharaohs in Egypt, knowledge, wealth and power continued to be amassed in the capital city of Alexandria, located at the estuary of the Nile River. Alexandria would later become the second largest city of the Roman Empire, after Rome itself. To expand the nation's power and to continue the achievements in the natural philosophy of Greece, Ptolemy I Soter began construction of the Great Library of Alexandria in the city in 290 BC. At first, the library collected various literary works, mainly those written in the Greek language and also books translated from other languages. The library was dedicated to the Muses, the nine goddesses of the arts. After Pharaoh Ptolemy II acceded to the throne in 285 BC, the library became widely known, as scientific research activities began. Alexandria came to be regarded as the capital of knowledge and learning, thanks to the generous funding provided by Ptolemy II and his father. With Egypt's bounty and riches, excellent remuneration for researchers, and favorable living conditions, the library attracted many outstanding scholars.

An atmosphere of burgeoning scholarly pursuit began to permeate the city when Demetrius of Phaleron (350–280 BC), the first director of the Library of Alexandria, brought a collection of books from Athens. This gradually transformed Alexandria into a center of scientific learning around the Mediterranean region. To continue expanding the library's collection, successive Ptolemaic pharaohs enforced the regulations that required all ships entering the port of Alexandria to surrender the original manuscripts of all the books on board before they were allowed to leave port. Copies of these manuscripts would then be returned to the manuscripts' original owners instead of the originals. This policy indeed enabled the library's collection to expand significantly. In its heyday, the size of the library's collection reached perhaps as many as ten thousand volumes, and a library catalog was produced for its patrons. The Library of Alexandria thus became a highly recognized repository of great knowledge, attracting visiting scholars from far and wide. Some of the greatest and most influential scholars of the era, such as Greek natural philosophers Euclid, Eratosthenes and Archimedes, whom we introduced in Chap. 1, all worked at the library, and during their tenure here they made significant contributions to scientific knowledge for future generations.

Although the Ptolemaic Kingdom inherited its land and regime from Alexander the Great's empire, Greek scholars in Egypt would no longer engage in the investigation of the theory of universals advocated since Aristotle's time. This theory of universals had been widely studied in Greece for hundreds of years, but it gradually degenerated into a kind of intellectual superiority focused on fantastical and meaningless pursuits, which was completely detached from reality. The scientists who came to Egypt, where practical applications were highly valued, began to respond to the demand of society and adopted a more pragmatic approach in their scholarly endeavors. They turned to research in technology and applications that would be more relevant to people's everyday life, and eventually, they achieved remarkable progress in various disciplines, such as optics, fluid mechanics and astronomy. This atmosphere of pragmatism was evident from the research of Strato of Lampsacus (335–269 BC), the second director of the Library of Alexandria. An excellent scientist, Strato observed that the distances between droplets from water falling from the eaves become greater when they get closer to the ground. From this phenomenon, he deduced that free-falling objects are accelerating downward. He was therefore a pioneer in the study of mechanics. Another scientist also living in Alexandria, Philo of Byzantium (280–220 BC), who was credited as a military engineer, wrote *Mechanike syntaxis* (*Compendium of Mechanics*) in about 250 BC. This work not only explained a number of hydrostatic phenomena that he had observed but also made concrete suggestions on harbor building and defense fortifications. It also explained siege tactics and siege equipment such as catapults, as well as other inventions of practical military applications.

Because of these leading researchers, fields of study pursued by scientists in Alexandria were no longer confined to the type of all-encompassing natural philosophy indulged by the Aristotelian school in Athens, or the philosophical idea of pursuing universals. Instead, it focused on a limited number of areas, including hydrostatics, optics and astronomy, in order to explore the scientific principles possibly hidden in natural phenomena and relevant to everyday life. These scholars transformed the idealistic and impractical Athenian research practice into an alternative that emphasized practical knowledge that would help to enhance the quality of life. This would then allow scientific research to spread rapidly from the ivory tower to the general populace, where ordinary citizens could apply the knowledge acquired from studying nature to create useful apparatuses. One notable case involves the Greek inventor Ctesibius of Alexandria (285–222 BC), the son of a barber in Alexandria. Around 250 BC Ctesibius invented a suction pump that used low pressure to pump water and also a water clock that kept accurate time.

This pursuit of practical research continued over the next 300 years or so when Hero of Alexandria (10–70) wrote in his *Catoptrica* that the paths of the incident and reflected light on a flat mirror traverse the shortest distance. Although Hero never offered an explanation for this hypothesis at the time, it was verified in the 17th century by Christiaan Huygens (1629–1695) when he worked out a wave theory of light, which would lend credence to Hero's highly developed intuition for natural phenomena. Because of his exceptionally intuitive mind, Hero managed to invent numerous instruments and mechanical contraptions based on natural phenomena. For example, he invented the surveying instrument called a theodolite by applying the optical principle that the angle of incidence equals the angle of reflection on a plane mirror. Hero also created siphons, catapults, and a steam-powered engine (called an aeolipile), which may be considered rudimentary when compared with similar devices invented in more recent times, yet Hero's works, including the design principles and blueprints, proved indispensable in inspiring modern inventions. Do not confuse the inventions of Hero mentioned above with children's toys or stage props. These were actual, workable machinery that could be used in everyday life or even in wars, and indeed they did find practical and well-received applications, especially in military and civil engineering.

2.10 Ptolemaic Astronomy

In this prevailing atmosphere of pragmatism, an astronomer and geographer by the name of Claudius Ptolemy (90–168) burst onto the scene in the city of Alexandria. His astronomical works were profoundly influential in shaping future generations of astronomers. Little is known about the life of this great scientist, but his given name Claudius indicates he possessed Roman citizenship, and his surname Ptolemy suggests he was of Greek descent. At the age of 27, Ptolemy came to Alexandria to study the various disciplines that were required to understand natural phenomena, and there he would remain for nearly a quarter of a century. In his work *Optics*, he extended Hero's reflection principle for plane mirrors to curved mirrors. He also explored the principle of refraction in situations where light passes through different media. The experiment designed by Ptolemy on refraction is deceptively simple. A stick would be inserted into the water, and it would appear to be bent under the water surface. He then inferred the relationship between the ray's angle of incidence, as it penetrates the water surface and the angle of refraction. Unfortunately, Ptolemy's research in optics remained limited to qualitative description,[17] as he devoted most of his research efforts in his true passion, astronomy, to which he made remarkable contributions.

---

[17]The actual law that governs the angles of incidence and refraction was not known until the 17th century, when Pierre de Fermat (1607–1665) proposed the principle of least time: the path taken between two points by a ray of light is not the one with the shortest distance but the one that can

The celestial configuration established by Ptolemy, like that of Aristotle, is based on a spherical Earth at the center of the Universe. Ptolemy's goal was to calculate the trajectories of the five planets known to humans at the time. To achieve this goal, the first challenge was to explain the unresolved problem of the retrograde motion of Mars. First proposed by Hipparchus and Apollonius of Perga over a hundred years before (see Sect. 2.8), this was a problem that had perplexed all astronomical observers since ancient times. It originated from the then yet unknown fact: all the planets in the Solar system orbit the Sun in the same direction along their respective orbits, but their tangential (orbital) speeds are all different. Therefore, when viewing the position of a planet from Earth, sometimes the planet would appear to stop briefly and reverse direction at certain times, due to Earth's orbital speed being different from those of other planets. After a while, however, the planet would resume its course in the original direction. This is what is known as planetary retrograde motion. When astronomers attempted to explain this phenomenon based on a geocentric theory, their efforts often ended up fruitless. Ptolemy took up this difficult challenge with confidence only because he had inherited a large quantity of observational data of the heavens left by Hipparchus and others. From historical data and his own observations, Ptolemy discovered that the trajectories of some planets did not follow a circular orbit, but instead returned to earlier positions along the original path, which was consistent with the deferent-and-epicycle model of Hipparchus. Thus Ptolemy proceeded to explain the epicycle model using mathematical tools and founded the geocentric Ptolemaic system.

In this geocentric model, which is based on a substantial amount of observational data and analytic results, Earth is a stationary sphere located at the center of the Universe. All celestial bodies move at uniform speeds along circular orbits centered on Earth. Among them, the Moon is the heavenly body closest to Earth, followed by Mercury, Venus, the Sun, Mars, Jupiter, and Saturn, with distant fixed stars located on the outermost sphere. Although the circular orbit hypothesis is consistent with the idea of the perfect circle revered by Aristotle, it was refuted by the observed data, the most prominent of which is the retrograde motion of planets widely acknowledged at the time.

In order to find a geocentric model that would fit the observed data and phenomena, Ptolemy proceeded to modify the cosmological model proposed by Apollonius and Hipparchus, which consists of the epicycle, deferent, and eccentric. The reason was that during the two centuries since the introduction of this model, astronomers remained unable to describe the planetary

---

be traversed in the least amount of time. This hypothesis was later developed into Descartes' (or Snell–Descartes) Law.

retrograde motion with any level of accuracy. Once his revision of the model commenced, Ptolemy continued to adopt a circular orbit and uniform orbital speed for each of the celestial bodies, but not for very long. Ptolemy noticed that the revised model still could not explain the observations, as all planetary data indicated that the speeds projected onto the celestial sphere were non-uniform. With this discovery, he deduced that planetary orbits should be non-circular. Now, to make the model fit the observed planetary data, Ptolemy repeated what Hipparchus had done, i.e., displacing Earth's position from the center of the celestial sphere, thus turning each circular orbit into either an eccentric or a deferent. When a planet arrives at the apogee (farthest point from Earth) of its orbit, its orbital speed slows down, but when it is at the perigee (nearest position from Earth), its speed picks up. Refer to Fig. 2.7.

To reproduce the retrograde motion of a planet, Ptolemy added an epicycle to each of five visible planets, but not the Sun and the Moon. This allows each planet to revolve around the center of its epicycle, which in turn revolves around the deferent to which the planet belongs. Once the movements of the epicycle and the deferent are combined, planetary retrograde motion can be observed from the stationary Earth, although this phenomenon is more pronounced at the perigee and less so at the apogee. To amplify the effect of the retrograde motion at the apogee, Ptolemy then added a point called the equant, so that the center of the eccentric is located midway between Earth's center and the equant. Although the center of the deferent revolves around

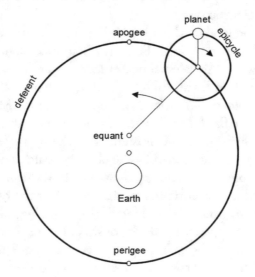

Fig. 2.7 The epicycle model of Ptolemy. Referenced from https://www.britannica.com/topic/Ptolemaic-system

the eccentric's circular orbit, its center moves at a uniform angular speed relative to the equant. In this system, the outermost layer is a stationary sphere on which fixed stars are located. Finally, the outermost shell is contained in another moving sphere, called the Primum Mobile ("first moved"), which provides a source of power to the entire system.

In the Ptolemaic system, the celestial body closest to Earth is the Moon, followed by Mercury, Venus, the Sun, Mars, Jupiter and, finally, Saturn. The Moon was considered to be the nearest object because other bodies were seen to be occasionally occulted (blocked) by the Moon (e.g. during a solar eclipse). The order of Mars, Jupiter and Saturn was determined by each planet's apparent orbital period (time to complete one revolution) around Earth,[18] with Mars, one of the inner planets, having the shortest orbital period. Ptolemy also calculated the ratio of the Earth-Moon distance to the Earth-Sun distance, and the results were quite similar to what Aristarchus had previously arrived at, so the Sun's position was determined to be roughly between the Moon and Mars. As for the order of the Sun, Mercury and Venus, since the observations of Mercury and Venus made by Ptolemy were all planetary transits across the Sun, he decided to place the two planets between the Sun and the Earth. However, this configuration was controversial as it could not account for portions of the observed data. There was yet another controversial aspect that exposed the shortcomings of the Ptolemaic system: when attempting to fit the model using actual data, Ptolemy was unable to derive the radii of each planet's epicycle and deferent; only an estimate of the two radii's ratio could be obtained.

To resolve the above controversy, Ptolemy applied the known orbital period data for the celestial objects repeatedly to his model and based on the error values obtained he made adjustments to model parameters, such as each planet's equant point, orbital speed around the eccentric, and the orbital radii of the epicycle and deferent. The idea was to obtain results that were consistent with the observed data using the adjusted parameter values. If the real problem is in fact a highly inaccurate model, no matter how much effort one puts into adjusting the parameters to fit the data, the model will ultimately be useless (lacks predictive power). Confronted with this fundamental issue and parameters too numerous to deal with, Ptolemy had to give up further attempts at improving his model. The Ptolemaic system, for all its flaws and deficiencies, remained unchanged for almost a thousand years, until the

---

[18]The orbital periods of the Moon and the Sun (both without an epicycle) are 27.5 days and 365 days, respectively. For Mercury and Venus, their orbital periods (along their own deferents) are 88 days and 225 days, respectively. Mars has an orbital period of 1.88 years, Jupiter 11.9 years and Saturn 29.5 years.

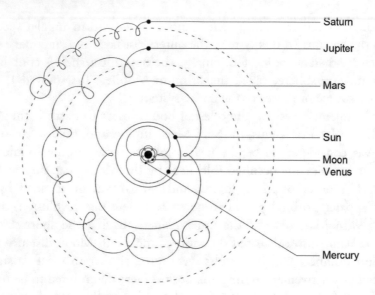

Saturn

Jupiter

Mars

Sun

Moon

Venus

Mercury

Fig. 2.8 A petal-like Ptolemaic cosmological model created under special conditions. Based on the information of the web: https://www.youtube.com/watch?v=EpSy0Lkm3zM\&t=68s

beginning of the Scientific Revolution of the 16th century. The unchallenged supremacy and longevity of Ptolemy's geocentric system can be attributed not only to its ability to describe a great number of astronomical phenomena but also to its perfect compatibility with Aristotle's cosmological view.

From the above discussion, we know that Ptolemy's geocentric system was constructed using geometric analysis on triangles and circles, based on observational data accumulated over a long period of time by his predecessors. In addition, by applying formulas and proofs of theorems and other mathematical techniques, Ptolemy was able to derive a number of demonstrative calculations and finally obtained the size of each planet's epicycle, the equant, the orbital angular speed relative to the equant, and the time required for completing a retrograde movement cycle. The model was almost successful in describing all observed astronomical phenomena. To produce the expected results with the model, special conditions must be specified. For example, appropriately adjusting the sizes of the epicycle and the deferent will produce the desired cycle for a planet's retrograde motion. Another example is that, for a given planet, since both the duration of retrograde motion and the time interval between successive retrograde events vary, it was necessary to make adjustments to the sizes of the epicycle and the deferent as well as the angular speed. Under certain conditions, one can even create a petal-like orbital model, such as that depicted in Fig. 2.8. In short, an artificially constructed, tailor-made orbit based on specific, carefully chosen data can only fit a limited number of observed scenarios and cannot be generalized or considered to represent physical reality.

In our discussions of the process with which Ptolemy constructed his cosmological model, it is evident that Ptolemy was profoundly influenced by the mechanism-causation relation advocated by Aristotle. This type of artificial systems, constructed only to explain a specific phenomenon with special conditions, could never hope to represent a harmonious state of affairs sanctioned by the forces of nature. It was unsettling at best and repulsive at worse. The best we can say about this process is its mathematical precision, not its physical validity. Unfortunately, these precise mathematics created an insurmountable obstacle for future generations of astronomers, and over the next thirteen centuries, the geocentric Ptolemaic system remained the only accepted astronomical model on which all research was to be carried out. In this vast time span, any revisions to the model were limited to fine-tuning the sizes of the epicycles or the positions of the eccentrics, yet such minor modifications would then be hailed, perhaps reluctantly, as "innovation." The final incarnation of the Ptolemaic model (before the dawn of modern science) contained over 80 epicycles, making it overly complex and difficult to understand.

2.11   *The Almagest*: A Great Astronomical Treatise

Despite its complexity and incomprehensibility, the Ptolemaic model's astronomical discourse, which is composed of epicycles, deferents and eccentrics, can be regarded as a type of rigorous mathematical proof. What's more, Ptolemy's model-building logic was not based on complete fabrication or mere personal preferences. Instead, he built his model of planetary motion based on a large quantity of observational data. This allowed him to confidently break away from the perfect spherical astronomy of the Aristotelian framework, which had endured for centuries before him. Incidentally, he also left behind incontrovertible evidence for the heliocentric theory, to be proposed by Copernicus 1,500 years later. In Ptolemy's system, he divided the 1,028 observed fixed stars into 48 constellations, which served as a summary to the study of the Sun and other celestial bodies in ancient Greek astronomy. Finally, at about the age of fifty, Ptolemy compiled a 13-volume mathematical and astronomical treatise containing his lifelong research, known as the *Almagest* (Great System of Astronomy), also later referred to as *The Greatest*.

In this masterpiece of astronomy, the first eight volumes contain the movements of the Sun and the Moon and also an augmented version of the star catalog created by Hipparchus. The last five volumes explore in detail the movements and trajectories of the planets based on the planetary model that he had created. In enhancing Hipparchus' star catalog, Ptolemy used the latitude and longitude measures of the invisible outermost sphere to describe the positions of the stars. He added more than 100 stars to the original list of 850 to expand the catalog to 1,028 stars. Ptolemy provided detailed descriptions of the changes in the sky from his observation over an extended period of time. He wrote extensively about the precession of celestial bodies and the corrections required for describing the refraction of sunlight and moonlight. He also included methods to compute the dates of solar and lunar eclipses. All of these represent exceptional contributions to astronomy. Although this great work was conceived based on the geocentric theory, the published observational data and the proposed research methodology remained authoritative over the next thousand years, and future astronomers such as Copernicus, Galilei and Kepler did not dare to discount or ignore it. The geocentric model depicted in Fig. 2.8 was named after Ptolemy by later generations of astronomers and is now known as the Ptolemaic system. This model would dominate the field of astronomy over the following thirteen centuries. The *Almagest* is an encyclopedia of astronomy. It was required reading for astronomers for a thousand years. Until the early 16th century, before Copernicus reintroduced the heliocentric theory in his *De revolutionibus orbium coelestium* ("On the Revolutions of the Heavenly Sphere"), the *Almagest* was never questioned or challenged. Ptolemy's masterpiece and Euclid's *Elements* are said to be the two works that survived the longest as textbooks in human history.

As described in Chap. 1, the lunar eclipse was in fact noticed and also documented as early as the time of Thales, around 2000 BC. After a thousand years of astronomical advancement, predicting the times that lunar eclipses would occur became knowledge that was much sought after. Astronomical data was compiled into ephemerides (singular ephemeris) and became indispensable for marine navigation. It was reported that this information helped save the life of Christopher Columbus (1451–1506) in the late 15th century on his voyage to the Americas. This legend goes as follows. When Columbus arrived at the New World, he landed on San Salvador Island in the Caribbean. Immediately after landing, Columbus and his crew were surrounded by the indigenous people, the Lucayan, and they feared for their lives. As the captain, Columbus had to improvise. He found an interpreter and warned the apprehensive and skeptical natives about the mortal danger the latter were about to face. This reason, according to Columbus, was that the Moon was going to turn red for three consecutive nights, starting from the following evening. The Earth would devour the Moon, and then the end of the world would come. Having successfully intimidated the natives, Columbus took advantage of the situation and threatened them further: If the natives refused to provide him, without compensation, with the supplies he required, then they would receive divine punishment and perish. The following night, the Moon in the sky indeed turned blood red, which terrified the natives. They knelt before Columbus, bowed down to worship him, and offered him food and water. What saved Columbus from an early demise was the ephemeris that he had brought with him. The ephemeris contained astronomical data recorded by astronomers over a long period of time, restructured to provide useful information to sailors in their nighttime navigation. Data regarding lunar phases across all seasons was also indispensable in navigation.

Appendix 2.1: Principle of Minimum Energy: Cases and Proofs

To explain why Mother Nature has selected these polyhedra (the Platonic solids) as the shapes for the crystallization of matter, scientists today have proposed the theory of the "natural tendency to require the least amount of energy," which states that among all the available shapes, the energy required to form these polyhedra are minimum. A popular (two-dimensional) example that supports this theory is the regular hexagonal shape of the honeycomb. The reason for the formation of this shape stems from the fact that for a given length, a regular hexagon with a perimeter equal to that length has the largest enclosed area among all regular polygons that tile the plane (without leaving gaps). Therefore, for a given amount of propolis (bee glue) produced by honey bees and the same amount of labor (or time), a honeycomb cell with a regular hexagonal shape will provide the largest amount of space for the bees. The proof of a regular hexagon having the "same perimeter but maximum area" is as follows.

1. Regular hexagon (Fig. 2.9: right):
   Each side is shared by 2 hexagons, and each hexagon has 6 sides, so a side corresponds to one-third of a hexagon. Suppose each side of the hexagon is of length R. Since a regular hexagon has an area of $\frac{3\sqrt{3}}{2}R^2$, each side

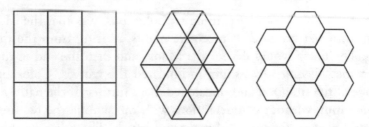

Fig. 2.9 Three different types of regular tiling: square/regular quadrilateral (left); equilateral triangle (center); regular hexagon (right)

corresponds to an area of:

$$\frac{1}{3} \times \frac{3\sqrt{3}}{2} R^2 = \frac{\sqrt{3}}{2} R^2 \tag{2.6}$$

2. Regular quadrilateral or tetragon (square; Fig. 2.9: left):
   Each side is shared by 2 squares, and each square has 4 sides, so a side corresponds to one-half of a square. Suppose each side of the square is of length R. Since a square has an area of R^2, each side corresponds to an area of:

$$\frac{1}{2} \times R^2 = \frac{1}{2} R^2 \tag{2.7}$$

3. Equilateral triangle (Fig. 2.9: center):
   Each side is shared by 2 triangles, and each triangle has 3 sides, so a side corresponds to two-thirds of a triangle. Suppose each side of the triangle is of length R. Since an equilateral triangle has an area of $\frac{\sqrt{3}}{4} R^2$, each side corresponds to an area of:

$$\frac{2}{3} \times \frac{\sqrt{3}}{4} R^2 = \frac{\sqrt{3}}{6} R \tag{2.8}$$

From the three proofs above, it is clear that the regular polygons can be ranked by the criterion "area per unit length of perimeter" as follows, in descending order: (a) regular hexagon; (b) square (regular quadrilateral); (c) equilateral triangle.

If we increase the number of sides of the regular polygon to 8 (an octagon) or 12 (a dodecagon), the tiling will leave gaps and much of the area would be wasted and fail to satisfy the minimum energy requirement. This highly efficient hexagonal shape is adopted not only by bees. Tens of thousands of columns of the igneous rock basalt, formed from underwater eruptions of magma and found in Northern Ireland, the northern coast of Germany and the Pescadores Islands of Taiwan, also have hexagonal cross-sections. Efficiency is a mysterious force of nature. Whether it is the living bees or the inanimate basalt columns, the nature tends to opt for the most efficient geometric shape in the formation of physical structures.

Another geometric shape considered the most efficient is the circle, or its three-dimensional counterpart, the sphere. They are found in abundance in the Universe, from objects as large as planets to those as tiny as water droplets or phytoplankton, all of which are spherical. Therefore, to form a soap bubble with the least amount of soapy water,[19] the shape of the bubble is bound to be spherical. A sphere is therefore regarded as the most efficient surface. If we join two soap bubbles of equal size together, the common wall separating them will be a flat surface as a result of surface tension. If six bubbles of identical sizes are joined in a symmetrical fashion, then the space in the middle will form a regular hexahedron (i.e. a cube). If twelve bubbles are joined symmetrically, the shape in the middle will be a regular dodecahedron, with all twelve sides being regular pentagons of the same size. This shape of the space that forms in the middle of the bubbles represents that which minimizes its volume.

As an application of regular polyhedra, the dice for tabletop games and gambling have been designed with the principle of minimum energy. When a die in the shape of a regular polyhedron is rolled on a table, the probability of each face turning up is identical. In addition, a rolled regular polyhedral die takes the least amount of time to come to rest compared with other non-regular polyhedral dice. This shortens the total amount of time required to determine the results of a game, as well as saving gamblers' mental and physical efforts. Although saving energy is not the factor gamblers are most concerned about, saving time is one way for casino owners to maximize their income. Therefore, using regular polyhedron-shaped dice in gambling is a win-win for both the casinos and the gamblers, and this has endured over a thousand years uninterrupted or with little change.

The idea that all matter is made up of Platonic solids is exemplified in today's vast computer-animated film industry, where multimedia figures are created by animators as computer memory capacity and computing power continue to grow beyond leaps and bounds. This technology was first implemented by engineers at the Boeing Company in the development of flight simulators. The software engineers who designed the simulators employed

---

[19]The thickness of soap bubble film is smaller even than the wavelengths of visible light, which is about 1/20000 the diameter of a human hair.

millions of triangles to represent the complex shapes of terrestrial surfaces that appeared in the background of simulated flight scenarios. This technology was later used by the computer animation movie studio Pixar to create animated feature films, where dozens of human characters and various animate and inanimate objects were rendered with tens of millions of basic geometric shapes. Capable of adapting in response to changes in the environment, these animated entities are transformed into life-like characters, creating stunning visual effects and revolutionizing the film industry of the 21st century.

Appendix 2.2: Derivation of Equations for Planetary Retrograde Motion

The retrograde motions[20] of planets baffled generations of ancient astronomers, who proposed a number of theoretical models to try to explain the phenomenon. Here we analyze their causes from a modern perspective and perform a few simple calculations.

When observing celestial objects in the sky from Earth, we not only see the Sun and the Moon but also a few what are known as "wandering stars," or planets, which move relative to a fixed background of distant stars. A planet normally moves westward relative to the fixed stars, and this is called prograde (or direct) motion. Occasionally, however, it would appear to move eastward relative to the fixed stars, which is referred to as retrograde motion. This rather strange planetary movement was a great mystery in astronomy for a long time until geocentrism was abandoned in favor of the heliocentric (Sun-centered) model, which offers a much simpler way to explain retrograde motions.

Consider the planetary system of Fig. 2.10. P is a planet that orbits the Sun and is superior to Earth (i.e., Earth's orbit lies inside that of P). According to Kepler's third law of planetary motion, Earth's angular speed is faster than that of P. For example, suppose Earth has completed half of its orbit around the sun, P may have completed only one quarter of its orbit. In the diagram below, five pairs of E and P's relative positions are shown: When Earth is at the E1 position and P at P1, P's projection onto the celestial sphere is A1; when Earth and P are at the positions $E2$ and the $P2$, respectively, P's projection is $A2$; and so on. For an observer on Earth, P's trajectory, as projected onto the distant background, appears to be $A1 \rightarrow A2 \rightarrow A3 \rightarrow A4 \rightarrow A5$,

---

[20]The diagrams and the data in the last paragraphs are referred to Wikipedia: https://en.wikipedia.org/wiki/Apparent_retrograde_motion.

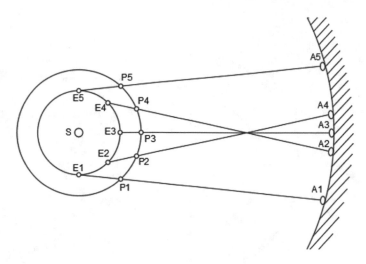

Fig. 2.10 Planetary retrograde motion

where the path $A1 \rightarrow A2$ is prograde (direct), the part $A2 \rightarrow A3 \rightarrow A4$ is retrograde, and the path $A4 \rightarrow A5$ resumes the original prograde direction.

We will now try to compute the time required for a planet to complete a retrograde motion relative to Earth. Three assumptions are required:

1. The masses of planet P and Earth are negligible (compared with the Sun), and the Sun can be regarded as a fixed point.
2. Planet P has the same orbital plane as Earth's and travels in a circular orbit (instead of an ellipse, which is more tractable mathematically).
3. Gravitational forces on P, Earth and the Sun from other Solar System bodies are ignored.

Next, we consider a simple model as shown in Fig. 2.11. Here we have a Cartesian coordinate system with the Sun S located at the origin. Let E be Earth and P be the planet under observation, and suppose the radii of their respective circular orbits around the Sun are R_1 and R_2. Let E and P have angular orbital speeds of ω_1 and ω_2, respectively. Select $t = 0$ as the time when planet P is in opposition to the Sun (i.e., the Sun, Earth and P are perfectly aligned in a straight line, with Earth between the other two bodies). Now all three bodies are located on the x-axis.

After a while, at time t, Earth's coordinates can be expressed as

$$(x_E, y_E) = (R_1 \cos \omega_1 t, R_1 \sin \omega_1 t) \tag{2.9}$$

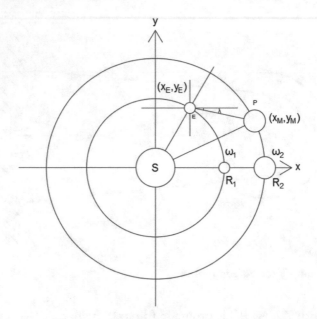

Fig. 2.11 Computing the various positions of a planet's retrograde trajectory

Likewise, planet P's coordinates (x_M, y_M) are

$$(x_M, y_M) = (R_2 \cos \omega_2 t, R_2 \sin \omega_2 t) \qquad (2.10)$$

We then define the bearing of P with respect to E to be

$$\lambda = \tan^{-1}\left(\frac{y_M - y_E}{x_M - x_E}\right) \qquad (2.11)$$

Notice that whether P is orbiting in a prograde or retrograde direction at time t depends on the sign of $d\lambda/dt$. If $d\lambda/dt > 0$, the movement is prograde, and if it is negative, the motion is retrograde. Substituting (2.9) and (2.10) into (2.11) we have

$$\tan \lambda = \frac{y_M - y_E}{x_M - x_E} = \frac{R_2 \sin \omega_2 t - R_1 \sin \omega_1 t}{R_2 \cos \omega_2 t - R_1 \cos \omega_1 t} \qquad (2.12)$$

Now we take the derivative of (2.12) with respect to t to get

$$\sec^2 \lambda \frac{d\lambda}{dt} = \frac{\omega_1 R_1^2 + \omega_2 R_2^2 - (\omega_1 + \omega_2) R_1 R_2 \cos(\omega_1 - \omega_2)t}{(R_2 \cos \omega_2 t - R_1 \cos \omega_1 t)^2} \qquad (2.13)$$

Note that at $t = 0$ (when the Sun, Earth and P are in syzygy), as Earth's angular orbital speed is higher, P will move in a retrograde direction.

Next, we will compute the time $t > 0$ when $d\lambda/dt$ becomes zero. Both sides of Eq. (2.13) vanishes when $t = t_S$ satisfies the following equation:

$$\omega_1 R_1^2 + \omega_2 R_2^2 = (\omega_1 + \omega_2) R_1 R_2 \cos(\omega_1 - \omega_2) t_S \qquad (2.14)$$

or

$$t_S = \frac{\cos^{-1}\left[\frac{\omega_1 R_1^2 + \omega_2 R_2^2}{(\omega_1 + \omega_2) R_1 R_2}\right]}{\omega_1 - \omega_2} \qquad (2.15)$$

Since $t = 0$ is precisely the time when P is halfway through its retrograde movement (as $t > 0$ and $t < 0$ are perfectly symmetric), the amount of time for P to complete a retrograde cycle is

$$T = 2t_S = \frac{2\cos^{-1}\left[\frac{\omega_1 R_1^2 + \omega_2 R_2^2}{(\omega_1 + \omega_2) R_1 R_2}\right]}{\omega_1 - \omega_2} \qquad (2.16)$$

We can apply the above equation to compute the retrograde constants of the planets in the Solar System relative to Earth. Substituting each planet's orbital data into the equation, we get:

- The number of days in Mars' retrogradation is 72 days for each synodic period of 25.6 months.
- The number of days in Jupiter's retrogradation is 121 days for each synodic period of 13.1 months.
- The number of days in Saturn's retrogradation is 138 days for each synodic period of 12.4 months.
- The number of days in Uranus' retrogradation is 151 days for each synodic period of 12.15 months.
- The number of days in Neptune's retrogradation is 158 days for each synodic period of 12.07 months.

The above results are close to those obtained from actual observations. The above data indicates that the farther away a planet is from Earth, the shorter the time interval between its two consecutive retrograde motions. This interval is the time interval for Earth to be located between the Sun and the planet in question for two consecutive times. For its relationship with the sidereal period (time required for a planet to complete one cycle around the Sun), refer to (3.3) in Chap. 3.

3

The Tumultuous Astronomical Revolution

朝聞道, 夕死可矣

Hear the Way in the morning,
and it won't matter if you die that evening,

-----------*Book 4 ("Humaneness")*, Analects, *by Confucius (551–479 BC).*
Translated by Burton Watson (1925–2017)

Although the astronomy of Ptolemy laid a solid foundation for future scientific research, Europe would see a period of severe scientific decline over a millennium following the astronomer's death. From the time in ancient Greece where science was regarded as a metaphysical discipline, scientists were convinced that that all things in the natural world adhered to the laws of nature, though the data or empirical evidence that would prove the veracity of these so-called laws was scarce, if not entirely nonexistent. Ptolemy's research mitigated the predicament of this lack of evidence, as scientists were finally able to calculate the sizes of the epicycles and deferents of the five planets, the location of the equant, and the angular velocities of their orbiting motion using Ptolemy's epicycle-deferent cosmological model. Furthermore, under certain special conditions, the time required for a planet to complete a cycle of retrograde motion could also be obtained accurately. This epicycle-deferent model, with its fantastical predictive capability, thus became the only theory recognized by astronomers for more than a thousand years, until the Scientific Revolution sparked by Nicolaus Copernicus at the end of the 15th century. Copernicus removed Earth from the center of the cosmos and replaced it with the Sun, after which astronomy took on an extraordinary new life as a scientific discipline. Below we provide an account of this millennium-long history of astronomy, which would begin with one of the darkest episodes in history.

3.1   Science of Medieval Europe

The type of scientific framework advocated by Ptolemy, supported by conjectures and hypotheses under the guiding influence of Aristotelianism for over a century, was not changed significantly until 146 BC, when Rome conquered Greece. This period of nearly two centuries was a time of relative stability in terms of scientific development, but with the rise of the Roman Empire, the progress of science began to enter a thousand years of decline, a period referred to as the Dark Ages by future historians of science. Initially Greek-style scientific research continued for nearly a century, but the focus was on the applications based on previous research, and little or no significant innovation was achieved. For example, in late 300, Theon of Alexandria (335–405) wrote commentaries on Ptolemy's *Almagest* and Euclid's *Elements*. A century later, John of Philoponus (490–570) also contributed to the commentaries on Aristotle. In his writings, Philoponus raised issues with Aristotelian kinematics, as many others had. The increasingly weakening development of Greek science, which had enjoyed a robust period of progress but was now in a precarious position, began to deteriorate even further beginning in the 5th century. Even commentary work was no longer carried out. What was most devastating in this turn of events was the Roman emperor's education policy, which precipitated the demise of ancient Greek science, and its repercussions would continue to be felt over the next thousand years.

3.1.1   Destruction and Redemption of Empires

The Roman rulers who conquered the Greek city-states by force were mostly military leaders who had built their fortunes from plundering. As administrators of the Empire, these rulers often showed their disdain for the natural philosophers, who formed cliques or factions among themselves and did nothing but philosophizing all day long. Only when there were bridges to be erected and palaces to be built would the rulers try to solicit the help of those who understood mathematics and geometry in order to conduct surveying and computational work. Other than these tasks, the natural philosophers were neither welcomed nor considered qualified even as entertainers at social events. For this reason, one would be hard-pressed to find a famous mathematician or physicist who lived under the rule of the Roman Empire, which dominated the Western world for nearly a thousand years. Neither could one identify a single scientific theory that would have been influential to scientists of later generations.

Under the imperial policy of prioritizing present needs and committing very little to the future, funds that should have been invested in institutions of scientific and philosophical research instead went to the construction of bridges, roadways and palaces. When the Roman emperors realized that the Christian Church was an ally in consolidating their imperial rule, they did not hesitate to provide large sums to church-run schools and monasteries but turned a blind eye to the survival of the academic institutes responsible for imparting general knowledge. The consequences were the preservation of knowledge that was only relevant to religion and the total disregard of general knowledge, science or philosophy. From this point on, academic institutes of ancient Greek tradition, already on the verge of collapse, were replaced by religious schools and monasteries wholesale. Now the advancement of knowledge and topics of research were required to be related to religion, and exploring natural phenomena had become entirely meaningless and devoid of value. This serious deficit of knowledge would continue for several more centuries, and when the Western Roman Empire fell in 476, what was left of it were little more than cities in decline, the rise of feudalism, and religious control of politics and public administration. This would eventually lead to closed societies and economic depression.

Following the collapse of the Western Roman Empire in the 5th century, Europe was plunged into turmoil and disarray, with the exception of the territories under the control of the Byzantine Empire. The populace was mostly uneducated and remained illiterate. Only a very small number of elites were able to receive a religious education and acquire knowledge in schools operated by the Church. Latin was generally used, and fewer and fewer were able to read Greek texts. A thousand years of intellectual and scientific heritage left by the Greeks, in the form of fragmented parchments, was collecting dust in the corner of the libraries of the monasteries. Just when the Arab world to the East was immersed in the natural philosophy from the Western world, Charlemagne (Charles the Great, 742–814) was busy unifying the fractured provinces, leaving behind devastated cities devoid of vitality everywhere. The people were generally uneducated, and even their lords were still being taught writing skills or learning to write their own names. This kind of considerable structural damage to society would continue for another few hundred years until the 11th century, at which time all knowledge and culture based on Greek science and philosophy were described by historians as having been wiped out from the European continent.

In this dark age of science spanning almost a millennium, knowledge would fortuitously continue to pass down through the generations, sadly not in Europe but in the Islamic world of the Arabs, where knowledge would be elevated and carried forward in a special way, even if only on a limited scale. The last emperor of the Western Roman Empire, Romulus Augustulus (460–507), was deposed in 476, and this marked the end of the Empire in the western provinces. Meanwhile, the Greek-speaking eastern half of the Empire continued to exist under the administration of what would be referred to as Byzantine Empire, which would in time even expand its territorial reach. Indeed the Byzantines defeated the Persian Empire in 627 at the Battle of Nineveh. It was during this period of upheaval with a power vacuum that the Arabs rapidly came to prominence. Under the leadership of Muhammad (571–632), this fierce and agile nomadic people, who had been quietly inhabiting the lands bordering the desert and the farmland, emerged at the end of the 6th century and began to spread their monotheistic religion from Mecca and to resist the rule of the Byzantine Empire, eventually expanding the Islamic territory to the entire Arabian Peninsula. Upon the death of Muhammad in 632, the empire that he had built was inherited by his four followers or relatives, but he himself remained highly revered by the people.[1]

In 636, when the Byzantines were still reveling in their victory at the Battle of Nineveh, the Muslim army managed to capture the province of Syria from the Byzantine Empire to the north of the Arabian Peninsula, and the Arab forces eventually came to occupy Mesopotamia, Persia and Egypt, putting nearly the entire southeastern region of the Mediterranean under their control. The fourth ruler Ali, however, was a weak leader who was murdered during his reign, and the Islamic world split into two factions, the Sunnis and the Shiites, a situation that has continued to this very day. The Sunni Muslims established the Umayyad Caliphate, with its capital at Damascus, and they would continue their military campaigns. At its peak the Caliphate stretched across present-day Morocco and the southern part of the Iberian Peninsula in the west, the entire southern shores of the Mediterranean Sea in the south, and the Western and South Asian countries such as Iran, Pakistan and Afghanistan in the east. In contrast, the territory of the contemporary

---

[1]The four leaders who were referred to as the Rashidun Caliphs (Rightly Guided Caliphs) by followers of Sunni Islam are: Abu Bakr, Omar, Othman and Ali.

Byzantine Empire consisted mainly of modern Turkey and a few loosely allied city-states in Eastern Europe.

3.1.2 Science in the Language of Arabic

Under Sunni rule,[2] Greek scientists who were previously persecuted under the rule of the Roman Empire were welcomed by the Muslims, as the Arabs were able to discover valuable scientific information in many Greek works, such as human anatomy and healing methods, engineering apparatuses, and design principles. Wealthy Arab merchants were therefore willing to spend a fortune on translating a large number of Greek works into the Arabic language. Consequently, the Islamic world gradually became fertile ground for scientific research, and its intellectual legacies successfully preserved ancient knowledge of the Greeks and Egyptians, which would in time provide a solid foundation for Europe's Renaissance in the 14th century and the Scientific Revolution in the 16th century. From 754 to 833, Muslims translated a large number of scientific works written in Greek, Persian and Sanskrit, including those written by Plato, Aristotle, Galen, Euclid and Ptolemy. Such large quantities of translation work were completed at the House of Wisdom in Baghdad, the capital of the Abbasid Caliphate.[3] This great Arabic institution, like the Mouseion of Alexandria in the Hellenistic age, boasted not only an extensive collection but also a multi-ethnic group of scientists who were responsible for performing translation work and also creating new works. Among them were the Arabs, Greeks, Persians, Jews and Turks. The translations they produced were also spread as far as Egypt and Iberia to the west as well as Persia and northern India to the east, allowing ancient Greek knowledge to be reintroduced to the Western world and reversing the catastrophic loss of scientific legacies under the rule of the Roman Empire.

Islamic culture flourished in the West for more than six centuries until 1258, when the Mongol cavalry from East Asia captured Baghdad and ended the rule of the Caliphate. The Abbasid collapsed and split into three regional powers: the Umayyad dynasty in southern Iberian Peninsula, the Fatimid Caliphate in Egypt, and the Almoravid dynasty centered in Morocco. In addition, the territories of the Abbasid Caliphate in Syria and Palestine were occupied by the Byzantine and later the Crusaders from Western Europe. Just like any other country on Earth, the political turmoil in the Arab world was

---

[2]See also the story illustrated in Chap. 6 of Mlodinow [23].

[3]Established by the descendants of Muhammad's uncle Abbas in 750, the Abbasid Caliphate was the second hereditary dynasty of the Islamic caliphates; it was later conquered by Hulagu Khan of the Mongol Empire in 1258.

equally detrimental to the progress of science. Financial resources that were originally earmarked for scientific research in Islamic schools were instead used to finance the wars that would determine the survival of the Caliphate. In the absence of the necessary funding, the preservation and transfer of knowledge in the Islamic academic institutions turned away from Greek science and philosophy and embraced theology and religious studies meant to nourish the soul.

Although science came to a halt as a result of wars and upheaval, sporadic research continued to be carried out in service of wars, ironically funded by Hulagu Khan of the "barbaric" Mongol Empire. This intrepid general with a sense of justice believed in divination, and during the siege of Baghdad he was aided by local practitioners of astrology and was eventually victorious against the Arabs. To reward these astronomers, he generously funded the construction of an astronomical observatory in Maragha in 1259. Despite the great turmoil, these astronomers, who had fallen into poverty, were understandably grateful and dedicated their work to the new ruler after having been treated with such benevolence. The observatory's first director, Nasir al-Din-al-Tusi (1201–1274), used the data collected by the new instrument to compile star catalogs and to revise Ptolemy's system of epicycles and deferents. The data collected with the observatory continued to accumulate, and a century later, Ibn al-Shatir (1304–1375) developed a system of planetary motion by modifying Ptolemy's eccentric model using epicycles based on this vast amount of data. The new model did not radically change the distance between the Moon and Earth and was able to remedy the shortcomings of Ptolemy's geocentric system. Due to the significant contribution of astrology, the observatory at Maragha continued to expand with financial support from the Mongols. The largest observatory was the one located at Samarkand, which was inaugurated in 1420. The sidereal year calculated at the observatory was 365 days, 5 h, 49 min and 15 s, which represents truly amazing precision.

Apart from their remarkable work on astronomical measurements, as explained above, the greatest contribution in the Arabic language may have been the mathematical works of the Persian scholar Muhammad ibn Musa al-Khwarizmi (780–850). He wrote *The Compendious Book on Calculation by Completion and Balancing*, which introduced the concept of algebra and described ways for solving quadratic equations (note that the word *algorithm* was derived from his name). In his book he also adopted three types of numeral systems: Roman numerals, the sexagesimal digits of ancient Babylonians, and the decimal system of India ("Hindu numerals"), the last of which later came to be called "Arabic numerals"[4] in Europe, as they were introduced to the West by the Arabs. In addition to mathematics, the research conducted by Al-Battani (858–929) in astronomy also led to fruitful results. Although Al-Battani produced mostly commentaries on Ptolemy's *Almagest*, he would occasionally obtain more elaborate and precise astronomical data, such as the angle of inclination of the ecliptic plane relative to the Earth's equatorial plane, which he measured to be 23.5° (attaining one decimal place), as well as the timings of the four seasons and the positions of stars. To measure stellar locations with higher precisions, Al-Battani adopted the trigonometric function *sine* from India to simplify calculations involving triangles. In addition, Al-Biruni (973–1048) attempted to estimate the radius of Earth, but his result remained quite far away from the true value, the principal error being from the unit of length used at the time, the "cubit", the actual value of which remains uncertain to this day. In optics, Al-Haitam (965–1040) suggested that refraction is the result of light changing speeds while passing from one medium to another, but he provided no empirical evidence and his descriptions remained only a hypothesis. In chemistry, Jabir ibn Hayyan (c. 721–c. 815) employed the techniques of evaporation, sublimation, and dissolution in his research on crystallization. In medicine, Al-Razi (854–925) studied smallpox and measles, but criticized the Romans' theory that the body possessed four separate "humors" (liquid substances) and their balance are the key to health. Ibn Sina, often known in the west as Avicenna (c. 980–1037) wrote *The Book of Healing*, a highly influential medical masterpiece in the Middle Ages. Ibn Rushd (1126–1198) confirmed the retinal function of the eye. All these works and contribution indicate that by this time the study of anatomy in medicine had already reached a sophisticated level.

3.1.3 Science in the Language of Latin

In the 10th century, people in European territories outside the control of Muslims were living under the shadow of the Latin (Western) Church's jurisdiction. Although there were fewer incursions by the barbarians, and agricultural productivity had increased with the rise of new technologies, almost all of the wealth created went to the Church's coffers, and the acquisition of knowledge remained a privilege enjoyed only by the clergy in religious schools. As the practice of knowledge transfer and cultural transmission went

---

[4]The digit zero in the Arabic numerals was introduced to Europe in the 12th century by Thierry of Chartres, who taught in Paris.

into decline, a scientific activity passed down from the Arab world emerged as the savior of European civilization: translation.

As mentioned earlier, at the peak of the Islamic Empire, Muslim scientists translated a considerable number of scientific works written in the Greek language, thus allowing Greek science to flourish around the Mediterranean in the Arabic language. When Europe gradually recovered from the wars and conflicts, the emerging agricultural technologies sparked the people's desire to learn, and the demand for scientific knowledge surged. During this time, most books on intellectual subjects were preserved in the form of Arabic manuscripts in monasteries, and translating them became a necessity. The first group of people who undertook this responsibility was the Monastery of Santa Maria de Ripoll in the Pyrenees, close to the modern-day French-Spanish border. The initial pace of the translation work was rather slow, and not until the early twelfth century did the speed of translation began to gradually pick up.

One of the most successful translators of this age was perhaps Gerard of Cremona (1114–1187), who worked as a teacher in the Kingdom of Castile (in modern-day Spain) and translated the Arabic texts of Ptolemy's *Almagest* and Euclid's *Elements* into Latin. His translations of Aristotle's works, including *On the Heavens*, *Physics* and *Meteorology*, were most influential to his contemporaries. The high demand for education necessitated a large number of translators, and the focus of their translation work was on Aristotelian works. Consequently, in the two hundred years from the 12th to the 14th centuries, Aristotelianism exerted much influence on the development of European science, but it also fueled a great deal of debate. The two rival religious orders to this debate were the Franciscans, also known as the Greyfriars, and the Dominicans, or the Blackfriars. Their confrontations would take place at newly established "universities". The emergence of universities brought a breath of fresh air to the otherwise stagnating scientific research and the deteriorating and endangered ideological innovation. Let us examine the hardships faced by European universities in their early days.

3.1.4   Universities of Medieval Europe

In continental Europe, the education system originally intended to impart knowledge was nearly destroyed under the rule of the Roman Empire, and the situation remained hopeless until the end of the 11th century. In 1085, Christian forces reconquered a significant portion of Iberia. This time, however, the soldiers did not focus on burning, killing and looting. Instead they transported a substantial quantity of Arabic scientific literature, translated from the original Greek, back to Rome, which allowed the Church in Rome to gain access to the superb wisdom and superior knowledge of ancient scientists and philosophers. As a result, these Arabic texts began to be translated into Latin or other vernacular languages. The most influential and widely circulated among the texts translated was a number of medical research papers by the Benedictine monk Constantinus Africanus (d. 1099). When it was discovered that the medical knowledge contained in these Arabic documents and their translations were so advanced and highly applicable, a race to seize the opportunity to translate these spoils of war began. Translating from Arabic into Latin became a popular business, and funding translation work became a sort of status symbol in high society. The wealthy would collect these Arabic works translated from ancient Greek as though they were collecting art and antiques, with a view to making a handsome profit. The wisdom of the ancients that suddenly re-emerged in European societies was not disappointing to their new investors, either. By applying this advanced knowledge and technology from the past, people's lives gradually improved, and the society as a whole began to accumulate wealth. Governments, which held all the real power over their citizens, began to shift their attention to educational reform.

Among the bureaucrats in the government, those in the know understood that to reform education, they had to first resurrect the educational system and institutions, and also to invite scholars who had scattered across the old empire to return and teach. As a result, starting from the 11th century, a number of universities in Europe were established, many well-known ones continuing to exist to this day. Examples are the University of Bologna (est. 1088) and the University of Padova (1222) in Italy; the University of Paris (1200) in France; and the University of Cambridge (1209) and the University of Oxford (1224) in England. These higher educational institutions provided a venue in which scholars would be able to teach and conduct research. In their early days, however, these pioneering universities were more of a loose alliance rather than well-funded organizations, as they lacked permanent assets or endowments. The universities were administered from buildings geographically scattered and they relied solely on donations, a rather unstable source of income, for their day-to-day operations. Classrooms were leased from private property owners or borrowed from the Church. Sometimes classes could even be conducted in brothels at the whim of the professors, who would be eager to endear themselves to the students. This was because a professor's salary was paid directly by the students, who held the power to decide whether a professor would go or stay. The University of Bologna once even allowed students to punish their professors. When they deemed a professor to be incoherent, or if the professor was absent, tardy, or leave class early without prior permission, he would be ridiculed or thrown at with stones by the students.

Despite all the hardships associated with running a university and the danger that came with teaching, natural philosophy was by now the focus of the activities of imparting knowledge. Religious philosophy, on the other hand, had become secondary in importance, and to be educated in this branch of knowledge one would have to visit a monastery. This seismic shift in what was being taught would over time make the transfer of knowledge better satisfy learners and would also allow universities to become the principal venue for promoting practical skills and carrying out the exploration of nature. Professors would therefore reluctantly put up with students' mistreatment in order to keep their academic positions and the resources available to them for conducting research. Among the significant changes, as well as struggles, experienced by many universities, the debate over Aristotelianism at the University of Paris and the reforms in the research on mathematical physics at Oxford stood out in particular. In the following we will explore the revolutionary innovations in natural philosophy that occurred at these two fine institutions.

3.1.5 Student Activism at the Sorbonne: Aristotelian Debates

The critical scrutiny of Aristotelian doctrines began at the University of Paris almost as soon as the institution was established. The reason was that to its critics Aristotelianism were full of empirically unproven and controversial claims. These claims were like iron chains that shackle the hands of God, rendering the revelation of truth impossible. On the other hand, the University of Toulouse, also located in France and established three decades after its counterpart in Paris, agreed to adopt the works of Aristotle in its curricula. The University of Paris soon followed suit and announced the lifting of the ban on Aristotelianism, and from that point on the ancient Greek philosopher's writings were elevated to constitute the core of education at all French universities. The scholars who led this movement were Albertus Magnus (1200–1280) and Thomas Aquinas (1225–1274), both doctors of theology[5] and Dominican friars, with the latter being the former's student.

---

[5]There were three types of doctorates conferred by the academia of Europe during this time: Doctor of Theology, Doctor of Law, and Doctor of Medicine. A scientist only received the Master's degree, known as Master of Natural Sciences. The full title of Albertus Magnus was Universal Doctor (or *Doctor universalis*) and that of Aquinas was Angelic Doctor (or *Doctor Angelicus*). Albertus Magnus was proclaimed the patron saint of natural scientists by the Pope in 1941.

Albertus Magnus taught in France and Germany and founded the University of Cologne. He was a moderate Aristotelian, but preferred Ptolemy's epicycle-deferent system to Aristotle's concentric spheres model. Magnus' pupil, Aquinas, expanded on his teacher's assertions and successfully integrated biblical theology and Aristotelian philosophy in his work *Summa Theologica*. Aquinas supported the concentric spheres model of Aristotle and believed that his system represents perfection and is consistent with God's original purpose of Creation. To him, even though the Ptolemaic model was able to explain certain observational data, it was also erroneous in a large number of phenomena, and for this reason accepting this model would not be appropriate. Aquinas emphasized that Christian doctrines and knowledge about nature should be reconciled with each other, and the ultimate goal of education is to understand the many reasons of the omnipotent God's Creation. This type of middle-of-the-road stance adopted by Aquinas helped make his works the syllabus for the graduation examination at the University of Paris.

Despite its acceptance at universities, the Aristotelian philosophy did not continue to propagate unopposed in Europe. In the four decades following the 1250s, Aristotelianism was banned by the Franciscan friars at the University of Paris and the University of Toulouse, and the fervent opposition even reached Pope John XXI (1215–1277), who ordered an investigation into the matter and decided to condemn Aristotelianism and the works of Aquinas. However, these official ban was reversed after the new Pope John XXII (1249–1334) came to power and canonized Aquinas in 1323. French and German universities followed suit and lifted the ban. In 1341 the University of Paris even went as far as requiring its graduates to swear an oath to the effect that they would be willing to teach Aristotelianism at universities before they would be conferred the Master of Natural Sciences degree. After nearly a century of debate, creative scientific research that gradually loosened itself from the shackles of theology finally emerged in Paris at the beginning of early 14th century. The person who played a leading role in this regard was Jean Buridan (1295–1363), who had served two terms as president of the University of Paris. He was opposed to the definitiveness of scientific principles, and he advocated that any scientific principle is required to be supported by an actual phenomenon or experience and not derived purely via logical reasoning. This doctrine, later referred to as empiricism, established itself as the foundation of experimental science: only through careful observation and detailed analysis are we able to draw a valid conclusion.

To illustrate the essence of empiricism, Buridan provided an example from Aristotelianism that he believed to be entirely incorrect. Aristotle's theory suggests that an object thrown upward into the air will continue to rise because the air underneath the object pushes it upward. Buridan refuted this assertion and countered that if Aristotle's theory was true, an arrow with a pointed tail should arrive at a smaller height than one with a blunt tail, because the latter's flatter tail has a larger area and thrust exerted on it by the air should also be greater. This reasoning, however, obviously contradicted what was actually observed: a more streamlined pointed-tailed arrow would continue to travel higher and longer, whereas and a blunt-tailed arrow would quickly lose its momentum and fall to the ground. This observation also confirmed another fact, that is, what air exerts on the arrow is in fact resistance and not thrust. Buridan was correct in this particular case, but empiricism is not the answer to everything and occasionally it does lead to seriously flawed conclusions. Buridan agreed with Aristotle's contention that vacuum cannot exist, as he believed that, for instance, air will naturally rush in when a pair of bellows is pulled opened, the reason of which is to avoid creating a vacuum. To Buridan, this suggests that nature detests vacuum, a rather preposterous argument.

Many of Buridan's errors were fortunately rectified by his student Nicole Oresme (1320–1382), the most famous one involving the arrow and Earth's rotation. Buridan argued that Earth does not rotate, because an arrow shot vertically upward will always fall to the same spot where it was launched. If Earth does rotate on its axis, the arrow should have fallen to a place deviated from its original position and in the opposite direction of Earth's rotation. Oresme, however, contended that when an arrow is shot, its initial motion has already incorporated the horizontal speed caused by Earth's rotation, so it will fall back vertically to its original position. Being able to propose this correction indicates that Oresme was someone with an intuitive understanding of physics and keen observational skills when it came to objects in motion. In particular, his discovery that the motion of the arrow is a combination of horizontal and vertical components allowed him to provide the physical interpretation. In fact, according to current psychological research, the concept of relative motion already exists in children. It is a type of innate intuition, possibly suppressed later in life by education or due to the environment. This suppression of instinct goes as follows: at first it is suspicion; then rejection follows. A child learning a musical instrument is a good example: if a mistake is initially made from misreading the musical score, to correct it later it may take several times the amount of time spent on learning the piece. This also applies to learning science, quite unexpectedly, so basic education at the university level is indeed very important.

3.1.6   Emergence of Kinematics: Mathematics at Oxford

The type of expositions[6] given on kinematics, as in the case of Buridan and Oresme, was rather common at the University of Paris at the time, and soon the study of kinematics spread to Oxford, England. The city of Oxford became a prosperous municipality on the upstream stretches of the River Thames in early 12th century even before the University of Oxford was established, and many students and teachers began to gather there. Officially the University of Oxford was founded in 1250, but the year Robert Grosseteste (1175–1253) was appointed the university's first Chancellor, which is 1224, has been taken as the institution's first year of existence by convention. Scientific research at Oxford University's early years focused on mathematics, which was a direct result of the influence of its first Chancellor, as Grosseteste contributed to the success of Roger Bacon (1214–1292), a remarkable intellect and passionate mathematician. In addition to being enthusiastic for mathematics, Bacon was equally full of fervor for optics and geography. The influence of this pair of teacher and student at Oxford University was passed down to the first residential college in England, Merton College,[7] established in 1264 by Walter de Merton (1205–1277), then Lord Chancellor of England.

From the outset the teaching of natural philosophy was the main objective of Merton College, and Aristotelianism was at the very center of its ideological foundation. Professors therefore still employed qualitative methods in scientific research methods and teaching was limited to the subjective logic of causality. This academic focus continued well into the 17th century, during Newton's times, because to break through the millennium-old intellectual framework that was Aristotelianism was nothing short of a daunting task. Merton College, however, was able to achieve brilliant results in mathematical research, who had been long been overlooked by Aristotelian philosophy. The four outstanding principal researchers in these endeavors were: Thomas Bradwardine (1290–s1349), William of Heytsebury (1313–1372), Richard Swineshead (fl. c. 1340–1354 in Oxford), and John of Dumbleton (d. 1349). Their research focused on using mathematics to model and explain mean velocity. This problem in kinematics might seem very simple from a modern

---

[6]Referenced from Evans [9] and Mlodinow [23].

[7]In England, the universities of Oxford, Cambridge and Durham started as residential colleges. As both students and teachers lived on campus, university buildings were scattered throughout the city. This residential college setting has been a successful model for hundreds of years and has now become a proud tradition of English universities.

point of view, but back then it was considered a profound and thorny philosophical issue. As the definition of speed is based on distance and time, two physical quantities whose definitions were still a bit abstract and controversial and therefore remained unclear at the time, these four mathematicians realized that first and foremost they needed to formulate hypotheses about these physical quantities and perform relevant experiments on them. They chose the most abstract and also the most important physical quantity, time, to embark on their journey of research in the world of physics.

Since ancient times human beings have described time in a variety of different ways, but these discussions were limited to abstract discourse or artistic conception. Concrete quantitative empirical results were few and far between. At the inception of Merton College, where the concepts of "minutes" and "seconds" were nonexistent, "time" might still have been a vague literary term. To define time as a measurable physical quantity was a novel and abstract undertaking. Anyone could come up with a far-fetched or implausible theory of time. The very first clock that measured time in hours reportedly did not appear until the 1,330 s. Prior to this, it was customary to divide daytime into 12 equal parts. However, these time divisions would vary greatly in length depending on the region and season. For example, each part of the 12 equal divisions of an area in high latitudes during the winter might only be half of its counterpart in the summer. For these reasons, attempting to understand speed or acceleration based on imprecisely defined concepts of time was obviously a difficult task. The scholars at Merton, however, applied their refined observational skills to describe the motions of objects, and they are credited with having derived the first principles of motion in history, which was later known as the Merton rule of uniform acceleration (or mean speed theorem). The rule states that the distance covered by a uniformly accelerated object, starting from rest, is exactly half the distance traveled by a body moving at a maximum uniform speed over the same period of time. This can be stated succinctly in modern mathematical notation:

$$\frac{1}{2}v_0 \cdot t_0 = \frac{1}{2}at_0 \cdot t_0 = \frac{1}{2}at_0^2 \tag{3.1}$$

Here v_0 is the so-called maximum speed, and v_0t_0 is the distance traveled at this maximum speed; $a = v_0/t_0$ is the uniform acceleration represented as a fixed value in the equation.

This principle, seemingly simplistic from a modern perspective, could only be described at the time with unfamiliar physical quantities and a lengthy verbal description, which would sure cause considerable controversy in the academic circles of Europe. To prevent further pointless arguments from such

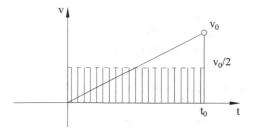

Fig. 3.1 Mean speed and acceleration

imprecise and awkward language, French natural philosopher and theologian Nicole Oresme (1320–1382) made important contributed to mathematics. Just like Newton, who invented calculus to prove the law of universal gravitation, and Einstein, who applied high-order tensors to prove general relativity, Oresme created brand new mathematical tools to explain his new physical law. In the process he confirmed that mathematics is an essential tool for making physics come to life. Oresme's mathematics describes how the velocity of a moving object changes with time via geometric figures. We illustrate Oresme's principle with geometry in Fig. 3.1. In the diagram, the horizontal axis represents time, and the vertical axis represents velocity. The left-hand side of (3.1) is the area of the shaded rectangular region, and the right-hand side is the area of the right triangle with v_0 as one of its vertices. The equal sign indicates that these two geometric figures have the same area, and these two areas also represent the distances traveled by the same object under two different motion scenarios. These were mentioned in the Merton rule of uniform acceleration, one being uniform acceleration starting from zero initial velocity (indicated by the right triangle), and the other being constant velocity at maximum speed v_0 (indicated by the rectangle). Oresme placed time and velocity on two mutually orthogonal axes, which was an attempt previously unheard of, and indeed it was quite difficult for the scientific community to understand or even accept this invention. It was therefore no surprise that Oresme's work failed to gain the recognition it deserved from his contemporaries.

3.1.7 The Printing Press Facilitates Knowledge Preservation

The trend continued by Oresme to employ diagrams to explain physical phenomena had already been widespread, and the fields of learning that benefited the most from it was in fact *not* the difficult subject of physics but the simpler and more practical engineering disciplines, such as mechanical principles and inventions: water wheels, windmills and printing presses. The first two mechanical devices were instrumental to the irrigation of farmland and the processing of agricultural products, such as increasing crop yields and boosting agricultural sales. The third piece of mechanical equipment, the printing press, also made groundbreaking contributions to science and made knowledge accessible to a wider audience. Movable-type printing had been invented by the Chinese around 1040, but because of the tens of thousands of different characters used in the Chinese language, movable-type printing did not quite fulfill its potential in China. When this technology was introduced to Europe around 1450, however, the entire publishing industry came to life. This is because the alphabet-based languages in the West can be written with only a few dozens of letters, and a page of a book is simply a combination of thousands of these highly repeating symbols. Movable-type printing quickly became a core technology for the dissemination of information and sharing of knowledge in Europe, and no publishing houses could survive without gaining access to this type of printing presses.

Generally speaking, a movable-type printer was able to produce thousands of copies of a book in the same time span as what would take a scribe to make one single copy. During the latter half of the 15th century, the quantities of books printed by the publishing industry may have exceeded the sum total of books produced in the previous one thousand years, not to mention the cost of printing was only a fraction of that of hiring a scribe to produce copies. This was a revolutionary technology, and an entire industry made up of professional copyists and lasting a millennium disappeared almost overnight. This is similar to the digital camera revolution of the 21st century, where digital cameras quickly replaced optical camera in the span of a few years. Photographic film cartridges that hold at most a few dozens of pictures per roll are today nowhere to be found, and they have instead been replaced by tiny silicon chips capable of storing thousands of images. The photographer no longer needs to think long and hard to make sure all settings are perfect before pressing that shutter release. Nowadays the clicking sounds of camera shutter can be heard anywhere, anytime, and photographers enjoy the freedom of artistic expression to the fullest extent. This must have been the case with the advent of movable-type printing, when publication and acquisition of books became commonplace with unprecedented ease and popularity.

With such magnificent printing machinery at the publishers' disposal, the industry flourished. Many writers and thinkers with novel insights but could not afford the traditional means of publishing books were now suddenly blessed with a tremendous space for disseminating their new ideas. Movable-type printing allowed more efficient and precise knowledge transfer and cultural exchange, while at the same time facilitated innovation across disciplines. The modern-day "smart economy," which everyone is proud of, actually became a way of life as early as the 15th century. Intellectual elites began to feel that the social distance between themselves and the nobility had narrowed, and the demand for new ideas and knowledge now gave them the desire to change and innovate. Thanks to the innovations driven by movable-type printing, it was now much easier for scientists to acquire a wide range of knowledge to face the challenges of Mother Nature. Artists were given more opportunities to be inspired and to become more creative. Sculptors were able to create life-like artworks that expressed their affections based on human anatomical structures found in medical literature. Painters learned projective geometry and perspective from mathematicians to create three-dimensional scenes never before seen. These worlds of their creation not only showcased the strength of muscles or the softness of the body but also expressed the characteristics of humanity in terms of postures and facial expressions. When applied to architecture, these artistic ideas completely transformed and renewed the skylines of European cities. New buildings incorporated many mechanical structures that adorned the palaces, where the princes and nobles lived, and also guest houses, various church buildings, and spiritual centers. Leonardo da Vinci (1452–1519) was one of the most representative figures in this era who pioneered the fusion of art and science.

3.1.8 Harbinger of the Astronomical Revolution

Movable-type printing not only brought Renaissance to Europe but also sparked the first wave of revolution in the history of science. According to Weinberg,[8] Europe at the beginning of the 15th century saw national governments consolidated and societal stability becoming a reality in France, under Charles VII and Louis XI, and in England, under Henry VII. The declining Byzantine Empire disintegrated after the fall of Constantinople in 1453, causing many Greek scholars in the empire to seek refuge in Western Europe, including the Italian Peninsula, France, England and the German states. The Renaissance, which was at its inception, not only aroused the people's desire

---

[8]Refer to Chap. 11 of Weinberg [32].

for the rejuvenation of the soul but also enhanced their quest for the exploration of natural phenomena. Following the arrival of Columbus in the Americas, Europeans were made aware that there were many things that were still hidden beyond what the Bible had revealed. The Protestant Reformation led by Martin Luther in the 16th century proposed rationalism and empiricism, which contributed to an environment of social values that was conducive to scientific exploration.

However, at the very beginning of the 15th century, Western Europe entered a period of darkness due to wars and diseases. The Hundred-Years' War between England and France, which erupted in 1337 and lasted 116 years, plunged the entire Western Europe into a turbulent, chaotic world. Meanwhile, the Black Death claimed the lives of nearly half of Europe's population. As a result, cities were in ruin, the population suffered, soldiers were powerless to defend their countries, and even the Church split into two papal authorities, one in Avignon, France and the other in the Vatican. Scientific research, being vulnerable to upheavals and the destruction of war, gradually declined in both England and France, but it grew and flourished in areas to the east of Central Europe.

At the beginning of the 15th century, almost all astronomical studies that were of importance were carried out in Germany, Italy, Poland, and the Czech lands. German astronomer Nicholas of Cusa (1401–1464) used vague language to describe the moving Earth and the infinite cosmos, which diverged substantially from the teachings of the Catholic Church. In Germany, Georg von Peurbach (1423–1461) and his student Johannes Müller von Königsberg (1436–1476; better known as Regiomontanus) also collaborated to expand on Ptolemy's epicycle-deferent model, the conclusions of which were later cited repeatedly by Copernicus. Another example is Alessandro Achillini (1461–1512) and Girolamo Fracastoro (1476–1553), both graduated from the University of Padova, the bastion of Aristotelianism, in Italy. They raised a number of issues with Aristotle's concentric spheres model. First of all, the Aristotelian model advocates that the distances between the planets and Earth never change, but this contradicts the observations that the brightness of a planet often changes. Secondly, if Ptolemy's equant model is used to observe Venus, we know that the farthest point of Venus from Earth is sixty times its nearest distance, yet the change in brightness does not reflect this huge difference, so the reasoning must be incorrect. It is now understood that the brightness of a light source is inversely proportional to the square of its distance from the observer, but the theory that brightness and distance are inversely proportional to each other was generally accepted at the time.

Of course we now understand that these contradictions or errors are the results of the Aristotelian model's incorrect assumption that Earth is at the center of the cosmos. The erroneous assumption, however, would have to wait until the 16th century before it was discredited by Galilei, when he made observations in the night sky with his own eyes through the telescope. Prior to that, German and Italian astronomers could only raise questions and live with the unresolved mysteries. Under this atmosphere of escalated and intensified scientific debate, the Polish astronomer Nicolaus Copernicus (1473–1543), who later proposed the heliocentric theory and sparked the Scientific Revolution of the modern era, was quietly but surely working at this time toward the demolition of geocentrism, a theory that had survived more than a thousand years, in the remote kingdom of Poland.

3.2 Copernicus: Pioneer of the Scientific Revolution

The Ptolemaic geocentric model that Copernicus was up against was not merely an astronomical theory; it also represented the most important philosophical and theological framework in his time. As God created human beings and made Earth the abode his creation, Earth had to be at the very center of the entire Universe, and other celestial objects had to revolve around Earth as well. Therefore, even though Ptolemy's geocentrism was unable to satisfactorily explain the change in brightness or the waxing and waning of the Moon, Earth's nearest celestial neighbor, or to perfectly match the observational results against the predictions of geocentrism, the theory was a perfect compromise. This was because, most importantly, geocentrism was welcomed by the Catholic Church despite its limited ability to make accurate predictions of celestial events. Geocentrism was therefore able to stand unchallenged for over a millennium with the support of the Catholic Church, a power that had been dominating Europe's political, economic and religious life. As the accuracy of astronomical observations gradually improved over time, however, astronomers became aware of the increasing inconsistency of the Ptolemaic model against actual observed data. In order to rescue this "perfect" system from being consigned to the dustbin of history, scientists began to pile more and more epicycles upon the existing geocentric model, making it even more confusing and unwieldy.

This chaotic development in astronomy exposed a fact. In the period of fifteen centuries between the advent of Ptolemy's *Almagest* and the publication of Copernicus' *De revolutionibus orbium coelestium* ("On the Revolutions of the Heavenly Spheres", or simply *De revolutionibus*) in 1543, the astronomical community in Western Europe failed to contribute to scientific knowledge of any significance, which allowed the cosmological system established by Aristotle and Ptolemy to maintain its foothold.[9] The artificial orbits constructed with epicycles, deferents and the equant, along with their meticulously tuned sizes and the angular velocities of their circular motions, were able to describe the apparent motions of the planets, as observed from Earth's surface, in a reasonable manner. These calculations could even predict the future positions of the planets, under certain circumstances, with errors within only one-sixth of a degree. Despite its limitations, this was considered a rare scientific achievement since the time of ancient Greece. In the course of more than a thousand years, although some Islamic and Indian astronomers in the East occasionally proposed modifications, however inconsequential, to the Ptolemaic model to improve its accuracy, many long-standing astronomical issues that had been vigorously debated remained unresolved. The lack of satisfactory resolutions prompted challenges to the Ptolemaic model to become increasingly severe, as reflected by observational results that contradicted the dominant theory. However, the mighty Church's stranglehold on academic freedom refused to yield and the Church was able to enforce the suppression of scientific evidence that sprang up here and there. Any academic debates that arose from time to time within the walls of the academic institutions would die down just as easily and posed no threat to the establishment or exerted minimal impact on society. The repressive atmosphere that represented the status quo was gradually challenged in the 16th century by the series of movements ignited by the Renaissance, which began in Italy. Advocacy for the freedom of thought spread like wild fire on the European continent, and changes were inevitable. Nicolaus Copernicus, a priest, physician and astronomer from the conservative Kingdom of Poland, emerged amidst such great changes. The Polish priest,[10] in defiance of the authority of the Catholic Church, bravely proposed that the Sun is actually at the center of the cosmos and openly claimed that Earth is one of the planets orbiting the Sun. The truth behind celestial motions was finally revealed to the world.

---

[9]After Thomas Aquinas incorporated Aristotelian philosophy into Christian doctrines in the 1200s, Aristotelianism formed the solid, unchallenged foundation of university education in Europe. This reason is that the natural cause-and-effect theory advocated by Aristotelian philosophy coincided with Biblical teachings (in fact the teachings of many other religions): the Creator is the cause, and His Creation, including living beings and their actions, are the effect.

[10]The beginning of the Christian era saw the Christian Church gradually taking control of the political and economic power in Europe and becoming the only institution capable of sponsoring academic research. Those who wished to do research and earn a living therefore had to become associated with the Church. Many researchers who dedicated their lives to science at the time were by profession priests or friars, with a small number of them being nobles.

Nicolaus Copernicus, the revolutionary and pioneer in astronomy, was born in the town of Toruń in 1473 to a merchant family from Kraków, the capital of the Kingdom of Poland. Copernicus' father died when he was ten years old, and he was raised by his maternal uncle Lucas Watzenrode the Younger, Prince-Bishop of Warmia, who became rich from serving in the Church. At the age of 17, Copernicus was arranged by his uncle to enter the University of Kraków and began his studies in the Department of Arts, a major center of astronomical and mathematical studies. At the age of 23, he went to the University of Bologna in Italy to study canon law and became an assistant to the astronomer Domenico M. Novara (1454–1504). There he commenced his career in astronomical observation and research. During his time in Bologna, he became a canon of the diocese Frombork (Frauenburg in German) with his uncle's referral. In this lifetime position Copernicus would receive a generous salary but was not required to perform any church duties. Copernicus, however, was never ordained a priest. Instead he went to the University of Padova in the Venetian Republic to study medicine. In 1503, he finally received the degree of Doctor of Canon Law from University of Ferrara in Italy at the age of 30, after which he returned to Poland to become a canon, for which he received a fixed income and a church castle. This allowed him to focus on astronomical research, his true passion, outside his official duties. Starting in 1506, therefore, he employed a simple instrument he had constructed himself for astronomical observations and collected dozens of observational data sets. Later in 1510, he returned to his hometown of Frombork and built a small observatory. He remained there until his death in 1543.

Not long after he settled in Frombork, Copernicus began to study Ptolemy's *Almagest* in earnest, like any other astronomer. The Ptolemaic model that were widely accepted at this time had been modified with numerous additional epicycles to improve its predictive accuracy over a period of a millennium, making it quite intractable and unwieldy to work with. Just when Copernicus became increasingly frustrated with the complicated Ptolemaic system, he discovered the heliocentric model presented by the ancient Greek astronomer Aristarchus of Samos (see Sect. 2.6 in Chap. 2) some 1,700 years earlier. Copernicus considered heliocentrism to be much simpler compared with the incomprehensible geocentric system and also to provide more clarity, and he began to explore the feasibility of substituting the Sun for Earth as the center of the cosmos. Copernicus was very cautious in the beginning, because placing the Sun at the center of the Universe with Earth revolving around it in a circular orbit in uniform angular motion. This is because it would require reinterpreting the available observational data collected over a thousand years by Ptolemy using the new dynamic coordinates of the moving Earth. Setting aside the extremely difficult and unprecedented attempt of viewing a moving planet from the perspective of moving coordinates, Copernicus had to deal with an enormous amount of data and complex phenomena of planetary motions under the new model. It is hard to imagine how intense the struggles of Copernicus must have been. In any event, Copernicus believed that Ptolemy's geocentric model had run its course and heliocentrism might be the only way forward.

3.2.1   Rise of Neoplatonism

The rise of heliocentrism did not happen overnight. Ptolemy's geocentric model had endured more than a thousand years almost unchallenged until the end of the 15th century, when the precision of astronomical apparatuses improved so significantly that observational data collected since the introduction of the Ptolemaic system had to be radically adjusted or rectified, thus shaking the very foundation of geocentrism and finally signifying the beginning of its end. One of the most famous challenges came from Peurbach (1423–1461) of the University of Vienna. He fitted the data collected by Arab astronomers with a theoretical model he had devised and was able to produce predictions of planetary positions and distances much more accurately than those from the Ptolemaic model. His student Regiomotanus (1436–1476) compiled this information and incorporated it into his work *Eptiome of the Almagest*. Later, A. Bruchaisky (1445–1497), a professor of mathematics and astronomy at the University of Kraków in Poland, provided comprehensive expositions to the astronomical theory of Peurbach based on this new data and produced the popular textbook *Theoricae novae planetarum*. In this new work in astronomy, the author pointed out many errors produced as a result of the Ptolemaic model, such as the ratio of the size of the Moon's epicycle to Earth's equant being too large, causing the Earth-Moon distances at full moon and at new moon to diverge significantly, which was inconsistent with Peurbach's results. These inconsistencies sparked the interest of one of Regiomotanus' students, Domenico Maria Novara, who began to work with his own student Copernicus to reexamine the astronomy of the Ptolemaic model with the reemerging Neoplatonism.

The so-called Neoplatonism considered mathematics to be a real entity, and a cosmological model can be governed by mathematical relations. Within this mathematical framework, the cosmos would not only become orderly and harmonious but also simpler than those complicated models constructed

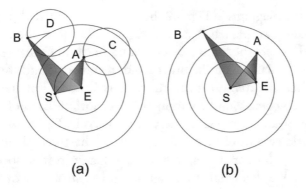

(a)   (b)

Fig. 3.2 Similarity between the **a** geocentric and the **b** heliocentric systems. Inspired by Fig. 3.7 of Hsiang et al. [15]

with layers and layers of geometric figures. The rise of Neoplatonism could be attributed to the proliferation of epicycles, which numbered more than 80 at one time, of various sizes in the Ptolemaic system that was constantly being modified. Successive generations of astronomers would become aware over time that this excessively elaborated model could not be further from physical reality. Mathematics, which had been highly valued by astronomers up to this point but was not utilized to its full potential, emerged as a highly useful tool in several applications for explaining astronomical models and quantifying the universe. This new view of mathematics ignited the hope of astronomers in applying mathematics to model everything in the Universe, especially those complex and intractable phenomena. Their motives for turning to mathematics were also quite simple: the Creator must have created the cosmos with minimalism and harmony in mind, so mathematics would naturally play a role in the Universe's order. One can therefore discover the hidden truth of the cosmos in mathematical structures.

Novara, who was teaching at the University of Bologna at the time, was encouraged by the rise of Neoplatonism and took the first step toward dismantling the bastion that was geocentrism. He described the geocentric model as being full of complexity and chaos, as well as being at odds with God's idea of simplicity and harmony and entirely incapable of reflecting the spirit of mathematics or the order it brings. The Ptolemaic model would from this point on be challenged relentlessly by Neoplatonism. To illustrate the unwieldy nature of geocentrism, we reproduce in Fig. 3.2 below and explain the comparison of the geocentric and heliocentric systems, as described in Hsiang et al.[11]

---

[11] Refer to Hsiang et al. [16], p. 61.

In each of the diagrams in Fig. 3.2 there are four celestial objects: Earth (E), the Sun (S), and two planets A and B. Figure 3.2a shows the Ptolemaic geocentric model, where the Sun revolves around Earth in circular motion. Each of the other two planets, on the other hand, revolves around the center of its epicycle in circular orbits. In turn, the centers C and D of the two planets' epicycles also revolve around Earth in circular motion. Each of these circular movements proceeds at the same angular velocity. In Fig. 3.2b, which shows the heliocentric system of Copernicus, each of the three planets (including Earth) orbits the Sun at the same angular velocity in its respective circular orbit. Suppose we deliberately arrange the celestial objects in the two diagrams in the following fashion: let the radii AC and BD of the two epicycles equal to the Sun-Earth distance SE, i.e. AC = BD = SE, and let these three line segments be mutually parallel to one another (as shown in Fig. 3.2). Now, From Earth's point of view, the relative positions of the Sun and the two planets are identical in these two systems. It should be clear from the foregoing that under specific configurations, the Ptolemaic geocentric system can be seen as equivalent to the Copernican heliocentric system. The difference is that heliocentrism is simpler and more harmonious, while geocentrism is complicated and unwieldy. Motivated by this reasoning, Copernicus became determined to conduct in-depth exploration of the heliocentric system.

3.2.2   Heliocentrism: *Little Commentary* and *on the Revolutions of the Heavenly Spheres*

From the two equivalent systems arranged in a particular manner above, we can see that the Ptolemaic model is valid as a theory to describe the planetary motions as viewed from Earth, but only when very specific conditions are met. Moreover, each planet in the Ptolemaic model requires an independent system to describe its relative position to Earth and the Sun. If the five planets known to scientists at the time are described along with the Sun and the Moon, the intricate petal-like diagram depicted in Fig. 2.8 could be formed, but it is difficult to imagine that it truly reflects the way the cosmos works, since observational data never confirmed the model's veracity. As shown in Fig. 3.1, if we swap the positions of the Sun and Earth and dispense with the epicycles, however, the data will still fit the observations as if epicycles were employed, but the now Sun-centered system will conform much better to the simplicity and harmony of God's creation. At this time Novara and Copernicus knew exactly what they should do: modify the Ptolemaic model by placing the Sun at the center of the cosmos and eliminate the epicycles (numbering more than eighty) from the system altogether. Copernicus then proceeded to put in words the exact cosmological model he had in mind into a short but important astronomical research paper in astronomy, *De hypothesibus motuum coelestium a se constitutis commentariolus* ("A Commentary on the Theories of the Motions of Heavenly Objects from Their Arrangements"), commonly referred to as *Commentariolus* ("Little Commentary").

(1) *Commentariolus*

Copernicus never published the *Commentariolus* during his lifetime, so this work was not as well-known as his masterpiece *On the Revolutions of the Heavenly Spheres*. Nevertheless, *Commentariolus* contains several important arguments that Copernicus used as a solid foundation to construct his heliocentric system. For example, he argued, among others, that Earth is not the center of the cosmos but the center of the lunar sphere, i.e., the orbit of the Moon around Earth. With the exception of the Moon, all other celestial bodies revolve around the Sun. The distance between Earth and the stars is much greater than the Earth-Sun distance. The apparent diurnal motion of the stars is caused by Earth's daily rotation. The retrograde motion of Mars is caused by Earth's orbital motion around the Sun. All these arguments presented by Copernicus were derived from his study of Ptolemy's *Almagest* and not from his own observations. However, each argument was made according to the same basis: Earth revolves around the Sun and the Moon revolves around Earth. On this basis, Copernicus was able to reinterpret Ptolemy's data and arrive at the conclusion known to modern scientists: the heliocentric system, where the six planets orbit the Sun in the order of increasing distance from the Sun: Mercury, Venus, Earth, Mars, Jupiter and Saturn. Their orbital periods are, respectively, 3 months, 9 months, 1 year, 2.5 years, 12 years and 30 years. Copernicus pointed out in the *Commentariolus* that although the hypothesis of the Earth revolving around the Sun was proposed by the Pythagoreans as early as 500 BC, it was pure speculation as no actual observations or evidence was ever offered to back up the claim. To observe new data based on the new heliocentric theory from a moving Earth was no easy feat, either. Frustrated and at his wit's end, Copernicus managed to get a copy of Ptolemy's *Almagest* and realized that to refute this great work of astronomy would require much more than the few rather weak hypotheses in his *Commentariolus*. Copernicus then proceeded on the basis of this shorter work and began his long journey of exploring the heliocentric system. After more than three decades of research and gathering of scientific evidence, he finally completed *De revolutionibus orbium coelestium* ("On the Revolutions of the Heavenly Spheres"), his seminal work on the heliocentric theory, in 1543.

(2) *De Revolutionibus Orbium Coelestium*

At the beginning of this work, Copernicus wrote emotionally that when he decided to treat Earth as a planet and took into account the movements of all planets (including Earth) together, not only could the issue of irregular planetary retrograde motions be settled decisively but the relative sizes and order of the planets' orbits could be reasonably explained as well. Copernicus further emphasized that the geocentric hypothesis had been the source of all contradictions in astronomy. Once Earth yielded its central position to the Sun, all these problems were quickly resolved. The end result is a cosmos that possesses the qualities of orderliness, simplicity and harmony, the premises of which is that the Sun occupies the center of the Universe.

In order to thoroughly trace his own thought process in arriving at the above conclusions, Copernicus divided *De revolutionibus* into six volumes or "books". Book I begins with a clear description of the heliocentric model of the cosmos and classifies Earth's motions into three types: rotation (or spin), revolution around the Sun, and (axial) precession. It also explains the principles of planetary retrograde motion and the formation of the four seasons, which are quite advanced and accurate. When Copernicus first suggested that Earth spins on its axis, one of the challenges he was confronted with was that one cannot feel Earth's movements at all. At the time it was almost impossible to prove Earth's rotation by scientific means, as indirect evidence would be required. We present two natural phenomena and an experimental device at the end of this chapter, "Appendix 3.1: Indirect Evidence of Earth's Rotation", to prove that Earth does rotate on its axis.

Book II provides a basis for the arguments used to demonstrate the principles of the heliocentric system's motions, while explaining the relationship between the spherical shell model and the phenomena of sunrise and sunset. Books III and IV describe the apparent motions of the Sun and the Moon, respectively. First, he explains how to calculate the length of the year and outlines the shape of Earth's orbit around the Sun. Then he discusses the movement of the Moon and the cause of lunar eclipses. As Copernicus believed these orbits to be circular, some of his calculations in this book contain significant errors compared with actual modern measurements. Books V and VI expand upon the heliocentric system and provide clear and close connections among the movements of the Sun, Moon and Earth as well as the other planets. The books also explore separately the "longitudinal" and "latitudinal" motions of the five major planets, namely Mercury, Venus, Mars, Jupiter and Saturn on the celestial sphere. This great work of Copernicus provides a cohesive analysis on the astronomical data he collected and discusses the long-standing, controversial issues in astronomy in his time using the theory he established. Copernicus demonstrated his superb geometric and mathematical skills in his calculations and reasoning, confirming the superiority of Neoplatonism, which was highly regarded by his teacher Novara. Written in Latin, *De revolutionibus* contains over 400 pages and deserves to be called the scientific masterpiece the astronomical community had been waiting for in Copernicus' time.

3.2.3   Modern Proofs of the Heliocentric Theory

Copernicus' masterful mathematical skills, as exemplified in *De revolutionibus*, consist of hundreds of pages of complicated geometric derivations and proofs, which are difficult to understand even to modern readers. In their work *Mysteries of the Ages*,[12] Hsiang and his coauthors attempted to illustrate the superiority and transcendence of the Copernican system with the following descriptions. The goal is to compute the orbital radius and orbital period of a planet revolving around the Sun via the relationship between the heliocentric system and retrograde planetary motion. A basic assumption for these derivations is that each planet orbits the Sun in uniform circular motion. In reality, all planets travel in non-circular elliptical orbits with varying angular speeds, so in Hsiang et al. own words, their results can be referred to as a simplified version of the Copernican heliocentric model.

In *Mysteries of the Ages*, Hsiang et al. [15] first derive mathematically the orbital periods and radii of the outer (or superior) planets, i.e. planets whose orbits contain Earth, which are Mars, Jupiter and Saturn. Suppose that Earth (E) orbits the Sun (S) in 1 Earth year (or simply "year"), in which case Earth's average angular speed is $\omega_E = 360°/1$ year. Suppose also that another planet P revolving around the Sun has an orbital period of T years. Then P's average angular speed is $\omega_P = 360°/T$ years. Hsiang et al. next consider the movements of E and P starting at the point in time where SEP forms a straight line, and they assume that E's angular speed is larger than P's. After E has revolved around the Sun once, it will catch up with P after having traveled a certain distance, at which point SEP again forms a straight line. This straight line and the one in the beginning forms an angle α, and assuming that the time taken by E to catch up with P is t years, we then have the following

---

[12]Refer to Hsiang et al. [15], pp. 66–74.

two equations:

$$\omega_E = \frac{360°}{1} = \frac{360° + \alpha}{t}, \quad \omega_P = \frac{360°}{T} = \frac{\alpha}{t} \tag{3.2}$$

Combining the above two equations and dividing by 360° simultaneously, we obtain the relationship between the orbital periods of Earth and planet P.

$$T = \frac{t}{t - 1} \tag{3.3}$$

Since the time t can be measured from Earth, we can readily calculate the orbital period T of P.

Hsiang et al. next derive R, the orbital period of plant P. As before, Earth's orbital radius is assumed to be one unit and calculations will begin at the time where SEP forms a straight line. When SEP becomes a straight line for the second time, E will have covered an additional angle θ to form a right triangle with S and P, i.e. $\angle SEP = 90°$. Assume it takes E τ years to travel this distance. We therefore obtain the following relationships

$$\frac{360°}{1} = \frac{360° + \alpha}{t} = \frac{\theta}{\tau} \tag{3.4}$$

We can express the above relationship of the right triangle $\triangle SEP$ with a trigonometric function

$$R = \sec \frac{360° \tau}{t} \tag{3.5}$$

Since the point in time where $\angle SEP = 90°$ is observable from Earth, τ is known. The orbital radius of each outer planet can therefore be calculated from the above equation. Similarly, the relationship between the orbital period and radius of each of the two inner (or inferior) planets, Mercury and Venus, can also be derived.

$$T = \frac{t}{t + 1}, \quad R = \cos \frac{360° \tau}{t} \tag{3.6}$$

Here t is the time (measured in year) elapsed between two consecutive instances of the Sun, planet P and Earth forming a straight line, whereas τ is the time elapsed between the time when the points SEP form a straight line and when they form a right triangle.

The above relationship between a planet's orbital period and radius has been derived using modern trigonometry. In Copernicus' *De revolutionibus*, however, the equations are proved with rather cumbersome and complicated geometric figures, although the end results are identical. From a modern perspective, the formulas in (3.3) to (3.6) appear to be quite ordinary, and the derivation process has also been straightforward. Yet the mathematical relationships are so simple and harmonious, which provided great encouragement to the followers of Neoplatonism in Copernicus' time. Furthermore, even though the above derivation is based on the assumption of a circular orbit, Hsiang et al. was able to apply the observations of two retrograde motions of Saturn and Jupiter between 2007 and 2008 to the formulas to calculate their orbital periods and radii, with an error within 5% compared with known true values. Of course the principal source of this error is the assumption of the two planets' circular orbits rather than their true elliptical orbits in this simplified version of the Copernican model.

The reasoning above not only allowed Copernicus to successfully place the Sun at the very center of all planetary orbits but also provided accurate calculations of the sizes of these orbits. Moreover, Earth assumed its rightful place among the planets, and any ambiguities regarding the order of these planets from the Sun were also resolved. From this point on, the Solar System's structure and composition became well known: the Sun is at the center, and the planets Mercury, Venus, Earth, Mars, Jupiter and Saturn orbit the Sun in circular motion, in the indicated order, each having its own distinct orbital period. After Copernicus introduced and verified the heliocentric system, astronomers began to seriously consider the relationship between mathematics and the Universe, and they even began to believe that anything that applies in mathematics must also apply to the Universe. This is also what Copernicus, a follower of Neoplatonism, had hoped. From his pursuit of a simple and harmonious cosmological system, with the assumption of all planets revolving around the Sun in circular orbits, to finally designating the Sun as the center of the Universe, where the orbital radii and periods of planets can also be calculated, it was indeed a triumph for Neoplatonism.

Compared with the millennium-old Ptolemaic system, Copernicus' heliocentric model is simpler and more aesthetically pleasing, which is nothing like the rather unpalatable contraption of epicycles and deferents in the geocentric system designed to explain stellar parallax and planetary retrograde motion while insisting that Earth should be at the center of the cosmos. The end result was a strange and complicated Universe extremely difficult to comprehend. On the other hand, Copernicus' assumption of planets having circular orbits resulted in predictions being at odds with actual observational data. For example, the different speeds by which Earth revolves around the Sun in its orbit during the summer and the winter cannot be explained with the Copernican model. To compensate for these shortcomings, Copernicus, who by now had exhausted all his means of making further improvements, surprisingly reintroduced Ptolemy's epicycles and equant scheme to his heliocentric system. The result was that the center of Earth's orbit is on an epicycle that revolves around the Sun at uniform angular velocity. In this modification of his model, Copernicus introduced as many as thirty epicycles at one point, and in the end he placed the Sun away from the center of all planetary orbits, which was apparently contrary to his original intention. As the corrections he made are related to the change in the orientation of Earth's rotation axis and the orbital plane, which are rather complex, we will not pursue it further in this book.[13] It was also because of the contradiction brought about by this compromise that after the publication of *De revolutionibus*, heliocentrism failed to replace geocentrism right away as expected, as the epicycle-augmented heliocentric model was no less complicated than the petal-shaped cosmos of Ptolemy.

3.2.4 How *on the Revolutions* Was Published: Inner Struggles

While Copernicus was busy developing his heliocentric model, the Protestant Reformation led by Martin Luther (1483–1546), whose teachings were later known as Lutheranism, became increasingly influential in the Catholic Church of Poland. Although Mauritius Ferber, Prince-Bishop of Warmia,[14] was a hardliner against religious reforms, Copernicus was able to avoid, for the time being, the persecution of the Church regarding the heliocentric theory he had been advocating, thanks to his status as a private physician to Ferber. Still Copernicus was not prepared to openly publish his new theory, the discussion of which was confined to his circle of friends, and he was unable to communicate with his colleagues in the astronomical community at all. Despite the purely private discussions, Copernicus' theory was subjected to attacks by certain members of the Church, although this time they did not come from the Catholic authority but the anti-Catholic Lutheran Church in Germany.

---

[13] For further information on how Copernicus developed his theory, refer to Chap. 7 of Hsiang et al. [15], pp. 178–190.

[14] The uncle who raised Copernicus was the bishop of this diocese, so Copernicus would have had a close relationship with Ferber, the person who succeeded his uncle as the new bishop.

Martin Luther, a seminal figure in the Protestant Reformation and Copernicus' contemporary, in one occasion publicly called Copernicus "that fool" and accused him of trying to prove himself unconventional and to seek fame and compliments, as well as to turn the entire discipline of astronomy upside down. In the end, Luther, like the Catholic hierarchy, placed his faith without question on the Bible, blindly quoting the passage that described how Joshua commanded the Sun, not the Earth, to stand still. Another Lutheran priest cited Ecclesiastes 1:5, "The sun also ariseth, and the sun goeth down, and hasteth to his place where he arose." (KJV) to attack Copernicus' heliocentrism, which advocated a stationary Sun. It was therefore understandable to Copernicus that in this rather hostile atmosphere, attempting to publish one's research findings in an academic publication comparable to the *Almagest* would be an exceedingly risky endeavor and insurmountable obstacle. Despite these adversities, the solid skills in theology, medicine and mathematics that Copernicus had acquired before the age of thirty allowed him to refine and perfect his heliocentric model with his clear thinking and accurate memory, as well as the access to a considerable amount of astronomical data.

Even though the general atmosphere of the time was rather hostile to heliocentrism, after more than two decades of dedicated research the theoretical and structural foundations of the Copernican heliocentric model gradually took shape. On the other hand, as Copernicus was, after all, a career canon who had sworn allegiance to and was regularly receiving salaries from the Roman Catholic Church, the decision of whether to publish a treatise of the new cosmological model he had spent thirty years to complete was nothing short of a painful battle within himself. In the end, Copernicus overcame the apparent contradictions between theology and science and decided to be on the side of truth. After a great deal of soul-searching and painful struggles, Copernicus proceeded to publish the results of his astronomical research in the book *De revolutionibus orbium coelestium.*

Just when Copernicus' work on the manuscript was nearing completion, however, his self-doubts and uncertainties resurfaced, the same way they arose when he first decided to write and publish the book. First of all, he recognized that the phenomena described by the heliocentric model seemed counterintuitive when applied to everyday life, and his new theory would not be easily understood or accepted by the public at large. Secondly, heliocentrism would disprove and supplant the millennium-old geocentric model, which would undoubtedly incur the wrath of and fierce opposition from the Aristotelians, whose teachings had long dominated academia. Even more seriously, however, was that heliocentrism infringes upon the unassailable theological beliefs that the Catholic Church had held consistently over a thousand years, with

Earth being at the center of the cosmos. The consequences of publishing this work could be self-destruction. To make things worse, most publishing houses were unwilling to undertake a project of such magnitude. Capable and willing publishers were simply nowhere to be found, if they existed at all.

Copernicus saw a silver lining, however, in the young mathematician Georg Joachim Rheticus (1514–1574) during the spring of 1539. Rheticus, from the Archduchy of Austria, had been appointed professor at a Lutheran university at the age of 22. Rheticus was passionate about astronomy, and he once traveled from Germany to the University of Kraków in Poland, hoping to study astronomy with Copernicus. Rheticus was awestruck by Copernicus' heliocentric model after reading the manuscripts of *De revolutionibus*. He was convinced this revolutionary theory should be published so that the errors of the astronomical framework that had been held as the truth for a millennium could be rectified. After having taken into consideration of Copernicus' apprehensions, Rheticus came up with a plan to publish Copernicus' work in two stages. First, he arranged for Franz Rhode, a publisher in Danzig (modern-day Gdansk, Poland), to publish the book's abstract, known as *Narratio Prima*, in early 1540. The purpose was to pique curiosity of the scientific community regarding cosmological systems. In late 1542, the manuscript was delivered to Johannes Petreius (1497–1550), a famous German printer in Nuremberg, the city known as the Athens of Germany.

Although Copernicus was dreading the prospect of drawing condemnation from the Catholic Church by publishing the book, his project was nevertheless supported by several important members of the Church hierarchy, including Bishop Giese and Cardinal Schoenberg. Prior to the official publication of the book's second stage, Copernicus purposely arranged to have his work dedicated to Pope Paul III (1468–1549), who was erudite and well read. When the book was published, however, an unsigned *Ad Lectorem* ("To the Reader") was added to the preface. The purpose was to emphasize that the theory expounded in this work was merely a hypothesis put forth to aid computation so that the results would be consistent with the observed astronomical data. The level of confidence and rigor of the language employed was quite different from that of the meticulously crafted arguments of the main text. The contemporary readers would therefore find these two contrasting tones confusing and wonder what Copernicus' original intention for writing the preface was. Not until a few decades later that Kepler discovered a short letter in a group of old documents that contained the same message as *Ad Lectorem*, which was signed by the Lutheran preacher Andreas Osiander (1498–1522), an acquaintance of Copernicus. It turned out that Copernicus had solicited Osiander's view on whether the publication of the book would provoke a backlash from the Aristotelians. Osiander might have added *Ad Lectorem* himself to the preface without Copernicus' consent in order to allay the author's concerns regarding the negative reaction by the Aristotelians. Judging from the circumstances under which Galilei would suffer a century later, Osiander's unauthorized action appears to have protected Copernicus from potential persecution. On May 24, 1543, Copernicus was handed a sample copy of his *De revolutionibus orbium coelestium*[15] from Nuremberg, Germany on his death bed. He reportedly passed away after touching its cover.

Regardless of whether this novel but subversive heliocentric theory was acceptable to the Catholic Church or the Lutheran Church, what was actually more crucial was its acceptance by the astronomical community at large. The first scholar who openly praised the Copernican theory was none other than Rheticus, pupil and admirer of Copernicus. Recall that Rheticus had published *Narratio Prima*, which was meant to be an introduction to *De revolutionibus*, and that he had submitted Copernicus' manuscript to the printer in Nuremberg for publication. Rheticus was supposed to have written the preface on behalf of Copernicus, but he was too busy at the time to do it, and that was how Osiander was able to covertly include his unsigned letter *Ad lectorem*. In 1551, Prussian astronomer Erasmus Reinhold recalculated, based on the new Copernican framework, the astronomical data from the 1275 *Alfonsine Tables* and also the observations accumulated over the centuries since its publication. The results were a new version of the *Prutenic Tables*, which allowed the old-style calendar to be improved significantly. This was recognition of Copernicus' contribution and substantiation of his heliocentric system.

The Copernican model places the Sun at the center of the cosmos, with the celestial objects revolving around it in circular orbits. The order of the planets according to this model, in increasing distances from the Sun, is as follows: Mercury, Venus, Earth, Mars, Jupiter and Saturn. The Moon, on the other hand, orbits the Earth. On the surface, this system appears not to be very different from Aristotle's geocentric model. If we remove the Moon from the geocentric model and exchange the positions of the Sun and Earth, the resulting model will be the Copernican heliocentric system. This ostensibly insignificant change, however, would result in errors of an enormous magnitude. Under the Copernican model, the retrograde motion of Mars, which

---

[15]The first edition of *De revolutionibus* has been a much sought-after item on the antique market. On June 17, 2008, at Christie's Auction in New York, a first-edition of the book, published in Nuremberg in 1543, fetched a staggering US$2.21 million. Eight years later, in 2016, at another Christie's Auction event in London, a copy of *Narratio Prima* published in Danzig was sold for US$2.41 million.

had been baffling the astronomical community for over a thousand years, could now be reasonably explained with clarity, completely dispensing with the cumbersome and confusing theory of epicycles and deferents. Regrettably, however, the Copernican system depends on a model that uses a circular (rather than elliptic) orbit for its description of planetary motion, which inevitably leads to serious discrepancies between data calculated based on the model and actual observations concerning the speeds of planetary motions and distances between celestial objects. Despite these errors, the Copernican model allowed a great many truth-seeking astronomers, who had been denied the opportunity to conduct genuine astronomical research for over a millennium, to gaze into the enigmatic night sky afresh. In the atmosphere of freedom and liberation of the Italian Renaissance, Copernicus' theory also inspired a plethora of research activities, from which a series of remarkable scientific breakthroughs would be made possible. Furthermore, the following century, the 1600s, was such a momentous period that the philosopher Alfred N. Whitehead (1861–1947) referred to it as a "Century of Geniuses" in the history of human civilization. Three of Copernicus' contemporaries were among these geniuses: Tycho Brahe, Johannes Kepler and Galileo Galilei. Below we will look at the life of Tycho and his connections to astronomy.

3.2.5   Tycho Brahe, the Noble Astronomical Observer

The study of the movements of celestial objects did not stop following the death of Copernicus. The reason was that astronomical observation was more than just pure science, as the astrological elements of which it was a part were of great interest to royal families around Europe. In this regard, Danish nobleman and astronomer Tycho Brahe (1546–1601), a proud and intelligent man, made invaluable contribution to astronomy. Tycho was born to an aristocratic family in northern Denmark in 1546, three years after the death of Copernicus. He had always been fascinated by celestial phenomena since childhood. At the age of 13 he began to study philosophy in Copenhagen. He later went on to study astronomy in Leipzig at the age of 16. While still in secondary school he used instruments that he had designed to observe partial solar eclipses. At Leipzig University, Tycho learned that astronomical knowledge accumulated since Copernicus could accurately predict the occurrences of solar eclipse, and his research on the *Prutenic Tables* revealed that over a millennium earlier Ptolemy had predicted the date of the syzygy of Jupiter, Saturn and Earth (i.e., the three planets forming a straight line) that occurred in 1536, which deviated from the actual date observed by him by only a few days. Tycho was both completely surprised and highly motivated by the results. What surprised him was the accuracy of Ptolemy's prediction, and what motivated him was the possibility of improving upon Ptolemy's yet-to-be-perfected model.

Tycho proceeded to purchase a copy of Ptolemy's *Almagest* and studied its star charts. He then designed a new quadrant to greatly improve the accuracy of the predicted positions of celestial objects. Tycho's very first important discovery was made in 1572 at the age of 26, when a very bright "nova" (new star) suddenly appeared in the sky of northern Europe. While the phenomenon was causing widespread panic among the populace, Tycho directed the quadrant he had constructed toward the area in the sky now known as the constellation of Cassiopeia and conducted thorough observation of the bright star. This "nova" is now understood to be the image of a thermonuclear explosion of a Type Ia supernova determined to be about 9,000 light-years from Earth using advanced astronomical telescopes. After several months of observation, Tycho concluded that the position of the star had remained unchanged night after night, and its location appeared to have been much farther away from Earth compared with the Moon. This discovery was completely at odds with the mainstream (Aristotelian) astronomy at the time: celestial objects other than the Moon were unchanging. With this extremely radical theory Tycho published the book *De Nova Stella*, which was unexpectedly circulated widely and did much to allay the public's fear of the mysterious phenomenon, and he became well-known throughout Europe.[16] To reward this heroic young man, King Frederik II (1534–1588) of Denmark immediately announced his decision to grant the Island of Hven (a small island in the North Sea, now part of Sweden) to Tycho as part of his land holdings as a nobleman. The king also built the most advanced astronomical observatory in Europe on the island, known as Uraniborg. Apart from the four state-of-the-art observing stations, this castle-observatory was also equipped with a laboratory, a printing house, and a paper mill. It also housed Tycho's residence, which allowed Tycho to conduct astronomical observation and research with minimal interference. Tycho's responsibilities included the interpretation of major astronomical phenomena with the collected observational data, and to provide his benefactor, the Danish king, with his astrological predictions about the kingdom.

With abundant funding and support from the king, Tycho began to construct larger, more powerful astronomical equipment based on instruments

---

[16]Tycho is perhaps best known for his prosthetic nose made with brass. He lost the bridge of his nose in a sword duel against a fellow nobleman.

he had studied for years in order to achieve higher stability and accuracy. Following years of construction work and calibration efforts, the island of Hven now boasted a powerful and precision instrument capable of collecting data on the positions of planets and stars over extended periods of time. This was quite different from what had been done previously, e.g., observations could be made only when planets came into view at observable locations. It can be seen from a mural still fixed on the wall in the Uraniborg Observatory today that the quadrant designed by Tycho had a radius exceeding two meters and a precision of $1/360°$. It was advanced astronomical instruments like these that Tycho employed to make accurate and regular observations and measurements of the Sun, Moon, planets and stars over a 20-year period. Soon after Tycho began to make such observations, however, he found that the star catalogs published by contemporary astronomers came in a variety of configurations and were often disorganized and inconsistent with facts. A better, more comprehensive star catalog was needed to remedy these shortcomings. Tycho believed that it was necessary to perform long-term astronomical observation and to keep faithful records of them. Therefore, with the help of the precision instruments available at Uraniborg, Tycho was able to complete a number of star catalogs that remain valuable to this day. The data contained in them were also recognized by later generations of astronomers to be the most accurate prior to the advent of the telescopes as well as the concrete evidence that convinced European scientists of Copernicus' heliocentric theory. The data also ultimately provided a solid foundation from which scientists such as Kepler and Galilei were able to launch their efforts to modernize astronomy.

Although Tycho's astronomical observations were rather accurate, his interpretation of the data was very different. In 1577, the year after Tycho began to conduct astronomical observation on Hven, he made close observations of a large comet that traveled across the skies of Denmark. Tycho discovered that the comet's orbit could not have been a perfect circle, and that the comet was obviously farther away from Earth than the Moon. Between the time the supernova was observed five years earlier and the time of the comet's appearance, all the evidence showed that the Aristotelian geocentric model could not explain the observed phenomena. After several years of reorganizing his work and careful deliberation, Tycho published the book *On Comets* in 1583, in which he proposed a theory that lied somewhere between geocentrism and heliocentrism, with the motionless Earth remaining at the center of the cosmos. The Moon, the Sun and the fixed stars, therefore, would all revolve around Earth in spherical orbits and the five planets would orbit the Sun. Why in the world would Tycho create a cosmological model that deviated so egregiously from the highly accurate data from his star catalogs? It had nothing to do with whether Tycho had any ideological issues with the Copernican theory, which was already well known in the astronomical community at the time. In fact, Tycho was attempting to describe the movements of the entire cosmological system based on data observed on a moving Earth, and only those with three-dimensional thinking could grasp the dynamic interaction among celestial objects. Kepler may have been one of these talented scientists, and Newton and Einstein after him were perhaps the cream of the crop.

Tycho eventually spent over 20 years living and working in the well-equipped and comfortable Uraniborg. In 1599, at the invitation of the Rudolf II, Holy Roman Emperor and King of Bohemia, Tycho left Denmark for Prague to become Rudolf's court mathematician and began to develop the *Rudolphine Tables*. There Tycho encountered Johannes Kepler's *Mysterium Cosmographicum* ("Cosmographic Mystery" or "Cosmic Mystery") and was impressed by the young man's unique insight into the cosmological system using perfect geometric reasoning. Tycho invited Kepler to Prague to be his research assistant the following year. As the health of the 53-year-old Tycho was worsening, the assistance of a bright, young person was needed to take care of the large amount of observed data. On October 24, 1601, Tycho contracted acute cystitis after attending a banquet and fell ill. He died the following day. On his deathbed, Tycho urged Kepler to complete the star catalog Emperor Rudolf II had commissioned him to compile, and he also asked Kepler to incorporate the Tychonian System into the catalog. Kepler was happy to oblige his teacher's first request. The second request, however, was an entirely different story.

3.3   Kepler: The First Mathematical Physicist

Johannes Kepler (1571–1630) was born to a poor family in Weil der Stadt, a Protestant city in the south of modern Germany. His father was a bad-tempered slacker, and his mother a rather eccentric character who was constantly worried. Kepler's parents had a total of seven children, but they did not do a very good job taking caring of them. Their father abandoned the family when he was hired as a mercenary. As a child Kepler nearly died from smallpox himself, and the experience would later have a detrimental effect on his health. He suffered from gastrointestinal diseases, skin rashes and frequent fever, not to mention severe nearsightedness, throughout his life. Though sickly and weak, Kepler was hardworking at a very young age, and at the age of 13 he became a preparatory student for a theological seminary through a very competitive examination process. The diocese in which Kepler lived placed emphasis on education, and as a result Kepler, with his excellent academic record, was able to receive a five-year rigorous education in theology on scholarship. In 1589, Kepler attended the University of Tübingen as the age of 18. While the first two years of his university life was in general education, in the last three years he was trained in theology.

Although principally trained in theology at university, Kepler also studied under the astronomy and mathematics professor Michael Maestlin (1550–1631), a scholar who was well-known throughout Europe. Kepler was impressed with the heliocentric system proposed by Copernicus, which he regarded to be profoundly elegant, simple and harmonious. The last decade of Kepler's life happened to coincide with the beginning of the Thirty Years' War (1618–1648), one of the most brutal and destructive conflicts in European history. The two hegemonic powers, supporters of Lutheranism and those of the Catholic Church, fought against each other in a series of large-scale and savage battles, the result of which was that almost half of Europe's population was wiped out. As a Protestant, Kepler grew up in a Lutheran-dominated diocese. Yet the country was ruled by a Catholic king and therefore he was often persecuted for his faith. Kepler was twice forced to abandon his estates and properties as well as being exiled. The resulting poverty and involuntary migration led to the untimely deaths of both of his first and second wives as well as the demise of half of his children. Despite these extremely difficult circumstances, Kepler never gave up on his work on astronomical research. He eventually published the three laws of planetary motions, thus pioneering the use of mathematics in modern astrophysics. Below we will examine the life of Kepler and find out about how he was able to discover these laws. Let us begin with the time he started his three-decade-long professional career after graduating from university.

In April 1594, a mathematics teacher at a Lutheran school in Graz, the diocese to which the University of Tübingen belonged, fell ill and passed away. The schoolmaster wrote to the university to request a qualified teacher to fill the position left vacant by the mathematics instructor. Kepler, who was highly regarded by the University, was appointed to the position and began teaching mathematics and astronomy there. While teaching astronomy, Kepler spotted a curious phenomenon: the ratio of the diameter of a circle inscribed in an equilateral triangle to the diameter of the circumscribed circle seemed to be identical to the ratio of the diameters of Jupiter's and Saturn's orbits around the Sun. A Protestant faithful and someone profoundly influenced by the perfection advocated by Aristotelianism, Kepler began to employ geometry to explore if any mysteries of a geometric nature existed among the orbits of the then known six planets. He thought of the five Platonic solids (see Fig. 2.1) and attempted to fit them within the six planetary spherical shells by inscribing each orbital shell in a platonic solid (where the sphere is tangent to all faces of the polyhedron) and circumscribing the shell about another platonic solid (where the sphere is tangent to all vertices of the polyhedron).

After making painstaking computations and careful rearrangements, Kepler placed at the innermost layer the sphere of Mercury, which was then inscribed in an octahedron. This regular convex polyhedron of eight faces was next inscribed in the sphere of Venus, which was in turn inscribed in an icosahedron. This regular convex polyhedron with 20 faces was next inscribed in the sphere of Earth, which was in turn inscribed in a dodecahedron. This regular convex polyhedron with 12 faces was next inscribed in the sphere of Mars, which was then inscribed in a tetrahedron. This convex polyhedron with four faces was then inscribed in the sphere of Jupiter, which was in turn inscribed in a cube (a regular hexahedron). Lastly, Saturn's spherical shell was the outermost sphere, and the cube was inscribed in it. This configuration is depicted in Fig. 3.3. Kepler was keenly aware that the greater number of faces a polyhedron has, the smaller the distance between two adjacent spheres, and vice versa. He applied this logic to arrange the positions of the platonic solids. Kepler's motivation, as described previously, was to seek out the true

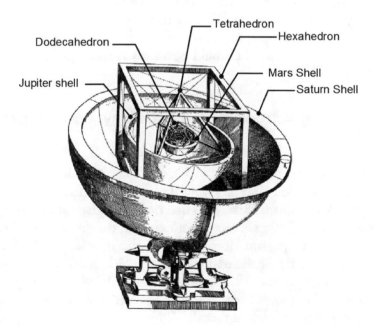

Fig. 3.3 Kepler's cosmological model constructed with the five Platonic solids (regular convex polyhedra). *Source*https://en.wikipedia.org/wiki/Platonic_solid

laws of physics using Plato's perfect geometric structures, indeed a culmination of applying the geometric doctrine that "only five convex regular polyhedra exist." The reasoning was that exactly five platonic solids were all that was necessary to construct God's perfect cosmos; one more would be completely superfluous and one fewer would be utterly inadequate. Despite this rather far-fetched idea that could be characterized as wishful thinking, Kepler made judicious adjustments of the thickness of each planetary spherical shell and the length of each platonic solid's side, the results of which were roughly consistent with the sizes of the planetary orbits that had been estimated by Copernicus earlier.

Young Kepler, passionate about astronomy, was gratified with his meticulously designed model of the cosmos, having brought together then mainstream heliocentric system advocated by Copernicus and the perfect geometry that God favored.[17] He therefore compiled these results into his very first published book, *Mysterium Cosmographicam* ("The Cosmographic Mystery"). After its completion, Kepler became aware that his cosmological model built with the perfect Platonic solids exhibited a great deal of inconsistencies with actual observed data. However, fully confident with the accuracy of his creation, Kepler blamed the discrepancies on poor accuracy of the astronomical apparatuses used to collect the data. Kepler then sent a copy of his book to Tycho, already a well-known scientist in the astronomical community. Kepler was hoping that Tycho would allow him access to the large collection of data in his possession. As a result, Kepler was hired by Tycho to work at the observatory at Uraniborg, where he would begin his amazing journey into astronomical exploration.

3.3.1   Unifying Astronomy and Astrology: Cosmological Model

Apart from teaching mathematics in Graz, Kepler was also responsible for producing astronomical almanacs as well as making astrological predictions regarding the kingdom's warfare, the good or ill of a given situation, weather patterns and the harvesting of crops. For example, the unusually bad weather in 1598 that led to peasant riots as well as the invasion of the Ottoman Empire were correctly predicted by Kepler, which made him not only a well-known astrologer but a highly valued one in Graz by the upper class of the society. However, his true passion remained the cosmological system of Copernicus. Kepler believed that the Copernican theory was built on two

---

[17]This is referred to as Kepler's Platonic Solid Model.

main core components: the Sun and mathematics. He was convinced that the simplicity and harmony exhibited by God's vast cosmos should be paired with a rigorous mathematical model to describe the motion of celestial objects. On the other hand, the Copernican model places the Sun at the center of the cosmos, around which planets revolve, so the Sun must be responsible for providing the power for planetary motion. The magnitude of the force exerted on a planet should also be inversely proportional to its distance from the Sun. Although this theory was different from the law of universal gravitation proposed later by Newton (which states that the force is inversely proportional to the square of the distance), this was a bold attempt by an astronomer to bring together physics and mathematics for the first time to explain celestial phenomena.

Such an advanced theory certainly required the incorporation of many different ideas and methods, and Kepler was indeed inspired by his teacher Michael Maestlin from their discussions of the Copernican model. When Kepler published his *Mysterium Cosmographicum*, Maestlin offered his assistance in the book's layout and proofreading with much enthusiasm. The book was finally published in 1596. However, misfortune often followed Kepler. In 1598, Archduke Ferdinand of Graz (the future Ferdinand II, Holy Roman Emperor) began to force Protestants to convert to Catholicism and imposed a deadline for the conversion. Those who refused would have to leave the city. Kepler held onto his Lutheran faith, and in September 1598 he left the theological seminary where he had been working for five years with his wife and children. However, this next phase of his career would lead Kepler to the famous astronomer Tycho Brache, who was older and had a personality very different from his. In fact, Kepler had begun correspondence with Tycho in his last years in Graz on issues related to astronomy.

Having read Kepler's *Mysterium Cosmographicum*, Tycho traveled in February 1600 to Benátky nad Jizerou, a town not too far from Prague, to meet with Kepler. The two astronomers, however, did not quite get along during the few months that they spent together due to their disagreements on a number of issues. Then Archduke Ferdinand began to expel Protestants from Graz, and Kepler had to temporarily abandon his ideological disagreements with Tycho and moved to Prague with his wife and children. There he began to construct the *Rudolphine Tables* with Tycho. Ten months after working in Prague, however, another tragedy struck. Tycho suddenly died of acute cystitis, which left Kepler with two important but tricky tasks: to continue with the work on the *Rudolphine Tables*, and incorporating the "Tychonian System" into the tables. Although the tasks that Kepler were assigned to complete were not exactly what he had in mind, the result was that Emperor Rudolf

II proceeded to appoint him as Tycho's successor and to take over the role of the imperial mathematician. Overnight Kepler was transformed from an assistant into the principal investigator, which allowed him unrestricted access to Tycho's vast collection of astronomical data and to conduct research on astronomy and mathematics in any manner and as thoroughly as he pleased. The most direct consequence, however, was that Kepler became much better off financially with the funding provided by the emperor. Taking advantage of all these unprecedented and serendipitous conditions, Kepler was favorably positioned to take his astronomical research to the next level. The first order of business for the loyal and honest Kepler was naturally to honor Tycho's last wish: establishing the cosmological model.[18]

3.3.2   Discarding Yesterday's Model and Focusing on Mars Data

A number of cosmological models had already been proposed by the astronomical community in Kepler's time. Examples include the concentric spheres model of the Eudoxan system described in Chap. 1 as well as the eccentric model of Apollonius and the epicycle-referent model of Hipparchus introduced in Chap. 2. Later, Ptolemy added the equant (refer to Fig. 2.6) to the Hipparchus model, so that for each planet there is a straight line connecting the center of Earth, the center of the epicycle and the equant. In addition, the planet moves along its epicycle in circular motion, whereas the center of the epicycle revolves around, also in circular motion, the deferent at non-uniform angular velocity while moving around the equant at uniform angular velocity. Although multiple restrictions and conditions are built into this model, the movement data of the five "wandering" planets calculated under the Ptolemaic system produced differences that were within 10 arc minutes of the observational data (1 arc minute $= 1/60°$ and a circle contains $360°$). This astonishing accuracy had allowed the Ptolemaic model to stand the test of time in the astronomical community over a period of over one thousand years. Copernicus, however, would subsequently discover that Ptolemy's equant model was causing the epicycle's center to move along the deferent unevenly, which he thought destroyed the simplicity and harmony of nature. The result was that Copernicus advocated that the Sun should replace the Earth as the center of the cosmos. This would allow Earth and all other planets to revolve around the Sun in circular motion, thus eliminating

---

[18] Parts of following discussion follow the reasoning in Weinberg [32].

the disharmony caused by the equant and achieving consistency between the model and observational data for a number of planets.

In an atmosphere where two entirely different cosmological models, one created by Ptolemy and the other by Copernicus, competed head-to-head, Kepler agreed with the latter astronomer's decision to place the Sun at the common center of the circular planetary orbits, including Earth's. However, the observational data of Tycho indicated that the angular velocity of a planet's motion around the Sun was not constant, which prompted Kepler to rethink Ptolemy's equant model as an appropriate representation of planetary motion. Kepler decided to correct the Copernican model, although the mathematical methods he employed was derived from the Ptolemaic tradition. Although the data thus obtained could reduce the deviations caused by the Copernican model, a great deal of discrepancies remained compared with the data collected by Tycho. These results deeply disappointed and frustrated Kepler.

This setback, however, did not hold Kepler back for too long. Realizing he was endowed with resources and a favorable research environment, Kepler decided to put his original priority of building a new cosmological model on the back burner for the time being. He began to conduct research on the large quantity of data left behind by Tycho. The first order of business was to study the data for Mars, for three reasons. First, Mars data was the most abundant. Secondly, when compared with the two inferior planets Venus and Mercury, which were often obscured by background light close to the horizon, Mars was often located high above the night sky and is therefore easier to observe. Thirdly, the speed of Mars' revolution around the Sun was much faster than those of the two superior planets, Jupiter and Saturn, and so Mars' observational data varied much more and offered a great deal more value as research data. What was even more fortunate for Kepler was that, according to later analyses of the data, the orbit of Mars has an eccentricity (deviation from being a perfect circle) greater than those of any other planets, with the except of Mercury. Otherwise it would have been difficult for Kepler to discover the first law of planetary motion, which states that planets move in ellipses with the Sun at one focus. In addition to the above three objective reasons and a fortuitous one, there may have been a "subjective" reason as well. That is, the nature of the retrograde motion of Mars had always been the primary problem confronting the astronomical community. Every single astronomer would eventually have had to deal with it, not least the relentless and tenacious Johannes Kepler.

Using the data for Mars to study planetary orbits was nevertheless another matter; it was a challenge that Kepler had to face. Since ancient times, measuring objects in the sky had always entailed measuring the angle between the lines of sight of celestial objects and their positions, or in other words the celestial latitudes and longitudes of the Earth-centered celestial sphere. As for measuring the distances between celestial objects, geometric analyses of Aristarchus and Hipparchus (see Sect. 2.6 in Chap. 2) were available, but they were limited to only three objects, the Earth, the Moon and the Sun. Measuring the distances between other planets was completely out of the question, until Kepler found a way, that is. Kepler's secret weapon to solve this problem was to take advantage of the periodicity of the planetary data. The periodic motions of the planets (including Earth) around the Sun all have different characteristics apart from orbital periods. Fixing Earth, the Sun and any given planet, one obtains a set of trigonometric relations among the three celestial objects. Once the temporal dimension is factored into these relationships, one can also generate an entire collection of trigonometric relationships at different points in time, which we can use to calculate various distances between the planet and the Sun (or Earth). With this basic concept, the first planet Kepler selected was none other than Mars.

When Kepler proceeded to perform calculations of Mars' orbit, he initially made the mistake of ignoring the complexity of the available data and believed that he was able to complete the analysis in a short time. Over the next five years, Kepler performed calculations on Tycho's data repeatedly using different models, and it is said that as many as 900 pages of such manual computations have survived to this date. After a series of detailed and lengthy calculations, Kepler discovered that the trajectory of Mars, if considered a perfect circular orbit, would produce a difference of eight arc minutes compared with Tycho's data. Although this small gap was quite commonplace in the astronomical community at the time, Kepler was highly confident about the accuracy of Tycho's observations and so he repeated the mathematical calculations several more times. He concluded that this suspicious gap had to imply something more substantial. Now we can consider this "luck" a gift from heaven, because had Kepler instead used the data for Venus, the resulting elliptical deviation would not have been eight but just one arc minute, which would have simply been treated as an acceptable error of measurement to Kepler rather than an underlying problem with the cosmological model. If this had happened the truth about the cosmos would perhaps have to wait for a few more centuries before it was finally revealed. Below we will review Kepler's astronomical research over the five-year period in which he conducted his astronomical research.

The first thing that Kepler wanted to accomplish was to verify whether the center of the deferent in the Ptolemaic model was located half way between the Sun and the equant, as claimed by Ptolemy. He selected four different dates,[19] on which the celestial longitudes of Mars emerged from the horizon at sunset were recorded. Then a straight line connecting one of these Mars positions (recorded in 1587) and the Sun was used as a baseline to examine the many triangular relationships formed from the four longitudinal positions with the Sun. Kepler's method involved constantly changing the position of the equant to check whether the motion of Mars under the eccentric model was consistent with the observational data. After shifting the equant and recalculating the data, the conclusion was that the center of Mars' deferent, the center of the Sun and the equant of Mars' orbit indeed formed a straight line. However, the center of the deferent did not fall half way between the Sun and the equant. These findings refuted Ptolemy's equidistant theory[20] and even rejected Copernicus' claim that the equant did not exist. They also provided guidance to Kepler when he made three important conclusions about the planetary model: (1) a planet revolved around the equant at uniform angular velocity; (2) the existence of the equant provided a higher likelihood for the Sun to be at the center of the planetary orbit; and (3) the force acting on a planet when it is farther away from the Sun is less than the force acting on the planet when it is closer to the Sun. Granted these three conclusions lacked rigorous proofs at the time, yet they had a profound influence on the development of the laws of planetary motion later. For these reasons the conclusions were referred to by later generations of astronomers as the vicarious hypothesis, or Kepler's zeroth law of planetary motion.[21]

3.3.3 Confirming the Law of Areas Using Earth Data

Kepler was only able to refute the Ptolemaic model with Mars data, which would then lead to the vicarious hypothesis, mentioned previously, as a conclusion, yet he was unable to articulate an exact model for planetary orbit. Kepler therefore asked himself: Is it possible to remedy this shortcoming with just Earth's data? More precisely, if the orbit of Mars contains an equant, does

---

[19] According to Hsiang et al. [15], the four dates chosen by Kepler were March 16, 1587; June 8, 1591; August 25, 1593; and October 31, 1595.

[20] The Ptolemaic model advocated that the center of a planetary orbit (then Earth but now the Sun), the center of the deferent and the equant form a straight line, and the center of the deferent falls half way between the orbital center and the equant. For this reason it was termed the equidistance theory.

[21] Refer to Hsiang et al. [16] for additional details.

Earth's also have one as well? To verify this hypothesis, Kepler pioneered a geometric analysis technique that was quite unprecedented. He used the four different dates above to estimate the four relative positions of Earth on its orbit. As the orbital period is 687 Earth days for Mars and 365 days for Earth, when Mars returns to the original position after having revolved around the Sun once, Earth's position will differ by $[(687 - 365)/365] \times 360 = 318°$ compared with that from a year before. At this time the Sun, Mars and the two positions of Earth together form the four points of a geometric figure. It is therefore possible to employ trigonometric functions to compute Earth's position in orbit (assumed to be a perfect circle). Since a single date corresponds to one position of Earth, we now have four different positions of Earth in orbit. From these four points, the center of Earth's orbit can be obtained, and likely angular velocities at these four positions can be used to derive the position of the equant. Under this logical framework, Kepler proved that the center of Earth's orbit is different from its equant after making extensive calculations and repeated comparisons of the data. This confirmed that, much like Mars, Earth indeed possesses an equant. Similarly, it indirectly confirms that Earth revolves around its equant at a uniform angular velocity.

In the process of demonstrating that both the orbits of Mars and Earth possess equants, a working assumption was that the orbits are perfect circles. As a result, many discrepancies existed between the results thus obtained and Tycho's data. Realizing that his model was inconsistent with the facts, Kepler was determined to seek the best, most appropriate orbital curve for planetary trajectory. The first requirement was to abandon the perfect circle. There were two reasons: first, a circular orbit led to results that were deviated from Tycho's observational data by eight arc minutes; second, as three (non-collinear) points can determine a circle, yet the circles formed by randomly choosing sets of three points from Tycho's Mars data were all different, and therefore the orbit of Mars could not be a perfect circle. Next, Kepler made a new assumption that the orbit was a top-heavy oval-shaped curve, but he quickly noticed that the data presented even more discrepancies. Out of options, Kepler revisited Tycho's collection of data in earnest and charted a total of 19 sets of Mars orbital data containing the positions of Mars relative to the Sun. These 19 trajectories, he realized, were positioned half way between the perfect circles and ovals. It finally dawned on him that the ellipse would be the perfect candidate for the planetary orbit.

Having arrived at these conclusions, the only thing left for Kepler to do was to recalculate everything from scratch under the assumption of an elliptic orbit and to compare the results with the observational data to see how closely they matched. Using the geometric relations defined by an ellipse and all the planetary data available to him, Kepler calculated the parameters associated with the elliptical orbit for each planet, such as the lengths of the major and minor axes and the eccentricity of the ellipse. Having confirmed that each of the planetary orbits did indeed match the characteristics of an ellipse, Kepler proceeded to proclaim his first law of the planetary motion (the "elliptic" law): the orbit of a planet is an ellipse with the Sun located at one of its two foci. The process of proving the first law was an agonizing and challenging endeavor. Not only did Kepler have to sift through mountains of data but he was also responsible for analyzing countlessly many geometric figures. The end results, however, sparked an evolution in modern scientific research. Kepler is now regarded as the first mathematical physicist in history and the first person to have offered explanations of physical phenomena using rigorous mathematics.

3.3.4 First Two Laws of Planetary Motion: Modern Derivations

During the period of more than five years in which Kepler was enjoying his unrestricted access to Tycho's astronomical data, the astronomer continued to produce meticulously written reports detailing his research findings and computations. He wanted to ensure that his work would be preserved for posterity after he was gone or at least when he was no longer active doing research due to old age. Kepler eventually published his work in the volume *Astronomia nova* ("New Astronomy"), which contains details of the derivations of the elliptic and area laws (first and second laws).[22] Although the book had been essentially completed in the winter of 1605, Kepler waited four years before it was finally published in 1609. Even then, the book was filled with complicated geometric analyses to explain how the author had applied Tycho's data and devised a planetary model to derive the results via a series of meticulous, detailed calculations. This made the book lengthy and difficult to understand. For this reason, we will use simple geometry, trigonometric functions and algebra of modern mathematics to replicate the geometric analysis exclusively conceived by Kepler. The shifts in the spatial triangle formed by Mars, Earth and the Sun are used to recreate Kepler's elliptic and area laws for Mars.[23]

First of all, we will use information already known at Kepler's time about the Sun, Earth and Mars to compute the relative positions of Earth and

---

[22] First law of planetary motion (elliptic law): Planets move in elliptical orbits with the Sun at one of the two foci; second law of planetary motion (area law): Planets sweep out equal areas in equal times.

[23] Detailed derivation of Kepler's laws can be seen in Chap. 7 of Morin [24].

Mars at any given time, from which the laws of planetary motion can be derived, see also Yao and Huang [33]. Based on available historical records, background knowledge possessed by Kepler may have included the following:

(1) Both Earth and Mars orbit the Sun, and their orbital planes are coplanar with the ecliptic.
(2) Earth's angular velocities around the Sun are different at different locations, something noticeable from the varying lengths of the four seasons.
(3) The orbit of Earth around the Sun is not a perfect circle; its angular velocity changes with its position in orbit.
(4) From the above it can be inferred that the orbit of Mars around the Sun is also something other than a perfect circle, and its angular velocity changes with its position in orbit as well.

In addition, as all astronomical observations were carried out from the perspective of Earth at the time, it was possible to determine the angles of motion of the Sun and Mars relative to Earth only by making comparisons with the distant stars in the background. These angles can then be used to obtain the ratios of distances between celestial objects. Under these constraints, the possible methods of measurement employed by Kepler at the time are the following:

(1) Earth's daily angular velocity when orbiting the Sun: At sunset (or sunrise), the position of the Sun relative to other distant fixed stars is measured to arrive at the Sun's positional changes relative to Earth, from which the Earth's angular velocity around the Sun can be determined. This is known as observation at dusk.
(2) The angle $E_1 S E_2$ (i.e., θ) in Fig. 3.4: By observing the position of the Sun relative to distant fixed stars on two different dates, one can calculate the Sun's angular movement relative to Earth between these two dates, which is also equivalent to Earth's angular movement around the Sun.
(3) The angle SE_iM (i.e., μ_i), $i = 1, 2$, in Fig. 3.4: Observe the relationship between the position of Mars in the night sky and its time. When SEM is collinear, Mars will appear at the highest point reached in the sky (the zenith), as viewed from Earth, at midnight (or 12 noon, but Mars will be out of view). On the other hand, when SEM forms a right angle, Mars will be at the zenith during sunrise or sunset.
(4) The ratio of the orbital (or sidereal) periods of Mars to Earth: Mars' orbital period expressed as the number Earth years, which can be calculated using Eq. (3.3) in this chapter.

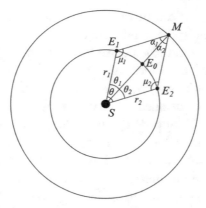

Fig. 3.4 The triangular relationship between the Sun (S), Earth (E) and Mars (M) used to derive Kepler's first (elliptic) law of planetary motion. Inspired by Fig. 4.5 of Hsiang and Chang [14]

In addition, it is also necessary to assume that the orbits of Mars and Earth around the Sun are closed. In reality, however, gravitational pull from other planets in the Solar System cause a planet's orbit to undergo slow precession (refer to Chap. 6 for the calculation of the precession of Mercury's perihelion based on the general theory of relativity). This precession rate is nevertheless so small that we will ignore it for the time being. The fact that Mars' orbital precession is so tiny was also why Kepler was so lucky. Based on the above information and assumptions, we can now carry out the following derivation. The very first task is to consider a planet's orbit, and Kepler's first order of business was to explore Earth's trajectory when it revolves around the Sun.

A. Earth's Orbital Trajectory around the Sun: The Elliptical Law

In this derivation the geometric shapes of Mars and Earth's orbits are assumed to be unknown, and there exist a number of factors that affect these shapes. It is therefore necessary to first fix a few variables before other variables can be explored. In other words, we will divide all the affecting factors into two categories: manipulating variables and control variables. Since we will first consider Earth's orbit, Mars' position will be fixed for the time being. A summary of the assumptions we have made are: Mars' orbit is closed, and if we make one observation per Martian year, Mars will appear to be at the same location. Consequently, we can regard Mars as being fixed. For ease of computation, therefore, we choose the first observation to be at the time when SEM is collinear, i.e. Earth is positioned exactly between the Sun and Mars, as shown in Fig. 3.4 when Earth is located at E_0. Then for each Martian year, measurement method (4) can be used to obtain the fact that 687 days have elapsed on Earth, while at the same time the angle SEM can also be obtained using measurement method (3). Then measurement method (2) can be used to obtain the angle ES E_0. Once these two angles are available, we will be able to derive the relationship between Earth's angle of revolution and the Sun-Earth distance. The detailed calculations follow.

Referring to Fig. 3.4, suppose Earth at the positions E_1 and E_2 are separated from its position at E_0 by an integer number of Martian years, then Mars will be at the same location M when Earth is at any of these three positions. Therefore we can measure all of the following: μ_1, μ_2, θ_1 and θ_2. This will in turn enable us to derive the angle $\alpha_1 = 180° - \mu_1 - \theta_1$ and $\alpha_2 = 180° - \mu_2 - \theta_2$. Setting d as the distance SM and r_1 as the distance $E_1 S$, and r_2 as the distance $E_2 S$, then by applying the laws of sine we have:

$$\frac{r_1}{\sin \alpha_1} = \frac{d}{\sin \mu_1} \tag{3.7}$$

and

$$\frac{r_2}{\sin \alpha_2} = \frac{d}{\sin \mu_2} \tag{3.8}$$

Dividing (3.7) by (3.8) we have

$$\frac{r_1}{r_2} = \frac{\sin \alpha_1 \sin \mu_2}{\sin \mu_1 \sin \alpha_2} \tag{3.9}$$

Note that the angles in (3.9) are all known, so we can calculate the ratios of different Sun-Earth distances whenever each Martian year has elapsed. Moreover, because the orbital periods of Mars and Earth do not form an exact integer ratio, different pairs of θ and its corresponding r can be obtained after a large number of observations have been made, and Earth's orbital trajectory around the Sun can be obtained. From this information, Kepler was able to deduce that Earth's orbit was indeed an ellipse. His calculations, however, required the use of a great deal of Tycho's data for each Earth position calculated, and for this reason it took Kepler nearly five years to complete the recording of Earth's orbital trajectory. Orbital charts were then created, and after several attempts at speculating about the shape of this orbit, Kepler finally concluded that the ellipse was the correct shape.

B. Mars' Orbital Trajectory around the Sun: The Elliptical Law Reaffirmed

Since Earth's orbit around the Sun is elliptical, it stands to reason that Mars' orbit also has the same shape. To prove or disprove this conjecture, we must figure out the relationship between the Sun-Mars distance d and Mars' angular motion around the Sun. The logic of this derivation will still be based on the triangular relationship shown in Fig. 3.4, but unlike the previous case, since Earth's orbit has been determined to be elliptic, we can use this fact to deduce the Sun-Mars distance d at different locations. Based on this characteristics, we can now propose a simple derivation method as follows.[24]

Suppose, as depicted in Fig. 3.4, the two alternate positions of Earth, E_1 and E_2, are exactly an integer number of Martian years apart in time. Mars (M) will then be located at the same position in both instances. Also suppose that the Sun-Mars distance is d during these instances. Note that the derivation that follows does not require E_0, so it is not necessary for E_0 to be an integer number of Martian years apart from E_1 or E_2. As in the previous case, the three angles θ, μ_1 and μ_2 generated by $E_1 S E_2$ can be measured via observation, and since Earth's orbit is now known, both r_1 and r_2 are also known. Furthermore, as the interior angles of any quadrilateral add up to $360°$, β can be deduced from the relation $\beta = \alpha_1 + \alpha_2 = 360° - \theta - \mu_1 - \mu_2$. First we rearrange (3.9) as

$$k = \frac{\sin \alpha_1}{\sin \alpha_2} = \frac{r_1 \sin \mu_1}{r_2 \sin \mu_2} \tag{3.10}$$

Therefore k is also a known quantity. Note that we have $\beta = \alpha_1 + \alpha_2$ and also $k = \sin \alpha_1 / \sin \alpha_2$, so α_1 and α_2 can be calculated. Now we employ the angle difference identity for sine in trigonometry

$$\sin \alpha_1 = \sin \beta \cos \alpha_2 - \cos \beta \sin \alpha_2 \tag{3.11}$$

and divide both sides of (3.11) by $\sin \alpha_2$ and substitute $k = \sin \alpha_1 / \sin \alpha_2$ into the expression to get

$$\alpha_2 = \cot^{-1} \left(\frac{k + \cos \beta}{\sin \beta} \right) \tag{3.12}$$

Once α_2 is available we can obtain d by substituting it back into (3.8):

$$d = \frac{\sin \mu_2}{\sin \alpha_2} r_2 \tag{3.13}$$

---

[24]Referenced from Jacobsen [18].

Using the above approach, we can deduce the Sun-Mars distance d with this specific position of Mars on two different dates one Martian year (687 days) apart. For example, using the data obtained on Day 1 and Day 688, one Sun-Mars distance d can be deduced. Likewise, using data obtained on Day 2 and Day 689, another value for d after Mars has revolved a small angle around the Sun can be obtained. Repeat this process until a complete Marts orbital trajectory is derived.

C. Angular Velocity and Orbital Data: Deriving the Area Law

At the same time Kepler was laboring to record Earth's orbital coordinates on a chart, he was also calculating the tangential velocity v of Earth at each point. Kepler discovered that Earth was moving faster when it was close to the Sun and more slowly when it was far away. Based on these findings, Kepler speculated that the product of the (varying) orbital radius (i.e., distance from the Sun) and the corresponding tangential velocity, namely rv, could be equal at Earth's aphelion (farthest point from the Sun) and perihelion (closest point from the Sun). After comparing multiple data points, this conjecture was confirmed. Kepler then proposed an important conclusion: when Earth has revolved around the Sun in its orbit by a small angle, the distance it has traveled $\Delta s = v\Delta t$ will project a length of $r\Delta\theta$ in the direction perpendicular to the orbital radius. If this projected length is treated as the base of an isosceles triangle and the other two sides are taken to be Earth's two (nearly equal) distances from the Sun, then the area of this approximate isosceles triangle is $\Delta A = r \times r\Delta\theta/2 = r^2\Delta\theta/2$. Kepler discovered that the area of the "triangle" swept out by the line joining Earth and the Sun in unit time stays constant at any point on Earth's orbit. This can be represented as follows:

$$\frac{\Delta A}{\Delta t} = \frac{1}{2}r^2\frac{\Delta\theta}{\Delta t} = \frac{1}{2}r^2\omega = C \qquad (3.14)$$

Here $\omega = \Delta\theta/\Delta t$ is the angular velocity and C is a constant. This "area law" is commonly known as Kepler's second law of planetary motion. The discovery of this law brought great inspiration and encouragement to Kepler, as the angular velocity of a planet in orbit can be measured directly, and the distance between the Sun and the planet at any point on its orbit has the following relationship:

$$\frac{r_1^2}{r_2^2} = \frac{\omega_2}{\omega_1} \qquad (3.15)$$

Hsiang et al. [15][25] took data collected between May 1948 and January 1984, which spans 30 Martian years, to check if formula (3.15) is valid. The results showed that the two ratios in (3.15) were indeed very close. Hsiang et al. also carried out computations on the elliptical characteristics of the Martian orbit and found that its eccentricity was about 0.093, which confirmed that elliptic law was indeed valid. The researchers also used data collected for Mercury, Venus and Jupiter to confirm that Kepler's elliptic and area laws (first and second laws of planetary motion) are applicable to all planets. In this section, our derivations of Kepler's first and second laws have been inspired by the work of Hsiang et al.

With the first two laws of planetary motion at his disposal, Kepler was now able to predict planetary positions. In the case of Mars, for example, the planet orbits the Sun once every 687 Earth days, and if its elliptical orbit has a semi-major axis a and semi-minor axis b, then the orbital area will be πab and the line connecting the Sun and Mars sweeps out an area of $\pi ab/687$ per day as Mars orbit the Sun. If we divide twice this area by the square of orbital radius on the given day, the result will be equal to Mars' angular velocity relative to the Sun on that day. Therefore, if we know the location of Mars on a certain day, we can predict the general location of the planet several days later. In 1629, Kepler used the method described above to predict the transits of Mercury[26] across the Sun on November 7 and December 6, 1631, both of which were later confirmed by the French astronomer P. Gassendi (1592–1655).

3.3.5 Discovering the Third Law

Two years after the publication of *Astronomia nova*, in 1611, Kepler was still basking in his accomplishments. Little did he know then that disaster was looming for him and his family. It was during this year that Kepler's long-time financial sponsor, Emperor Rudolf II, was forced to abdicate and later succeeded by his younger brother Matthias. Kepler himself was dismissed as the imperial mathematician. In that same year, Kepler's three children all fell seriously ill with smallpox, and his beloved eldest son succumbed and passed away. His wife later died of typhoid fever, the result of the turmoil of war.

---

[25] Refer to Chap. 5 of Hsiang et al. [15].

[26] A transit of Mercury across the Sun occurs when the Sun, Mercury and Earth form a straight line, at which time the black shadow of Mercury will appear across the surface of the Sun. On average, there are about 13 transits of Mercury each year, which are visible only during daytime. To view the phenomenon properly, it is necessary to project the image from the telescope onto a piece of white paper. The black dot of Mercury will traverse across the projected solar image.

The following year Kepler left Prague with his two surviving younger sons to escape an even worse disaster. Having been uprooted and drifting from place to place, Kepler finally settled in Linz, Austria. In late 1613, Kepler married the 24-year-old Susanna Reuttinger, who was 18 years his junior. She came from a poor family, but she was able to give Kepler a happy life. Apart from raising her two stepchildren with dedication, Susanna also gave him seven children during the marriage, although five of them did not survive to adulthood.

Despite the ongoing war and a life of destitute, Kepler continued to work tirelessly in the following years on the orbital data left behind by Tycho on Earth, Mercury, Venus, Mars, Jupiter and Saturn, the six planets known to astronomers at the time. In Kepler's cosmological view, he firmly believed the cosmos to be a harmonious entity and each object that existed in the world to be associated with a harmonious quantity. Based on this firm belief of his, Kepler, who turned 47 in 1618, discovered yet another law, more important than previous two, from his data: the harmonic law. This law expresses at the outset a profound relationship between a planet's orbital radius and its orbital period: *the square of a planet's orbital period is directly proportional to the cube of the semi-major axis of its orbit.* Differing substantially from the previous two laws, which are qualitative in nature, the third law describes the relationship between two physical quantities mathematically. This relationship enables us to understand that an extremely rigorous mechanical system governing the Sun and its orbiting planets operates steadily and harmoniously in the Universe.[27]

How was Kepler able to deduce this perfect relationship? One theory postulates that Kepler still relied on the Platonic solids that he had employed earlier in his work and also made use of the musical scale concepts advocated by the Pythagoreans to construct his harmonic law for a given planet. He then made repeated calculations using Tycho's Martian data to try to figure out how the semi-major axis was related to the orbital period. A more believable theory, however, suggests that Kepler calculated the orbital radius (semi-major axis a) and the orbital period (T) of each planet based on the elliptic law and observational data.[28] From Eq. (3.9) Kepler first deduced the orbital periods of the six known planets via observation and then employed the method outlined in Sect. 3.19 B to obtain their orbital trajectories. Once the orbital

---

[27] Referenced from Kozhamthadam [19].

[28] From the subtitle given by Kepler to his *Epitome*, which was "A description of the movement of Mars based on an astrophysical framework", it stands to reason that it was Mars data that Kepler used in his discovery of the third law.

Table 3.1 Orbital radius (semi-major axis) and orbital period of each of the six major planets in the Solar System known in Kepler's time. The figures shown are consistent with the predictions of the third law. The orbital period of Earth is defined as 1 unit (Earth year) and its orbital radius as 1 AU (astronomical unit). The corresponding measures of the other five planets, along with Earth's, are shown in the following table (Refer to Chap. 7 of Polya [28].)

Planet	a (AU)	T (Earth year)	T^2/r^3
Mercury	0.38710	0.24085	1.0001
Venus	0.72333	0.61521	0.9999
Earth	1.00000	1.00000	1.0000
Mars	1.52369	1.88809	1.0079
Jupiter	5.20280	11.8622	1.0010
Saturn	9.54000	29.4577	1.0010

data for all the planets was collected, Kepler was able to deduce the relationship between semi-major axis a of each planet and its orbital period T. Next, Kepler once again utilized his superb computational skills and patience to tabulate the output (similar to the figures contained in Table 3.1), from which he checked the data against the (incorrect) relationships $T \propto a$ and $T \propto a^2$ to see if any of them held. The results of Kepler's computations suggested that the true relationship between T and a should lie somewhere between these two, so after several additional attempts with other formulas, Kepler finally settled on $T^2 \propto a^3$, the now-celebrated third law of planetary motion or the harmonic law.

To Kepler, the discovery of the third law reaffirmed the harmonious nature of the many phenomena observable in the Universe, which he had compared to the beauty of geometry and music. Kepler collected these findings in his *Harmonices Mundi* ("The Harmony of the World"), which was published in 1619. This was also the year of the eruption of the Thirty Years' War, one of the most devastating wars in European history. Compared with the simplicity and harmony of the Universe, the chaos and savagery displayed by the parties involved in the conflicts was a sort of spiritual persecution and physical torture to all truth-seeking scientists. During this period, Kepler was able to earn only a meager income and he and his family never stayed in one place for very long. Kepler later attempted to secure a teaching position at the University of Tübingen, his alma mater, but his request was declined due to the University's concerns over the perceived religious views of Kepler. At Linz, he was also accused of being unfaithful to the church and later expelled by local Lutheran authorities.

Facing such harsh treatment and adverse circumstances, Kepler's work on astronomy, however, remained uninterrupted. Between 1617 and 1621, Kepler published his final magnum opus, *Epitome astronomiae Copernicanæ*

("Epitome of Copernican Astronomy", or simply the *Epitome*), a three-volume work that extended the applicability of his three laws of planetary motion on Mars to the remaining five known planets of the Solar System. In the book even the Moon and the four satellites of Jupiter were analyzed, making *Epitome* a kind of encyclopedia in astronomy. One salient feature of the book was that there was no mention of the epicycles, deferents or the equant of the Ptolemaic system. His reasoning was straightforward: the simple and harmonious cosmos simply could not be comprised of planets moving on complicated, petal-like orbits, such as those depicted in Fig. 2.8. The physical implications just could not be explained away, as it was inconceivable that nature had a powerful mechanism to maintain such overly complex mechanism of planetary motion.

3.3.6   Giants of the Scientific World

It is not difficult to see that Tycho's astronomical data was vitally important to Kepler's development of the three laws of planetary motion. To summarize, Kepler first determined the elliptical orbit of Mars' orbit. He then employed the calculations in his book *Astronomia nova* to propose the area law, a quantitative law that was unprecedented in the astronomical community: the line connecting a planet orbiting the Sun in an elliptical orbit sweeps out equal areas during equal intervals of time. From another perspective, this law can also be described as: the planet travels at a higher tangential speed when it is close to Sun than it does when it is far away.[29] This law completely refuted the theory that had been in existence from Aristotle in ancient Greece to the time of Copernicus: planets move at constant speeds in circular orbits. Finally, Kepler argued that since the motion of a planet is related to its distance from the Sun, it stands to reason that the Sun must be exerting a certain force on the planet. The force turns out to be the universal gravitation proposed by Newton a century later. This so-called "non-contact force", in modern usage, had become a widespread concept in the scientific community during Kepler's time due to the discovery of magnetism[30] earlier. Although these concepts

---

[29]If the equant of the Ptolemaic model is placed on one of the two foci of the elliptical orbit, then Kepler's model will be nearly equivalent to the epicycle-deferent model when describing planetary motion.

[30]In 1600 William Gilbert (1544–1603) published the book *De Magnete* ("On the Magnet"), which provided a systematic description of Earth's magnetic field. Gilbert's book was once described by Galilei as a work so great that it was the envy of the astronomical community.

are somewhat similar, interpreting the gravitational pull of the Sun as magnetic force was a mistake.[31] The most influential law, however, was Kepler's third law of planetary motion, which Newton later employed to derive his law of universal gravitation. The law states that the ratio that exists between the orbital periods of any two given planets is equal to the ratio of the 2/3th power of the semi-major axes.

Kepler's three laws pushed the study of planetary orbital motion to an unprecedented level of perfection. Under his first two laws, the complicated and nearly incomprehensible astronomical frameworks of the previous generations, consisting of epicycles and deferents, were replaced with Kepler's system of "planets traveling in elliptical orbits at non-constant speeds", and the position of any planet can be accurately predicted using these laws. Kepler's third law governs a planet's orbit, orbital speed and orbital period with a precise quantitative relationship. This pioneering work represented not only a breakthrough in astronomy but also the entire physics community (which at the time consisted of only a handful of disciplines such as astronomy and optics). Moreover, even though the "force-distance" relation proposed by Kepler was not consistent with the known facts, it nevertheless paved the way for future scientists, most notably Newton and also others, to model physical phenomena using mathematical tools.

Kepler laid the foundation for future research by combining experimental data with theoretical models, thus creating a challenging but hopeful future for further scientific research. Finally, he delivered a very clear physical image of the cosmological system: the motion of celestial objects is so simple and harmonious. It was easy for the world to understand and accept. In the following centuries, this image would continue to reverberate in the physics community, first with Newton's law of universal gravitation and later Einstein's theory of relativity. Both modern theories benefited greatly from the inspiration and conceptual framework of Kepler's three laws of planetary motion. This reflects on Newton's comment that if he had seen further in the field of celestial mechanics it was by standing on the shoulders of Giants in the scientific world, and one of these Giants to whom Newton referred was Johannes Kepler.

Despite Kepler's academic and research career being full of mysticism, religious fanaticism and also philosophical reflection, his great discovery in planetary motion not only augured the death knell of the geocentric system that

---

[31] English physician William Gilbert (1540–1603) proposed a geomagnetic theory to explain why planets orbit the Sun in elliptical orbits. He made the assumption that both the Sun and Earth have magnetic fields in their north and south poles. When Earth is orbiting the Sun, both gravitational and repulsive forces are generated to propel Earth's movement around the Sun. Earth's orbit becomes elliptical because of the differing magnitudes of the two opposite forces.

had dominated for nearly two millennia but also provided the basis on which Newton would develop his law of universal gravitation. Compared with the law of universal gravitation, however, Kepler's laws represented the conclusion drawn inductively from a large compilation of observational data. Scientific investigation via rigorous logical reasoning would have to wait until Newton's time, when his law of universal gravitation was derived and confirmed. Kepler's laws describe the movement of planets around the Sun, while the law of universal gravitation can be used to compute the force between celestial bodies as well as to describe the relationship between force and motion. The law of universal gravitation can also be used to calculate the orbits of planets and comets, which may be parabolic or hyperbolic under different gravitational conditions, something not derivable from Kepler's laws.

After having completed his *Epitome*, Kepler continued to work on honoring Tycho's other dying wish: the compilation and publication of the *Rudolphine Tables*. This star catalog (with planetary tables) was finally published in 1627. *Rudolphine Tables* contained substantially refined data based on earlier tables and by applying Kepler's first and second laws, the accuracy of which was more than a hundredfold higher than the data compiled with the Copernican model. It was also more comprehensive than any previously published star catalog. As a result, it was widely adopted by the astronomical community as a standard catalog for over a century following its publication. Although the *Rudolphine Tables* was eventually completed, the publication process was fraught with peril. First, it was the issue of publishing costs.

After bargaining with the new Holy Roman Emperor and King of Bohemia, Ferdinand II, Kepler paid for two-thirds of the costs out of his own pocket before finally receiving royal consent for the book's publication. While the book was being printed, a peasant rebellion broke out in 1626, and all the books that had been printed and stored at the printing house in Linz were completely destroyed by fires. Fortunately, the manuscript miraculously survived. Near the end of that year, Kepler relocated his family to Ulm (incidentally the birthplace of Albert Einstein) in the south of Germany. With the cooperation of the local printer, the 586-page first edition of *Rudolphine Tables* was finally published in September 1627. Kepler spent the last three years of his life fighting religious persecution, escaping the ravages of war, and trying to secure his livelihood. On November 15, 1630, having lived a life of sorrow but also glory, Kepler died at the age of 60 from high fever while at a friend's house in Regensburg, a city situated along the Danube River.

3.4   Galilei: Founder of Experimental Physics

Galileo Galilei (1564–1642) has been described as one of the greatest scientists in history, along with the likes of Isaac Newton, Charles Darwin, and Albert Einstein. The refracting telescope that he invented fundamentally changed how observational astronomy was done, and his research in kinematics also set a new model for modern experimental physics. Galilei was the first scientist to have publicly and boldly advocated that the heliocentric model was superior to the Ptolemaic system. In his later years, this action, however, would lead to his persecution and the great humiliation of having to recant his life's work on physics and astronomy, as well as his being sentenced to life imprisonment. For this reason, although Galilei's scientific achievements were his greatest contribution to humankind, he is perhaps best known today for his conflict with the Roman Catholic Church and his subsequent sentencing by the Inquisition.

Galileo Galilei was born in the Renaissance era. This was an age of flourishing ideas, and thinkers and philosophers from the many city-states of the Italian peninsula competed fiercely with one another. Social and economic developments were in full swing. During this period when all segments of the population were endeavoring to succeed, however, two unchallenged, authoritarian powers remained deeply entrenched in the society. One of them was the Roman Catholic Church, which continued to dominate in religious life and social institutions. The other was Aristotelianism, a system of philosophy that monopolized mainstream science at the time. It was during this period, near the conclusion of the Renaissance, that Galilei came into the world. Galilei was born in the city of Pisa to a noble family (incidentally, English playwright William Shakespeare was born two months after Galilei was born). He was the eldest of the seven children in his family. His father Vincenzo was a nobleman without land, and the family was not at all wealthy. Vincenzo worked as a lutenist and was a creative composer, although many of his compositions did not conform to the musical norms of his time.

Galilei inherited his father's nonconformist views and developed keen insights into things around him. At the same time, he was confident and possessed the qualities that would make him an explorer of the mysteries of the Universe. Galilei was also regarded as the father of experimental physics who was simultaneously a mathematician, physicist, inventor, entrepreneur, and also an accomplished debater. He was enthusiastic, witty, and was highly respected in academic circles, churches and courts, and he had a large network of friend in the arts, sciences, and in the religious community. Despite his connections and his status as a respected member of society, Galilei went through two religious trials, once in 1616 and a second time in 1633, on matters related to science. In the end, this great elocutionist was forced to yield to the Catholic Church and Aristotelianism and to turn his back on the results of his lifelong research efforts, in exchange for house arrest for the remainder of his life, having to be confined to his former residence in the countryside to avoid actual imprisonment.

Due to his father's influence, Galilei began to develop a fascination with music and mathematics at a very young age, and he was particularly fond of creating drawings and poetry. What is even more unique was his excellent skills at designing and constructing mechanical apparatuses. With his father's blessings, Galilei enjoyed learning with great freedom until his was 12 years old, when his family moved to Florence. Then everything changed. Galilei's new school, a Benedictine Abbey, discouraged free thinking, and he received a rigorous education and religious training. There Galilei advocated against the teachings of Aristotelianism. He later studied medicine at the University of Pisa, but he was so critical of the human anatomy introduced by first-century Roman physician Galen of Pergamon Glenn that he was dismissed from the University without having finished his degree. Galilei then began working as a mathematics tutor. At the age of 23, he attempted to apply for a mathematics lectureship at the University of Pisa, but the position was offered to a 32-year-old university graduate.

During the time when he was unemployed, Galilei concentrated on the research of kinematics, a subject he was very fond of since childhood. He proposed the concept of an object's center of gravity, treating an object in motion as a point mass centered on the object's center of gravity. While observing the swinging motion of a cathedral's chandelier, Galilei suggested that the pendulum's cycle was not related to its amplitude. He also studied Archimedes' theory, from which he constructed the hydraulic balance (or hydraulic scale). With his creative research efforts, he was appointed to a mathematics lectureship at the University of Pisa in 1589, teaching geometry and astrology. He also often took the opportunity in his academic career to earn extra income, for instance practicing astrology and casting horoscopes for medical students so that they would know the appropriate times for carrying out bloodletting (as therapy) on patients. For such services he charged 12 silver coins, which represented one-fifth of the annual salary he was earning as a professor. Galilei was also a gambler, and he often bluffed his way to winning bets by employing mathematical tricks against his students.

3.4.1   Research at University: First Attempt at Kinematics

After becoming a mathematics lecturer at the University of Pisa, Galilei realized that other professors were still teaching the Aristotelian school of natural philosophy from two thousand years before. From a modern perspective, some of these millennia-old philosophical ideas that dominated the world are now rather absurd. For example, Aristotelian scholars explained the reason for the slowing motion of an object as fatigue rather than the result of friction, and objects fall not because of gravity but because they "desire" to be one with the Earth. These are all instances of teleological statements advocated by Aristotle. Galilei felt uneasy or even pain whenever confronted with these theories or philosophical interpretations. He believed that an object falls not because of its "desire" but rather because it is guided by natural factors. He therefore put forward new perspectives on the motion of objects, although his main ideas remained constrained by the Aristotelian doctrines. For example, in his published work *De Motu* ("On Motion"), Galilei argued that if Earth indeed rotated on its axis, an object released from a high tower would land at a horizontal distance hundreds of yards from the tower. This phenomenon was, in fact, not observed in reality, so he concluded that Earth did not rotate. During his career as a professor at Pisa, although Galilei made no substantial findings on physical principles, in De Motu he did elaborate on his scientific view on the interplay between experimentation and theory, and he emphasized that scientific research is the attempt to identify simple and harmonious relationships that have already existed in complex natural phenomena. This pioneering concept was further developed and enhanced by Newton and Einstein later: Truth is characterized by two qualities, simplicity and harmony.

Even though Galilei was enthusiastic about observing natural phenomena and made attempts to understand and uncover the truths hidden behind them, these truths were often used by him as a tool to argue with members of other schools of thought in the scientific community. His passion for seeking the truths and debating them earned him more than a few enemies. For example, from his observation of the motion of free- falling objects,[32]

---

[32]Legend has it that Galilei performed the famous free fall experiment at the Leaning Tower of Pisa. However, Galilei never mentioned this alleged event either in his classroom or in his written research reports. In fact, no historical records ever existed that described this particular experiment. On the other hand, there are records which show that the Italian Simon Stevin (1548–1620) was the first person to have conducted a free fall experiment. In his 1586 notes on experiments, Stevin wrote that two objects of different weights were simultaneously released from the top of a tower. Only one instance of an object hitting the ground could be heard.

the speed with which an object falls does not depends on its weight. His claim was a stark contrast with the Aristotelian theory, which states that the greater the weight of the object, the faster it falls. As another example, Galilei talked about how Kepler had observed changes in distant stars and determined that celestial bodies beyond the Moon did indeed undergo changes. His observation thus refuted the Aristotelian model that the Universe beyond the Moon was eternal and unchanging. Galilei tended to provoke the Aristotelian scholars in their gatherings with these arguments, which caused great annoyance among them. Eventually their sentiment prompted them to plot against Galilei.

In 1592, his three-year contract with the University of Pisa expired, and Galilei was fully aware that his teaching contract would have no hope of being renewed given the animosity between him and the Aristotelians. His next move was to seek teaching positions elsewhere. Through the recommendation of a friend, Galilei was unexpectedly offered a mathematics professorship at the University of Padova (Padua) in Venice. Although Galilei's transfer to Padova was out of necessity and not of his own choosing, he found himself thriving and the relocation was in fact a blessing in disguise. Padova offered Galilei a great deal of academic freedom, allowing him to freely observe natural phenomena without any restrictions. Galilei was also endowed with a pair of dexterous hands that let him fully demonstrate his skills with machinery and also his strengths in mechanics. He would soon engage in the study of the free-fall motion of objects, something he had contemplated doing but did not have the opportunity to accomplish at the University of Pisa. At Padova his first order of business was to perform the inclined plane experiment, one of the most famous in the history of science.

This classic experiment, far-reaching in the history of physics, was carried out in 1603 when Galilei was teaching at Padova. At the beginning of the experiment, Galilei was concerned that in a free falling setup the object would be falling too rapidly to be observed closely enough. In order to extend the total traveling time of the object's downward motion (or, equivalently, to slow down its speed), Galilei designed an alternative experiment with inclined planes instead. The platform for the motion experiment had a simple configuration: two inclined planes and a level surface, together with a bronze ball. The inclined planes were placed on either side of the level surface. The angle of inclination on the left plane was fixed, whereas that of the one on the right was adjustable. The bronze ball was at first placed at the top of the left inclined plane and was allowed to roll down. The surfaces of the inclined planes were lined with parchment to ensure they were as smooth and polished as possible. A groove about the width of a thumb was carved out along the center of the inclined plane to allow the bronze ball to descend in a straight line. When the angles of inclination of the two inclined planes were the same, the bronze ball would roll down from the left plane, travel along the level surface and then climb up the right inclined plane before reaching the same height as when it was first set in motion. If the angle of inclination of the right plane was reduced but its length increased, the ball would still ascend to the original height, although it would also travel a longer distance on the plane. Finally, Galilei found that if the right plane's angle of inclination was reduced to zero (i.e., made horizontal), then the bronze ball's rolling action on the right plane, after traveling from the left plane to the level plane, showed no sign of stopping. In other words, if the right plane was sufficiently long, the ball would be scrolling at a constant speed indefinitely.

In carrying out the experiment, Galilei carved equidistant shallow markers along the inclined plane for the purpose of measuring the changes in the bronze ball's speed when it descended along the inclined plane. When the ball rolled over each marker, a sound would be produced, and its time of occurrence would be recorded and subsequently used to calculate the ball's changes in speed. At first Galilei timed the ball's movements with his own pulse, but when this proved inadequate he employed the water clock that he had invented. Apart from keeping time, there were also two important experimental parameters that needed to be adjusted: the first was the angle of inclination of the left inclined plane, which determined the speed of the bronze ball, and the second was the distance between markers on the right plane, which were required for time measurement. According to Galilei's only description of the experiment, the initial angle of inclination was set at $2°$ and the inter-marker distance was about 1 mm. However, the small angle caused the ball to travel too slowly and the noise made by the ball's contact with the markers were too frequent, so it is not too difficult to image that the ball's trajectory was being distorted to a great extent, which resulted in serious experimental errors. Galilei therefore expanded the inter-marker distances progressively in a series of experiments using a water clock with an enhanced timing mechanism, so that the movement on the inclined plane became smoother and the accuracy of the time measurement time was greatly improved.

Using an inclined plane in place of a free-fall setup in the experiment was indeed an ingenious idea on the part of Galilei. A bronze ball rolling on an appropriately inclined plane allowed Galilei sufficient time to observe its dynamics and to record the change in speed during the movement. Galilei discovered that the ball would continue to accelerate as it rolled down, and the degree acceleration would increase as the left plane's angle of inclination was increased. If this angle was set to close to $90°$ (i.e., the plane is at a near-vertical position), then the acceleration would presumably be close to that in the free-fall situation. After having conducted a series of experiments with different angles of inclination, Galilei arrived at an important conclusion: free-fall motion has uniform (or constant) acceleration. He also discovered that when the bronze ball in his experiment rolled down from the inclined plane and entered the level surface, it would continue to move indefinitely at a constant speed. The second conclusion, therefore, was that in the absence of external forces, the ball would continue to move at a constant speed in rectilinear motion. These two conclusions were later refined by Newton and transformed into his, respectively, second and first laws of motion.[33] However, Galilei did not investigate further into these two important findings at the time. Nor did he explain why the bronze ball would rise to the same height on the right plane or produce any reports on these observations. One explanation was that he was actively engaged in the enterprise of research, construction and sale of scientific instruments that had military applications.

3.4.2   First Astronomical Telescope in History

In the 18 years that he lived in Venice, Galilei introduced a number of mechanical instruments that were of notable commercial value. For example, based on Archimedes' principle (which the philosopher reportedly employed to determine the purity of gold based on the density of water), he designed a hydrostatic balance to measure the density of any object. He also combined pairs of dividers and rulers to create military compasses, one purpose of which was to calculate the weights of cannonballs required to bombard enemy camps. Based on his research, Galilei taught military defense engineering and surveying classes to select audiences, and these classes were well received by young European aristocrats, which earned him considerable extra income. Galilei's greatest and most influential invention was what would later be referred to as the Galilean telescope, which he invented in 1609, the same year Kepler published his *Astronomia nova*. As a result, 1609 was wide considered to mark the beginning of the modern Scientific Revolution.

The invention of the telescope had its beginnings in our knowledge of glass spheres having the effect of magnifying images long before the instrument came into being. In 1267, the English philosopher and Franciscan friar

---

[33] In his *Philosophiæ Naturalis Principia Mathematica* ("Mathematical Principles of Natural Philosophy", often referred to as simply *Principia*), published in 1686, Newton attributed the discovery of the first law to Galilei. Judging from Newton's long scholarly disputes with Robert Hooke and Gottfried Wilhelm Leibniz, this kind of professional acknowledgement was indeed quite rare.

Roger Bacon (1214–1292) had a series of discussions on this topic in his book *Opus Majus* ("Greater Works"). The actual assembly of spherical lenses into a functional telescope, however, was still a mystery. Some historians credited Hans Lippershey (1570–1619), a Dutch maker of eyeglasses, as the inventor. One version of the account says that one day in 1597, while Lippershey's two children were playing with lenses in his spectacles shop, they commented on how they could make a far-away weather vane on the top of a church building seem closer when looking at it through a combination of two lenses. This prompted Lippershey, according to Galilei's account, to experiment further with various combinations of lenses. He eventually put together a type of monocular (then referred to as a spyglass) with a convex lens and a concave lens, but the magnification factor was only about 2× to 3×. Almost a decade later, in 1608, Lippershey filed for a patent for this image-amplification technology but his application was not approved. Apparently a similar device had already been in existence for centuries.

When Galilei heard about this invention in May 1609, he showed no interest in it at first, until his friend Paolo Sarpi urged him to pursue it after he realized the instrument's commercial potential. Sarpi believed that if the telescope was to be improved by enhancing its magnification capability, the Venetian Navy could use it spot enemy vessels from a distance. Galilei instantly embraced this idea, as it made sense to him that Venice, a city-state built on an archipelago of more than a hundred islands and undefended by walls could use such instruments to defend against foreign invasion via an early warning system. Galilei immediately agreed to take on the project and began to procure supplies and equipment necessary to build telescopes.[34] Several months later, he successfully assembled a refracting telescope. In his design, he placed a *plano-convex* lens (flat surface facing front, convex surface facing back) at the front of the telescope and a *plano-concave* lens (concave surface facing front, flat surface facing back) in the back. The distance between the two lenses was configured to be exactly the difference between the two lenses' focal lengths, and the magnification would then be equal to the ratio of convex lens' focal length (larger) to the concave lens' focal length (smaller). This first telescope constructed by Galilei had an 8× to 9× magnification. Galilei later made the detailed design of this invention public and claimed to have

---

[34]The story of how Galilei built his first telescope in 1609 has been confirmed by a procurement list discovered by historians. This list is believed to have been owned by Galilei when he lived in Venice. Galilei's first task was to find the right glass and lapping (grinding) machines for creating spherical lenses. He later used iron cannonballs as tools for grinding and polishing concave lens of different focal lengths.

independently invented the refracting telescope.[35] Since that time, all telescopes constructed with this combination of lenses have been referred to as Galilean telescopes. See "Appendix 3.2: Principles behind Galilei's Telescope Designs."

On August 23, 1609, then still a poor mathematics professor, Galilei transported the telescope he had constructed to Venice and demonstrated it to the navy there. His telescope allowed an observer to spot the flags of distant ships two hours in advance to determine whether they were an invading fleet. Galilei installed the telescope on the top of the St. Mark's Campanile, located at the Piazza San Marco of Venice, and made demonstrations of the instrument's observational capabilities to the admirals, who promptly commissioned Galilei to construct a sizable number of telescopes for them. The Doge of Venice (the city-state's chief magistrate and leader) was so pleased with Galilei's invention that he raised the inventor's annual salary to 1,000 silver coins and awarded him the honor of a tenured professorship. In December of that year, Galilei made further improvements to his instrument, producing a telescope that has a magnification factor of 20. Later, he constructed a super telescope with a magnification factor of 30, the highest possible magnification that could be achieved with a monocular refracting telescope at that time (due to limitations imposed on the lens size). Galilei was not only a seasoned craftsman in experimental apparatuses, but he was also skilled in using his instruments. He applied his $20\times$ telescope, for the first time, to observe the mysterious, and at times controversial, night sky. Galilei's observations naturally brought him fruitful results in astronomical research, which helped to establish him as a prominent figure in the astronomical community.

3.4.3 *Sidereus Nuncius*: Jupiter's Moons

In March 1610, Galilei published a 60-page book entitled *Sidereus Nuncius* ("Sidereal Messenger" or "Starry Message"), which sent a shockwave through the world of astronomy and made Galilei famous overnight. As this work recorded a large number of important developments and discoveries in astronomy, each of which could be observed live with the instrument he invented. First of all, Galilei pointed out that the surface of the Moon was as rugged as Earth's due to the many peaks and valleys that could be seen using the telescope. Farther away in the depths of the night sky were a great

---

[35]In 1611, Kepler used a convex lens as the eyepiece to enlarge the field of view and to increase its eye relief distance, but this produced an inverted image. This design does increase the magnification, but it also requires a higher convex lens focal length ratio to eliminate the visual distortion caused by the object lens.

number of fixed stars, including nearly 50 in the Pleiades, more than 500 in the Constellation Orion, and countlessly many in the distant Milky Way Galaxy. As the shapes of fixed stars were not observable using the telescopes available at the time (flickers resulting from very faint light passing through the atmosphere makes them unresolvable), Galilei suggested that stars must have been very far away from Earth, and therefore no stellar parallax could be observed from Earth's annual parallax[36] when revolving around the Sun. He also pointed out that Venus also exhibited phases like the waxing and waning of the Moon (from new moon, gibbous moon, to full moon), a phenomenon which fit perfectly with the Copernican theory's advocacy that Venus revolved around the Sun. The most important and most convincing evidence for heliocentrism, however, was Galilei's realization that the four new "stars" Copernicus had discovered near Jupiter were in fact its four satellites (or moons),[37] each of which was orbiting Jupiter in its own orbit and orbital period.

To verify that these four celestial objects were in fact Jupiter's satellites, however, it would require several nights of careful observation of the starry sky and very detailed recordkeeping. This was indeed what Galilei did, as described in his *Sidereus Nuncius*. On the first night, Galilei observed two objects on the east side and a third one on the west of Jupiter. On the second night he spotted all three objects having moved to the west side. A few days later, Galilei discovered a fourth object, and he also confirmed that the orbital planes of all four satellites were coplanar with the ecliptic plane. Based on these detailed observations, Galilei concluded that the Moon was also a satellite of Earth, although its orbital plane was inclined at a small angle relative to the ecliptic plane where the all the planets are located. This discovery provided strong evidence for Copernicus' heliocentric theory, as the Jovian system represents a microcosm of the Solar System. One year later, Galilei measured the orbital periods of Jupiter's four moons: Io (1.18.30; 1 day, 18 h, 30 min), Europa (3.13.20), and Ganymede (7.4.0), and Callisto (16.18.0). The measurements made by Galilei are almost identical to what are known today. In addition to the superior planets such as Mars and Jupiter, Galilei also observed the inferior planets, such as Venus. Like the Moon, Venus also exhibited waxing and waning phases, which also confirmed the fact that Venus orbits the Sun.

---

[36]In theory, when Earth orbits the Sun during the year, the positions of nearby stars, as viewed from Earth, would change relative to the more distant stars. However, Galilei failed to notice any changes in the stars with his telescope, so he proposed that fixed stars were extremely far from Earth.

[37]The Jovian moons are: Io, Europa, Ganymede and Callisto, all of which were named after four of the many lovers of Greek deity and king of the gods Zeus (or it Roman equivalent, Jupiter).

Finally, Galilei boldly conducted observations of the Sun, although not directly with a telescope but with a solar image projected onto a wall through a telescope. On the solar surface Galilei was able to observe what are now known as sunspots, which would move all the way across the Sun. Through his *Letters on Sunspots*, he communicated that when the sunspots were in the middle of the Sun's disk, their shapes appeared to be cloud-like, just like patches of cloud observed on Earth. When these sunspots were near the edge of the solar disk, however, their shapes would elongate and become slanted. From these observations he concluded that the sunspots were moving along the solar surface. Galilei also used this evidence to refute the Aristotelians, whose theory suggested that the Earth did not revolve around the Sun because neither the Moon nor the clouds were "left behind" by Earth's movement. By the same line of reasoning, Galilei now argued that since the sunspots on the Sun's surface were not left behind by the Sun, it was also not possible for the Sun to revolve around Earth. This illustrates once again that Galilei was an eloquent and accomplished debater.

All these great astronomical discoveries helped to transform Galilei from a purely academic mathematician into an astronomer endowed with a physical grounding in philosophy. Galilei had been dissatisfied with his being merely a university mathematician, so now that he no longer felt constrained by his designated role he was ready to do even greater things. In 1610, he named the four Jovian moons the *Medicean Stars*, a reference to his former pupil and the current Grand Duke of Tuscany, Cosimo II de' Medici. This showed that Galilei had considerable finesse when it was necessary to advance his interests in the world of politics. In return, Cosimo rewarded him with the position of chair of mathematics at the University of Pisa and the title of court mathematician and philosopher for the Grand Duchy of Tuscany, which offered no better salary than his professorship at the University of Padova but were of much higher prestige. Galilei was delighted to accept these appointments, as he could then return to Florence, which was the capital of Tuscany and not too far away from Pisa, his birthplace. His new designation as the court philosopher also represented a much improved academic status. The most practical benefit for Galilei was that he was no longer required to teach at the University of Pisa, which allowed him to concentrate on his research and expand his network in political and religious circles. The downside to this new reality, however, was that Galilei had to leave the academic environment in Venice that had allowed him tremendous freedom to think and perform his work. Now he was returning to a calculating and chaotic Italian city-state society, which was riddled with political and religious intrigues.

3.4.4 Return to Florence: First Trial by the Roman Inquisition

In September 1610, Galilei arrived in Florence to assume his new post, but he soon fell ill for several months. He reportedly contracted the so-called "French disease" (syphilis), which was untreatable at the time. Fortunately, Galilei recovered from the disease a few months later. After regaining his strength and returning to his active self, this scientist, a big and well-built gentleman, traveled to Rome to meet with the astronomers of the Society of Jesus (the Jesuits), hoping to expand his sphere of influence to the entire Italian Peninsula. In Rome, he employed his 20× telescope to conduct live demonstrations in front of the public, allowing Jesuit priests to catch a glimpse of the dynamic cosmos through Galilei's telescope. These Jesuits, who were ardent supporters of Aristotelianism, were indeed impressed with what they saw, and they suddenly found themselves more receptive to new ideas and ready to change their minds. They even made Galilei a member of the Accademia dei Lincei (Lincean Academy), a science academy founded by Prince Federico Cesi (1585–1630). Galilei was now an important member of the astronomical community in Rome. Soon afterwards, Galilei was granted an audience with Pope Paul V and also met with Cardinal Maffeo Barberini (1568–1644, later Pope Urban VIII). By now Galilei had established himself as an important and respected figure in the religious world as well.

The astronomical phenomena observed with Galilei's telescope and carried out in a series of important activities contributed to the near collapse of geocentrism in the scientific community. Nevertheless, Galilei's numerous enemies in academia were reluctant to admit the truth or defeat. Some refused to employ telescopes to conduct research, arguing that the images obtained with the instruments were merely illusions caused by the scattering of light. Others tried to turn these academic debates into religious issues with political overtones. Despite these circumstances, the Catholic Church at first neither condemned nor prohibited Galilei's criticism of Aristotelianism or his covert efforts to promote the Copernican model. Thanks to the social atmosphere created by Galilei, the Church decided to take a step back for the time being to watch how things would develop before passing judgment. As an example of the Church's stance, in a letter written to a friar in Naples who supported the Copernican theory, the Jesuit Cardinal Roberto Bellamine (1542–1621) urged the recipient of his letter to treat Copernicus' heliocentrism simply as a mathematical tool for explaining astronomical phenomena and not the absolute truth (i.e., regarding the Sun as being located at the center of cosmos and Earth being on the third spherical shell around the Sun). Cardinal Bellamine also reminded the friar that this type of thinking might lead to an erosion of the faith and compromising the authority of the Holy Bible. Galilei, who had been one of the chief advocates of the scientific revolution, nevertheless held onto his discoveries and never yielded. He quoted the famous saying of the Librarian of the Vatican, Cardinal Caesar Baronius (1538–1607), which stated that "The Bible teaches us how to go to heaven, not how the heavens go." With this statement Galilei was defending his view that the Bible had not been written to explain the structure of cosmos, nor were the results of scientific research, including his own, in any way in conflict with the Bible's teachings.

Galilei's arguably weak and disingenuous attempt at pleasing those in power by quoting authoritative sources certainly did not sit well with his Aristotelian adversaries, who had been fighting him for more than two decades. In late 1611, the Florentine philosopher Ludovico delle Colombe (1565–1616) gathered a group of Aristotelian scholars from Pisa and Florence to form a secret society known as LIGA and subsequently published a book in opposition to Galilei's heliocentric model of the cosmos, thereby commencing a long-term struggle against Galilei. However, in the series of public academic debates held before the end of 1614, Galilei was able to thoroughly defeat his opponents over various issues, from Archimedes' floating bodies to the sunspots discovered by the Jesuit priest and astronomer Christoph Scheiner (1575–1650), rendering the scholarly attacks launched by LIGA completely ineffectual. In response, LIGA adopted a new strategy to shift the debate to the sensitive and more dangerous realm of theology. To counter this wave of attacks, Galilei's strategy was to emphasize that God had given believers two "books", one being the Bible, which teaches people how to go to heaven; and the other being Nature, which teaches people about the cosmos created by God. Galilei further stressed that the truths from these two sources could never contradict each other. The Bible, therefore, should not be taken by natural philosophers to support their faulty reasoning. Galilei, who had gained the upper hand in the open debates, now decided to intensify the conflict by sending the proceedings of these debates to Grand Duchess Christina, the mother of Grand Duke Cosimo II, in the form of a letter. This action greatly enraged the Jesuit priests of LIGA, as they wished to have all their debates with Galilei confined to academia and not accessible to the upper echelons of society.

In February 1615, the onslaught of LIGA reached its highest point. Jesuit priest Niccolò Lorini (1544–1617) wrote to the Grand Inquisitor of the Roman Inquisition, Cardinal Millino, falsely accusing Galilei of, among other alleged offenses, attempting to reinterpret the Bible in his letter to the Arch-duchess at the expense of Aristotelianism, and Lorini believed Galilei's action had a detrimental effect on the Christian faith. Cardinal Millino treated these accusations as a very serious matter and convened a committee of theologians to investigate the two central issues in Lorini's letter: Is the Sun located at the center of the cosmos? Does Earth revolve around the Sun and also rotate on its axis daily? After receiving a summons from the Inquisition, Galilei traveled to Rome alone, as before, to defend Copernicanism. Pope Paul V had appointed a committee of theologians to review the Copernican the-ory. However, lacking the scientific knowledge to assess the veracity of the two statements about the cosmos, the theologians turned their attention to the two "hypotheses" and focused on whether they contradicted the Church's interpretation of the Bible, and whether faith of the believers in Christ would be jeopardized.

Facing the impending trial, Galilei was confident he would prevail, but he was also distracted and ended up misjudging the situation. He spoke with the eloquence of a university professor and proceeded to educate those in court who were powerful figures in the Catholic Church about the principles of heliocentrism, while at the same time shredding his opponents' self-esteem with razor-sharp arguments. He even went as far as to imply that the Pope and his advisors were all stubborn imbeciles. Sure enough, having been ridiculed, the Inquisition came up with a verdict in only two short weeks. Coperni-canism was deemed "foolish and absurd in philosophy, and formally heretical since it explicitly contradicts in many places the sense of Holy Scripture". In February 1616, Galilei was given two judgments from the Inquisition. One signed judgment ordered him to abandon the Copernican opinions and not to defend them. Another judgment, which was unsigned, further prohibited him from teaching the Copernican doctrine. With regard to Copernicus' *De revolutionibus*, as it was useful for the revision of the calendar, it was not for-mally banned but merely withdrawn from circulation, pending "corrections" that would clarify the theory's status as hypothesis. The book was therefore effectively censored by the Catholic Church, until the ban was lifted in two centuries later in 1835. Following this major setback, Galilei returned to Flo-rence to recuperate and decided to never advocate Copernicanism ever again. He then turned his attention to kinematics, although from time to time he would continue to criticize Aristotelianism.

3.4.5   Galilei's *Dialogue Concerning the Two Chief World Systems*: The Work that Proved Fateful

The research on the far less controversial discipline of kinematics allowed Galilei to continue his work in Florence for several years without incident. In August 1623, his friend Cardinal Barberini was elected as Pope Urbano VIII, and Galilei thought that the chance to reverse his condemnation by the Roman Inquisition seven years earlier had come. Galilei wrote to the new Pope, hoping to travel to Rome to congratulate him in person and to show him the new findings in his research on kinematics. Pope Urbano VIII had been an admirer of Galilei's work, so he granted Galilei's request and invited him to Rome to demonstrate his work on the motion of Earth's tides, a topic Galilei had been studying for a number of years. After Galilei explained to the Pope the workings of tides and what caused them, he asked the Pope to allow him to publish this work openly. The Pope was willing to grant Galilei the permission to do so, but Galilei was required to adhere to one condition: as Galilei's tidal theory was formulated based on the hypothesis of a self-rotating Earth, so as long as Galilei treated heliocentrism simply as a mathematical tool rather than a physical fact, the Pope would be willing to permit him to publish his research findings. Thanks to the Pope's support and the encouragement from his friends at the Lincean Academy, Galilei rediscovered his passion for astronomy. Starting in 1624, Galilei spent the better part of the next eight years working on his book, the *Dialogue Concerning the Two Chief World Systems* (often abbreviated the *Dialogue*), which was written in the Italian language. It turned out that Galilei largely ignored the condition imposed by the Pope when he wrote the *Dialogue*, intentionally or otherwise. His insistence on giving more weight to the Copernican position would eventually lead to his second trial by the Inquisition.

Why did Galilei choose the Ptolemaic and Copernican systems to articulate his cosmological view? To recap, there were four major astronomical systems recognized by various schools of thought in Galilei's time: (1) Aristotelianism, which believed that the Earth was at the center of the cosmos, and all celestial bodies, including the Sun, orbit Earth on their own concentric spherical shells; (2) Ptolemaic system (second century), which stated that Earth is at the center of the cosmos, and the celestial bodies, including the Sun, revolve around Earth on complex orbits consisting of epicycles, deferents and the equant; (3) Copernican system (15th century), which believed that the Sun is the center of the Universe, and all planets, including Earth, orbit the Sun on their respective deferents; (4) Tychonian system (16th century), which advocated that the Sun and the Moon orbit the stationary, motionless

Earth, but other celestial bodies (the planets) revolve around the Sun. However, as knowledge of astronomy was confined to academia and the Church, trying to explain these cosmological systems clearly to the public was no easy task. Galilei chose the Ptolemaic and Copernican models for the purposes of debate and comparison, as these two systems were diametrically opposed to each other, one advocating Earth as the center of the cosmos and the other championing the Sun as having that honor.

Moreover, in order to allow the public to appreciate his work (which had always been Galilei's policy), he decided to write the *Dialogue* in Italian rather than Latin, the latter being the common language among academics in his time. More interestingly, to maintain the impression of the narrative's impartiality toward the theories under discussion, as well as to comply with the Pope's order not to regard the Copernican system as the absolute truth, Galilei employed three fictional characters in his book whose names begin with the letter "S": Salviati, Galilei's alter ego (named after his friend Filippo Salviati); Simplicio, which represented the followers of Aristotle (named after a friar, but it was suspected the name actually meant "simple-minded fool"); and Sagredo, a wise arbiter named after Giovanni F. Sagredo, a friend Galilei knew from his time in Venice. Because the book was written in the vernacular Italian tongue, Galilei succeeded in rendering a subject as difficult as astronomy comprehensible to the public by using familiar language. During the course of the dialogue, however, Galilei argued each topic with language infused with a hypothetical tone commonly employed at the time, which often conveyed a kind of ruthless irony, sharp criticism and witty humor. Galilei also peppered the book with ideas from the cosmological view that he supported, making it nothing less than an elaborate and powerful statement in astronomy.

The dialogues between the three central figures of the book took place in a span of four days. On the first day the conversation focused on the criticism of Aristotelianism's fallacy with respect to the definitions of celestial bodies and matter, and on the second day, the topic turned to Earth's rotation. Earth's orbital motion around the Sun was discussed on the third day, and tidal phenomena were the main topic on the fourth. The discussions systematically touched on the divergences between the two models, and knowledge in mechanics was employed to argue for the Copernican model's superiority and against the Ptolemaic doctrine's deficiencies. One of the most famous insights, drawn from mechanics, was a scientific idea that would become important later: motion is not change; it does not lead to growth or annihilation. This idea not only overturned the millennia-old Aristotelianism but also became an integral part of the philosophy of mechanics. It would theorize that Earth's constant motion was a process initiated naturally and therefore involved no special purposes. When the *Dialogue* was completed, Galilei had entirely forgotten about the preconditions imposed by Pope Urban VIII on his publication eight years earlier: Earth's motions should be treated as nothing more than a tool and not a fact, and the book's focus should be on Earth's tides. The reality was that tides, the supposed main topic, did not appear in the *Dialogue* until the fourth day, and the first three days were all about heliocentrism, covertly or otherwise. No doubt Galilei had apparently broken the promises he had made to the Pope.

As Galilei finally completed the book and was ready to publish it, his friend Prince Cesi passed away, and the plague that was affecting Europe started to sweep across the continent. The *Dialogue* therefore was not officially published until February 1632 in the city of Florence. When Pope Urban VIII read his copy of the *Dialogue* and found that Galilei was defending the Copernican opinions, hence in violation of papal orders, and, more seriously, read about what he had told Galilei all those years earlier from the mouth of Simplicio (the intellectually inept fool), Pope Urban VIII became absolutely furious. He immediately ordered that the book be banned from public sale and demanded that the perpetrator of this abomination be brought to justice. It was too late to remove the copies, however, as they had been completely sold out. Apparently this new work of astronomy was so popular in the Italian society that there had been intense ongoing discussions about its contents.

3.4.6 Last Judgment

On October 1, 1632, Galilei received a summons from the Roman Inquisition, ordering him to go to Rome to stand trial on suspicion of heresy within the next thirty days. Having been suffering from arthritis and hernia for an extended period of time, the 69-year-old Galilei suffered a stroke upon hearing the news. Galilei, however, did not shirk his responsibility as long as he still drew breath, so he traveled from Florence to Rome once again. He set off on January 23, 1633 and finally arrived in Rome after a grueling three-week journey. Due to his close personal friendship with the Pope (since the latter was still Cardinal Barberini), Francesco Niccolini, Florence's ambassador to the Holy See, arranged for Galilei to stay at the Roman residence of Ferdinando II de' Medici (1610–1670), Grand Duke of Tuscany, in a comfortable suite equipped with amenities and servants. The trial began on April 12, and Galilei, who had gradually regained his physical strength, argued that he never insisted that the heliocentric theory of Copernicus was the correct doctrine; instead he had elaborated on the theory's weaknesses. Galilei again quoted from the classics a number of times, emphasizing that the Bible's role was to teach people how to go to heaven and not how the heavens operate.

In this court of the Inquisition, which was accustomed to passing judgment even before the trial was held, Galilei's eloquence and hard-hitting rebuttals exerted tremendous pressure on the panel, who, being infuriated with the humiliation, argued that Galilei was merely mocking them by flaunting his wealth of knowledge in science. The Inquisition handed down its verdict only a few days later: Galilei was found in agreement with Copernicus' heliocentrism, thus not only breaking the vows he made in 1616 but also violating the Bible's teachings and undermining the faith of Christ. The judgment was unequivocal, yet there were differences in opinions as to how Galilei should be sentenced. After lengthy deliberations, the court agreed to offer Galilei a chance to plead guilty and confess his sins. On April 30, 1633, Galilei admitted his guilt openly in court and stated that the *Dialogue* contained numerous errors, and that the book's publication was deeply regrettable. However, staunch conservatives among the Jesuits opposed this compromise, and they pressured Pope Urbano VIII into imposing a more severe penalty on Galilei. Finally, the Inquisition issued a harsh judgment on June 21: *Dialogue* was to be banned (placed on the Vatican's *Index of Forbidden Books*) and Galilei was sentenced to life imprisonment; Galilei was also required to recite seven penitential psalms weekly for two years. After the judgment was passed, the Aristotelian scholars breathed a sigh of relief, and the Jesuit priests celebrated joyously. With this victory the Aristotelians had successfully turned a scientific manner into a theological question with their tactic, and the Jesuits were now able to maintain their grip on the loyalty of the masses toward the Church, that is, with Earth continuing to sit comfortably at the center of the cosmos.

However, Pope Urbano VIII was, or at least had once been, a close friend of Galilei's after all, and he was a true admirer of Galilei's scientific work. In December 1633, the Pope permitted Galilei to serve his sentence as a guest of the Archbishop of Siena for a period of time. Galilei was soon allowed to return to his villa at Arcetri near Florence and would continue to receive his special allowances. During the final eight years of his life under house arrest in the countryside, Galilei was able to complete his final book *Discourses and Mathematical Demonstrations Relating to Two New Sciences* (or simply *Two New Sciences*), a great scientific testament covering his work in physics spanning thirty years. *Two New Sciences* contained references to what would become Newton's first and second laws of motion in kinematics[38] a

---

[38] In his *Philosophiæ Naturalis Principia Mathematica,* later referred to by others as the greatest work in the history of science, Isaac Newton publicly credited this work by Galilei as the original source

few decades later. Galilei went completely blind in 1637, due to his unceasing efforts to observe the Sun and planets appearing nearby with the telescope. *Two New Sciences* was finally completed with the help of his son and pupils. Unsurprisingly, the book was judged to be heretical by the Vatican. Galilei, despite his persecution, never left the Church. In fact, he often prayed before God into the wee hours and also invited friends to pray on his behalf. On January 8, 1642, Galilei passed away at the age of 77, with his family and two of his pupils by his side.

3.4.7 Persecution of Science by Religion

Religious persecutions, such as those suffered by Galilei, were heard from time to time. The reason was that the Roman Catholic Church was beginning to lose its supremacy and grip on power in European societies. This was caused by anti-Vatican political events that had first occurred in Germany and later in England. The year 1520 saw the beginning of the Protestant Reformation led by Martin Luther (1483–1546), who proclaimed that the essence of the Christian faith was justification by faith alone and not by buying the "indulgences" from the Catholic Church. Luther was excommunicated by the Pope, and all the books he had written were ordered to be burned. The German Church therefore split from the Vatican and thus born the Protestant Lutheran Church. Elsewhere in Europe, King Henry VIII (1491–1547) of England wished to have his marriage with Catherine of Aragon annulled, as his queen failed to bear him a male heir. However, the Vatican disagreed with such unprecedented royal request, as it contradicted the teachings of the Church. King Henry VIII decided to officially break with Rome and initiated the English Reformation with the creation of the Anglican Church (Church of England). These political events caused the Vatican to become particularly sensitive to and be no longer tolerant of anti-Catholic incidents based on scientific or theological grounds. Galilei was unfortunate to have lived in the aftermath of these religious battles and become embroiled in the subsequent power struggles involving the Church, the state and the scientific community.

In reality, 33 years before Galilei's scientific work was declared heretical by the Vatican, the Church had publicly executed another supporter of Copernicanism by burning him at the stake in the city's central square. The victim was the brilliant and honest Italian mathematician and philosopher Giordano Bruno (1548–1600). An advocate of an infinite universe later in his

---

for his first law of motion. Throughout his career Newton rarely shared his research findings with other scientists, and even less frequently did he give credit to them; this could have been the only time he had done so.

life, Bruno entered the Dominican Order at the monastery of San Domenico Maggiore in Naples at the age of 17 to study theology. Born with a rebellious spirit and a voracious appetite for knowledge, Bruno found his way to a large collection of banned books about the cosmos at the monastery library and was delighted to read them. This was a time when the Catholic Church was still in strict control of the political and economic life in Europe. One of the books that caught Bruno's eye was *On the Nature of Things*, written by a Roman poet and philosopher Lucretius who had lived about 1,500 years before. In the book, the writer was convinced that the Universe was infinitely large, a belief which Bruno, a believer of an omnipotent God, could wholeheartedly agree with. The Almighty God could only have created an infinite cosmos. When Bruno began to talk about his discoveries with other friars, however, the Church immediately accused him of preaching heresy and expelled him from the monastery.

Having been banished from the Church and unemployed, Bruno now began an itinerant life, spreading his ideas about the infinite universe and its connection with the omnipotent God. After leaving Naples, he first traveled to the Vatican and then to Switzerland to visit Calvinist and Lutheran churches, but he was less than welcomed there. He finally fled to France and openly supported the struggle of King Henry III against the Catholic Church. Bruno traveled to the University of Oxford in England, on the French king's recommendation, to advocate Copernicanism. Armed with the stamp of approval from French royalty, Bruno began to take advantage of his lectureship at Oxford to openly teach his controversial views on the infinite cosmos, emphasizing that the Copernicus' heliocentrism was true and that every star in the night sky constituted a Sun, possibly with its own Earth-like planets filled with different lifeforms. These radical arguments, with no supporting evidence whatsoever, were understandably ridiculed by the scholars at Oxford, and Bruno was eventually dismissed. Exhausted and desperate, Bruno decided to return to his native Italy.

After returning to Venice in 1591, Bruno became a private tutor to the nobleman Giovanni Mocenigo. Italy was at the time considered to be the strictest country in Europe in terms of religious control, with a number of inquisitorial courts under the Inquisition having been established in major Italian cities such as Rome and Venice. Their purpose was to combat religious dissent and to put pagans and heretics on trial. A year later, Bruno's advocacy for the infinite cosmos was condemned by Mocenigo, who promptly reported him to the Venetian Inquisition, and Bruno was subsequently sentenced to an eight-year prison term, again for his heresy. In 1600, Pope Clement VIII pronounced that all of Bruno's works would be destroyed publicly in St. Peter's

Square, and Bruno would also be burned at the stake there. A decade after Bruno's sacrifice for his belief in the infinite cosmos, Galilei looked at the night sky with the telescope he had invented, and he was surprised to find Bruno's boundless universe filled with countlessly many Sun-like stars.

3.4.8 Relationship with Kepler

Finally, we will examine Galilei and Kepler,[39] two great astronomers who were contemporaries of each other, and find out why their paths barely crossed in their work on astronomy and the exploration of the cosmos. The two scientists probably made contact for the first time in 1597. When Galilei was teaching at the University of Pisa, the Ptolemaic model was the standard theory taught in his astronomy classes. Kepler had been following Galilei's work and accomplishments in observational astronomy and his experiments in kinematics, but he was wondering why Galilei still held onto the Ptolemaic doctrine in his teaching. So Kepler asked someone from Germany to bring a copy of his scientific debut *Mysterium Cosmographicum* to Galilei. Upon receiving Kepler's book, Galilei replied that he was also a follower of Copernicanism, but he also admitted that he was reluctant to openly admit his beliefs lest he would be persecuted by the astronomical community much as Copernicus himself had been. However, he also said that when he started to use the 20× telescope he had invented to observe the mountains and valleys on the Moon's surface in 1609, as well as the four satellites that were orbiting Jupiter, the phases of Venus, the rings of Saturn, and the sunspots, all he could see was phenomena that the Aristotelian cosmological view failed to explain or even outright contradicted. Galilei had no choice but to abandon Ptolemy's geocentrism and begin to embrace Copernicanism. In 1613, Galilei eventually acknowledged Copernicus' heliocentric theory openly in his *Letters on Sunspots* when the pamphlet was published.

Also in 1609, however, Galilei read about Kepler's area law and elliptic laws in his *Astronomia nova*, but he offered no public support for them. This was probably because Galilei still held the belief that planets orbit the Sun in uniform circular motion. This could also explain why, much later in 1632, when Galilei was sentenced to life imprisonment by the Roman Inquisition for supporting Copernicanism, he remained skeptical of Kepler's three laws of planetary motion and was reluctant to adopt them in his defense. It might have been because Galilei believed that Kepler's laws were merely a special

---

[39] Referenced from Hummel [17].

case of derivation in geometry and could not provide a comprehensive physical interpretation of the nature of planetary motion. It was Galilei's own fault that he failed to take advantage of Kepler's work in his trial, but perhaps Kepler himself should also shoulder the blame for Galilei's fate, as he never truly attempted to provide any physical interpretation as to what the three laws he discovered truly meant. This is also why Galilei was praised and appreciated by later generations of scientists, as he always insisted on the principle of using experiments to analyze physical phenomena, which he believed to be the only right way to truly attack a problem and then try to construct the mathematical relationship hidden behind it.

3.4.9 Accomplishments in Kinematics

In Galilei's time, the study of celestial motion was gradually evolving from observing the movements of celestial bodies to scientific research based on kinematics and observational data. The protagonist of this transformation was none other than Galileo Galilei, whose work became the sources of inspiration for Newton's research in universal gravitation. These sources included Galilei's achievements in kinematics, the logical reasoning approach that he established, namely combining different elements such as experiments, induction and deduction to establish the relativity of uniform linear motion and the constancy of gravitational acceleration. Still other sources were Galilei's question regarding the mechanism that allows a planet to revolve around its host star in a circular orbit. In the final years of his house arrest, Galilei applied the theory of motion on objects, which he had not had a chance to complete in his earlier in his career, to explain and elaborate on the results from his observations with mathematical reasoning considered quite sophisticated at the time. Fortunately, before he suffered from total visual impairment, he had completed two reports, one on *Statics* and the other on *Dynamics*. The former report explained the forces exerted on a stationary object, whereas the latter report clearly described those forces acting on an object in motion. Finally, Galilei combined these two works and published his final work, *Two New Sciences*, for posterity. The completion of this important work in kinematics left behind a priceless scientific legacy, upon which Newton and later Einstein would continue the research on gravitation. Below we will give and account of Galilei's contribution to the study of mechanics.

According to Einstein's commentaries, Galilei was a great experimental physicist who routinely invented new instruments with specific features and functionalities to verify the conjectures embedded in his brand new theories.

For example, he invented the water clock[40] for keeping time in his experiments. The clock was made of a water container with a small hole at the bottom. When a timekeeping session began, the small hole at the bottom was opened to allow water to drain to a bucket beneath it. At the conclusion of the session, the water collected at the bottom was weighted. The proportional relationship of the water's weight relative to the length of time measured was then used to calculate the amount of time required to collect the water. Another example is that when Galilei studied how an object would behave in a free-fall setting, he found it difficult to observe or measure the object's position or time, since it was traveling too fast. Galilei therefore invented a contraption with the bronze ball and the carved inclined planes to conduct the experiment. Recall that when the ball rolled over each carved shallow marker, it would produce a sound. Galilei would then measure the increase in distance traveled by the ball relative to time, or change of speed.

Although Galilei was quite experienced at conducting experiments, some of them happened to be exceedingly difficult to perform. For example, since a projectile experiment does not involve a rectilinear motion in a free-fall configuration, a simulation based on an inclined plane would no longer be appropriate. Also, the trajectory of a projectile is a constantly changing parabola, and the object's position at each point in time is difficult to record visually. As a physical experiment was not possible, Galilei then carried out the simulation of the experiment entire in his mind, which was later referred to as a "thought experiment". Galilei imagined the following situation on a ship (in contrast, Einstein's thought experiment occurred on a train, which did not yet exist in Galilei's time. A ship would work equally well if it was sufficiently stabilized and free from random movements, assuming strict conditions applied). A goldfish was swimming in a fish bowl, and a child was throwing a ball on the ship's deck. The resulting speed of the ball and that of the fish were each the sum of its own speed and the ship's speed. Galilei was therefore able to explain, using the conclusions obtained from the inclined plane experiment, that when an object was projected forward, the horizontal component of its movement was uniform linear motion, as there was no external force in that direction. The vertical component of its movement, however, would be in

---

[40] Historical records show that the earliest water clock, then known as clepsydra, was invented by royal court officials in ancient Egypt around 1500 BC. The main principle was to observe the extent of the decrease in water level to estimate the time elapsed. Millennia later, during Galilei's time, the water clock had undergone numerous improvements and modifications to accommodate various purposes. The water clock used by Galilei had a gear-and-pointer design. Rising water levels would drive the gear assembly and push the pointer to indicate the correct time; it was already quite practical and sophisticated.

free-fall motion due to gravity. A projectile motion therefore consists of the combination of movements in these two directions.

The results of these experiments, which are of great significance in the history and development of physics, were recorded by Galilei in his *Two New Sciences*, which was completed in 1635 while he was under house arrest in his villa in Arcetri. The three protagonists from the *Dialogue* reprised their roles in this new book. Although the discussions remained critical of the views of Aristotelians, the physics being studied was kinematics, the scope of which was entirely on Earth's surface and was entirely divorced from the Bible's teachings. Although *Two New Sciences* was eventually banned from publication by the Catholic Church, several copies were smuggled out of Italy and reprinted in 1638 by a publisher in Leiden, South Holland. Like its predecessor, the book was also divided into four parts, each representing the dialogue that occurred on one day.

On the first day, discussions focused on the landing of objects with different weights. The result was that all objects landed with identical terminal velocity, which directly contradicted the Aristotelian prediction that heavier objects should travel faster. The second day addressed the question of the strength of materials of objects with different shapes, which could be interpreted as the earliest exposition on the mechanics of materials, although it remained a preliminary discussion with no concrete results. Returning to the science of motion, the third day discussed uniform and naturally (uniformly) accelerated motion. Free fall was described as a type of uniformly accelerated motion. Although Galilei's objective was simply to describe this type of motion at the time, it unintentionally brought kinematics into an unexplored area not entirely foreign but sufficiently important: Newtonian dynamics. On the fourth day, Galilei discussed projectile motion, which was essentially the results of the thought experiment he conducted in 1608, which was mentioned in a previous section. In this masterpiece of kinematics completed in his final years, Galilei put forward 58 propositions, including uniform linear acceleration, horizontal projectile motion and projectile motion with various launch angles. Galilei presented each proposition in three stages: description of the physical phenomenon, process of logical reasoning and analysis, and derivation of the mathematical model. His work helped to establish a standard for many future works in physics.

3.4.10 Father of Experimental Physics

Galilei's influence on the research of subsequent generations of physicists concerning his object motion experiments includes two main steps: (1) establishing a set of experimental apparatuses to simulate and observe the physical phenomenon under study; and (2) creating a theoretical model for the phenomenon in order to explain a variety of phenomena and principles in nature. Before Galilei, natural philosophers often observed a natural phenomenon passively and then attempted to extract and explain the patterns contained in the complex and incomprehensible manifestations. The end result was, more often than not, disappointing if not entirely fruitless. Since Galilei, physicists have employed patterns derived from observations to construct satisfactory mathematical models, from which the principles behind a physical phenomenon can be explained. Refer to "Appendix 3.3: The Work of Scientific Revolutionaries." As Galilei himself said, he conducted an experiment for two purposes: to verify his own ideas and to measure the physical quantities involved. At a time when the qualitative nature of Aristotelian research dominated scientific thoughts, Galilei's quantitative approach was much ahead of its time and therefore quite incompatible with contemporary mainstream scientific ideas. However, Galilei's research in physics led to an amazing evolutionary development: a rapid shift from qualitative observations to measurement-based quantitative research. The objectives of scientific research also shifted from understanding the workings of specific natural phenomena, often with practical applications, to establishing a set of physical models to reveal otherwise hidden natural phenomena. Subsequently, an individual mathematical model is constructed to carry out analysis and computation, and to make predictions.

Take free fall motion as an example. The exploratory model for physics established by Galilei can be roughly divided into the following four steps:

(1) First, Galilei provided a description of the phenomenon and any relevant logical discussion: In *Two New Sciences* he pointed out the fallacy of the Aristotelian assertion that two objects, one heavier than the other, would not land on the ground at the same time when released at the same height.[41] Then Galilei posed the question: If the heavier object does

---

[41]The assertion that the weight of a free-falling object does not change the amount of time required for it to land was made by Galilei when he taught at the University of Padova in 1610. The event was not recorded until Galilei included it in his *Two New Sciences* when his was working on the book during his house arrest after 1638.

indeed falls at a higher speed, then if we tie both objects together as a single unit, would it fall faster or more slowly than the heavier object alone? According to the Aristotelian doctrine, the combined unit, which has a greater weight than either object, should fall at a faster speed. On the other hand, since the heavier component of the combined unit would be slowed down by the lighter counterpart, the speed of the combined unit should therefore lie between the individual speeds of the two objects when released separately. Galilei therefore arrived at the following conclusion: If a contradiction results from attempting to explain a phenomenon using the same theory (in this case Aristotelianism), one must conclude that the theory in question is erroneous.

(2) Secondly, Galilei was attempting to define the motion's mathematical relationship with its velocity[42]: He understood that in free fall motion the velocity is related to both distance and time, so he first assumed that the speed v of the falling object was proportional to the vertical distance traveled s, which can be expressed as $v \sim s$.

Suppose that as the object falls from an initially stationary state, its velocity becomes v_1 after traveling a distance of s_1. If the overall speed during this period of time is represented by the average speed $v_1/2$, then the above three quantities can be related using the formula $t_1 = 2s_1/v_1$. As the object continues to fall for a further distance of s_1, the total distance traveled at that point is $2s_1$, and the velocity is increased to $2v_1$, with an average velocity of v_1. The total time required for the object to travel the two segments of distance will then be $t_2 = 2s_1/v_1 t$, which is the same as t_1, the amount of time to travel the first segment, an obviously contradictory result.

(3) Since the initial hypothesis of $v \sim s$ was determined to be incorrect, Galilei proposed the new hypothesis,[43] $v \sim t$, and assumed that the falling body was traveling in uniformly accelerated motion. He therefore formulated the relation $v = gt$, g being the acceleration due to gravity, a constant.

(4) In order to prove this relationship, Galilei designed the inclined plane experiment described earlier in this chapter. He measured the change in distance traveled by the bronze ball as time progressed, after which he converted this measure into the change of velocity over time. After having conducted a series of experiments with different angles of inclination,

---

[42]Refer to Yao and Yu [34].

[43]This hypothesis had been proposed by Merton College, Oxford more than three centuries earlier. That is, the concepts of mean velocity and uniform acceleration are the same (see Fig. 3.1). Galilei probably derived his concept from the research conducted at Merton College.

Galilei discovered that the increase in velocity per unit time was always constant. Therefore, if velocity v (on the vertical axis) is plotted against time t (on the horizontal axis), the graph is a straight line (see Fig. 3.1), and the slope of this straight line is proportional to the angle of inclination of the inclined plane. Galilei also discovered that the area of the triangle under the straight line was the distance traveled by the bronze ball, i.e., $s = vt/2 = gt^2/2$. This is the famous law of motion, which states that the distance traveled by a falling object is directly proportional to the square of time.

As the principal investigator of this experiment, Galilei was honored by Albert Einstein as the father of experimental physics. The experiment helped to create a mode of thinking that became profoundly influential on the development of physics: First observe the phenomenon and eliminate any unreasonable conclusions. Then assume a mathematical model (e.g., $v = gt$) and design a experimental target to be verified (e.g., $s = vt/2 = gt^2/2$). Finally, conduct the experiment to validate it. In the four-step exploratory model for physics outlined above, the most remarkable part is that Galilei employed mathematical equations to express the details and the objectives of the experiment. In the scientific investigations before Galilei's time, mathematics played a very insignificant role, which had to do with the fact that Aristotelianism had been the dominant natural philosophy for such a long time.

Aristotle criticized his teacher Plato for placing too much emphasis on mathematics, because he believed the results would lead to many erroneous conclusions. Aristotle therefore claimed that mathematics was merely an abstract concept with no concrete meaning, as it often led to results that were inconsistent with actual experiences in real life. The blame should not be placed squarely on Aristotle, however, as mathematics in his time was not based on any well-defined, concrete physical quantities, so it was difficult to apply quantitative methods to explain complex phenomena such as motion. It was also impossible to account for the details of complex motion explicitly with mathematical equations or geometric relations. Even in the experiment conducted by Galilei, as described earlier, the attempts at establishing a mathematical model to describe the object's motion could lead to contradictory outcomes. All these factors contributed to the absolute dominance of the so-called theories advocated by Aristotelians in the scientific world based only on empirical rules.

However, constructing mathematical equations to model physical phenomena was no small feat. Let's consider objects in free fall motion again. There are a number of physical factors that affect an object's motion in free fall, such as its weight and shape, and the friction of air surrounding it. The limited mathematical knowledge available in Galilei's time, however, was inadequate to accommodate these factors. What was special about Galilei's approach is that he was able to identify the crucial factors and ignore less significant ones. In the end, Galilei managed to unravel the mystery by isolating the most relevant factors in the complex phenomenon and clarify them one by one, and eventually he was able to derive a simple and harmonious mathematical formula. Galilei's in-born ability to weed out the undesirable elements and extract the essential ones allowed him to not only examine subtle changes in motion but understand the cause that gave rise to the phenomenon as well. He could then proceed to simplify the process before a satisfactory model was successfully constructed. As a perfect example for Galilei's unraveling strategy, air resistance in the free fall experiment could explain why different objects landed at different times, so if it was ignored, the objects should hit the ground at exactly the same time. In the mathematical relationship that Galilei was eventually able to derive, namely $s = gt^2/2$, he transformed mathematics into a precise tool for depicting physical phenomena, thus allowing physicists of future generations to follow his footstep: describing new physical phenomena necessitates new mathematical models.

The Scientific Revolution begun by Copernicus did not end with the death of Galilei. Following the series of unprecedented discoveries made by these great scientists, Aristotelianism, which had dominated the world of science for nearly two millennia, was on the verge of collapse, and its survival was tied to the extent of the Catholic Church's grip on power in an age when the Church's influence was waning. The pace at which European science was developing at a particular geographic location would be directly proportional to the square of its distance from Rome. The farther away it was from Rome, the faster its scientific development and the greater the fruits of the scientists' labor. One of these places far away from the control of the Vatican was Great Britain, and the protagonist of the next phase of the Revolution was Sir Isaac Newton.

Appendix 3.1: Indirect Evidence of the Earth's Rotation

We live on a spherical planet that rotates on its axis once every 24 h, but we have no idea that the ground is actually moving. Our bodies have been physiologically adapted to this dynamic environment, and therefore the fact that Earth is rotating is hidden from us. Since the time of Copernicus, scientists have proposed new theories to describe Earth's motion. Now we know that Earth revolves around the Sun and rotates on its axis, and that Earth's equatorial plane is inclined to the ecliptic plane at an angle of about 23.5°. Scientists have been exploring ways to prove, right here on the surface of Earth, that the planet is indeed rotating. Below we present two natural phenomena and an experimental device. By analyzing their principles of operation or interpreting their mechanisms, we will be able to verify Earth's rotational motion.

(1) Tropical Cyclones: Counterclockwise Rotation in the Northern Hemisphere

Suppose a typhoon of moderate intensity, located at 30° N on the Pacific Ocean, has a radius of 300 km and an eye of radius 30 km. The typhoon's maximum wind speed is typically about 80 m/s, which occurs at the eyewall, as shown in the following diagram. Below we will derive a wind speed distribution model and then apply it to estimate the maximum wind speed around the eye.

Suppose a typhoon has formed at 30° N. Figure 3.5 shows a magnified diagram of a part of the typhoon. We can create plane polar coordinate system with the center of the cyclone's eye as the origin and use it to describe the motion of any fluid particle within the storm, with its speed expressed as

$$\vec{v} = v_r \hat{i}_r + v_\theta \hat{i}_\theta = \dot{r}\hat{i}_r + r\omega \hat{i}_\theta \tag{3.16}$$

Here $\dot{r} = dr/dt$ and $\omega = d\theta/dt$. Now, let's assume that Earth rotates at an angular velocity of Ω, which equals to 15°/h or $15 \times 2\pi/(360 \times$

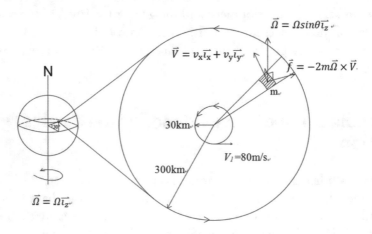

Fig. 3.5 Dynamic structure of a tropical cyclone on Earth's surface

$3600) = 7.27 \times 10^{-5}/s$. Since the typhoon's point mass is located at $30°$ N, the angular velocity of Earth's rotation at this point has the following orthogonal component: $\Omega \sin 30° \, \vec{i}_z = 0.5 \,\Omega\, \vec{i}_z$. At this time, the point mass is subject to the effect of the Coriolis force and has a magnitude of

$$\vec{f} = -2m(0.5\Omega)\hat{i}_z \times \vec{v} = m\Omega(r\omega\hat{i}_r - \dot{r}\hat{i}_\theta) \qquad (3.17)$$

We substitute the data given above into Eq. (3.17), which expresses a relationship in mechanics, as follows. The change of angular momentum of the point mass $d(mr^2\omega)/dt$ should then be equal to the torque created by the Coriolis force exerting on the point mass in its the tangential direction $\vec{r} \times \vec{f} = -m\Omega r\frac{dr}{dt}\hat{i}_z$, thus

$$d\left(mr^2\omega\right) = -m\Omega r dr \qquad (3.18)$$

Assuming that the typhoon's flow field is rotationally symmetric, the velocity of the fluid particles at a distance of r from the typhoon center can be regarded as the wind speed at that location. Therefore, if we integrate the expression (3.18) from $r = r_2$ (the typhoon's outermost radius) to $r = r_1$ (the boundary of the typhoon's eye, or the eyewall's radius), then we obtain the following relationship:

$$\omega_1 = \frac{\Omega}{2}\left(\frac{r_2^2}{r_1^2} - 1\right) \qquad (3.19)$$

In the process of arriving at the integral (3.19), we have assumed that the angular velocity of the eye's point mass is ω_1, and the angular velocity of fluid particles in the periphery of the typhoon $\omega_2 = 0$. We now substitute $r_1 = 30$ km and $r_2 = 300$ km into the formula to obtain $\omega_1 = 3.6 \times 10^{-3}/s$. Since this is a positive quantity, it indicates that the rotation is counterclockwise.

Next we will estimate the wind speed around the eyewall. The tangential velocity of a fluid particle on the eyewall is $r_1 \times \omega_1 = 30 \times 10^3 \times 3.6 \times 10^{-3} = 108$ m/s. This quantity is about 20% greater than the observed (typical) value of 80 m/s mentioned earlier, which suggests that the Coriolis force that drives the typhoon constitute a rather significant factor. Other factors that affects the typhoon's movement include the friction between the storm and the sea's surface, evaporation and rise of water vapor at the eye, precipitation within the typhoon's chamber, and the obstruction of the typhoon's periphery, all of which having contributed roughly 20% to the observed speed.

(2) **Atmospheric Circulation**

In the absence of Earth's rotational motion, the air closer to the equator would rise after being heated by the Sun. As this heated air reaches a certain altitude, it would drift northward and southward to the polar regions until it cools down, at which time the air would drift back to the equatorial region, forming a north-south atmospheric circulatory system. Due to Earth's rotation, however, the atmospheric structure is a much more complex situation. At the upper reaches of the Pacific Ocean and the Atlantic Ocean, we can designate an atmospheric circulatory zone bounded by the $30°$ and $60°$ latitudes (either in the Northern or Southern Hemisphere). We will then employ a simple point- mass motion model to estimate the wind speed of atmospheric circulation at $30°$ latitude in order to confirm the importance of the Coriolis force generated by Earth's rotational motion on atmospheric circulation.

Again let's refer to Fig. 3.5. Suppose a point mass m, located at latitude θ and longitude φ in the upper atmosphere, is in motion along a direction parallel to Earth's surface. The motion generated by the Coriolis force under Earth's rotation satisfies the equations

$$m\frac{dv_x}{dt} = (2m\Omega\sin\theta)v_y \qquad (3.20)$$

$$m\frac{dv_y}{dt} = -(2m\Omega\sin\theta)v_x \qquad (3.21)$$

Here the velocity along the latitude is $v_x = R\cos\theta\frac{d\varphi}{dt}$ and the velocity along the longitude is $v_y = R\frac{d\theta}{dt}$. Substituting these two velocity expressions into (3.20) and (3.21), we obtain

$$\frac{d}{dt}\left(\cos\theta\frac{d\varphi}{dt}\right) = 2\Omega\sin\theta\frac{d\theta}{dt} \qquad (3.22)$$

$$\frac{d}{dt}\left(\frac{d\theta}{dt}\right) = -2\Omega\sin\theta\cos\theta\frac{d\varphi}{dt} \qquad (3.23)$$

Rearranging (3.22), we have

$$\frac{d}{dt}\left(\cos\theta\frac{d\varphi}{dt} + 2\Omega\cos\theta\right) = 0 \qquad (3.24)$$

Integrating (3.24) with respect to time, we obtain

$$\cos\theta \frac{d\varphi}{dt} + 2\Omega\cos\theta = C_1 \tag{3.25}$$

Here C_1 represents the constant of integration. Substituting (3.25) back into (3.28), we have

$$\frac{d}{dt}\left(\frac{d\theta}{dt}\right) = -2\Omega\sin\theta(C_1 - 2\Omega\cos\theta) \tag{3.26}$$

Now we multiply both sides by $d\theta$ to get

$$\frac{d\theta}{dt}d\left(\frac{d\theta}{dt}\right) = 2\Omega(C_1 - 2\Omega\cos\theta)d(\cos\theta) \tag{3.27}$$

Integrating (3.27) with respect to time, we have the expression

$$\frac{1}{2}\left(\frac{d\theta}{dt}\right)^2 + C_2 = 2\Omega\cos\theta(C_1 - \Omega\cos\theta) \tag{3.28}$$

Here C_2 is another constant of integration. When $\frac{d\theta}{dt} = 0$, the direction of the wind returns to the latitude's east-west orientation. In other words, the wind no longer blows toward the north or south pole. The latitude θ^* at this time represents the highest latitude the wind is able to reach, and the following applies:

$$C_2 = 2\Omega\cos\theta^*\left(C_1 - \Omega\cos\theta^*\right) \tag{3.29}$$

Now, we will need boundary conditions in order to compute the two constants of integration. First of all, if we assume that the wind along the equator only blows toward the north or south pole (i.e., along the longitude's north-south orientation), then its velocity v satisfies the relationship $\left(\frac{d\theta}{dt}\right)_0 = \frac{v}{R}$. On the other hand, the velocity of the wind blowing in the easterly or westerly direction (i.e., along the latitude) should be 0, or $\left(\frac{d\varphi}{dt}\right)_0 = 0$. Substituting this condition into (3.25), we get $C_1 = 2\Omega$, where Ω is the angular velocity of Earth's rotational motion. Substituting this constant into (3.28), we arrive at the following

$$-\left(\frac{d\theta}{dt}\right)_0^2 = 4\Omega^2\left(2\cos\theta^* - \cos^2\theta^* - 1\right) = -\left[2\Omega(1 - \cos\theta^*)\right]^2,$$

(3.30)

from which we obtain

$$\theta^* = \cos^{-1}\left[1 - \frac{v}{2R\Omega}\right]$$

(3.31)

Since we know that $30°$ N represents a boundary of the atmospheric circulatory zone, i.e., $\theta^* = 30°$, if we substitute this expression into the above, we can conclude that the wind speed at this particular latitude should be $v = 448\,\text{km/h}$.

According to what we know today, the atmosphere at this latitude contains a jet stream (alternatively, subtropical jet). The jet stream's speed is generally between 200 and 300 km/h, and can increase to 400 km/h occasionally. Our calculations above result in a value larger than the usual estimate. The reason is that actual atmospheric circulation involves a robust, three-dimensional vortex, which is subject to depletion from the friction of viscous fluids and the phase transition between liquid water and water vapor. These complexities are not modeled by our smooth, two-dimensional configuration with non-viscous fluids. However, the simplified model discussed here emphasizes that, although inaccurate, the estimated value computed from the Coriolis force generated by Earth's rotation is nevertheless sufficiently close to the actual value measured. Such Coriolis force has a considerable effect on atmospheric circulation. To conclude, the calculation of atmospheric circulation also provides indirect evidence of Earth's rotational motion.

(3) Foucault Pendulum

The principles behind the Foucault pendulum, which we will employ to confirm Earth's rotation, can be explained using one of the following two approaches: the method of mechanics[44] and the geometric approach.[45] Now we will explore the method of mechanics as follows.

Consider a simple pendulum placed at latitude φ. Assuming that the pendulum swings at a very small amplitude, the motion of the bob's point mass

---

[44] For the explanation using mechanics, consult https://en.wikipedia.org/wiki/Foucault_pendulum.

[45] Refer to the geometric approach here for additional details (in Chinese): http://highscope.ch.ntu.edu.tw/wordpress/?p=46671.

can be regarded as that on a level surface parallel to the latitude. An xy (two-dimensional) rectangular coordinate can therefore be superimposed on the location with the origin being the bob when the swing amplitude is zero. Assuming also that the length of the pendulum is l, then the resultant force (force of gravity plus the tension of the pendulum's string or rod) can be decomposed on the xy Cartesian coordinates as $-mg \times \frac{x}{l}$ and $-mg \times \frac{y}{l}$, respectively. Now, the vertical component of the angular velocity Ω of Earth's rotation on the Cartesian plane is $\omega = \Omega \sin \varphi$. Substituting this into the Coriolis force formula, we obtain the effective Coriolis force component

$$\vec{F} = -2m\vec{\Omega} \times \vec{v} = 2m\omega(\dot{y}\hat{\imath}_x - \dot{x}\hat{\imath}_y) \tag{3.32}$$

In the above formula, $\hat{\imath}_x$ and $\hat{\imath}_y$ are the unit vectors in the x and y directions, respectively. The motion equations are therefore

$$m\ddot{x} = -mg \times \frac{x}{l} + 2m\omega\dot{y} \tag{3.33}$$

$$m\ddot{y} = -mg \times \frac{y}{l} - 2m\omega\dot{x} \tag{3.34}$$

Adding (3.33) to (3.34) multiplied by i, we can take advantage of the complex polar coordinates (r, θ), $z \equiv x + iy = re^{i\theta}$, and combine the two expressions

$$\ddot{z} + 2i\omega\dot{z} + \frac{g}{l}z = 0 \tag{3.35}$$

The general solution to (3.35) can be written as

$$z = e^{-i\omega t}\left(Ae^{i\sqrt{\omega^2+g/l}\,t} + Be^{-i\sqrt{\omega^2+g/l}\,t}\right) \tag{3.36}$$

From (3.36), we can see that when the Coriolis force is absent, the solution (3.36) becomes

$$z = Ae^{i\sqrt{g/l}\,t} + Be^{-i\sqrt{g/l}\,t}, \tag{3.37}$$

which is the general solution to the ordinary single pendulum problem, and the angular frequency is $\sqrt{g/l}$. In the case where the Coriolis force is present, the angular frequency becomes $\sqrt{\omega^2 + g/l}$, and $e^{-i\omega t}$ as a multiplicative factor is added to the front. This implies that in addition to swinging, the plane on which the pendulum swings will also rotate gradually in a counterclockwise direction, and this angular speed of the rotation will be $\omega = \Omega \sin \varphi$, where $\Omega = 360°/\text{day}$ is the angular speed of Earth's rotation. Substituting Ω into the expression we obtain

$$\omega = \frac{360° \sin \varphi}{\text{day}} \tag{3.38}$$

If the location is the equator, $\varphi = 0$, and the plane of the single pendulum's motion will not rotate. At the north or south pole, $\varphi = \pi/2$, and the plane of the single pendulum's motion will rotate once per day. The reason for its rotation is the Coriolis force generated by Earth's rotation. Again the fact that Earth rotates is indirectly confirmed.

Apart from solving the differential equation directly, Somerville [30] applied a brilliant geometric method to directly observe the angular velocity of the plane of the pendulum's motion, see please the Fig. 3.6. As explained by Somerville, one may consider the case where the single pendulum is located at point C on the latitude φ. Suppose Earth has rotated an angle of $\delta \alpha$ after a period of time. During this time, the position of the pendulum has moved from C_2 to C_1. Consider the angle $\angle C_1 A C_2$. If this angle is viewed from the plane of the vertical pendulum's swinging motion, namely the plane where the triangle $\triangle C_1 B C_2$ lies, then the swinging direction of the pendulum on this plane should move in parallel. The swinging direction of the pendulum

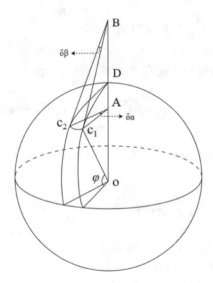

Fig. 3.6 Geometric analysis of Foucault pendulum's motion. Based on the information of Fig. 3.5 of Somerville [30]

at C_1 will therefore still be parallel to C_2B, i.e., the angle of $\delta\beta$ will form between C_2B and C_1B. From the geometric relationship shown in Fig. 3.6, we will therefore obtain

$$\overline{C_1C_2} = \overline{AC_1} \times \delta\alpha = \overline{BC_1} \times \delta\beta \qquad (3.39)$$

Since $\frac{\overline{AC_1}}{\overline{BC_1}} = \sin\varphi$, we have $\delta\beta = \delta\alpha \sin\varphi$. After integrating this expression, a rotation of $\alpha = 360°$ per day by Earth implies a rotation of $\beta = 360° \sin\varphi$ per day by the Foucault pendulum.

Appendix 3.2: Principles behind Galilei's Telescope Designs

The Galilean telescope employs a convex lens with a focal length of f_1 as the object lens and a concave lens with a focal length of f_2 as the eyepiece. The principle behind the formation of an image with this configuration is illustrated and explained in the diagram below. Assuming that the effective radius of the object lens is r_1 and the effective radius of the eyepiece is r_2, the following relationship applies

$$\frac{f_1}{f_2} = \frac{r_1}{r_2} = M \qquad (3.40)$$

Here M is also the magnification of the telescope. Since an image often appears blurry around the lens' edges, they are excluded from the calculation of the effective radius. Consequently, the effective radius of a lens will be smaller than its actual physical radius (Fig. 3.7).

Now we consider the first telescope constructed by Galilei to spot the ships far away from the Venetian coast. It had a magnification factor of $9\times$, or $M = 9$. Suppose the radius of the pupil of a human eye is 3 mm, and the effective radius r_2 of the eyepiece is twice that of the pupil, or 6 mm. Substituting these values into the above formula, we obtain $r_1 = 9 \times 6 = 54$ mm. Assuming that the actual radii of the object lens and the eyepiece are twice their respective effective radii, then this telescope is equipped with an eyepiece of actual radius 12 mm and an object lens of actual radius 108 mm. The telescope's actual length, in this case, has nothing to do with the focal length. Now suppose that the telescope's body is not excessively long for the purposes of transportation and installation. Then we can assume that its focal length is $f_1 = 1$ m, in which case $f_2 = (1/9)$ m. Here f_1 is the minimum

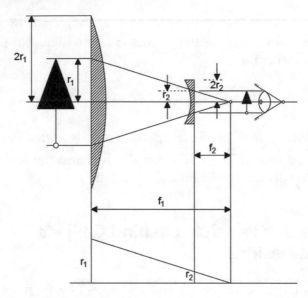

Fig. 3.7 A Galilean telescope's structure

length of the telescope's main body (the tube). If shading parts in front of the object lens and behind the eyepiece are taken into consideration, the maximum length would perhaps be 1.1 m. Galilei reportedly used spherical cannonballs to grind and polish his concave lens, so $f_2 = (1/9)$ m could have been the cannonball's radius, which is reasonable considering that the inner diameter of the artillery barrel in Galilei's time was comparable in size. Taking into account of these factors, the design specifications of the telescope described above appear to be reasonable.

Now we consider the telescope Galilei employed to observe Jupiter and the Moon. It had a magnification factor of $20\times$. Again we adopt the same specifications for the eyepiece, as in the previous telescope, with $r_2 = 6$ mm and $f_2 = (1/9)$ m. The corresponding object lens would have the specifications $r_1 = 20 \times 6 = 120$ mm and $f_1 = 20/9$ m. The length of this telescope would therefore be over 2 m, with the diameter of the object lens approaching 25 cm. This would have been a rather large and bulky telescope, and with its size and weight it would have been best regarded as a telescope for an indoor observatory rather than a mobile or transportable instrument. A large-diameter objective lens would also be better suited for astronomical observations, as it provides a field of view as wide as 30 arc minutes, which is required to observe the Moon.

As Galilei might have used cannonballs to grind and polish his concave lenses, we can speculate that the quality of lenses could not have been too high. The scattering of light caused by the lenses' rough surface and the chromatic aberration caused by white light passing through the lens would have produced blurred images. In addition, as shown in the photograph of a Galilean telescope below, the diameter of the eyepiece is perhaps close to 123 mm, as mentioned earlier, but the diameter of the object lens is obviously much smaller than 120 mm. It is therefore reasonable to believe that the images seen by Galilei were probably quite blurry and the field of view could not have been too wide. What Galilei saw were probably only portions of the objects he tried to observe and not their entirety.

Appendix 3.3: The Work of Scientific Revolutionaries

Continental Europe in the 16th century witnessed the fervent cultural, artistic, political and economic growth of the Renaissance and the Scientific Revolution, which brought about rapid improvement in the quality of spiritual and material civilization in the history of humankind. Traditional manufacturing methods were no longer adequate to meet the needs of society, and a variety of new inventions would soon become reality. One important invention was vacuum-producing machinery, a practical invention providing an effective substitute for human and animal labor. Vacuum machines were used for pumping and extracting groundwater and for driving simple farming machinery. Even military equipment for besieging cities found use for them. The originators of vacuum engineering were from Italy, the birthplace of the Renaissance.

All well-drilling operators in Florence in the 16th century were aware that the suction pumps they had been using to extract groundwater could reach no deeper than 32 feet (or 18 cubits) underground. Evangelista Torricelli (1608–1647), one of Galilei's last students, conjectured that this limit was the result of the maximum thrust the weight of the air could exert on the well water. To prove his theory, Torricelli designed an experiment in which a vertical glass tube filled with mercury had its upper end sealed off and its lower end exposed to air. Due to atmospheric pressure, the mercury column remained in the tube, but the maximum height of the mercury column reached only 30 in. (760 mm). Any mercury above this height would flow out of the tube via its lower end. Dividing the weight of the mercury column by the cross-sectional area of the glass tube gives what is now called one atmosphere (a unit of pressure).

Subsequently, Frenchman Blaise Pascal (1623–1662) extended Torricelli's experiment and reasoned that if the mercury tube was placed at a location of a higher altitude, such as the top of a mountain, then the height of the mercury column would be reduced, as it was common knowledge that air is thinner at higher altitudes (the lighter the air, the lower the pressure). Pascal conducted a series of experiments between 1648 and 1651 to prove his conjecture. At about the same time, Robert Boyle (1627–1691), the son of Anglo-Irish nobleman Richard Boyle, 1st Earl of Cork, was obsessed with science, even though he had put himself in extreme danger by choosing to side with parliamentarian forces in the 1640 s when the English Civil War broke out. Though perhaps a bit eccentric with his fascination with science, Boyle was a lifelong, devout Christian, so whenever he discovered a new phenomenon he would attribute it to the Creator. Boyle continued the vacuum experiments of Torricelli and Pascal, using an improved vacuum pump designed by his student Robert Hook (1635–1703) to extract the air entirely from the glass tube. He discovered that the height of the mercury column in the pressure gauge connected to the container was reduced. From the relative changes between the height reduction and the container's volume, he proposed what would be later referred to as Boyle's law: the pressure of a gas in a container multiplied by its volume is a constant.

Apart from practical principles concerning vacuum, geometry was also one of the greatest facilitators of the Scientific Revolution, having existed for over two millennia and had a much wider scope of practical applications. One of the most famous geometers was the philosopher René Descartes (1596–1650). Descartes was born into a French aristocratic family and studied law at the University of Poitiers. He joined the Dutch States Army of Maurice, Prince of Orange, during the Dutch War of Independence (Eighty Years' War) against the rule of Habsburg Spain. After the war, Descartes settled in the Dutch Republic in 1628 and devoted himself to scientific research. In 1649, Descartes was invited by the Queen of Sweden to Stockholm to be her tutor. He contracted pneumonia early in the following year due to the cold weather and died days later. Interestingly, Descartes only spent the last 20 years of his entire 53 years of life engaging in scientific research, but he managed to make notable contribution in a number of disciplines, including meteorology, mechanics, astronomy, mathematics and optics. As a result, he was regarded as one of the key figures in the Scientific Revolution.

The greatest contribution of Descartes is the creation of analytical geometry, a mathematical method that can be applied to describe geometric curves or surfaces and to compute the point(s) of intersection of two curves or the curve(s) of intersection of two surfaces. Descartes' achievements opened the door to using equations to describe relationships between quantities, as well as computational processes. The days of using purely verbal descriptions for these purposes were numbered, and the classical Euclidean approach for drawing geometric diagrams was no longer the only method available to make geometric arguments. In physics, Descartes described the relationship between the angles of incidence and refraction when light passes through a boundary between two different isotropic media. He further used this relationship to elaborate on the formation of rainbows. This was a time when the fact that white light is in fact a combination of lights of different (roughly seven) wavelengths in the visible spectrum was not widely known. Nevertheless, Descartes was able to apply the relationship between the angles of incidence and refraction to calculate the angular differences between the seven colors. The fact that Descartes was able to come up with values quite close to what are currently known is remarkable indeed. Overall, Descartes was lauded by the philosophical community for his approach to scientific research, but opinions regarding his practice of science were less favorable. As an example, France's delayed acceptance of Newton's universal law of gravitation for decades can be attributed to the many French scientists' adherence to Descartes' physical theories that did not agree with Newton's.

At last, there is another natural philosopher who remained an influential figure during the Scientific Revolution, so we are including him on the list of scientific revolutionaries: Englishman Francis Bacon (1561–1626). Bacon was the son of Sir Nicholas Bacon (1510–1579), Lord Keeper of the Great Seal of England. After graduating from Trinity College in Cambridge, Francis Bacon became quite successful in law, politics, and diplomacy, as his father had been. He was created Baron Verulam in 1618 and was promoted to the office of Lord High Chancellor at the same time and later Viscount St. Alban. In 1626, Bacon died of pneumonia, reportedly after having contracted the condition while studying the effects of freezing on meat preservation. Bacon's philosophy in science focused on empirical research; some would even characterize this emphasis as a kind of extreme empiricism, in stark contrast to Plato's philosophical approach. He not only disapproved of Aristotelianism but also rejected both the Ptolemaic and Copernican cosmological views, since these doctrines offered no pragmatic value to him. Instead he believed that scientists ought to collect all the useful knowledge and phenomena hidden in nature as comprehensively as possible, so that humans would be able to recover their God-given abilities to control nature which had been lost since Adam's expulsion from the Garden of Eden. In the 17th and 18th centuries, Bacon's scientific accomplishments were compared to those of Plato and Aristotle. In retrospect, however, no scientific achievements can be unequivocally attributed to Bacon, let alone his impact on future generations of scientists.

4

The Well-Ordered Newtonian Mechanics

萬有是如此簡約
引力是這般和諧
啊!萬有的引力
因有你
這個世界變得井然有序
我們也安然悠游於反覆更新的變化中

So harmonic is the universe
And so simple is the gravity
Ah, it's all thanks to the universal gravitation
The world thus becomes an orderly space
And the universe reaches eternal through inertia and conservation

Just like many scientific giants, Issac Newton (1643–1727) was also a weirdo. He was anti-social and aggressive, unique in thinking as well as actions, at the same time refusing to yield to authority. Born in the county of Lincolnshire, he later moved to Cambridge and London for research and government work, all the while living in his isolated world both mentally and physically, almost without any contact with real people, not to mention getting married. Though he spent his whole life in the long, narrow area bounded by these three places and never actually saw the ocean with his own eyes, he was able to describe the tidal phenomenon and its causing factors. Even more surprising, though, was how healthy he was. In fact, rumor has it that

© The Editor(s) (if applicable) and The Author(s), under exclusive
license to Springer Nature Switzerland AG 2020
F. Chen and F.-T. Hsu, *How Humankind Created Science*,
https://doi.org/10.1007/978-3-030-43135-8_4

he only shed one single tooth from cradle to grave. Equally amazing was how he managed by integrating astronomy, physics, and mathematics in an unprecedented way to explain a number of celestial and earthly movements with physical concepts and calculate changes in the natural world with mathematical approaches, thereby solving certain astronomical mysteries that had perplexed natural philosopher for more than 2,000 years. With achievements that almost no one in the history of physics could ever emulate, he debunked Aristotle's scientific hypothesis, which had been widely accepted for thousands of years and was still wielding significant influence on many scientists even in the 17th century. It's no wonder the famous economist John Maynard Keynes (1883–1946) hailed him not only as a magician who bridged the gap between ancient natural philosophy and modern physics but also a model in present-day scientific research worthy of our tremendous respect.

In addition to his research in physics, Newton also spent quite some time on unorthodox subjects during midlives, such as the miracles and divine wonders described in the Bible. According to this holy book of Christianity, God promises to reveal His will to those who bear true faith to Him, which led Newton to believe that prophets like Moses, Daniel, and even ancient philosophers, such as Plato and Pythagoras, were all aware of the law of universal gravitation because of God's enlightenment. Such a belief further prompted him to apply his mathematical talent to calculating the possible dates of important events in the Bible, including God's creation of the world and Noah's construction of the Ark. Besides, he even proposed that Apocalypse would take place between 2060 and 2344 based on the prophetic visions in the Book of Revelation. Despite his position as a professor at the Trinity College and interest in Biblical studies, his doubtful attitude towards the divinity of Jesus Christ gave rise to his opposition to the theological theory of trinity, putting him in a dangerous and ironic situation which luckily didn't cause another tragedy where science was suppressed in the name of religion. In addition to his dedication to holy matters, he was addicted to alchemy as well, spending nearly 30 years on relevant research and investing a good amount of budget to build labs and libraries for the sake of it. A hundred years after his passion for the mysterious practice cost his life, scientists concluded by analyzing his hair that the lead, arsenic, antimony, and other heavy metals in his body were four times the amount compared to normal people. As for mercury, it was a shocking 15 times higher. In a Sotheby's auction in 1936, Newton's 650,000 and 1,350,000-word manuscripts on alchemy and religion were put on sale, where his loyal fan Keynes purchased part of the former, probably hoping to grow rich overnight with his secret tips for turning basic substances into gold.

4.1 Teenage Prodigy

Newton's alienation from and hostility to the outside world might be products of his unhappy childhood. Born on Christmas Day of 1642 at the Woolsthorpe Manor in Lincolnshire, he didn't get a chance to meet his father, an illiterate farmer who died a few months prior to his birth. As for his mother, she just couldn't bring herself to like the tiny, premature newborn, so soon after marrying a rich priest twice her age when Newton was three, she sent him to the countryside to stay with his grandma, but then, the year of Newton's tenth birthday saw the death of his stepfather, who'd always disliked him, so he was able to reunite with his mother for a short while, only to be sent away at twelve again, to the Puritan-founded King's School in Grantham. Though he continued to act eccentric and isolated at school, his homestay host, pharmacist/chemist William Clark, was appreciative of his ideas and encouraged him to exercise his creativity, letting him scare neighbors with lanterns tied to kite tails, happily sprinkle medicinal powder into the air to test wind speeds, and spin toy windmills by making mice run on wheels. The good times, unfortunately, didn't last too long. Five years later, Newton's mother asked him to return home and take over the family farm, but he turned out to be a lousy farmer who didn't at all bother to weed his farmland or look for the poultry and livestock that went astray. Luckily, as if God had been reluctant to see this mathematical genius buried in haystacks and animal poop, the president of the King's School wrote a recommendation letter for him in 1661 and opened his door to the famous Trinity College of the Cambridge University. Over the 35 years following his admission, therefore, Newton was able to make use of the resources at the academic sanctuary and engage in his long-term research in mathematics, optics, and mechanics that profoundly influenced the development of human civilization.

The University of Cambridge was, though the ultimate academic heaven, a living hell for Newton at the same time because of the exorbitant tuition fees, which he couldn't possibly afford with the meager £10 given by his mother every year as allowance. Feeling the pinch of life but still striving to keep his spot, he chose to become a "sizar," entitled to lower fees but required to stay after school to help rich students groom, do their laundry, shine their shoes, clean their chamber pots, serve beers, and what not to pay the still-expensive tuition and basic expenses. During his first two years at Cambridge, he had to attend to his wealthy classmates besides studying Aristotle's long-standing theory day and night so he could maintain a minimal life. But such a demanding environment was exactly what helped cultivate his perseverance and patience for detailed thinking, which were indispensable characteristics for conducting the groundbreaking research in physics that he achieved later on. He worked 18 hours a day and seven days a week since the beginning of college life, all the years until he left Cambridge. It was indeed the decades of intense and sustained effort that made him who he is in the history of science.

In 1664, the just-turned-22 Newton started reading the publications of Galilei, Kepler, as well as Descartes (1596–1650) and started his own research plans. When he finished his university studies in 1665, he was also granted a public fund, which could've secured him four more years in Cambridge. Sadly, August of the same year saw the outbreak of black death, which took London by storm and afflicted one fourth of the city population (roughly 100,000 people) with dark, swollen lymph glands, eventually causing all patients, alive or dead, to be tossed into plague pits. With the entire country devastated by the horrifying disease, Cambridge, though 90 km away from London, was struck as well. Therefore, Newton had no choice but to go home since the school was suspended. Yet his two-year stay at the Woolsthorpe countryside was surprisingly a tranquil period of time, giving him the chance to engage in profound thinking and obtain significant results from his pioneering studies, such as the colors of light, the Binomial theorem, calculus, the law of universal gravitation, etc. In fact, these research findings from the short but prolific couple of years before he returned to Cambridge in the spring of 1667 were precisely what firmly laid down the foundation for his remarkable achievements in physics. That said, the early explorations were nothing but the beginning of his enlightenment. His real, thorough research didn't start until he returned to Cambridge. In the fall of 1667, the university was set to choose nine fellows to stay at the institute for research work, and Newton, now 25, was among those considered. Although many of the other candidates were strong competitors, some even backed by senior officials or handpicked by the king, he still won by a small margin, which saved him from the destiny of spending the rest of his life pasturing cattle and handling sheep poop at home.

4.2 Early Research

According to Newton's *Research Log*[1] (which was, to be honest, more like a day-to-day journal containing a bunch of trivial matters), he didn't dive right into the field of dynamics, which he had long been studying before, right after he was chosen to be a fellow. Instead, he was focused more on mathematics and optics. In fact, his research in any of the three fields from 1664 onward could easily land him a spot in the science hall of fame. In the following sections, we will take a look at his dedication to and achievements in optics, mathematics, and mechanics.

4.2.1 Optics

In the field of optics research, Newton had, in his student days, read up on the studies of Robert Boyle (1627–1691) and Robert Hooke (1635–1703), the former a renowned chemist and the latter an attentive experimentalist working as Boyle's assistant, familiar with all his theories. Venturing into the world of optics, Newton found that Descartes had long before calculated the angle variations of the seven-colored lights by analyzing the angles of incidence and refraction 20 years before but didn't realize white light was a combination of the colors. Prompted by his exceptional scientific intuition, Newton started to study the relationship between light and the seven colors on the visible spectrum before further designing a prism for white light to be split into red and blue constituents. He then continued to direct the red light into a second prism, only to find out it couldn't be further divided, but when he set up yet another pyramid-shaped piece of glass for the red and blue rays to enter, they exited as white light again. A few experiments later, he concluded that white light was composed of the seven-colored rays listed by Descartes and that these were primary colors that could no longer be spread apart.

After understanding the constitution of light, Newton set off to modify Galilei's long, thin refracting telescope, which would blur the edges of objects after two refractions due to the chromatic aberration induced by the prism in it. Therefore, he replaced the original concave lens with a curved mirror and added a flat secondary mirror for light to be reflected again into the eyepiece to eliminate the aberration effect and produce clear images where objects could be magnified up to 40 times. This telescope that he made was

---

[1]All Newton's research at the countryside was detailed in a blank notebook, left behind by his late stepfather, a.k.a. the rich priest. It was all thanks to this hard-to-come-by notebook (paper was a luxury back then) that he was able to write down his thought development, giving us a chance to look at his research during the several years after 1664.

merely about 15 cm, 1/3 the length of Galilei's predecessor; besides, it also inspired the invention of optical telescopes for astronomical use. With such achievements in the realm of optics, it's no wonder he was appointed the Lucasian Professor of Mathematics[2] two short years later at the young age of 27.

The 40x super telescope received widespread recognition among members of the royal family in London as well as political and academic elites. In 1671, Henry Oldenburg (1619–1677), honorary President of the Royal Society, invited Newton to submit his research to the Philosophical Transactions of the institution (which continues to see the publication of important articles today). When the study was actually published the next year, however, several influential figures stepped forward to question the originality and significance of it. Among the huge wave of criticism, the most unrelenting was nobody but Hooke,[3] who was then the curator of experiments at the Royal Society. According to him, Newton, who he claimed had previously read his papers, simply repeated the experiments on lights and colors that he had conducted in his own lab. In other words, he was basically accusing Newton of plagiarism. In the end, he even openly expressed his disagreement with Newton's theory that white light was the combination of seven colors out of indignation and indiscretion, igniting a debate that lasted for quite a while in academia. After the incident, Newton started being precautious about publishing articles[4] and isolated himself from academia by refusing to engage in any kind of

---

[2]The post of Lucasian Professor was founded by the Cambridge University's Member of Parliament, Henry Lucas, (1610–1663), who made a donation before he died in February of 1663 for the university to establish the chair of mathematics, with Issac Barrow (1630–1677) as the first Lucasian Professor. As for Newton, he had worked as Barrow's assistant for years before eventually being appointed himself.

[3]The curator was responsible for modifying and supplementing experiment ideas proposed by other members. With the sole exception of summer, Hooke had to submit evaluation reports to the Royal Society every single week, which lasted for 40 years before he passed away. It's said that he was a pale old guy with a bad hunch on his back, but his talent in invention made him a welcomed figure in the British upper class. With his endless curiosity, he was constantly able to bring up questions that surprised those present. In fact, he was credited with the discovery of cells, which he found in a piece of cork using the compound microscope he invented by himself. Besides, he also contributed significantly to techniques of astronomical observation by enhancing the resolution of Galilei's telescope, discovered the law of spring movement (later called "Hooke's Law"), and even proved the stimulating effect of weed with hands-on experiments. Yet respected as he was as President of the Royal Society, there wasn't any portrait of him left behind. Why? Because when Newton took over the position after Hooke died in 1703, orders were given to destroy all paintings of him in the institution due to their feud over academic issues. In fact, Newton even considered burning all Hooke's publications, but was fortunately talked out of if by his colleagues, who, by doing so, stopped yet another incident of academic persecution.

[4]It wasn't until 1690 did Newton compile his optical research into *Opticks* and published the book in English. By then, he had long been held as a prominent figure in the physics world thanks to *the Principia* (short for *Philosophiæ Naturalis Principia Mathematica*, Latin for *Mathematical Principles of*

academic exchange. While he was just slightly over 30, the hair at his temples was already turning gray.

4.2.2 Mathematics

Mathematical research wasn't the witness of Newton's most exceptional achievements, yet he wasn't short of some excellent results in this field. Sadly, such success also made him the victim of unpleasant academic battles which lasted even longer. To take it from the beginning, mathematics was where Newton first devoted his effort after returning to the countryside due to the plague. Deeply influenced by Barrow's studies, he was obsessed with the concepts of maximum, minimum, and infinity. The notion of "limit" in Euclid's method of exhaustion described in Chap. 1 (see Sect. 1.8) is a good example of what attracted him. Yet when it comes to the abstract construction of mathematical thinking, he still preferred to seek help from physical phenomena in the natural world. In his *Research Log*, Newton detailed how he understood and explained the theorem of calculus. Regarding the concepts about differentiation and integration that he put forward, below is a two-part summary established from the perspective of modern calculus.

Let's first suppose an object is located at $P(t)$ at a given moment t, and moves to $P(t + \Delta t)$ after a minimum time interval Δt, which brings us to $t + \Delta t$. To calculate the instantaneous velocity (v) of the object at t, we need to divide the displacement that took place during Δt by the time interval, as shown below:

$$v = \lim_{\Delta t \to 0} \frac{\Delta P(t)}{\Delta t} = \lim_{\Delta t \to 0} \frac{P(t + \Delta t) - P(t)}{\Delta t} \tag{4.1}$$

If we draw a graph with $P(t)$ and t on the vertical and horizontal axes respectively, then v would be the tangent slope at the point $(t, P(t))$. Then, let's consider a simple example where $P(t) = t^n$ and n is a real number. With this equation, we can derive the following result:

$$P(t + \Delta t) = (t + \Delta t)^n = t^n \left(1 + \frac{\Delta t}{t}\right)^n \approx t^n \left(1 + n\frac{\Delta t}{t}\right)$$

$$= t^n + nt^{n-1}\Delta t \tag{4.2}$$

Natural Philosophy), his other masterpiece published by the London Royal Society in 1684, and was free of the worries about incurring attacks on his studies.

Please, note that we need to apply binomial approximation when developing (4.2), i.e. when $|x| \ll 1$, $(1 + x)^n \approx 1 + nx$. As such, (4.1) can be simplified as:

$$v = \lim_{\Delta t \to 0} \frac{P(t + \Delta t) - P(t)}{\Delta t} = nt^{n-1} \qquad (4.3)$$

Equation (4.3) was named by Newton the "fluxion" of $P(t)$, which is what we call "derivative" today and has become the primary rule of differentiation in modern calculus.

In the second part of his study, Newton applied the concept of minimum distance to calculate the area enclosed by a curve and two vertical lines passing through both ends of it. Suppose there's a curve in a Cartesian coordinate system, and we can define it as $y = nx^{n-1}$. To calculate the area below the curve whose two endpoints are represented by $x = x_1$ and $x = x_2$, we need to divide it into an infinite number of rectangles along the direction of the x-axis, each of them defined by a width and height of Δx and y. When Δx is so small that it approaches 0, each small rectangle would be $y\Delta x = nx^{n-1}\Delta x$. Therefore, to calculate the total area under the curve, we can simply add up all the rectangles between $x = x_1$ and $x = x_2$:

$$\lim_{\Delta x \to 0} \sum_{x=x_1}^{x=x_2} y\Delta x = \lim_{\Delta x \to 0} \sum_{x=x_1}^{x=x_2} nx^{n-1}\Delta x \qquad (4.4)$$

Based on the previously mentioned formula of differentiation, when $\Delta x \to 0$, $nx^{n-1}\Delta x = \Delta(x^n)$. With the substitution of this equation, (4.4) can be rewritten as:

$$\lim_{\Delta x \to 0} \sum_{x=x_1}^{x=x_2} \Delta(x^n) = \int_{x_1}^{x_2} nx^{n-1}dx \qquad (4.5)$$

In fact, values calculated from (4.4) and (4.5) are no different. In the latter, the expression on the left side of the equal sign is the total amount of change in x^n, which equals $x_2^n - x_1^n$. As for the expression on the other side, the symbol is what we commonly use in integration now. As this calculating process is a reversion of the previously demonstrated induction of differentiation, Newton called it inverse fluxion, a.k.a. the integration today.

4.2.3 Universal Gravity

According to Newton himself, universal gravity was something he'd started thinking about as far back as in 1666, inspired by Galilei's motion experiments. Focusing on how to describe instantaneous velocity and acceleration with the concept of limit that we've discussed before, he sought to replace "average speed," an attribute often employed by scientists back then, with the two quantities and was eventually led by the research to the invention of calculus. When trying to test related theories that he developed, he was smart in choosing the hottest topic at the time, i.e. why do planets in the solar system circle around Sun following elliptical orbits, thereby popularizing the discipline of kinematics. However, to answer the question using a kinematic approach, he had to calculate not only the trajectory and speed of planetary motion but also the gravity between planets and Sun, which lies at one of the foci of any orbit.

How does one measure the gravity between celestial bodies, though? *The reason behind the falling apple was probably the Earth's force that drew it downward, but it happened too fast for me to observe the change in speed, not to mention measuring the vaguely existing gravity because the tree's just a couple meters tall*, thought Newton to himself. Yet just when the tricky situation was bogging him down, Galilei's ramp experiment (see Sect. 3.21 in Chap. 3) surfaced in his mind. He started suspecting that it was due to the offsetting effect between the vertical gravity of Earth and the opposite reaction from the inclined plane that Galilei's copper ball moved in linear motion. From Newton's perspective, although friction does exist between the ball and ramp, it's probably small enough to be disregarded, so it's reasonable to assume that the total external force exerted on the copper sphere was zero at the time of experiment, which allowed it to remain in uniform motion in a straight line. Through this insightful observation, Newton discovered the hidden gem in Galilei's experiment and established his first law of motion: an object at rest stays at rest and an object in motion stays in motion with the same speed and in the same direction unless acted upon by an unbalanced force.

Meanwhile, he noticed the copper ball's gradual acceleration after released from top of the inclined plane or thrown over the edge of a table due to the gravity of Earth, which drew it toward the ground, and the phenomenon that the level of acceleration seemed to be inversely proportional to the ball's mass when the amount of external force exerted on it was not considered. In other words, the larger the mass, the slower the acceleration, and vice versa. Following this rationale, he defined the physical quantity of "momentum" as the product of an object's mass and speed, contending that the change in momentum over time equaled the external force affecting the object. And this is Newton's famous second law of motion.

After these two laws dawned on him, Newton looked up at the Moon in the night sky one day, and started thinking about the gravity that affected planets: *why wouldn't they fall off just as the apple did?* In seeking an answer to the question, he again resorted to Galilei's experiment, where the copper ball hits the table after accelerating along the ramp and starts rolling at a steady speed toward the edge, where it then flies horizontally into the air before taking a parabolic fall. Out of experience, Newton considered the concept of limit once again and wondered: *if the speed at which the ball flies off the table is increased to a maximum, perhaps it can stay afloat instead of falling down, just as the Moon circles around Sun following a fixed orbit?* And that, was when he decided to shift his focus on the copper ball to the Moon and compare the gravitational forces on Earth's surface and Moon's orbit by calculating the ratio of them because Jean-Félix Picard (1620–1682) had already measured the Earth's radius at 6,370 km in 1669, and the distance between the Moon and Earth was also widely known to be 384,400 km[5] (i.e. the radius of the lunar orbit was 60 times that of the Earth). In addition, it was generally agreed that the Moon's revolving period around Earth was about 27.3 days, so Newton had all the necessary values for his calculation.

With all these numbers in hand, it was no hard task for him to calculate the Moon's revolving speed (v) by dividing the circumference of the lunar orbit by its orbiting period, which can then lead us to the Moon's centripetal acceleration, v^2/r, where we can further substitute the corresponding values mentioned in the previous paragraph to obtain the result of 0.002728 m/s^2. As for the gravitational acceleration on Earth's surface, Giovanni Riccioli (1598–1671) had long before determined the value at 9.81 m/s^2 (or 32.1 ft/s^2) in 1659 via a free-fall experiment. As $0.002728/9.81 \cong 1/3600 = 1/60^2$, where the number 60 represents the ratio of the Moon-Earth distance to that between his apple and the Earth's center, Newton was able to prove that the centripetal force (i.e. the gravity exerted by Earth on the Moon and apple) was inversely proportional to the square of distance. Obtaining such concrete evidence, he was probably as thrilled as a kid in a candy store, sitting quietly at his lab desk yet struggling to hide his excitement.[6]

---

[5]The radius of the lunar orbit can be measured with the diurnal parallax of the Moon. For details, please see Sect. 2.6 of Chap. 2.

[6]A difference source points out that the values Newton acquired regarding the mass of Moon and Earth as well as the distance between the two orbs were actually incorrect and therefore caused his results to be incompatible with the theory that gravitational force was inversely proportional to the square of distance since he used the wrong values in the equation of universal gravitation. For details, see Chown [5].

After proving the gravitational force between two celestial bodies was inversely proportional to the square of their distance, Newton took his research a step further to consider how such bodies moved in the universe and naturally thought of Kepler's three laws of planetary motion since they were the only theories at the time that described the movements of planets using a quantitative, structured approach. He first accepted the proposal about elliptic orbits in Kepler's first law before trying to prove it, and then, he looked deeply into the second law as it clearly states a line segment joining a planet and the Sun sweeps out equal areas during equal intervals of time. In other words, the area swept by the line between any planet traveling in an elliptical orbit and the Sun, which sits at one of the foci of the ellipse, can be written as $rv/2$, with r representing the Sun-planet distance and v the velocity component of the planet on the direction perpendicular to r. If $rv/2$ is a fixed value, so should the angular momentum acting on the planet (mrv), and as the vector sum of all torques acting on a particle is equal to the time rate of change of the angular momentum of that particle, a fixed angular momentum implies that the rates of change and torque are both zero. Simply put, the gravitational force acting on planets wouldn't produce any torque, meaning such force must be directed toward the Sun.

Having determined the direction of gravity using the theory about equal areas, Newton went on to measure gravitational force with the help of Kepler's third law, which suggests if the trajectory of a planets is approximated as a circle, the square of its revolving period would be directly proportional to the cube of the semi-major axis of its orbit, i.e. $(T_1/T_2 = 8)^2 = (r_1/r_2 = 4)^3$. In fact, this law enjoys an exalted status in the development of physics as it explicitly regulates the mathematical relationship between the physical quantities of r and T, which not even a physicist so brilliant as Galilei was able to do, so Newton told himself out of instinct that the equation must be an indispensable element in establishing a formula to calculate the amount of gravity. Below we demonstrate Newton's derivation process, where v represents the revolving speed of a planet and r the radius of its orbit, meaning the distance traveled after one revolution is $2\pi r$, while the time needed is $2\pi r/v$, which can also be understood as the orbital period (T). As indicated in Kepler's third law, $(2\pi r/v)^2$ and r^3 are directly proportional, so we know by multiplying the reciprocals of both values by r that v^2/r and $1/r^2$ are proportional to each other as well. As such, v^2/r can be written as:

$$\frac{v^2}{r} = \frac{\left(\frac{2\pi r}{T}\right)^2}{r} = \frac{r^3\left(\frac{2\pi}{T}\right)^2}{r^2} \tag{4.6}$$

When multiplied by the mass of Moon, v^2/r yields a result representing the gravitational force exerted on the Moon by Earth. Plus, the numerator of the rightmost expression in (4.6) can be confirmed as a constant since $r^3 \sim T^2$, which also allowed Newton to ensure again that the gravitational force between the Moon and Earth was inversely proportional to the square of their distance, and guess what, this mathematical relationship is applicable to the phenomena of falling apples and planetary revolution around Sun as well. Such discovery is exactly the core concept of the law of universal gravitation.[7]

The concise equation above was derived effortlessly thanks to Newton's exceptional scientific intuition, yet he still decided to prove it with a well-rounded approach out of discretion to make sure the formula wasn't flawed in any possible way. Otherwise, opponents like Hooke would probably jump at the chance to humiliate him mercilessly again. In his quest for proof, Newton first thought of Galilei's observation of Jupiter's four moons and his logs of the orbital radius and period of each. However, Jupiter was fairly far from the Earth, so it wouldn't be surprising if the measurement made using a not-so-advanced telescope and other astronomical instruments were inaccurate to some degree.[8] Still haunted by the criticism he'd suffered before, Newton was too hesitant to draw a definite conclusion although the results he got by calculating the data of Jupiter did suggest his equation was correct, so he eventually decided to stash the research away, keeping it unpublished for more than 20 years.

4.3 Contemporary Research on Solar System and Gravity

As mentioned earlier in this book, several groups of scientists in Europe had started studies on questions left unsolved by previous researchers about the solar system and gravity during the same period when Newton was developing his law of universal gravitation, and these fellow scientists also obtained important data such as the Earth's radius and distance from the Moon, etc. When it comes to gravitation-related issues, for example, the orbital radius of each planet and how strong the gravitational force is on Earth, there was

---

[7]In fact, Huygens already had the equation derived in 1659, but Newton had not yet realized how to conduct such research back then.

[8]In 1676, the Danish astronomer Ole Christensen Rømer (1644–1710) found out unexpectedly that the speed of light wasn't an indefinite value while he was observing a moon of Jupiter's. However, it turned out that the light speed calculated based on the velocity and distance of this moon was 30% deviant from the actual value, which shows clearly that the astronomical data back in those days was erroneous to a certain degree.

also an abundance of research results, which we will further discuss in two sections below:

4.3.1 Cassini's Calculation of the Radius of Earth's Revolving Orbit

In the world of 17th century's astronomy, the theory put forward by Aristarchus of Samos (310–230 B.C.) in 280 B.C. that the Sun-Earth distance was 20 times that of the Moon from Earth was still a mainstream belief, yet the fact that such important data was dictated by some rough estimate 2,000 years before not only left the French astronomer Giovanni Domenico Cassini (1625–1712) fairly disappointed but also gave him the aspiration to come up with an accurate calculation. In his time, detailed data about orbital periods of the planets in the solar system wasn't something uncommon in many scientists' star catalogues, so as long as he could acquire the radius (R) of a certain planet, the same attribute of the other planets and even the size of the entire solar system could all be calculated without much hassle. Of course, such potential aroused his interest, and he decided to determine the radius of Earth's revolving orbit first to use as his reference.

Since it was impossible to directly measure Earth's orbital radius, he decided to start from a geometric analysis by taking advantage of the parallax effect, which would require him to first acquire the distance between Earth and Mars. Therefore, he sent a colleague to Cayenne, a city located on the north-eastern tip of South America, to measure the angle between the lines drawn respectively from Mars and a distant star to where the observation was based, while he himself stayed in Paris performing the same measurement. When two sets of data were both collected, the orbital radius of Earth (R) was calculated using the aforementioned "parallax" effect as follows:

In Fig. 4.1, P_1 and P_2 represent where two separate groups of researchers are located to observe Mars. While P_1 is where the line between Mars and the

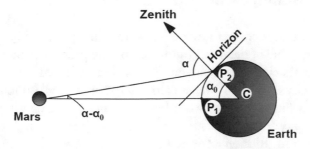

Fig. 4.1 Illustration of how Cassini measured the distance between Mars and Earth

Earth's center intersects with its surface, P_2 is where the angle (α) between the lines drawn respectively from Mars and a distant reference star, Zenith, to the point itself is measured. Suppose the distance from Mars to Earth is L, and the angle between P_1C and P_2C is α_0. When the radius of Earth is R, the geometric relationship illustrated in Fig. 4.1 can be represented as follows with the law of sines:

$$\frac{L}{\sin(\pi - \alpha)} = \frac{R}{\sin(\alpha - \alpha_0)} \tag{4.7}$$

From (4.7), the distance between Mars and Earth (L) can be calculated. Please note, however, that the equation is only valid when P_1 is where the line between Mars and Earth's center intersects with its surface. If not, related calculations would need to be adjusted for accuracy.

With L confirmed, Cassini developed the below equation with Kepler's law about orbital periods, with R representing the supposed radius of Earth:

$$\frac{T_e^2}{R^3} = \frac{T_m^2}{(R + L)^3} \tag{4.8}$$

Just as in (4.7), this equation only works when Earth, Sun, and Mars are on the same line. Under this circumstance, R can easily be calculated from (4.8) since the orbital periods of Earth (T_e) and Mars (T_m) as well as L are all known values. Although the R Cassini calculated (140,000,000 km) was 7% shorter than the actual distance, it was the first time the size of the solar system could possibly be estimated since the orbital radius of all the other planets could also be obtained following the same equation. Back in those days, nevertheless, Saturn was known to be the outermost planet in the system. It wasn't until two hundred years later when Uranus and Neptune were discovered in tandem.

4.3.2 Measuring Earth's Gravity

In Newton's time, there were two other important publications centered on Earth's gravity. One of them was published by the previously mentioned Giovanni Battista Riccioli from Bologna, Italy, who conducted his free fall experiments at the city's Asinelli Tower, an almost 100-meter-tall building originally constructed in the 11th century but has ever since seen a continued tilt that's over 5° now. In order to get an idea of how much time it took an object to free fall to the ground, Riccioli exhibited a maximum level of patience in counting how many times Galilei's pendulum swung a day to calculate its period.

Dividing the 86,400 s of a day by the oscillation count of the pendulum (imagine counting tens of thousands of times!), he calculated how long each period lasted in seconds, with an accuracy to three or four decimal places. Then, by multiplying the pendulum's period by the count of swings taking place during a free fall, he successfully came up with the time needed for an object to drop to the ground. According to Riccioli's experiment logs, one free fall from the 312-Roman-pes Asinelli Tower (equal to 323 ft, about 98.5 m) was completed over the course of 4.55 s, which, when coupled with Galilei's law of falling bodies, would set the gravitational acceleration at 32.1 ft/s^2, a jaw-dropping result that's incredibly accurate.

The other major research publication was authored by the Holland-born Christiaan Huygens (1629–1695). After entering Leiden University in 1645 as a math/law major, he started shining in fields like astronomy, statics, kinematics, optics, and of course, mathematics. Observing the sky through Galilei's telescope, he discovered the largest moon, Titan, of Saturn and proved that it wasn't just Earth and Jupiter that had natural satellites. Then in 1664, he was chosen to be a member of the French Academy of Sciences and moved to Paris, a city he called home for 20 years. Due perhaps to the well-paying position at the academy and living conditions that improved significantly with it, Huygens' Paris-based research was exceptionally successful, with his 1678 publication, *Traitè de la Lumiere* (*Treatise on Light*), giving shape to the wave theory of light.[9] As for his experiment on gravitational acceleration, it was inspired by one of Galilei's theories.

About half a century before, Galilei found out in his experiment that the time needed for a pendulum to complete one oscillation (period) was irrelevant of the angle it swung away from its equilibrium position (amplitude), yet Hyugens modified this theory by noting that the irrelevance was only valid when the amplitude was extremely small. In 1673, Huygens further proposed in his book *Horologium Oscillatorium: sive de motu pendulorum ad horologia aptato demonstrationes geometricae* (Latin for *The Pendulum Clock: or geometrical demonstrations concerning the motion of pendula as applied to clocks*) that the ratio of a pendulum's period to the time needed for an object to fall from the same pendulum's mid-point to bottom equaled the ratio of

---

[9]In his 1678 publication *Traitè de la Lumiere*, Huygens defined light as a type of disturbance wave traveling at a limited speed in luminiferous aether. Though today we know light to be electromagnetic waves that don't need any medium to transmit, which is slightly different from but similar to his understanding, there's no denying that his achievements are still very much worth noting. Back in his time, he was already able to use the orbital period of Io, which varies because of the ever-changing relative locations of Earth and Jupiter, as well as the revolving speed of Earth, already calculated by Kepler, to reach the conclusion that light needed 11 min to travel the distance of Earth's orbital radius, not much different from the 8.32 min that we know today.

the perimeter to the diameter of a circle whose radius was the length of the pendulum, and this ratio was 3.14159, exactly the π we know today. This mathematical relationship can now be expressed with the following equation: $T = 2\pi\sqrt{l/g}$, where T represents the period, l the length of the pendulum, and g the gravitational acceleration of the free-falling object. In order to prove his hypothesis correct, Huygens had to calculate the value of g, which was why he conducted an experiment to measure the gravitational constant. According to several related passages in *Horologium Oscillatorium*, he performed a free-fall experiment, where he found the distance traveled by the falling object to be 15.08 "meters" (which equals 16.1 ft or about 4.9 m now as the "meter" back then was defined as 1/10,000,000 the length of the meridian). If back-calculated with Galilei's theory that the square of an object's falling time was proportional to the distance it dropped, Huygens' findings would set the gravitational acceleration at 32.2 ft/s^2 (9.81 m/s^2), a reasonably accurate result surprisingly close to the actual value.

Although Huygens and Riccioli were successful in measuring the gravitational acceleration, both their research approaches were based on Galilei's law of falling bodies and pendulum theory. The former proves the distance traveled by an object is proportional to the square of its falling time, while the latter states that the period of a pendulum is irrelevant of its length, which explains why the device can be used as a time-measuring instrument with proper modification. To be fair, Galilei's greatness lies in the fact that with the foundation he laid in kinematics during his lifetime, especially concepts like average speed and acceleration that he proposed, he abandoned Aristotle's habit of describing motion based on mere "feeling" and transformed related studies into a modern scientific discipline well organized and supported by evidence. On the other hand, Huygens was no less exceptional considering he was able to derive the precise results mentioned above before calculus or advanced calculation tools were invented, which couldn't possibly be pulled off without his thorough, careful observation or accurate measurement, some of the key factors that allowed him to become an outstanding scientist when the scientific revolution was just gaining steam. Specifically, the gravitational constant that he calculated wielded an important influence on Newton's development of the law of universal gravitation, and his pendulum-based hypothesis that objects would also accelerate in curvilinear motion provided a solid bedrock for Newtonian mechanics as well.

4.4 Hidden Discovery

After deriving his first two laws of motion from Galilei's ramp experiment, Newton went on to develop the theory that the gravity exerted on an orbital planet was inversely proportional to the square of its distance from the force based on Kepler's three laws of planetary motion, thus integrating his studies on the movements of Earthly objects as well as celestial bodies and proving both groups to be prescribed by the same law. In other words, the force drawing planets toward the Sun and the gravity on Earth were supposed to be the same thing. Significant and groundbreaking as the discovery was, Newton didn't publish it until almost 20 years later, which inevitably aroused the curiosity of scientific historians as to the reason behind his choice, and most of them attributed the delay to insurmountable root causes like his anti-social personality, violent temper, unpredictable behavior, etc.

Other scholars, however, argued that it was Newton's competitive but prudent attitude towards knowledge that led to his dissatisfaction with two details in his developing process of the law of universal gravitation. For one thing, although the diameters of Moon and Earth as well as the distance between the two orbs were already established, the measurements weren't exactly that accurate. He did use the existing values to prove the inverse proportion between gravitational force and the square of distance, but what if the numbers had been wrong from the very beginning? For the other, perhaps he was even more concerned about the fact that his calculation of Earth's gravity was based on the hypothesis that all the gravitational force was completely concentrated on Earth's center. If the apple he observed had been fairly far from the ground, say, as far as Moon from Earth, his hypothesis might not stir too much controversy, but that's exactly where the problem was: the apple was nothing but a couple of meters from Earth's surface and could easily be drawn by the force of plants, mountains, oceans, and pretty much everything on the massive planet, so how was he supposed to prove the gravity to be centered on one single point only? Despite his initial frustration, the hypothesis turned out quite easy to prove, which Newton himself also explained in his *Philosophiæ Naturalis Principia Mathematica*. Below, let's validate his assumption using some arithmetic rules in today's integration.

Suppose there's a spherical shell whose mass and center are M and O respectively, while a ball with a mass of m is situated in a distance and r represents how far the ball's center is from O, as illustrated in Fig. 4.2. If the radius of the shell is R, the area of one specific tiny strip on the surface of it can be written as:

$$dA = 2\pi R sin\theta \cdot R d\theta = 2\pi R^2 sin\theta d\theta \qquad (4.9)$$

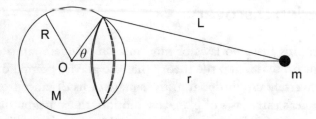

Fig. 4.2 Geometrical relationship between the ball and spherical shell

If we take the density (mass per unit of area) of the shell surface, ρ, into consideration, the mass of the strip would be:

$$dM = 2\pi\rho R^2 sin\theta d\theta \qquad (4.10)$$

Then, the law of universal gravitation that we're familiar with now can help us calculate the gravitational force between the spherical shell and the ball (m) to be:

$$F = m \int \frac{GdM}{L^2} \cdot \frac{r - R\cos\theta}{L} = m \int_0^\pi \frac{2G\rho\pi R^2(r - R\cos\theta)\sin\theta d\theta}{\left(R^2 + r^2 - 2Rr\cos\theta\right)^{\frac{3}{2}}}$$

$$= m \int_{-1}^1 \frac{2G\rho\pi R^2(r - Ru)du}{\left(R^2 + r^2 - 2Rru\right)^{\frac{3}{2}}} = m\frac{4G\rho\pi R^2}{r^2} = m\frac{GM}{r^2} \qquad (4.11)$$

Please note, that $\cos\theta$ was replaced by the variable u in the integral calculations of (4.11), where the last equation shows that the gravitational force between the shell and ball is inversely proportional to the square of r, which represents the distance between O and the center of m. Simply put, we can safely assume that the shell's mass is concentrated on its center O. Plus, as a solid orb can be understood as several concentric shells stacked together, the calculations above help prove the hypothesis that Newton made when trying to develop the law of universal gravitation.

Although Newton had already invented calculus when he proposed the law, all his arithmetic rules were based on empirical observations only rather than logic. Therefore, it was in fact some extremely lengthy geometrical analyses that he adopted in *Philosophiæ Naturalis Principia Mathematica* to explain the concept of integration and differentiation. As calculus was an unprecedented mathematical discipline entirely new to the public, the process of

making it understandable and accepted might actually be even more complicated than explaining the law of gravitation itself, so proving his hypothesis
about Earth's center using notions of calculus probably seemed to be a no
go for him. Meanwhile, as he was also keen on research in other fields, such
as optics, mathematics, Torah, alchemy, etc., plus the data about Earth and
Moon remained unconfirmed, it's no wonder he didn't publish the research
results until 1684. Why did this particular year see the publication of his
brainchild, then? Because in 1664, Halley's Comet lit up the night sky with
a jaw-dropping blaze.

4.5   Unearthing the Buried Law of Universal Gravitation: Halley's Comet

Historical accounts have shown us that through the past thousands of years,
the appearances of comets trailing tails of varying colors through the night sky
would always evoke people's fear around the world, making them believe the
phenomenon was an inauspicious sign from the heavens. For example, the
Maasai, Zulus, Aka peoples in Eastern, Southern, and Western Africa were
convinced that comet strikes were ill omens of famine, war, and pandemic
respectively, while as far east as in China, cometary orbits started to be kept
track of since 1,400 B.C., and the shapes and lengths of comet tails were even
used to predict what kind of disaster might be lurking, as shown in Fig. 4.3.
For example, comets whose tails covered the largest areas were represented

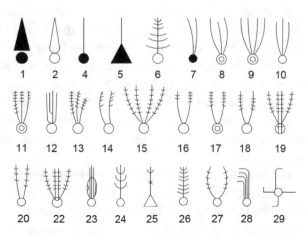

Fig. 4.3 Ancient China's rendition of different types of comets. All icons have
been recreated based on the picture of following website—http://aeea.nmns.edu.
tw/comets/mawangdui%20silk%20comets.html

by four lines and deemed portents of sweeping diseases. In contrast, the icon of three lines represented the second largest comets, which were regarded as signs of imminent national wars.

Nowadays, we know comets are nothing but the remnants left over from the formation of celestial bodies, which are small rocks mixed with ice formed by condensation. When comets passing by the solar system approaches the Sun, the ice contained in them gets melted by the warmth and releases gases that take away the dust among the rocky particles, thereby forming the radiating tails often seen with moving comets. After heated thousands of times by the Sun, comets would diminish in size, their trajectories deformed as well. Some of them would be gone from the solar system forever, while some others disappear after hitting other planets, becoming meteorites that can be seen on Earth. Out of the unknown number of comets in the universe, however, there are still some that circle as usual on their orbits and pass by the Sun once in a while, and these, are what result in the cometary activities that we see from Earth.

Nevertheless, comets are not all in the solar system but interspersed in the vast universe. Some only approach the Sun once every million years, while others can complete an orbital period in merely decades. According to astronomical estimates, at least 100,000 comets came close to the Sun in the past 100,000 years, while all humans did was panic about the phenomenon due to the lack of any knowledge about it. Luckily, this changed in 1682, when the secrets about comets were uncovered by a couple of scientists at the University of Cambridge after one illuminated the night sky of Europe. However, the decade before their research results were made public was, as coincidence would have it, plagued with all kinds of disasters and misfortunes, such as the notorious Black Death, a roaring fire that burnt down half of London, as well as the Anglo-Dutch Wars that sent ripples through the political sphere of Europe, so the comet, which trailed a blood-colored tail, was naturally seen as an inauspicious sign and became one of the hottest topics discussed all over England. Yet for the young and curious Halley, (Sir Edmond Halley, 1656–1742), the comet strike wasn't so much of an evil omen as an optimal chance for him to check out the enigmatic celestial body using the telescope given by his dad as a gift.

Halley was born in a wealthy family, with a father who was happy to inspire and indulge his curiosity. This loving dad didn't just bring home advanced scientific instruments but even sent Halley to the seas south of the equator in an exclusive research ship as well for his then 20-year-old son to draw a star map of the constellations in the southern hemisphere's night sky over the course of an entire year. His reason? Because with the newly-acquired map as well as existing knowledge about the stars in the northern hemisphere, British sailors could navigate all the major oceans of the world without getting lost. In addition to sending his son on the epic map-drawing trip, Halley's father was also persistent on collecting evidence about comet strikes, such as related documentation from ancient times as well as accounts of contemporary witnesses, and learned that the first documented comet was seen by a Byzantine monk in Constantinople, in June of 1337. Like father, like son. Halley later inherited his dad's research and started organizing the astronomical records of cometary activities established between 1472 and 1698. Sorting through a large amount of data, he carefully compared and contrasted the time interval between each strike as well as the similarity in cometary orbits, thereby grouping appearances that he deemed belonging to the same comets.

Eventually, Halley achieved his research objective and concluded that comets, just like planets in the solar system, revolved on elliptical orbits, with the major axes also pointing to the Sun. Basing his arguments on these findings, he claimed that the strikes in 1531, 1607, and 1682 all came from the same comet, which he said passed through the solar system periodically with an orbital period of approximately 76 years. In other words, 1758 should be the next time it came by. Then, just as he'd predicted, the comet again lit up the night sky of Europe 50 years after, exactly when the public was expecting it. And guess what, that particular strike wasn't following by any disastrous events, either, as if Halley's prediction had helped break the curse. People all marveled at this achievement, cheering for his success, and the comet was therefore named after him, becoming what we know as "Halley's Comet" today. This orb of ice, rocks, and other substances also got to shake off its notoriety as an ill omen and became a source of magical light that most people can only witness once in their lives.

Indeed, Halley secured his spot in the scientific history with the discovery of Comet Halley, but his other contribution may actually be of a greater significance, although it was thanks to Newton's research. So, the story all started with a bet made at a coffee shop in London. In the 17th century, coffee shops began springing up like mushrooms in London and attracted many to gather and talk about new knowledge, entrepreneurship, society, and politics over a coffee. Although it was an era firmly bound by the shackles of social class, coffeehouses were permeated with a fresh atmosphere where all felt equal, which also spread to France later on. In fact, it was a coffeehouse in Paris where rioters first gathered before heading over to bring down Bastille in the 18th-century French Revolution. Anyway, in the summer of 1684,

Hooke, Halley, and Christopher Wren[10] (1632–1723) met at a coffee shop imbued with a revolutionary spirit and raised a question while discussing the social unrest cause by a comet strike two years earlier: why is it the Sun that planets revolve around?

The orbit of a planet is an ellipse with the Sun at one of the two foci. A line segment joining a planet and the Sun sweeps out equal areas during equal intervals of time. The square of the orbital period of a planet is directly proportional to the cube of the semi-major axis of its orbit. Yes, these three simple, elegant laws that Kepler developed using data from Tycho's precise measurements were well known to the academia, but he never really explained what caused such planetary motion or if the laws were applicable to all planets and comets. In the scientific community of London back then, people generally believed that it was the Sun's gravity toward planets that made them revolve around it and that the amount of gravity exerted on a certain planet was inversely proportional to the square of its distance from Sun. Such belief, however, wasn't based on any detailed, logical reasoning but rather a simple notion that gravitational force should be equally distributed to each planet. So suppose F is the gravity exerted by Sun on surrounding planets, and this force is evenly disseminated towards all directions around it. Under this hypothesis, when the force is transmitted to an orbiting planet, the amount of gravity allotted to any point on its spherical surface would all be $f = F/4\pi r^2$, where r is the orbital radius and $4\pi r^2$ the tangent plane of the orbit and the planet. Therefore, f and r^2 are inversely proportional.[11]

The mathematical relationship of inverse proportion is straightforward and easy to understand, but how do we prove it? If the theory's correct, what's the underlying force, and how can it be measured? The discussion among the three scientists grew heated with the stimulation of caffeine, but a clear answer was still nowhere to be seen. Hoping to break the stalemate, Wren, who got a brilliant idea all of a sudden, proposed to buy the answer with 40 Shillings, which Hooke confidently claimed to be his right on the spot, bragging that he already knew the key to solving the question. Yet a few

---

[10]Wren was one of the founders of the London Royal Society, but in fact, he was not just an astronomer but one of the most renowned architects in England as well. He was in charge of the city reconstruction after the Great Fire of London and received the honor of knighthood in 1673. In 1681, he was appointed the President of the Royal Society and held the position for two years.

[11]The Irish missionary Johannes Scotus Erigena (815–877) started researching the same issue about gravity as far back as in the ninth century and claimed that the force from Earth that caused an object to fall was inversely proportional to its height. In the 17th century, the French priest Ismaël Bullialdus (1605–1694) also proposed that the force keeping a planet on its orbit was inversely proportional to the square of its distance from Sun.

months passed. Hooke still didn't come up with any results, so out of curiosity and impatience, Halley visited Newton in Cambridge in August of the same year.

At that point, Newton had been locking himself in his lab at the University of Cambridge for as long as 13 years because Hooke accused him viciously and violently of plagiarism when he published his research on the seven primary colors of light. Out of anger, the young and proud Newton decided to isolate himself in the College of Trinity at Cambridge to prevent contact with anyone he deemed evil manipulators of academic politics. During this period of total isolation, he tried indulgently to interpret the Bible code, hoping to pinpoint the time of the Second Coming, while dedicating himself to the study of alchemy, harboring the expectation of turning lead or mercury into gold. However, he had nothing accomplished in these fields after more than ten years went by, only to find himself wasting the golden years of life. When Halley visited him in August of 1684, the question about the planetary revolution around Sun was like a blow that knocked the 42-year-old Newton out of his little daydream. He told this visitor right away that he was well aware of the answer as he had conducted related studies as far back as 20 years before. Unwilling to be hoaxed again, Halley asked for his calculations immediately, but Newton just couldn't find the manuscript of the research even after turning his lab upside down, so at the end, he had no choice but to promise Halley that he'd send the materials over in a month after he sorted them through.[12]

Unlike Hooke, as it turned out, Newton was a man of his words. Three month after returning disappointedly to London, Halley really received a short nine-page essay from Newton, titled "De Motu," where he used the laws of motion and universal gravitation, which he developed 18 years before, to explain Kepler's second and third laws of planetary motion (as discussed in Sect. 4.2) while providing three possible types of planetary orbits: (a) planets that didn't move fast enough to break free from the Sun's gravity would stay on their elliptical orbits forever, (b) planets that were fast enough to break free from the gravity would move away from the Sun following hyperbolic trajectories and never turn back, and (c) planets whose speeds were balanced with the Sun's gravity would motion following parabolic tracks.

Newton's meticulous way of research was fully reflected on the well-structured essay. He started "De Motu" with a long and detailed explanation about several fresh, unprecedented definitions in the field of kinematics he'd been mulling over, which he then followed with several simple formulas and

---

[12] See also the story illustrated in Chap. 7 of Mlodinow [23].

quite a few geometrical shapes. Although he was already able to use his self-invented calculus plus some extra formulas to explicate how he developed the law of universal gravitation from Kepler's three planetary laws, he still chose to describe and prove his thought process with the relatively-complex geometrical approach in order for his theories to be accepted more easily by the academia, where people were then more familiar with geometry. Thanks to Halley's curiosity, Newton finally realized the importance of his research on gravity and was again motivated to showcase his results. At the end of the essay, he used the interaction between the gravitational forces of Moon and Earth as an example to endorse the validity of his law of gravitation. Just as Newton said himself, all these research materials were initiated during the two years at Woolsthorpe after he returned due to school suspension 18 years before. Though conducted centuries before our time, his research continues to hold an important status now, so below, let's take a look at how he examined Kepler's three planetary laws using his theory of universal gravitation.

4.6 Examining Laws of Planetary Motion with Universal Gravitation

With the law of universal gravitation developed, Newton then used it to verify if Kepler's laws of planetary motion were actually correct. Now, let's try to derive the three laws with calculus by basing our calculations on the law of universal gravitation. Details on the process will be provided later in this book.[13]

4.6.1 Kepler's First Law of Planetary Motion: Elliptical Orbits

To prove this law, which states that a planet's revolving orbit is elliptical in shape, let's first consider the impact of the Sun's gravitational force only and disregard all the possible disturbance that might be caused by the gravity of other planets. As the mass (M) of Sun (S) is much larger than that of the planets (say, the Earth) in the solar system, S is a fixed point which P, the planet in Fig. 4.4 whose mass is m, revolves around following a circular or elliptical orbit, as shown below. Suppose the planet is r away from the Sun in distance, and that the gravity working between the two orbs is $F(r)$.

---

[13]There's a different approach other than the one demonstrated here that can be used to derive the laws of planetary motion. For details, please see Hsiang and Chang [14].

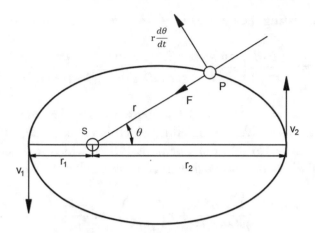

Fig. 4.4 How a planet is affected by gravity when motioning on an elliptical orbit

Under this circumstance, the kinematic equations of P on the radial and axial directions can be written respectively as follows in the polar coordinate system:

$$m\frac{d^2r}{dt^2} - mr\left(\frac{d\theta}{dt}\right)^2 = F_r \tag{4.12}$$

$$2m\frac{dr}{dt}\frac{d\theta}{dt} + mr\frac{d^2\theta}{dt^2} = F_\theta \tag{4.13}$$

In (4.12), F_r represents the universal gravity, but please note that the angular momentum of a planet is a fixed value when it's revolving on an orbit, so the gravitational force acting on P wouldn't generate a torque on it, which can be observed from (4.13). This equation, in fact, can also be rewritten as $\frac{d}{dt}\left(mr^2\frac{d\theta}{dt}\right) = F_\theta r$, and when $F_\theta = 0$, $mr^2\frac{d\theta}{dt}$ has to be a non-changing value, which again tells us that the angular momentum is fixed. Simply put, when it comes to the direction of gravity, we can confirm that it acts along the line between the Sun and the planet.

To solve (4.12), we need to simplify it with the equation $u = 1/r$, which will lead us to:

$$\frac{dr}{dt} = -\frac{1}{u^2}\left(\frac{du}{dt}\right) = -\frac{1}{u^2}\frac{d\theta}{dt}\frac{du}{d\theta} = -\frac{L}{m}\frac{du}{d\theta} \tag{4.14}$$

By differentiating the time (t) in (4.14), we can get the below result:

$$\frac{d^2r}{dt^2} = -\frac{L}{m}\frac{d}{dt}\left(\frac{du}{d\theta}\right) = -\frac{L}{m}\frac{d\theta}{dt}\frac{d^2u}{d\theta^2} = -\frac{L^2}{m^2r^2}\frac{d^2u}{d\theta^2} \tag{4.15}$$

In (4.15), $L = mr^2(d\theta/dt)$ represents the angular momentum of the planet P when it revolves on the orbit, which is a fixed value, so if we substitute (4.14) and (4.15) into the kinematic (4.12), which represents the force on the radial direction, the result will come out as below:

$$\frac{L^2}{m^2r^2}\frac{d^2u}{d\theta^2} + r\left(\frac{L}{mr^2}\right)^2 = \frac{-F_r}{m} \tag{4.16}$$

As the law of universal gravitation dictates the inverse proportion between F_r and the square of distance, it's safe to assume that $F_r = N/r^2$, while the negative sign in the rightmost expression of (4.16) signifies that the gravity points to the Sun. While $N = GMm$ = a non-varying value, G represents the gravitational constant. After substituting parameters where necessary, we can rewrite the kinematic equation as:

$$\frac{d^2u}{d\theta^2} + u = \text{constant} \tag{4.17}$$

The general solution to this equation of differentiation is $u = 1/r = A + B\cos(\theta + \theta_0)$, which can also be used to describe conic sections in the polar coordinate system, including parabolas, ellipses, and hyperbolas. Therefore, when a planet is affected by the gravitational force of a central body in its orbit, which is inversely proportional to the square of its distance from that body, the planet's trajectory must be a conic section, which can be an ellipse, parabola, or hyperbola, depending on its initial location and speed, as mentioned in the second last paragraph of Sect. 4.5. However, as planetary orbits have to be closed curves, the only possibilities are circles or ellipses.

4.6.2   Kepler's Second Law of Planetary Motion: Equal Areas

According to this law, a line segment joining a planet and the Sun sweeps out equal areas during equal intervals of time. In order to prove it right, let's first start with the conservation of angular momentum. Suppose a planet is r away from the Sun in distance, and the angle it sweeps through during a certain period of time is $d\theta$. Under this situation, the sector area with the angle $d\theta$ would be $r^2 d\theta/2$. In other words, the area swept out by the planet during the time unit dt is $r^2(d\theta/dt)/2$, while the angular momentum of its revolution can be written as $mr^2(d\theta/dt)$. Based on the law of angular momentum conservation, we know the area swept by the planet over the same period of time would be the same value as well.

4.6.3 Kepler's Third Law of Planetary Motion: Orbital Periods

This law states that the square of the orbital period of a planet is directly proportional to the cube of the semi-major axis of its orbit. As shown in Fig. 4.4, the planet P, which sits on its elliptical orbit with the distance r_1 from the Sun has an angular momentum of mr_1v_1, while its kinetic energy and gravitational energy are $mv_1^2/2$ and $-GMm/r_1$, respectively. When the distance between P and S changes to r_2, the three values mentioned above become mr_2v_2, $mv_2^2/2$, and $-GMm/r_2$. To calculate the gravitational energy, we can resort to the work done by the universal gravity. If we suppose the gravitational energy (U) is zero at an infinite distance, then the parameter can be calculated from the below formula of integration as:

$$U = -\int_{\infty}^{r} F(r')dr' = -\int_{\infty}^{r}\left(-\frac{GMm}{r'^2}\right)dr' = -\frac{GMm}{r} \tag{4.18}$$

Applying the laws of angular momentum and energy conservation, we can then retrieve the following equations:

$$mr_1v_1 = mr_2v_2 \tag{4.19}$$

$$\frac{1}{2}mv_1^2 - \frac{GMm}{r_1} = \frac{1}{2}mv_2^2 - \frac{GMm}{r_2} \tag{4.20}$$

Next up, we simplify (4.19) to be $v_2 = (r_1/r_2)v_1$ and substitute it into (4.20), which will lead us to this result below:

$$v_1 = \sqrt{\frac{2GMr_2}{(r_1 + r_2)r_1}} \tag{4.21}$$

From (4.21), we can further calculate the areal velocity of the Sun-revolving planet to be:

$$\frac{1}{2}v_1 r_1 = \sqrt{\frac{GMr_2 r_1}{2(r_1 + r_2)}} \tag{4.22}$$

As the following step, we divide the areal velocity from (4.22) by $\pi\sqrt{r_1 r_2}(r_1 + r_2)/2$, which is true for any elliptic area. This division will help us pinpoint the planet's orbital period at:

$$T = 2\pi\sqrt{\frac{\left(\frac{r_1 + r_2}{2}\right)^3}{GM}} \tag{4.23}$$

As shown in (4.4), if r_1 and r_2 combined equal the length of the major axis of the orbital ellipse (that is, when the planet sits on the either of the apsis of Sun), then the length of the semi-major axis would be $R = (r_1 + r_2)/2$. Therefore, (4.23) can be rewritten as:

$$\frac{R^3}{T^2} = \frac{\left(\frac{r_1 + r_2}{2}\right)^3}{T^2} = \frac{GM}{4\pi^2} \tag{4.24}$$

Since $GM/4\pi^2$ is a constant, plus R represents the semi-major axis, Kepler's third law is therefore proved: the square of the orbital period of a planet is directly proportional to the cube of the semi-major axis of its orbit, and the ratio between the two values is in proportion to the mass of Sun as well.

4.7   *Philosophiæ Naturalis Principia Mathematica*: A Scientific Masterpiece

Halley was extremely psyched after reading the carefully-structured essay as he knew the compelling arguments in it constituted an unprecedented theory in the field of astronomy. Therefore, he headed immediately to Cambridge again and asked Newton for permission to publish the essay in the Royal Society's scientific journal that he himself was in charge of so that the academia could share the pioneering results. However, the proposal was rejected as Newton was hoping to complete his research fully before having it published, partly because of the lessons he'd learned from the controversy about his optics studies 20 years ago. Meanwhile, the tract, which addressed the relationship between planetary orbits and gravity, was in part inspired by Galilei's ramp experiment, from which he derived his first law of motion, i.e. an object in motion stays in motion with the same speed and in the same direction unless acted upon by an unbalanced force. For a planet to stay on an elliptical orbit, in other words, there must be a force working on the radial direction, and this force is the gravity exerted by Sun. Unfortunately, the idea of drawing inspiration from Galilei's studies was in fact first given by Hooke in his letter to Newton, so he was probably worried that the essay would trigger his long-time enemy's harsh criticism and again trap him in an unescapable quagmire of academic persecution.[14] Therefore, he decided to hold off the publication until he could clarify all his doubts and research-related details. Yet from his conversation with Halley, Newton also realized how important his research topic was, so he was determined to pick up his studies from 20 years before and conducted a thorough and full-scale examination. In the 18 months that followed Halley's second visit, Newton disregarded all non-research work and dedicated himself wholeheartedly to the trivial but rigorous process of proving his arguments about universal gravitation. After days and nights of repeated theory development and countless searches for relevant calculation examples in past literature, *Philosophiæ Naturalis Principia Mathematica*, which is often referred to simply as *the Principia*, was finally born and went on to be hailed as the greatest scientific publication that redefined the progress of human civilization.

The 550-page, three-book masterpiece dives into several issues raised but left unsolved by natural philosophers since ancient times and reexamines planetary motion as well as ocean tides through the law of universal gravitation, established by the author Newton himself. Needless to say, the manuscripts sent from Cambridge simply blew Halley's mind. His scientific instinct told him that the law of universal gravitation would teach people an unforgettable lesson that some of the most perplexing phenomena in the natural world were regulated by nothing but a simple mathematic equation comprised of merely two factors: distance and mass, and that it was exactly the gravitational force between two masses that dictated the movement of

---

[14]Back in 1679, Hooke had openly explained his hypothesis about the motion of objects in class materials that he used for lectures. Unless acted upon by external forces, according to him, the direction of an object's motion wouldn't change, and the degree to which the object was affected depended on its distance from the force. Hooke once sent these hypotheses to Newton but didn't receive any response, possibly because the arguments weren't backed by any substantial evidence, or because Newton didn't see much innovation in Hooke's ideas since contemporary scientists generally held the same assumptions.

Earthly objects as well as celestial bodies. Overwhelmed by the epic master-piece, Halley traveled to Cambridge again to get Newton onboard for the publication of *Philosophiæ Naturalis Principia Mathematica*, and this time, he was given approval.

So, Halley took the 500 + page manuscript that he believed unveiled the mystery of the universe to the Royal Society, looking to have it published, but was given cold shoulder due to the society's financial predicament caused by the sluggish sales of its previous publication *De Historia Piscium* (Latin for *The History of Fish*). Deeply aware of how important Newton's research was, however, Halley decided to take charge of the editing work and even offered to provide the necessary capital for the publication. Whatever it took, he was resolved to make public the mighty law of universal gravitation. It's not that Halley was in any comfortable financial situation, though. In fact, he was also feeling the pinch because the Royal Society had been paying him only the meager sales profit from *De Historia Piscium* as his salary for months, but thanks to his unrelenting efforts in editing and fundraising, *the Principia* was finally presented to the public.

After successfully pulling off the publication, Halley embarked on three sailing trips under the order of the King of England to help solve the naviga-tion problems faced by the British navy. Halley first completed a map mod-eling the behavior of Earth's magnetic field, then one reflecting the surface winds in the Atlantic oceans. The special symbols he employed to signify wind directions are still in use today. In addition, he also initiated research in the demographic field by comparing the population distributions of Paris and London as well as metrics like birth/death rates, marriage statistics, and population density of the two cities. Through these studies, he came to the conclusion that if half of a city's adults were incapable of reproduction while the offspring of the other half could live to child-bearing age, then average speaking, each couple must have at least four children to prevent the popu-lation from decreasing. To give you an idea of how determined and detail-oriented Halley was when it comes to scientific research, he was actually willing to walk through a whole city and calculate the total distance with his number of steps due to the lack of effective measuring tools in order to determine the how big the city area was. Nevertheless, it wasn't just his hard work that's worth celebrating. His scientific ingenuity also helped him invent a brilliant approach to calculating the Sun-Earth distance by measuring the time needed for Venus to make a solar transit accurately, which he then cou-pled with the already-known orbital periods of Venus and Earth as well as the diameter of Sun to confirm the distances among the three bodies using the concept of similar triangles. The famous Captain Cook did actually, by following Halley's way of measuring Venus' time of solar transit, calculate the distance between the Earth and Sun to be 150,000,000 km when he was in Tahiti, a result extremely close to the actual value. As discussed earlier in Sect. 4.3, however, the French astronomer Cassini on the other side of the English Channel had long before calculated the Sun-Earth distance to be 140,000,000 km with the Earth-Mars distance measured in both South America and Paris as well as Kepler's third law of planetary motion. Cassini's calculation made full use of large-scale measurements as well as Kepler's law on orbital periods and became a great example of scientific analysis for later astronomers. In fact, Halley's stunningly precise calculation was very likely to be based on Cassini's precedent.

However, Halley's most renowned achievement was how he proposed the hypothesis that the comet striking in 1456, 1531, as well as 1607 was the same one predicted to visit again in 1758 with the help of the law of universal gravitation. When his prediction came true more than 60 years later, people didn't just understand and admire the Newtonian law of gravity even more but hailed Halley for his precise calculation and profound observation as well. Therefore, the comet was named after him as a tribute to his related research and to give him credit for encouraging Newton to publish *the Principia*. We don't know for sure how he calculated the orbital period of Halley's comet back then, but with Kepler's laws of planetary motion and Newton's law of universal gravitation already established, we can assume that he got some insight from the research results of the two fellow scientists and therefore conduct a simple model simulating how he delved into the details of the cometary orbit, as discussed below.

First, let's suppose the orbit of a comet is an ellipse based on Kepler's first law and only consider the impact of the Sun's gravity by disregarding all the possible disturbance that might come from the gravity of other planets. Under such circumstances, the Sun (S), whose mass is represented as M, sits unmoved at one of the foci of the elliptic orbit followed by the Comet C, whose mass is m. When C is r_1 and r_2 away from the Sun in distance, its corresponding tangential velocities[15] are v_1 and v_2, respectively. As the law of angular momentum conservation is valid for a comet that orbits on a fixed trajectory, we can confirm the equation $mr_1v_1 = mr_2v_2$ and substitute it into energy conservation (4.20), which would yield (4.21). As Kepler's third law dictates that T^2/R^3 is a fixed value for all planets, we can then derive (4.24) and develop the equation below, which represents the relationship between

---

[15]A more accurate way of phrasing it would be "the velocity that's perpendicular to the line joining the Sun and comet", which is too long and complicated. Therefore, "tangential velocity" will be used to refer to the concept in this book.

Earth and the comet:

$$\frac{(1yr)^2}{(1\text{AU})^3} = \frac{(76yr)^2}{\left(\frac{r_1+r_2}{2}\right)^3} \tag{4.25}$$

In (4.25), yr refers to a year (equal to the orbital period of Earth), and AU an astronomical unit (the distance between Sun and Earth), which is approximately 1.496×10^8 km. As Halley assumed that the orbit the comet revolved on was elliptic in shape and proposed the orbital period to be roughly 76 years after analyzing historical data of observation, he was able to infer that the radius of the comet's orbit the length of the semi-major axis of the ellipse. Meanwhile, we can calculate from (4.25) above that $r_1 + r_2 = 35.88$AU, where $r_1 = 0.586$AU, which is the comet's distance from Sun when at perihelion and it's tangential velocity is $v_1 \approx 39$ km/s.

Judging from the results above, when Halley's comet passes close to the Sun, i.e. when we're able to see it with bare eyes, it's actually closer than us to the fireball and moves at a speed of about 40 km/s. Once passing perihelion, the comet will gradually move away from Sun at a slowing velocity, sometimes even leaving the solar system. For the actual proportion of the entire system and the cometary orbit, please refer to Fig. 4.5. Based on the analysis data above, we can see a surge and a gradual decrease in the comet's speed when it travels close to and moves away from Sun respectively, which is the result of the strong solar gravity and perfectly complies with the rules stated in Kepler's law of planetary motion on equal areas. In fact, planners of space exploration

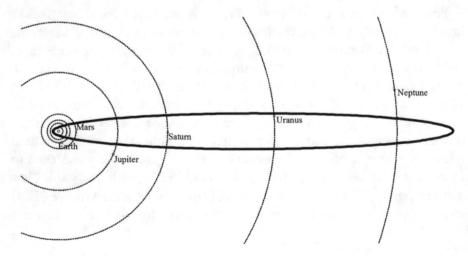

Fig. 4.5 The relative location of Halley's comet with solar system planets

would even take advantage of the Sun's gravity to accelerate small spaceships that cannot be loaded with enough fuel.

4.8 *The Principia*: Written by Standing on the Shoulder of Giants

When composing the classic *Philosophiæ Naturalis Principia Mathematica*, Newton integrated calculus, which he invented 20 years before, with the concepts of tangential and radial speeds of objects moving in circular motion before introducing the relationship of inverse proportion between gravity and orbital radius in order to explain that the Moon's revolution around Earth and the free fall of objects were both caused by the Earth's gravitational force. Then, he went a step further to study Jupiter and Saturn, wondering if they followed the same law in orbiting around the Sun. To solve this question, he wrote to several astronomers and asked for data about the planets' changes in speed and location when approaching the Sun between 1680 and 1684. After some fairly complicated calculations, he was finally able to confirm that all planets orbited on their trajectories following the law of universal gravitation. Yet his research didn't stop just there but continued with his attempt to prove that the law of gravitation was also applicable to other substances. Therefore, he performed free fall experiments with materials like gold, silver, lead, glass, sand, salt, wood, etc. and concluded that it was irrelevant what substance an object was made of. He also managed to get the values of Earth's and Moon's radius as well as their distance to help him calculate the Moon's degree of deviation from its orbit due to solar gravity. Such an experimental and exploratory research style gave the law a most thorough verification and served as the best endorsement of its credibility. Below, let's take a walk through the main topics in *the Principia*.

If you open this book, you'll be surprised how few formulas and how many geometric shapes the author uses to explain the transient movement of objects, though indeed with some help of calculus and the notion of limit. From an intuitive point of view in physics, his research can basically be understood as two main concepts: speed and acceleration, which are the distance traveled/speed change over an extremely short amount of time, as we have discussed earlier in Sect. 4.2. Simply put, the minimum time rate of change can be regarded as the amount of change happening over a transient span of time, and this is the underlying idea of Newton's calculus. To approach the notion of limit from a spatial point of view, on the other hand, we can borrow the concept brought forward by Apollonius of Perga two thousand years ago that the minimum circumference can be seen as a chord length (for details, please refer to the method demonstrated in Fig. 1.6 of Chap. 1). The theory of limit isn't particularly hard to explain, and the use of operators in calculus would make it even easier, but somehow Newton chose not to apply rules of differentiation but rather opted for the complicated geometric description to expound his theory. Why, we still don't know for sure, but will briefly go through possible reasons later in Sect. 4.9.

Before starting to derive the law of universal gravitation, Newton first lists eight basic elements in the motion of objects, including mass, momentum, inertia, external force, action, and three other items related to centripetal force, each explicated incisively with references to authoritative, recognized research studies. For example, he defines "mass" as the product of the density and volume of an object and differentiates it from the concept of weight. In addition, momentum (i.e. the amount of movement) is defined as the product of the speed and mass of an object, inertia (internal force) the resistance of any physical object needed to maintain its velocity/direction or motionless state, and action (external force) the force exerted on an object to change its state of zero motion or moving with the same speed and in the same direction. When explaining the concept of acceleration, Newton makes it a point to highlight the absoluteness of time and space. In other words, the acceleration of an object is, according to him, only affected by the change in its speed over time but has nothing to do with its temporal and spatial environments.[16]

After defining some key concepts, he continues to introduce what we call the three Newtonian laws of motion in mechanics. According to the first law, an object at rest stays at rest and an object in motion stays in motion with the same speed and in the same direction unless acted upon by an unbalanced force, which wasn't easy for people to accept as it went against the common, intuitive notion that came from everyday observations, such as horizontally moving objects gradually slowing down or even stopping entirely after traveling for a while or objects dropping on curved lines after thrown out. The second law, which pertains to the most widely applicable concept in mechanics, states that the momentum of an object will change over time on the direction through which the object is affected by an external force. In fact, what's really emphasized behind this law is the equation between external force and the change rate of momentum, i.e. the resultant force and the change rate of

---

[16]This notion of absolute time and space was overthrown more than two hundred years later by Einstein with his theory of relativity, which, however, also helped construct the physical mechanism of why two objects would draw each other close with gravitational force. In other words, he basically helped explain Newton's law with a theory that denied Newton's hypothesis, and for many, this exactly why science has been so enchanting over the past thousand years.

momentum lasting to exert the same effect on an object are inversely proportional. For example, when excluding the frictional effects from the ground and air, the force needed to accelerate a car at rest to 1 km/h within a second is 100 times that required to achieve the same effect in 100 s. However, this notion was also hardly acceptable to the general public back then as people tended to think intuitively that accelerating the same object to the same speed should take the same amount of force and failed to understand that time was a key factor as well. As for the third law, it sets a broad rule for the world of mechanics by stating that for every action, there is an equal and opposite reaction. Generally speaking, this last law can be understood as the conservation of all momentum in the universe. To put it simply, the total momentum across the whole universe is a fixed value since momentum is something that can be transferred among objects. On the other hand, this theory can also be interpreted in a narrow sense based on common phenomena in daily life, such as the facts that guns would recoil when bullets are fired, our heads would fall back after sneezes, and that the cue ball would rebound after hitting other balls on the pool table.

With the three laws of motion explicated, Newton starts going through academic theories about universal gravitation and their corresponding derivation processes with three volumes of comprehensive explanation. In Chap. 2 of the first volume of *the Principia*, he proves Kepler's second law of planetary motion and proposes that planets can revolve on conic sections (ellipses, parabolas, or hyperbolas) through calculations (see (4.17)). In addition, he supplements the condition that a planet would only orbit on an elliptical trajectory when affected by a centripetal force that's inversely proportional to the square of its distance from the Sun (i.e. the solar gravity working on the direction of the line joining the Sun and planet) and that it's only under such circumstances would this particular line sweep out equal areas in equal times. Meanwhile, he also provides an analysis of why Kepler's third law about the square of the orbital period of a planet being directly proportional to the cube of the semi-major axis of its orbit is true, which we have also discussed in detail in Sect. 4.6. Then in the second volume, Newton focuses on objects' movements in fluids to develop rules of how transmission media can cause obstruction to motion, while in the third volume, he extends the application range of the law of gravitation, which is the main topic of Volume 1, to all the celestial systems in the universe and emphasizes that his self-derived theories of mechanics can be used to prove that Kepler's all three laws are applicable to the six major planets in the solar system and explain why the Moon revolves around Earth as well as the cause behind the cometary strike in 1680. In other words, Newton's theory on the force that drives planetary motion (i.e. what we call "gravity" now) is irrelevant of the nature of the affected objects and only considers the direction and amount of centripetal force, which is quite different from the already-discovered magnetic force as the latter is closely influenced by whether objects themselves are magnetic or not.

It wasn't till about a hundred years later did Henry Cavendish (1731–1810) from the University of Cambridge finally transform Newton's theoretical analyses into the equation of universal gravitation that we widely use now: $F = Gm_1m_2/r^2$, with m_1 and m_2 representing the masses of two objects, r the distance between them, and G the gravitational constant. However, it was impossible for Cavendish to confirm the value of G as the masses of the Sun, Earth, Moon, and other planets were still unknown back then,[17] but to calculate F, just the ratio of two celestial bodies' masses would suffice[18]; as for G, it wasn't a required value. Luckily, data about the movements of Jupiter and Saturn around Sun was ready and complete, so it was no difficult task to calculate the ratio of mass between the Sun and Jupiter/Saturn and pinpoint the values of F for both sets of orbs. Yet when it comes to the gravitational force between Sun and Earth, the lack of the correct distance between the two bodies caused a deviation rate of over 100%[19] in Newton's calculation in Volume 3 of *the Principia*.

The ratios of mass mentioned in the previous paragraph can be analyzed using Kepler's third law of planetary motion with Newton's theory of gravity, and the results can then be coupled with existing data regarding planetary mass to calculate the same quantity for planets whose mass was yet unknown,[20] which has been demonstrated in detail in Sect. 4.6, where (4.24) shows that the ratio between the cube of the semi-major axis of a planet's orbit and the square of its orbital period is proportional to the mass of the celestial

---

[17] Cavendish eventually managed to measure the value of G in a 1797-experiment, with a result fairly close to 6.67408×10^{-11} m^3 kg^{-1} s^{-2}, the correct value we know today.

[18] The practice of representing physical quantities with ratios rather than absolute values started in Galilei's time as there was no standard system of measurements like length, time, mass, etc. In fact, it wasn't until 1742 did England start promoting for the development of such a system. After numerous negotiations among scientists from various countries, metric measurements were finally established in 1799 before dawn of the 19th century and have been gradually popularized till today.

[19] Regarding Newton's earlier calculations of the Earth-Moon distance, please see Sect. 2.6 in Chap. 2, but none of the values was quite accurate. The more precise results were actually obtained by Jean Richer and Giovanni Domenico Cassini in 1672 (see Sect. 4.3), when they calculated the distance to be 140,000,000 km, only 7% deviant from the 149,598,500 km we know now. It's possible that Newton wasn't aware of the French duo's findings when he started composing *the Principia* and therefore used his old values, which, at one point, caused the deviation rate in his results to be as high as 100% and frustrated him greatly. When he finally got his hands on Cassini's data, however, all the results generated therein were more than satisfying.

[20] It wasn't until 1797 did Henry Cavendish from the University of Cambridge get to confirm Earth's density with a self-designed experiment and use the already-known value of Earth's diameter to calculate its mass. For details, please see Sect. 4.12 of this chapter.

body it revolves around. With (4.24) and other known data in the field, it was already possible in Newton's time to calculate the ratio of mass between the Sun and its surrounding planets, which is exactly what he did and obtained the following results: (a) $R^3/T^2 = 2.51 \times 10^{19}$ km^3/days^2 (Earth's orbital period: 365.2425 days; average distance from Sun: 149,597,871 km), (b) $R^3/T^2 = 2.40 \times 10^{16}$ km^3/days^2 (Io's semi-major axis: 421,700 km; orbital period: 1,769,137,786 *days*, according to Galilei's research in 1610), and (c) $R^3/T^2 = 7.18 \times 10^{15}$ km^3/days^2 (Titan's semi-major axis: 1,221,870 km; orbital period: 15,945 days, according to Huygens' research in 1655). With the three values above, we can divide the value of R^3/T^2 in (a) by that in (b) and determine the ratio of mass between the Sun and Jupiter at $2.51 \times 10^{19}/2.40 \times 10^{16} \approx 1050$, and again, by performing the same calculation with (a) and (c), the mass ratio between Sun and Saturn can be calculated to be $2.51 \times 10^{19}/7.18 \times 10^{15} \approx 3500$. With both results in line with the actual values, a heavy load was taken off Newton's mind, making him more certain than ever of the infallibility of his law of universal gravitation.

4.9 Exploring Long-Standing Mysteries with the Law of Universal Gravitation

In Volume 3 of *the Principia*, Newton calculates two phenomena in the solar system that no one could ever explain since ancient times with his law of universal gravitation: ocean tides and the precession of Earth's axis. As the diameter of Earth's equatorial plane is longer than the distance between the North and South Poles, the gravitational forces from Sun and Moon aren't concentrated at the center of the orb, therefore causing axial precession and top-like spinning. The precession period calculated by Newton with the law of universal gravitation was 26,000 years, so close to the actual value that even he himself was probably quite surprised, but of course, pleased as well. On the other hand, he also provides in the book a voluminous analysis of the impact of the gravity of Sun and Moon on the ocean tides observed on Earth's surface. According to his results, although the Sun is 2.7×10^7 times larger than the Moon in terms of mass, it's also 390 times further away from the Earth. As the tidal force increases with the mass of a celestial body but is inversely proportional to the square of its distance from Earth, the Moon's gravitational pull on the oceans is twice as strong as that of Sun. Tidal movements, however, are also affected by a commonly known fact that the Moon always faces Earth with the same side (i.e. gravitational locking), but Newton never looked into this phenomenon. Therefore, it's a good chance for us to discuss the reasons behind tidal locking, how long it takes for the Moon's revolution to be synced with Earth's rotation, as well as Earth's rotation period when the synchronization happens. The below sections cover details about the topics mentioned in this paragraph.

4.9.1 Example 1: Precession Speed of the Earth's Axis

Roughly six billion years ago, the solar system was a cloud of gas and dust, which were then pulled together by gravity and started rotating around a central axis. Dust at the edges of the vortex clumped together due to the shorter radius of its spinning movement, while the particles on the center plane perpendicular to the rotation axis expanded outward because the centrifugal force was stronger than gravity. Going through such slow but consistent changes over several hundred million years, the originally ball-shaped nebula finally collapsed to become a flat disk. As for the debris of dust, they were partially drawn together due to continued spinning, so at the end, more than 99% of the dust formed the Sun at the center of the nebula, while the < 1% left behind became several planets and started rotating the Sun on the disk-shaped plane following the same direction. And this, explains why ancient people could observe five planets in the sky, i.e. Mercury, Venus, Mars, Jupiter, and Saturn, all moving on the same plane, which is what we call the "ecliptic" today.

As explained in Sect. 2.6 of Chap. 2, the concept of the ecliptic quickly revealed the fact that there's a 23.5-degree angle between the plane and Earth's rotation axis. In addition, the axis itself also rotates clockwise at a plodding speed around an axis perpendicular to the ecliptic, and this is the phenomenon of "precession," which Hipparchus discovered in 127 B.C. Although his calculation of the axial rotation period (36,000 years) was deviant from the actual value by about 10,000 years, it was already an exceptional achievement considering his times.[21] After coming up with the law of universal gravitation, Newton also saw precession as a potentially effective way of proving his new-found theory and decided to conduct relevant research as a result. According to his conclusions, the rotation of Earth is fairly fast, with the tangential velocity near the equator as high as 1,670 km/h, which causes the radius of the equatorial plane to be longer than the polar

---

[21]According to the calculation in Chap. 10 of Morin [24], the precession period of the equinoxes is 23,000 years.

counterpart by 23 km.[22] Due to this equatorial bulge, which is perpendicular to the rotation axis, the gravitational forces from Sun and Moon aren't concentrated at Earth's center, causing its axis to spin like a top, and this is the phenomenon that Newton referred to as "precession."

After clarifying the reason behind axial precession, he continued to calculate its period with the law of universal gravitation and obtained the surprisingly accurate result of 26,000 years, which is nearly the same as the actual value. In "Appendix 4.1: Calculating the Period of Earth's Precession with the Law of Universal Gravity", we've also attempted to calculate the same value using his theory. Considering the equatorial bulge, the Earth looks like an oblate spheroid, and therefore, the gravitational forces of Sun and Moon would each generate a torque on Earth, causing its axis to spin while the entire sphere's rotating as well. We then consider the interaction between gravitational energy and acting torques to derive the relationship between the frequency and angle of Earth's axial precession, thereby setting the period at about 25,500 years (less than 1.2% deviant from the actual value of 25,800 years) and proving that Newton's law of universal gravitation does capture the complicated patterns of planetary motion on revolving orbits.

4.9.2   Example 2: Marine Tidal Movement

Newton's discussion of the Earth's and Moon's gravity was based on the hypothesis that the gravitational force of Earth was concentrated at its center, an assumption that we'd probably dismiss as bold but sloppy today since the mountains and oceans on Earth's surface are not just different in mass but their distances from the Moon as well; plus, seawater has been observed to deform under the external forces generated from Earth's rotation, so Newton's hypothesis seems very far from convincing. This is probably why he dedicated so much volume in the three-book *Principia*, first published in 1687, to detailed, complex geometric shapes trying to prove his assumption. In addition, he also dedicated almost a whole volume to discussing how the Moon's gravity could impact the oceans on Earth, i.e. the phenomenon of tidal movement much-documented since ancient times.

The first documentation of tidal waves might be created in 330 B.C. by the Greek philosopher Pytheas of Massalia (350–285 B.C.), who observed that tides were the highest during Full Moon and New Moon when he was sailing from the Mediterranean into the Atlantic Seas and therefore inferred

---

[22] Please refer to "Appendix 4.6: Calculating the Difference between Earth's Equatorial and Polar Radiuses" at the end of this chapter.

the relationship between marine tides and the Moon. In a Shakespearean play, a witch also warns the Roman emperor Caesar to be particularly careful with his ships if he plans to disembark on the shores of England on March 15th, a scene that's clearly written with some reference to real-world facts because besides the high tides during Full Moon, the seawater near the British coasts chosen by Caesar for landing would also be a lot more turbulent due to the special landforms. Such a phenomenon can also be observed at the estuaries of the River Severn and the Qiantang River, where Cardiff, the capital of Wales, and Hangzhou, the capital of China's Zhejiang province, are located. Both tidal mouths are triangular in shape with the river courses tapering inland, which tends to cause the water levels to rise quickly and trigger spring tides after water floods in. While the tidal range of River Severn can go over 15 m, the Qiantang River has seen a range of more than 30 m. Both the records happened during Full Moon at the spring or autumn equinox, so it's certain that marine tides are closely related to Moon.

If strong enough, tidal movements can decide the outcome of the war and even put people's lives in danger. Therefore, theories attempting to explain the easily observable phenomenon hadn't been lacking since ancient times, but no one really conducted any systemized research. In Newton's era, even influential intellects who initiated the modern scientific revolution, such as Kepler and Galilei, tried to explain the reason behind ocean tides. While Kepler proposed that high and low tides were the results of the Moon's magnetic impact on Earth, Galilei mocked his theory and contended that tidal movements were caused by the combination of Earth's self-rotation and revolution around Sun, which was, to be honest, just as ridiculous. Newton, however, was different. After deriving the law of universal gravitation, he realized that ocean tides were definitely related to the Moon's gravity acting on Earth and came up with the hypothesis that high tides happened in oceans facing or on the opposite side of the Moon, which explained why both high and low tides took place twice a day.

Here's how he approached the phenomenon: the oceans on Earth are under stronger lunar gravity than centrifugal force when facing the Moon, while the situation becomes the other way around when the same ocean circle to the opposite side of Moon. As the Moon revolves around Earth in the same direction as the latter's rotation, a certain point on Earth wouldn't be directly facing the Moon again in 24 h but has to wait for the Earth to rotate a further 1/27.3 of a complete turn,[23] which takes about 53 min. This means

---

[23]While the Moon completes a circle around Earth every 27.3 days, the Earth's rotation period is 23.93 h. By dividing the latter value with the former, we can come to the result of 0.877 h, which can then be converted to 52.6 min.

the period between two high tides would be 24 h and 53 min, which is exactly what happens in real life and further proves the close connection between marine tidal movement and lunar gravity. In addition, Newton also analyzed the Sun's impact on Earth. Though the Sun is 2.7×10^7 times larger than the Moon in terms of mass, it's also 390 times further away from the Earth. As the tidal force increases with the mass of a celestial body but is inversely proportional to the square of its distance from Earth $(f \sim m/r^3)$, we can confirm that the Moon's gravitational pull on the oceans is twice as strong as that of Sun by calculating the Sun/Moon values with this formula.

If you still remember, the derivation details above have all been explained in "Appendix 4.2: Measuring the Influence of Moon's Gravity on Earth Tides with the Law of Universal Gravitation," where we easily prove with Newton's theory that the lunar tidal force is inversely proportional to the cube of the Moon-Earth distance. We also consider the fact that the effective force working on the Earth's oceans is comprised of the Moon's gravity and the centrifugal force generated from Moon-Earth co-orbiting and calculate the influences of both on Earth tides, thereby tracking how tide height changes along with latitude and reaching the rough results of 0.8 and 0.3 m for high and low tides. It's true that actual tide heights are still susceptible to coastal landforms and monsoon directions, but our result does reflect the average tide height under normal situation, which is reasonably accurate.

4.9.3 Example 3: Captured Rotation—The Phenomenon Left Out by Newton

It's widely known that the Moon always faces Earth with the same side, a stable phenomenon that persists due to the long-term interaction between Earth tides and Moon's gravity that we commonly refer to as "tidal locking". Although Newton did analyze the reason behind ocean tides, he never got to the part about tidal locking, so in below paragraphs, let's take a look at why Earth would capture the Moon's rotation and how long it takes for the Moon's revolution and Earth's rotation to be synchronized (i.e. co-orbiting).

The tidal locking relationship between the Earth and Moon is demonstrated in Fig. 4.6. Generally speaking, this phenomenon arises from the tidal force of A (natural satellite) that draws up tidal bulges on the surface of B (such as the oceans on Earth or Jupiter's surface fluid). When such tides are pulled up, A, B, and the two highest points of B's bulges are on the same line. However, as B is self-rotating while A orbits around it at a different speed and the angular momentum of B is larger, the line joining A and B and the one between the bulges of B would grow apart after a while, causing

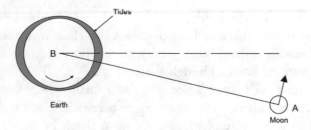

Fig. 4.6 How the tidal force of Moon (**A**) influences Earth (**B**)

A to accelerate following its revolving direction due to the pull-force of B's tides. The gravitational force between A and B would also change and transfer constantly because A's orbiting and B's tidal movements are asynchronous, causing the two bodies to experience a certain amount of back and forth, but eventually, a gravitational balance would be reached between them, and that's when tidal locking takes place. One thing to note is that the so-called "tidal bulge" doesn't just happen with the oceans on Earth. Even the Moon's crust experiences periodical bulging as well, which is why it always faces the Earth with the same side. Details will be explained in the following paragraphs.

When it comes to the co-orbiting of Moon and Earth, the lunar gravity doesn't simply cause ocean tides to rise but influences the groundwater level on Earth as well, therefore inducing periodic changes for wells and springs. Yet in fact, these changes are not directly triggered by the gravitational force but by the swelling of rocks caused by it. During high tides, the geological formation would loosen as rocks are pulled apart and swell due to the lunar force, which in turn widens well diameters and makes water levels decrease, precisely the opposite direction of tidal movements.[24] In fact, such swelling can also explain the decline in Moon's speed of self-rotation. As the Earth is about 80 times larger than Moon in terms of mass, the gravity it exerts would also cause rocks on the satellite to shrink and swell periodically, with the geological formations facing or on the opposite side of Earth swelling and those on the plane perpendicular to the line joining Moon and Earth shrinking.

However, as the diameter of the Moon is just 1/4 that of Earth, the magnifying effect on rocks facing or on the opposite side of Earth is only 1/4 as strong. In other words, the rocks on Moon's surface would swell 20 times as much as those on Earth and extend the orb's diameter by ten meters periodically, which gradually shortens the Moon's self-rotating period and generates peculiar flashes on its surface from time to time. According to the research

---

[24]For details, please refer to Chap. 3 of Chown [5].

of Professor Arlin Crotts at the University of Columbia, nearly 3/4 of the documented 1,500 flashes of all time happened near craters where crusts are seriously dismantled,[25] which indirectly proves that rocks on the lunar surface do swell. As the bulging side would point to the Earth and reduce the Moon's rotating speed, eventually the lunar rotation would be captured by the Earth, causing the rotating period to be aligned with its 27.3-day revolution. In other words, the Moon only faces the Earth with the same side ever since and no change will ever happen in its rotating behavior again due to the locking effect.

Although the conditions for tidal locking to happen can be derived from the law of universal gravitation, Newton didn't address research in this field, so it's a good chance for us to take a look at how the gravitational systems of Earth and Moon actually contributed to the phenomenon. Let's first suppose that when the total mechanical energy (i.e. momentum plus potential energy) in the Earth-Moon system hits a minimum, the Moon would be tidally locked to the Earth and rotate in exactly the same time as it takes to orbit the Earth. This hypothesis is proposed considering the fact that ever since the formation of the Earth-Moon system, the two bodies' gravity and tidal friction have been interacting constantly, which would cause the total energy in the system (the momentum of revolution and rotation plus potential energy) to die down gradually, hit a minimum, and eventually enter a stable state. After some calculation, we come to the conclusion that the below equation is the requirement of tidal locking:

$$L \geq \sqrt[4]{\frac{256G^2M^3m^3I}{27(M+m)}} \tag{4.26}$$

In (4.26), M and m represent the masses of Earth and Moon respectively, G the gravitational constant, I the Moon's moment of inertia during rotation, and L the total angular momentum of the Earth-Moon system. Regarding the detailed derivation process of this equation, please see "Appendix 4.3: Exploring the Requirement for Tidal Locking in the Earth-Moon System."

In addition to Earth and Moon, systems of duo astronomical bodies where the phenomenon of tidal locking also exists include Jupiter and its moon Io, as well as Pluto and its moon Charon. Table 4.1 contains the data of all three pairs of bodies, which all satisfy (4.26). Since Earth's gravity captures the Moon's rotation, it seems reasonable to suppose that the same mechanism would work the other way around as well. Following this rationale, the

---

[25] For details, please see the reference of the previous note.

Table 4.1 Data regarding systems of co-orbiting astronomical bodies where tidal locking has been observed

	Earth-Moon system	Jupiter & Io	Pluto & Charon
M (kg)	5.97×10^{24}	1.90×10^{27}	1.30×10^{22}
m (kg)	7.35×10^{22}	8.93×10^{22}	1.52×10^{21}
R (m)	3.84×10^{8}	4.22×10^{8}	1.96×10^{7}
ω (1/s)	2.66×10^{-6}	4.11×10^{-5}	1.14×10^{-5}
I (kg m^2)	8.90×10^{34}	1.18×10^{35}	2.23×10^{32}
$\sqrt[4]{\frac{256G^2M^3m^3I}{27(M+m)}}$ (kg m^2/s)	2.70×10^{33}	5.98×10^{34}	1.5×10^{30}
L (kg m^2/s)	2.85×10^{34}	6.54×10^{35}	5.96×10^{30}

rotation of Earth and the revolution of Moon should eventually be synchronized, i.e. the time needed for both bodies to complete a circle should be the same, for the gravitational interaction between them to gradually become stable, which is precisely what we call co-orbiting. Based on rough calculations, when the phenomenon actually happens, the rotation period of Earth will extend to 47 days compared to the one day right now. The problem, though, is that this isn't likely to happen till 10 billion years from now,[26] which is probably longer than the lifespan of Earth or even the solar system and basically negates the possibility of the Earth-Moon synchronization. So, does this mean it's impossible for two astronomical bodies to mutually lock each other? In other words, is this phenomenon a false proposition from the very beginning? And how did it come into the picture anyway? To answer this question, we have to trace it all the way back to a clay tablet found by archaeologists in Mesopotamia, just as we did in Chap. 1.

Written on this tablet is a sorcerer's prophecy regarding the full solar eclipse that happened at 8:45 am on April 15th in 136 B.C., when the Moon blocked the entire Sun and plunged Babylon into complete darkness. After learning about this event, modern scientists input the date into a computer program that simulated the motion of the solar system, only to find that Babylon was in fact not in the totality zone yet but still 1/8 of Earth's circumference away, which they converted into time using the Earth's rotation period and concluded that the eclipse should've happened 3.25 h later than the documented 8:45 am, in other words, 12 pm at noon. If the program was correct, the only reason that can explain the roughly 3-hour difference was the continued deceleration of Earth's rotation over the past 2,000 years, and with these numbers, the scientists went through related calculations and concluded that a day would be 1.7 ms longer every 100 years.

---

[26]For details, please see the source of the previous note.

The bold assumption was, however, followed by a modification, which scientists proposed after finding out when analyzing how Moon's gravity impacted satellite orbits that the deceleration of Earth's rotation should be 2.3 ms per hundred years rather than the 1.7 counterpart obtained from the simulation of the full eclipse in Babylon. In order to explain the 0.6-millisecond difference, they came up with another theory, contending that it was the slowly-melting ice at higher parts of the glaciers on Earth that reduced the orb's moment of inertia and accelerated its rotation. With the result of another simulation program, it was confirmed that the decline should be 0.6 ms every hundred years.

Now that it's confirmed that the co-orbiting of two astronomical bodies isn't impossible, the issue becomes, how long does it take for this to happen? To answer this question, we need to first look at the torque effect generated by tidal changes. As Earth rotates in the same direction the Moon revolves around it and has a higher angular velocity, the line between the tidal bulges on Earth would move ahead of and form a small angle with the Earth-Moon line. If we see the bulges as two equivalent masses, the Moon would exert its gravity on both of them, but as the bulge facing the Moon is acted upon by a stronger force, the Earth would then be affected by a torque opposite to its direction of rotation and decelerate due to loss of angular momentum. Analyzing the changes of motion in the Earth-Moon system with the principle of angular momentum conservation, we reach the conclusion that Moon's revolution and Earth's rotation would be synchronized in 15.2 billion years and that the Earth's rotation period would extend from one to 47.2 days. In addition, based on the findings of Project Apollo that the Moon-Earth distance sees an incremental increase of 3.8 cm every year, we also expect that one day on Earth would become 2 ms longer every hundred years due to the deceleration of its rotation speed, a result extremely close to the previously mentioned 1.7 and 2.3 ms. For details about related calculations, please see "Appendix 4.4: When Will We See Earth's Rotation and Moon's Revolution Synchronized?"

4.10 Bitter Controversy: Who Invented Calculus?

As we've mentioned earlier, Newton didn't use his self-invented calculus to elaborate on the formulation of the theories in *the Principia*, but rather resorted to complex geometrical analyses and took the trouble of explaining with one graph after another his ideas and research basis, such as the concept that chord length can be regarded as the shortest possible arc of a circle, which resonates with the method of exhaustion proposed by Apollonius of Perga (see Chap. 1), but such complicated explication makes his overall discourse extremely hard to understand and often too daunting for people to even try to read at all. It's possible that Newton intentionally chose the abstruse geometrical method to verify his theories after being bashed for his optical research in the early stage of his academic career in case he had to deal with the senseless criticism unleashed by those ignorant of advanced mathematical models. Another possibility, however, is that calculus was probably not an easy notion to clarify with geometry since it was an unprecedented discipline, and it might be even more time-consuming to get the public to understand and accept the theorem of calculus than the law of gravitation itself. Whatever his motive was, Newton would later regret taking the detour and find it unworthy because he wasn't the only person deeming himself the father of calculus. In fact, Gottfried W Leibniz (1646–1716) also claimed he invented calculus first.

The fight for the credit of inventing calculus caused confusing chaos where both sides stuck to their own version of the story, yet nobody knew for sure which was true. Here's how it happened. In 1669, Barrow passed Newton's calculus-related essays to the mathematician John Collins (1625–1683), who showed the manuscripts to Leibniz seven years later in London. To Leibniz's surprise, Newton's results were similar to the calculation rules that he himself had established in 1675. To make sure his credit wouldn't be snatched out of his hands, he published two related theses in 1684 and 1685, officially naming his theorem "calculus". In fact, both the differential and integral symbols we use today could already be seen in his research, while the former was essentially the antisymmetric form of the F-clef, likely to be his attempt to insinuate that calculus was an invention of his homeland Germany.

On the other hand, Newton, whose *Principia* had already brought him considerable fame by then, felt the need to fight back after reading the two essays and therefore made it a point to explicate his self-developed calculation rules in *Opticks*, published in 1704. Although Newton was already the President of the London Royal Society and enjoyed widespread fame, Leibniz was still reluctant to give in after reading the book and published another article right away the next year, where he suggested that Newton plagiarized his essays to develop relevant rules. This article, however, was submitted anonymously due to his fear of Newton's power as a heavyweight in the academia, yet the academic clan controlled by Newton still staged an organized series of fierce fightbacks, starting with the 1709 article by John Keill (1671–1721) in the *Philosophical Transactions* of the Royal Society championing Newton as the foremost inventor of calculus, which provoked an angry protest from Leibniz in 1711. As a response, the Royal Society convened an investigation committee to solve the controversy, the process of which ended in 1715 with the confirmation that Newton was the real inventor of calculus. Despite the result, the authoritative Royal Society's involvement basically muted the voice of Newton's opponents, so the truth behind the incident remained opaque in the international community of mathematics ever since even after the result of the official investigation was announced.

About two hundred years later, the truth finally came out thanks to later scientists' inspection into the Royal Society's meeting minutes. It turned out that the 1712 investigation committee was composed entirely of Newton's supporters, while the final report was even drafted by himself before published in the committee's name. This shouldn't come as a surprise, though, based on what we know today about his style of handling such disputes. In fact, Leibniz was lucky that Newton didn't handle him the way he persecuted Hooke, yet after several hundred years' discussion of what actually happened, scholars nowadays generally agree that calculus was invented by both Newton and Leibniz around the same period of time but independently; while the former started his research ten years earlier, the latter published his results one year ahead. Despite Leibniz's earlier publication, Newton was the one who first applied calculus to physics and solved the questions regarding planetary movements that had bothered scientists for more than a thousand years. Most valuable of all, he also established a perfect series of mechanics laws that can at once explain objects' motion on Earth's surface and planetary motion in the solar system, creating a connection between the two seemingly irrelevant phenomena using rules of calculus. Considering these achievements, it definitely makes more sense to credit Newton with the title of the inventor of calculus. Nevertheless, his outrageous attitude and arrogant behavior still make people can't help but sympathize with the oppressed Leibniz.

4.11 Malicious Criticism: Academia's Corrupted Norm

Newton's law of universal gravitation was met with much ridicule and criticism right after published. Some people said *the Principia* was nothing but a collection of discussions about geometry containing zero elements of physics, while others bashed him for arguing against Aristotle's theory and only talking about the effect of universal gravitation (how strong the force was) but avoiding the cause of it (how gravity came to be). There were yet some others who criticized him for trying to deify himself and assume the power of controlling the mechanism of the universe while he was merely a mortal British dude. These far-fetched attacks didn't cause much harm, but when it comes to the bombardment from his academic nemeses, he was cautious in building his defense.

Unsurprisingly, Hooke fired the first shot by emphasizing that *the Principia* contained nothing innovative but what he had long before told Newton in his letter 20 years earlier. Leibniz also seized this chance to leash out at Newton, questioning his motive for not discussing what actually caused gravity and contending that simply some formulae were by no means enough to prove that gravity could be transmitted through the vacuum of the universe and generate transient forces. In fact, his argument epitomized the mainstream belief in the European academic circle as well as Aristotle's key idea: the theory of causality. Newton did calculate the amount of force working between two objects, but why would gravity exist in the first place? He never explained.[27]

Joining the battle later on was Huygens, who came up with the gravitational constant. Although he did at first recognize Newton's achievement in extending the boundaries of physics with the law of universal gravitation, he also wondered why the period of a pendulum would be longer in equatorial areas. According to him, such a phenomenon existed not because Earth was an oblate spheroid but because the centrifugal force from Earth's rotation was the strongest near the equator and therefore mitigated the gravitational effect. The reason why he claimed so is that he'd once proposed a theory in his 1673 book *Horologium Oscillatorium*: the ratio of a pendulum's period to the time needed for an object to fall from the pendulum's mid-point to its bottom equals the ratio between the perimeter and diameter of a circle whose radius is the length of the pendulum, and this ratio is 3.14159, exactly the π we know today. If presented in the form of an equation, the theory can be written as:

$$T = 2\pi \sqrt{\frac{l}{g}} \qquad (4.27)$$

While T is the pendulum's period, l and g are its length and gravitational acceleration respectively. With (4.27), we can derive the following equation

---

[27]The answer didn't come out until 1915, when Einstein published his research in general relativity: the observed gravitational effect between masses results from their warping of spacetime, and the force of gravity is proportional to the degree of curvature.

to determine how T differs when in equatorial and polar regions:

$$\Delta T = 2\pi \left(\sqrt{\frac{l}{g_1}} - \sqrt{\frac{l}{g_2}} \right) \qquad (4.28)$$

The derivation process is described in detail in "Appendix 4.5: How the Theory of Pendulum and Its Period Change When in Equatorial and Polar Regions?" at the end of this chapter, where we will also analyze the reason why Huygens questioned Newton's law. Using (4.27) as a basis, we calculated how the centrifugal force from Earth's rotation would impact the period of a pendulum near the equator, and judging from the results, the influence of Earth's shape is almost 30 times stronger than that caused by the centrifugal force, which shows that the factor of rotation can be disregarded.

After the effect of Earth's rotation was proved too minor to be considered, Huygens came up with an even more difficult question: when the mass of an object is small to a certain degree, does the law of universal gravitation still apply? Just like the issues raised by Leibniz, who harbored malicious intent of sabotaging Newton's reputation, Huygens' question also posed a challenge that neither Newton nor any contemporary scientists could possibly solve, but it also gave later researchers an idea of what needed to be studied further, which was why related topics could be clarified in the 20th century.

In fact, before Newton published his law of universal gravitation, Descartes had already addressed the topic of gravity between the Sun and surrounding planets by applying Galilei's concept of inertia to planetary orbits. According to him, there was supposed to be an action force pushing planets toward the Sun during their movements; otherwise, they'd deviate from their orbits following the tangential direction. As for where the force came from, he devised the theory of vortices, in which he proposed that planetary motion would generate a whirl in the luminiferous aether of universe and that it was the forces from this whirl that pushed planets toward the Sun. During the Descartes era, this theory was widely accepted among astronomers and even used to argue against Aristotle's ideology. After *the Principia* was published in 1687, there were still many in the scientific community who held a reserved attitude towards the law of universal gravitation because of Descartes' influence. Luckily, Newton was well aware that the theory of vortices was something to be taken seriously, which is why he proved against it with Proposition 52 in Volume 2 of *the Principia* and managed to protect himself from widespread skepticism.

Apart from the theory of vortices, the British doctor William Gilbert (1540–1603) also tried to explain the reason behind planetary motion around

Sun and why the orbits were elliptic but cited Earth's magnetic field as the driving force. He based his arguments on the hypothesis that both Sun and Earth had North and South magnetic poles, which was why both bodies would attract and repel each other, and due to the difference in their magnetic forces, the orbit of Earth was distorted into an ellipse. However, such a claim was not just as groundless as the theory of vortices but couldn't be proved using any reasonable model of physics with mathematical calculation, either. Therefore, Gilbert's theory was simply disregarded by Newton.

Due perhaps to the extreme importance of Newton's discovery and the large extent of its influence, there had been no lack of questioning or even opposition to his theory. Even when Einstein introduced the general law of relativity that gravity can affect the direction of light in the 20th century, the Times was so quick to announce that Newton's research might be fallacious, but the truth is, Newton's theorem is based on absolute time and space, while what Einstein discussed is simply an exception in the relativity of the two elements. In fact, even Einstein himself said that Newtonian laws were 100% correct in systems where objects traveled slower than light, an ultimate endorsement of Newton's theories.

4.12 A Great Theory: Needs Rigorous Examinations

Despite the overwhelming criticism leveled against Newton, his opponents still struggled to cause any harm to the credibility of his theories because they had been endorsed by a series of scrupulous examinations, which pushed his achievements in the field of physics to an unprecedented height. In fact, he had already braced himself for possible attacks before publishing *the Principia*; therefore, he made it a point to include a range of self-examinations in Volume 3 of the book, such as the examples in Sect. 4.8. Unfortunately, not everyone was knowledgeable enough to understand his approach, which is why some decided to test the Newtonian theories using different methods.

The first examination was performed by Académie française (the French Academy) of France, where the academia refused to accept Newton's know-it-all theory due to the influence of Descartes' argument for the universal vortex and wasn't forced to take Newton seriously until 1735, when researches that could prove his laws were introduced one after another. Therefore, the Academy chose the theory stating that Earth was an oblate spheroid from Volume 3 of *the Principia* and sought to prove it, hoping to verify whether it was true that the high tangential velocity (1,670 km/h) near the equator triggered by the centrifugal force generated from Earth's rotation would cause the crust near the equator to expand outward and therefore result in the 23 km difference in the equatorial and polar radiuses, just as Newton had claimed. In fact, they considered this proposition worth pursuing partly because they knew a way of verification. So, they dispatched two groups of scientists to a Peruvian town near the equator and one in Norway near the North Pole to measure the meridian length at both locations. After taking measurements and collecting data for more than three years, the scientists found that the meridian was indeed longer at the North Pole than near the equator. In other words, it's true that the equatorial radius is longer than the polar counterpart, which causes the Earth to be shaped like a flat tangerine.

In fact, the difference between equatorial and polar radiuses can also be calculated using the law of universal gravitation. As shown in "Appendix 4.6: Calculating the Difference between Earth's Equatorial and Polar Radiuses," change in the length of the equatorial radius caused by Earth's rotation (h) can be calculated on the basis of energy conservation. During the rotation process, there are two types of potential energy generated on Earth's surface, i.e. the gravitational energy caused by gravity (mgh), and the energy of the centrifugal force from Earth's rotation $\left(-m\omega^2 x^2/2\right)$, where x is the distance between any point on Earth and its axis. At this point, the surface of Earth should be equipotential; otherwise, the surface mass would continue to be redistributed and cause the surface height to keep changing, which is why the sum of the gravitational energy and the energy of the centrifugal force has to be a constant. By supposing the change in the height of Earth's surface is minimum, we can derive $x = R \sin\theta$, which can then be substituted into the formulae of the two types of energy. Adding the two results, we can obtain the function $h(\theta)$, which represents how the degree of Earth's surface deformation changes with latitude:

$$h(\theta) = \frac{\omega^2 R^2}{2g} \sin^2\theta + C \tag{4.29}$$

In the function above, ω is the angular speed of Earth's rotation, R its radius, g the gravitational constant, and θ the angle of any circle of latitude.

Based on the result of (4.29), the difference between the equatorial and polar radiuses is about 11 km, registering a 50%+ deviation from the actual measurement of 23 km. As for the reason behind such a significant difference, there are a couple of factors that can explain the result. First of all, let's not forget that this proposition is a simplified model that only considers the influence of the centrifugal force from Earth's rotation. The other possible explanation is that the deformation of Earth's surface is in fact highly irregular, with some people claiming the Earth is a potato-shaped body with unevenly-distributed density. Both these factors could have contributed to the deviation, but the measurements of the French Academy and our calculation can both prove the validity of Newton's theory that Earth is a flat orb.

The second examination was conducted because of the publisher of *the Principia*, Halley. As previously discussed in Sect. 4.5, in 1705, Halley sorted through data on the cometary strikes observed in 1531, 1607, as well as 1682, concluded that the comet followed a quasi-parabola orbit, and predicted the next comet sighting to happen at the end of 1758 or beginning of 1759. Yet 50 years before Halley's prediction came true, the French mathematician Alexis-Claude Clairaut (1713–1765) decided that his theory was worth examining and therefore established a mathematical model in 1757 that included not only the solar gravity exerted on comets but the possible disturbance caused by Jupiter and Saturn to cometary trajectories as well. After six tedious months of calculations, he accurately predicted Halley's Comet to return in mid-April of 1759, only about one month later than the comet's actual visit on March 14 of the same years. As such, the validity of the law of universal gravitation was once again verified.

The third examination, however, wasn't nearly as smooth and probably caused a cold sweat for supporters of Newton's theories. This examination, though jointly initiated in the 18th century by French and British scientists, wasn't executed one hundred years later in 1846 and ended up leading to the discovery of Neptune, the eighth planet in the solar system. Nevertheless, were it not for the observation that the orbit of Uranus wasn't a perfect ellipse, mathematicians and astronomers wouldn't have realized it was actually the gravity of Neptune that disturbed its trajectory in their joint efforts. Therefore, we have to first talk about the discovery of Uranus.

Uranus was discovered as early as in 1690 by the British astronomer John Flamsteed (1646–1719) but mistakenly categorized as a star and cataloged as 34 Tauri. It just so happened that almost a century later in 1781, the German musician William Herschel (1738–1822) saw Uranus climbing near Gemini through his self-installed telescope when lounging in his backyard. After consulting existing celestial data, he confirmed the climbing body to be the "star" discovered by Flamsteed at the end of the previous century, except that it was in fact a planet which orbited on an elliptic trajectory. His discovery sent a ripple through the scientific sphere in Western Europe because the seventh planet (excluding Earth) of the solar system was finally discovered nearly 4,000 years after the Babylonians documented the five planets they knew of in 2000 BC. Therefore, it was only natural that the newly-discovered orb became the focus of people's heated discussion. After some twists and turns, it was finally named "Uranus."[28]

Uranus quickly became the focus of all spotlight in the astronomical circle right after its discovery, with almost every telescope you could possibly think of directed at this freshly found planet. However, after half a century's obser-vation, it was generally agreed that its deviation from the elliptical orbit grew more and more obvious, yet nobody would go so far as to question the valid-ity of Kepler's first law of planetary motion, so people started to consider the possibility that there might be another unknown planet affecting the shape of Uranus' trajectory. The hypothesis was easier said than done, though, because it was very hard to measure the influence of any planet other than the Sun on the orbit of another planet. Even Newton disregarded all the inter-planetary impacts and only considered the effect of the Sun's gravitational force on its surrounding planets when trying to prove Kepler's first law with his theory of universal gravitation, which was an inevitable compromise because although one can calculate the motion of two celestial bodies under all circumstances with data about their mutually-affected moving patterns, which is what we know as an "analytical solution", it's impossible to get a credible result by applying this approach to any situation where three bodies affect one another. A more reliable method nowadays is to convert simultaneous equations that regulate the forces among planets into a computer program through numer-ical computation and introduce the initial conditions to calculate a planet's motion and the force exerted on it at the subsequent point in time before inputting the results into the program again as a new set of initial conditions to calculate the necessary data at the next point in time, a process that has to be repeated over and over until the required values are obtained.

It's true that deviation can rise from the exclusion of planetary gravity, so why was Newton able to prove Kepler's first law when he considered Sun the only source of gravitational force? Well, because his calculations, though rough, were backed by facts. Let's take his calculation of Earth's orbit as an example. Excluding the Sun, Venus and Jupiter are the planets that exert the most gravity on Earth because they are the planets in the solar system that are closest to us and the largest in mass respectively. When these two bodies come to perigee, however, their gravitational forces are still only 1/8,000 and

---

[28] Herschel, who migrated to England in 1757 originally planned to name the planet Georgium Sidus (George's Star) to show his allegiance to King George III but was met with strong opposition from French astronomers, who argued that it should be named after the discoverer as Herchel's Star. After quite some disputes, Uranus, name of the Father of Sky in Greek mythology, was finally accepted by the academia after proposed by German astronomers.

1/16,000 of the solar gravity, which are little enough to be neglected. There-fore, it can be safely inferred that there would only be a very slight difference between the results from considering and excluding the impact of Venus and Jupiter.

Yet when it comes to Uranus, which is fairly far from the Sun, does such rough calculation still make sense? While many astronomical observers sim-ply assumed that Uranus didn't follow an elliptical orbit, the British math-ematician John Couch Adams (1819–1892) and French astronomer Jean-Joseph Leverrier (1811–1877) went against the tide and performed calcula-tions almost at the same time towards the hypothesis that Uranus couldn't stay on its elliptical trajectory due to the disturbance from some unknown planet. While Adams' conducted some extremely complicated calculations that're rarely used as a reference, Leverrier approached the issue by proposing several hypotheses, some of which proposed that this unknown planet was similar to Uranus in mass and not too far from it in distance, and that it also orbits the Sun on the ecliptic plane. Using these propositions as a start-ing point, Leverrier was able to greatly simplify his calculations and iden-tify the possible location of this new planet at the time of research. So on September 18th of 1846, he sent his results to Johann Galle (1812–1910), who was working for the Royal Observatory in Berlin back then, while five days later, Galle received the letter where Leverrier detailed why he suspected there was a planet with a mass 24.5 times that of Earth which completed a circle around Sun every 60,238 days and 11 h after calculating related val-ues based on Newton's law of universal gravitation as well as Kepler's laws of planetary motion and requested him to confirm if this new planet actually existed at the celestial coordinates provided in the letter as soon as possible. A long-time admirer of Leverrier, who taught at École Polytechnique (the Poly-technic school), for his astronomical research, Galle immediately pointed the Fraunhofer refractor telescope at his disposal to the coordinates, and guess what, with the instructions given by Leverrier, he spotted in the night sky within one hour this fresh new planet that had never been documented in any star catalog ever before.

The discovery of the eighth planet in the solar system, which was later named Neptune, sent another ripple through but carried a different meaning for the astronomical circle because Uranus was an unexpected gem spotted by Herschel while roaming the night sky with his telescope, Neptune was the product of an exceptional research that deeply explored an astronomi-cal anomaly on the basis of scientific theories. Therefore, this discovery cre-ated as big of a splash as in the previous century when Halley's prediction that the comet he studied would visit Earth again 76 years later came true and still elicited people's heartfelt admiration for Newton's law of universal gravitation. The only difference was that Halley's prediction took more than 50 years to be proven right, while Galle only spent five days on confirming the existence of Neptune thanks to the law of gravity, which Leverrier counted on to predict the never-seen-before Neptune and establish the coordinates where Galle spotted the planet. Such discovery didn't just prompt scientists to celebrate the magic of the law of universal gravitation yet another time but showed humans the predictive power of science as well, which is why scientific research has from then on become our way of exploring the unknown and foreseeing the unseen.

The large-scale examinations described above all helped elevate Newton's law of universal gravitation to a school of thoughts and revived the ancient idea that nature could be understood through observation and inference, which debunked Aristotle's theory of purpose, which dominated studies on the universe for more than 2,000 years. It's no wonder that the much impressed French mathematician Pierre-Simon de Laplace (1749–1827) would boast that with Newton's theories, nothing couldn't be predicted in the vast universe, and the world's future could be easily foreseen with human eyes.

4.13 Cavendish's Measurement of Gravitational Constant

Though the law of gravitational gravity was endowed with an unquestionable status after a whole series of meticulous examinations, Newton left in his theory an unknown value, G, i.e. the gravitational constant of the universe, which triggered many scientists' curiosity, yet none of their studies came to any concrete conclusion. In 1797, the professor at the Cambridge University of England, Henry Cavendish, conducted an experiment that led him to the gravitational constant in the equation of universal gravity, $F = GMm/R^2$. However, this wasn't what he set out to do as the original purpose of this experiment was to figure out the mass of Earth. As previously mentioned in Sect. 4.8, scientist could already calculate the ratio of mass between two planets using Kepler's third law and Newton's law of universal gravitation, so as long as one could measure the mass of any planet in the solar system, the same physical quantity could be calculated therefrom for all the other planets. As for which planet to measure first, the most practical choice was undoubtedly the Earth because related experiments are the most feasible and the results would be the most useful.

Yet how does one measure the mass of a giant ball like Earth? Back then, the Royal Society of London was planning a pendulum experiment that was historically called Schiehallion where scientists sought to find an evenly-shaped mountain whose volume was easy to estimate to measure its impact on the inclination of the pendulum that they had previously set up before converting the results into the density of Earth using the pendulum theory. After several experiments, the results showed the Earth's density as 4.5 times that of water, but the experimenting mechanism didn't seem discreet enough for Cavendish, so inspired by the geologist John Michell, he designed a set of precision equipment to perform a torsion pendulum experiment (see Fig. 4.7) to calculate the density of Earth again. What Cavendish did was compare the average gravity of the big and small balls with the gravitational force exerted on an orb on Earth's surface before using the ratio to calculate the mass and density of Earth, whose diameter was already known to be 6,370 km back then. As it turned out, his results showed the density of Earth to be 5.45 times that of water, 20% higher compared to the Royal Society's conclusion.

Despite the noticeable difference in results compared to the experiment executed by the authoritative Royal Society, Cavendish's version was widely thought to be the most accurate and successful ever in terms of measuring the mass of Earth thanks to his discretion and attentiveness in designing the

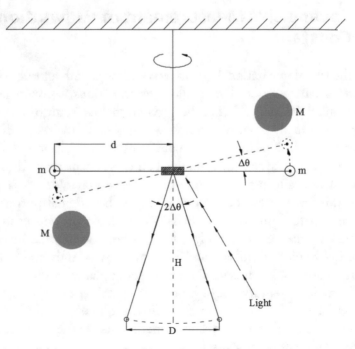

Fig. 4.7 Cavendish's apparatus for his torsion pendulum experiment

apparatus and running the experiment. When it comes to the design, he suspended the small balls with fine threads made of quartz so they could twist responsively upon the gravitational influence of the big balls. In addition, he added an optical magnifying instrument for incident rays to bounce off a reflective cylinder so that he could measure the angle between the incident and reflected rays, which was twice that of the small balls' range of swing. He also made it a point to carefully check the datum points of the small balls under the experiment environment, i.e. where they stayed fully at rest without the influence of the bigger counterparts. Then, when the experiment officially started, all staff had to be evacuated from the laboratory, while the big balls were controlled using a sliding equipment that had been installed outside the window beforehand. As for changes in the angles of the threads, they were observed remotely by scientists using a telescope by the window by measuring the distances traveled on the optical caliper by the reflected rays. In other words, the suspension system for the small balls and related instruments were completely isolated in the lab to cut off all external contact and avoid the impact of environmental conditions such as changes in airflow and temperature (they weren't so lucky to have thermostats back in those days, after all).

In addition to the apparatus design and related preparation, the measuring process was no less detail-oriented. Cavendish began the experiment by moving the big balls (M) towards the small ones (m), which would cause the torsion pendulum to start swinging due to the drawing force of gravity and the resistance of m, whose amplitude and period of twisting movement had to be measured now. Here are the results: before M was introduced, the period of the torsion pendulum was $T = 852$ s, while when the distance between m and M is $R = 230$ mm, the deflection caused by the twisting movement of m was $d\Delta\theta/2 = 3.725$ mm. Besides, the mass of the big balls was $M = 158$ kg. As the most difficult challenge of this experiment lay in measuring the tiny amplitude of the small balls' rotation caused by gravity, Cavendish designed an optical magnifying system where a source of light was directed towards the suspension point of the pendulum, above which was a cylinder that could reflect all incident rays toward the light collecting panel set up where the experiment staff was stationed. Since the reflective rays would be projected to different spots on the panel due to the rotation of the cylinder, it was possible to calculate $\Delta\theta$ with the location change D with the equation below:

$$\Delta\theta = tan^{-1}\left(\frac{D}{2H}\right) \tag{4.30}$$

In addition to the twisting angle of the pendulum ($\Delta\theta$), the corresponding period also needs to be measured. To do so, we can follow the equation of motion[29] for torsion pendulums, shown as follows:

$$I\ddot{\theta} = -k\theta \tag{4.31}$$

In (4.31), $I = \mu d^2$ is the moment of inertia of the lever from which the small balls are suspended, while μ and d represent the equivalent mass of the pendulum and the length of the lever. When the mass of the lever is so small compared to that of the small balls (m) that it can be ignored, we can obtain the result $\mu = 2m$, which is a standard equation of simple harmonic motion and allows us to calculate the pendulum period as:

$$T = 2\pi\sqrt{\frac{I}{k}} = 2\pi\sqrt{\frac{2md^2}{k}} \tag{4.32}$$

In (4.32), k is the torsion constant of the pendulum, whose relation with Γ, the torque working on the lever where the small balls are suspended, can be represented as:

$$\Gamma = 2Fd = 2\frac{GMm}{R^2}d = k\Delta\theta \tag{4.33}$$

In (4.33), F is the gravitational force between the big and small balls, which allows us to derive the result $k = 2GMmd/R^2\Delta\theta$, an equation we can then substitute back into (4.32) to calculate the pendulum's period T and obtain the below equation:

$$T = 2\pi\sqrt{\frac{dR^2\Delta\theta}{GM}} \tag{4.34}$$

Then, from Eq. (4.34), the value of G can be calculated as:

$$G = \frac{4\pi^2 d\Delta\theta R^2}{T^2 M} = \frac{4\pi^2 \times 3.725 \times 10^{-3} \text{ m} \times (0.23 \text{ m})^2}{(852s)^2 \times 158 \text{ kg}}$$

$$\approx 6.78 \times 10^{-11}\frac{\text{Nm}^2}{\text{kg}^2} \tag{4.35}$$

---

[29] For details, please see https://www.youtube.com/watch?v=WGHEXoCGXVY.

Based on the derivation process detailed above, one doesn't need the mass of the small balls (m) and the torsion constant of the pendulum (k) to conduct the experiment, while accurate measurements of the pendulum's period (T), the distance between the big and small balls (R), the amplitude of the big balls' swinging movement ($d\Delta\theta$) and their mass (M) are all necessary. Apart from M, which could be easily measured beforehand, all the other values were products of extremely accurate measurements at the time of the experiment, which was only made possible with Cavendish's precise instruments. In fact, there have been quite some scientists in modern times who tried to calculate the value of G with more advanced apparatus, but their conclusions were all quite close to Cavendish's result. With the value of G confirmed, Newton's formula of universal gravity can then be applied to all mechanical systems and calculate the attracting force between any two objects. In addition, this great law also inspired the development of fluid mechanics, solid mechanics, mechanics of materials, and dynamics, all contributing to the splendid era that saw a giant leap in human's material civilization from the 20th century onward.

4.14 *The Principia*: An Abstruse Masterpiece

In the composing process of *the Principia*, Newton patiently resorted to lots of complex geometrical shapes and tedious descriptions to prove his research so that contemporary readers, who didn't know much about calculus, could actually see his theories proved. Purposely or inevitably, however, all he left readers with was confusion and a sense of loss, because the book was simply too hard for ordinary people to understand. Even when the time had progressed to the 20th century, scientists still had to try really hard to simplify this masterpiece to make it more understandable to the general public, among which were two heavyweights in the world of physics. The first was the Indian scholar, Subrahmanyan Chandrasekhar (1910–1995), who won the Nobel Prize in Physics by proving that white dwarves can't be over 1.4 times the mass of the Sun.[30] As a well-cultivated researcher attempting to open for readers the door to the world of greatness and transcendence in *the Principia*, he mustered a great amount of patience in writing a book of

---

[30]When stars decompose into white dwarves, it's the electron degeneracy pressure (the main force that contributes to the formation of white dwarves as it doesn't allow two fermions to stay in the same energy state) working against the compression of gravity that prevents them from collapsing. Reversely calculated from the maximum gravity supported against, the maximum mass for white dwarves is 1.4 times that of the Sun. If exceeding the limit, they wouldn't be able to exist.

nearly 600 pages,[31] hoping to guide readers through the nooks and crannies of Newton's work. This publication, however, was deemed fairly complicated still and failed to serve the purpose of actual simplification, so another scientist, Richard Phillips Feynman (1918–1988), who won the Nobel Prize in Physics as well in 1965 for his achievements in quantum dynamics, also tried to prove the inverse proportion between gravitational force and the square of distance by establishing his own set of geometrical graphs. However, he never had his manuscripts published,[32] probably finding it lacking in terms of academic value, but in fact, the geometrical approach he adopted to demonstrate the elliptical orbits of planets is among the many studies where his wisdom shines through. To introduce readers to his ingenious research, we're including "Appendix 4.7: Feynman's Geometrical Approach of Proving Planetary Orbits to be Elliptic" in this chapter. To be honest, though, judging from the studies of the two authoritative figures in the 20th-century academia of physics, there really is no shortcut to proving the elliptic shape of planetary trajectories.

That said, efforts in reproducing Newton's derivation process of the gravitational law based on Kepler's laws of planetary motion using simple geometry or elementary calculus haven't been lacking in the modern era. In 2008, for example, Wu, Hsiang and Chang[33] also reinterpreted the abstruse *Principia* by using simplified but on-point geometric relationships as well as middle school-level calculus to (1) determine the direction of gravity with Kepler's second law (Propositions 1 & 2 in *the Principia*), (2) derive the law that gravitational force is inversely proportional with the square of distance (Proposition 11 in *the Principia*), (3) prove with the previous two results that planets circle on elliptical orbits where the Sun occupies one of the foci (Proposition 17 in *the Principia*), and (4) confirm that any working force acting on a sphere whose mass is evenly distributed can be considered concentrated on the barycenter of that sphere (Proposition 71 in *the Principia*). Their clear, straightforward derivation style served as a bridge between the puzzling *Principia* and general readers to popularize the Newtonian school of thought, which indeed makes their publication worth passing down. In fact, according to Hsiang and some other scholars, although Newton did illustrate the concept of barycenter in Proposition 71, he never really tried to prove his hypothesis that both the Earth and Sun could be regarded as point masses,

---

[31] Please see Chandrasekhar [2].

[32] The manuscripts were organized and published after Feynmann died by his friends David & Judith Goodstein as a dedicated book. For details, please see David and Judith Goodstein [11].

[33] Please see Hamilton [13].

which he simply formulated with bold but precise intuition. Yet as it turned out, it was indeed a wise assumption.

Upon its official publication in 1678, *the Principia* immediately triggered widespread discussion across the scientific community, even prompting some leading figures in mathematical studies to foray into the world of universal gravitation. Then, about a century passed. After numerous experimentation, observation, and derivation, Newton's gravitational law was finally accorded the highest status in the world of science. It's no wonder that, impressed by such a mighty achievement that had weathered all the challenges and meticulous examinations, the French and German mathematicians Joseph L. Lagrange (1736–1813) and Johann C. F. Gauss (1777–1855), as well as several famous scientists in the contemporary era, all hailed *the Principia* as an ultimate pearl of wisdom that served to elevate humans' mind and thought while recognizing Newton to be one of the most significant figures of all time whose scientific success can't possibly be paralleled.

4.15 Newton—A Scientific Giant Almost Impossible to Emulate

Newton's law of universal gravitation laid down a whole new definition of "scientific law" for later generations: a simple set of mathematical principles that can accurately explain phenomena consisted of complicated factors. The greatness of the gravitational law lies in the fact that it not only regulates the motion of celestial bodies in the universe but can describe precisely the motion rules of common objects on Earth as well. With this almighty law, we can calculate Earth's axial precession, cometary orbits, planetary motion, and tidal movements when applying the theory to astronomical bodies; as for the application on Earth, the law can be used in far more scenarios beyond predicting if an apple would be smashed after free falling to the ground. In fact, it inspired the development of disciplines that drove the advancement of human beings' material civilization in the 20th century, including solid mechanics, fluid mechanics, linear elasticity, mechanics of materials, etc. When magnificent architecture like the Brooklyn Bridge in New York and the Eiffel Tower in Paris lit up both sides of the Atlantic Ocean at the end of the 19th century, the world's admiration and trust for Newtonian mechanics had been elevated to an unprecedented height. Some critics even went so far as to say that all that was left for later physicists to do was confirm the physical parameters in newly-explored fields rather than try to discover unknown laws.[34] Simply put, it's their contention that with everything on Earth regulated by the law of universal gravitation, there's no question in the sphere of science that we can't find an answer to as long as related conditions are readily available.

It was on May 19th of 1686 when the Royal Society of London agreed to publish *the Principia*, but with the condition that Halley had to pay for all related expenses due to the tight financial situation caused by the sluggish sales of the society's previous publication. No one actually expected Newton's book to take England and the entire European continent by storm, as it turned out. *The Principia*, at once extensive in the range of topics and extremely detailed in addressing each of them, caused a buying rush and sparked enthusiastic discussions in academic institutes and even coffee houses. As for the good-willed Halley, he probably garnered a surprisingly big fortune for publishing the book, which we could also say is God's reward to him for helping Newton make public his major research. Despite the widespread sensation and admiration across European academia, there were still a mere few who unleashed malicious attacks. For example, the German polymath Leibniz claimed that calculus was something that he invented but plagiarized, the fellow British scientist Hooke contended that Newton stole his original idea that the gravitational force between two objects was inversely proportional to the square of their distance, which he revealed in a letter to Newton five years before the law was published, while there were yet some people who bashed out at Newton for his "mystical" assumption that gravity can be transmitted through the void of universe, which they said shouldn't be easily believed without any rational proof proposed.[35] However, such criticism and ridicule were triggered mostly by personal feuds, and at the end of the day, the great law of universal gravitation still stands unscathed.[36]

The tremendously important law of universal gravitation served as a selection mechanism that eliminated fallacious, unfit theories developed during the scientific revolution, which ran through the course of more than 160 years over the 16th and 17th centuries, and left only the ones reflecting the truths of nature. Essentially, the law was almost like a second scientific revolution,

---

[34] In fluid mechanics and mechanics of materials, for example, one simply needs to determine the stress-strain relationship (i.e. the composition) of a certain fluid or material and substitute the parameter into the formula of Newton's second law of motion to solve related issues.

[35] Regarding this issue, Newton's theory is indeed flawed to some degree since he contends in the book that gravitational force travels at an infinite speed, an ungrounded argument that was denied by Einstein more than 200 years later with the theory of general relativity. With tensor, a complex algebraic object in mathematics, Einstein proved through a complicated derivation process that gravity is the curvature of spacetime which travels at light speed.

[36] See also the story illustrated in Chap. 7 of Mlodinow [23].

revealing the real though cruel fact to later researchers: there's no way Mother Nature would show her true colors to those who do not pay close attention; scientists, if clever and alert enough, should see the natural world as an experienced secret keeper and count on careful observation and accurate experimentation to uncover the remarkable subtleties around us. Newton himself can be said to be the personification of detail orientation and meticulous observation, but even a great scientist like him once wrote humbly in a 1676 letter to Hooke that "if I have seen further than others, it is by standing upon the shoulders of giants." In fact, these giants that he mentioned, including Copernicus, Kepler, Galilei, and Descartes, had also been part of a hundred-year-long academic debate to slowly dismantle Aristotle's theory, which dominated the field of natural philosophy for nearly 2,000 years, yet it was Newton who, through strictly-organized analyses, finally summed up their research with a practical mathematical formula that can actually be used to calculate and predict natural phenomena. By then, Aristotle's theory had been fully debunked and basically collapsed.

Although Newton was humble enough to attribute his achievement to earlier researchers, his invention of calculus actually put him well ahead of the "giants" whose studies he used as references. When mentioning calculus, a critical milestone in his scientific career, we simply can't help but wonder if there was anyone back then who could remotely rival his command of mathematics, and the answer is most likely no. If you look back at how he derived the law of universal gravitation, you'll find it hard to overlook the fact that he adopted a model of mathematical calculations to develop a law of physics, an approach that Galilei had applied as well. However, Galilei wasn't able to clarify the details about his laws of planetary motion from the perspective of physics or derive accurate mathematical formulas to predict physical phenomena under different conditions. What he failed to do, Newton managed to achieve thanks to his scientific intuition, which somehow made him wonder if gravity was the driving force of all motion, and his serious effort in exploring the nature of this force and establishing a mathematical model for universal gravitation. Such seamless integration of mathematics and physics later became part of his legacy and an indispensable ability that researchers who'd like to engage in studies of physics have to equip themselves with.

Newton and Galilei were similar in the sense that they could both sort through complicated factors contributing to natural phenomena and disregard the secondary or irrelevant ones to develop simple regulatory laws describing the most critical factors. However, they were extremely different at the same time when it comes to the role that science played in their lives. While Galilei sought to transform his knowledge in mechanics, optics, and astronomy into army weapons, spying equipment for the navy, and bestselling books, basically treating science as a commodity to amass fortune, Newton was stubborn enough to dedicated ten of the golden years of his life to the studies of the Bible code and alchemy, only to throw himself into the grave due to mercury poisoning. Fortunately, he left the magnificent law of universal gravitation to the world. Otherwise, he wouldn't have been able to enjoy the royal-like status in the upper class of London.

Though somehow materialistic, it's a pragmatic way of evaluating the achievements of Newton and Galilei based on how economically beneficial their studies were, and honestly speaking, the cold, hard fact is that a scientific research can be incredibly fruitful or come to no avail based on how the topic is chosen, what approach is adopted, and how the results are interpreted. Newton's fields of interest, such as the Bible code and alchemy, were both unlikely to yield any substantial results no matter however hard he tried with his ingenuity. As for Galilei, he was smarter in transforming his research in mechanics into financial well-being and social status, yet he was also the one to be incarcerated due to his principle as a scientist. Looking back at the life stories of both of them and several revolutionary scientific figures mentioned in this book, we can't help but ask, is science a dangerous undertaking? The answer, however, seems to depend on how each and every researcher perceive their pursuit.

Severely affected by the profound and widespread influence of the law of universal gravitation, the Roman Church decided to preach the astronomical model proposed by Copernicus as a response, but the European society still saw the rise of anti-authoritarianism, which prompted people to question Aristotle's longstanding theory, hereditary monarchy, the authority of religion, etc. Such a change of mindset eventually pushed England to establish constitutional monarchy with the Act of Settlement and formulate quite some rules regulating the power balance between the King and the Parliament. As for France, the people of Paris stormed the Bastille on July 14th of 1789 and started a bloody revolution that lasted as long as five years. In the end, the feudal system in France, which traditionally favored royals and aristocrats, was dismantled, and what followed was the first-ever democracy in the world. In other words, Newton's law of gravitation didn't just initiate a glorious era for humans' material civilization but produced a fundamentally revolutionary effect as well, and Newtonian mechanics is also deeply grounded in various aspects of our life, with his terms of physics used frequently in phrases like the inertia of habit, life energy, and work momentum.

4.16   Life Summary: Awkward in Society But Sharp in Science

When looking back at his life at a later age, Newton described himself as a kid picking up cobblestones at the seaside, as in how little of nature's truth he'd managed to discover. In fact, Newton spent his entire life in the narrow area bounded by Cambridge, London, as well as Lincolnshire and had never seen the ocean with his own eyes. He had no close friends or family, not to mention anyone who he could remotely call a lover. For him, socializing was just as painfully difficult as it was for pets to play scrabble, which was probably why most people thought of him as mean, serious, arrogant, and always harboring vile criticism. Although he ended up winning the world's recognition and admiration, most time in his life was spent arguing with academic enemies, while the joy of his success was hardly shared. That said, perhaps it was such shy, antisocial nature that contributed partly to his scientific achievements since people who spend too much effort trying to mingle around are normally incapable of exploring nature's truth with a fierce determination and sharp, independent thinking.

In fact, Newton's resolute spirit was exactly the reason why he never gave up on the quest for nature's mechanism in an era dominated by theology while many still believed in magic, and as it turned out, the law regulating the world was just so surprisingly but elegantly simple. Over that past few hundred years, the scientific community never ceased to be thankful for his contribution and kept on celebrating his great ideas and rational style of research, even hailing him as a magician who unveiled the secrets of the universe, which shows that his sharp observation, well-structured thinking, and ability to transform obscure phenomena into simple laws have been very much missed and applauded over centuries of time.

The Principia, which was published in 1686, did win Newton tremendous fame but changed his life course as well. Abandoning his once-isolated lifestyle, he now actively dedicated himself to academic debates and bashed mercilessly out at his opponents like Hooke and Leibniz, who he wouldn't even stop persecuting after they passed away; as for those who ridiculed his theories as "mystical", he criticized them vehemently as too retarded to understand the phenomena of nature. When the political dispute between the University of Cambridge and James II of England broke out,[37] our once

---

[37]The political conflict arose when James II tried to declare Roman Catholicism the state religion and ordered the University of Cambridge to grant a degree to a Benedictines monk without proper examination, which the University rejected for violation of its several-hundred-year-old rules.

socially-awkward Newton also decided he should be part of it and was therefore appointed to be a member of the Parliament, representing the Cambridge University. In 1696, he was handpicked by the royal family to take charge of the re-coinage program for the Royal Mint, a job that paid him £400 a year, so he quit the teaching position at Cambridge and welcomed the promotion to Head of the Mint due to his success in running the program. At this point, the salary of his newly-acquired position (£1,650 per year) was up to 16 times that of his paycheck as a professor. Therefore, he left Cambridge permanently and relocated to London as a new base. In 1703, the year Hooke died, Newton took over his role as President of the Royal Society. During his 23-year term of office, he ruled with an iron fist and didn't really care to take up any advice. Such a tyrannical way of handling matters inevitably cast him in a negative light and became part of his legacy. In 1705, he was knighted by the Queen of England. Then, 21 years later, the 84-year-old Sir Isaac Newton died of serious pneumonia and kidney stones. He was buried in Westminster Abbey, his funeral somber but glorious.

Newton left a box to the world that contained manuscripts totaling almost a million words. Among them are the results of his alchemy experiments, his prediction of the date of Jesus' Second Coming, architectural calculations of Solomon's Temple, and a study that greatly humiliated the University of Cambridge and the Church of England where he strongly denied the divinity of Jesus Christ. Judging from the voluminous documents, he probably devoted no less time to alchemy, the Bible code, and theology than to scientific disciplines like optics, mathematics, and mechanics. In fact, he also spent 28 years conducting research about how to regulate England's national currency, and even tracked down criminals for counterfeiting coins as Head of the Royal Mint!

Newton's pursuits in so many different fields might make many think he's as alien as extraterrestrial creatures. And true, he's indeed bizarre in his own way, but what he left behind isn't mysterious geometrical patterns on farms but the great law of universal gravitation, which reveals the secrets of universe, and research logs of several million words that weren't just extremely valuable in a Sotheby's auction but provided an exceptional example of research commitment for later scientists as well.

Appendix 4.1: Calculating the Period of Earth's Precession with the Law of Universal Gravity

Since the Earth is an oblate spheroid with slight bulges in the equatorial area, the gravitational forces of the Sun and Moon would each generate a torque on it and affect the angular momentum of its rotation. Meanwhile, the Earth works like a giant top that not just rotates but also revolves slowly with its axis forming a 23.5-degree angle with the ecliptic plane. This phenomenon, as demonstrated in Fig. 4.8, is called axial precession in mechanics, also called the precession of equinoxes in astronomy.

To determine the period of Earth's precession, let's first suppose there are two astronomical bodies whose masses are M and m, as shown in Fig. 4.9. If we set the origin of the xyz coordinate system at the barycenter of m, the center of M would fall on the x axis with a position vector of $\vec{R} = (R, 0, 0)$. Now, let's again suppose there's a small point mass dm in m whose position vector is $\vec{r} = (x, y, z)$ and forms the angle β with $\vec{R}$. Considering that r, the average radius of m is far smaller than R, the distance between the two astronomical bodies, their combined gravitational energy can be represented as:

$$V = -\int \frac{GM dm}{\sqrt{R^2 + r^2 - 2Rr \cos \beta}}$$

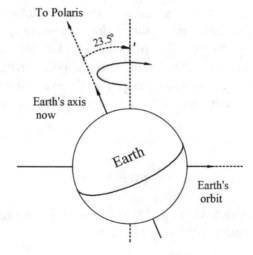

Fig. 4.8 The axial precession of the precession of equinoxes of the Earth

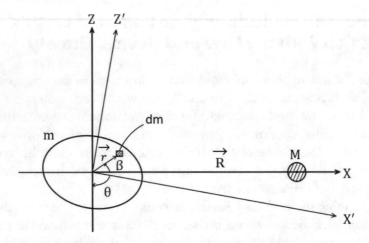

Fig. 4.9 The relationship between the two astronomical bodies to determine the period of the Earth's precession

$$\approx -\frac{GMm}{R} - \frac{GM}{2R^3} \int r^2 \left(3\cos^2\beta - 1\right)dm$$

$$= -\frac{GMm}{R} - \frac{GM}{2R^3} \int \left(2x^2 - y^2 - z^2\right)dm \qquad (4.36)$$

Just like the Earth, as you can see in Fig. 4.9, m is also an oblate spheroid in this proposition and rotates on its axis of symmetry z'. If we represent the moments of inertia from m's circling on z' and x' as C and A respectively, we can confirm that $C > A$ because m is flat and oblong in shape while z' is its axis of rotation. Now, if we rotate the coordinate system by an angle of $(\pi/2) - \theta$ along the y-axis, z' still stays on the xz plane but will form the angle θ with the x-axis. Please note, that the post-rotation variant y' still points toward the paper and will overlap with the original y-axis. Meanwhile, the spin of the coordinate system would help us derive the equations below:

$$x = x'\sin\theta + z'\cos\theta, \qquad (4.37a)$$

$$z = -x'\cos\theta + z'\sin\theta \qquad (4.37b)$$

Then, by substituting x', y', and z' for x, y, and z in (4.36) with (4.37a) and (4.37b), we'll come to the result below:

$$V \approx -\frac{GMm}{R}$$

$$-\frac{GM}{2R^3} \int \left[\begin{array}{c} \left(2\sin^2\theta - \cos^2\theta\right)x'^2 - y'^2 + \left(2\cos^2\theta - \sin^2\theta\right)z'^2 \\ + 6x'z' \sin\theta\cos\theta \end{array} \right] dm \tag{4.38}$$

Due to the symmetry of x' and z', the result of integrating the last item in (4.38), which contains $x'z'$, over m would be 0. Therefore, just a little simplification of the items of trigonometric functions in the above equation can help us rewrite it as:

$$V \approx -\frac{GMm}{R} - \frac{GM}{2R^3} \int \left[\left(1 - 3\cos^2\theta\right)\left(x'^2 - z'^2\right) - y'^2 + x'^2 \right] dm \tag{4.39}$$

In the meantime, we know of the two equations below because of the definition of the moment of inertia:

$$C = \int \left(x'^2 + y'^2\right) dm, \tag{4.40a}$$

$$A = \int \left(y'^2 + z'^2\right) dm = \int \left(x'^2 + z'^2\right) dm \tag{4.40b}$$

Derived from Eq. (4.40a) and (4.40b) is (4.41):

$$\int \left(x'^2 - z'^2\right) dm = C - A \tag{4.41}$$

While the symmetry mentioned before gives us (4.42):

$$\int \left(-y'^2 + x'^2\right) dm = 0 \tag{4.42}$$

Then, the substitution of (4.41), (4.42) into (4.39) would yield:

$$V \approx -\frac{GMm}{R} - \frac{GM}{2R^3}\left(1 - 3\cos^2\theta\right)(C - A) \tag{4.43}$$

To calculate the torque generated by M on m, we can differentiate the potential energy V over θ in (4.43):

$$\tau \approx -\frac{\partial V}{\partial \theta} = \frac{3GM}{R^3}(C - A)\sin\theta\cos\theta \tag{4.44}$$

From (4.44), we can see that the torque gives the angle θ a tendency to approach $\pi/2$. In other words, z' will slowly deviate towards z, meaning the direction of the torque points to $-y$.

While Earth's axial precession is affected by the gravity of both Sun and Moon, here, let's suppose the interaction between the two bodies is small enough to be disregarded, which is why we can evaluate the influence of each separately. If we follow the approach demonstrated above, the equations that are derived should be the same for both Sun and Moon; as long as we substitute two different sets of values as the parameters into the equations, we should be able to determine how the two bodies impact Earth's precession. So, let's first consider a system consisted only of the Sun and Earth, as shown in Fig. 4.10. If the z-axis is perpendicular to Earth's revolving plane (i.e. the ecliptic plane), then its axis of rotation (ON) would always form the angle φ with z. Then, if we again suppose the unit vector along ON is $\hat{n}$ and set its origin at point O, as time progresses, N will start motioning along the circle C continuously, as you can find illustrated in Fig. 4.10. In the next step, we need to mark a point (A) on C that would make $\overrightarrow{CA}$ parallel with $\hat{x}$ (the unit vector on the direction of x) and represent the angle $\angle NCA$ as α.

Now, let's look at a rotating coordinate system centered on the Sun where its angular velocity is equal to that of Earth's revolution. While the centers of both bodies are in a stationary state, the Earth would, due to is a revolution, be affected by a centrifugal force that's proportional to its revolving radius and points away from Sun. However, no net torque would be generated since centrifugal forces on Earth are always symmetric, yet when it comes to the Sun's torque on Earth, it's a varying value due to the continued circling of Earth's rotation axis around z, which causes the angle α to be changing constantly. Therefore, we need to develop the mathematical relationship between

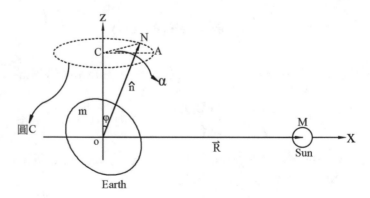

Fig. 4.10 A schematic description to determine how the two bodies impact the Earth's precession

this torque and α and calculate the integral of α to determine the result of $\langle\cos^2\alpha\rangle = \frac{1}{2}$, which is the average value of the torque. Please note that it is the direction of $\hat{x} \times \hat{n}$ that the torque points to. If we represent the angle between ON and the x-axis as θ, then we can also use (4.44) to calculate the torque value, which can eventually be written as:

$$\vec{\tau} = \frac{3GM}{R^3}(C - A)\sin\theta\cos\theta\frac{\hat{x} \times \hat{n}}{|\hat{x} \times \hat{n}|} \qquad (4.45)$$

As the angular momentum of Earth's rotation points to the direction of $\hat{n}$, the direction along which the momentum changes (i.e. the direction of the torque) should be perpendicular to the plane defined by the points O, C, and N, which can also be understood as the direction of $\hat{e} = (\sin\alpha, -\cos\alpha, 0)$. Therefore, it's only the torque on the direction of $\hat{e}$, which can be represented as $\vec{\tau} \cdot \hat{e}$, that contributes effectively to the precession of Earth. As for torques that point to other directions, they do cause φ, the angle between the axis of rotation and z-axis to change, but only to so small of a degree that the influence can be disregarded in this proposition.

Now, we can substitute all the angles and vectors in Eq. (4.45) with equations of φ and θ. Since related vectors $\hat{x}$ and $\hat{n}$ can be written as $(1, 0, 0)$ and $(\sin\varphi\cos\alpha, \sin\varphi\sin\alpha, \cos\varphi)$ respectively, we can derive $\cos\theta = \hat{n} \cdot \hat{x} = \sin\varphi\cos\alpha$, which leads us to the equation of the effective torque, as shown below:

$$
\begin{aligned}
\vec{\tau} \cdot \hat{e} &= \frac{3GM}{R^3}(C - A)\sin\theta\cos\theta\frac{\hat{x} \times \hat{n}}{|\hat{x} \times \hat{n}|} \cdot \hat{e} \\
&= \frac{3GM}{R^3}(C - A)\sqrt{1 - \sin^2\varphi\cos^2\alpha}\,\sin\varphi\cos\alpha \\
&\quad \frac{(0, -\cos\varphi, \sin\varphi\sin\alpha)}{\sqrt{\cos^2\varphi + \sin^2\varphi\sin^2\alpha}} \cdot (\sin\alpha, -\cos\alpha, 0) \\
&= \frac{3GM}{R^3}(C - A)\sqrt{1 - \sin^2\varphi\cos^2\alpha}\,\sin\varphi\cos\alpha\frac{\cos\varphi\cos\alpha}{\sqrt{1 - \sin^2\varphi\cos^2\alpha}} \\
&= \frac{3GM}{R^3}(C - A)\sin\varphi\cos\varphi\cos^2\alpha \qquad (4.46)
\end{aligned}
$$

After integrating the parameter α in (4.46) and adopting the average, we can derive:

$$\langle\vec{\tau} \cdot \hat{e}\rangle = \frac{3GM}{2R^3}(C - A)\sin\varphi\cos\varphi \qquad (4.47)$$

Please, note that $\langle \cos^2 \alpha \rangle = \frac{1}{2}$ was used in the derivation process above. Therefore, (4.47) is exactly the effective torque generated by the Sun (or Moon) on Earth.

Then, as the angular momentum of Earth's rotation is $\vec{L} = C\omega\hat{n}$, where ω is the angular velocity of its rotating movement, we can again derive the below equation:

$$\frac{d}{dt}\vec{L} = (\vec{\tau} \cdot \hat{e})\hat{e} \tag{4.48}$$

As a result, the angular velocity of Earth's axial precession can be written as:

$$\frac{d\alpha}{dt} = -\frac{3GM}{2R^3}\left(\frac{C-A}{C}\right)\frac{\cos\varphi}{\omega} \tag{4.49}$$

Note: In (4.49), the negative signifier means the precession is clock-wise if we look down from the top of $\hat{z}$-axis.

One thing that should be noted here is that the gravity causing Earth's axial precession comes from both Sun and Moon, and the total amount of torque also consists of the torques generated separately by the two bodies. Similarly, the total angular velocity of precession exerted by the Sun and Moon on Earth's axis of rotation is also the combination of their respective effects. Therefore, we need to calculate the angular velocity given by each with Eq. (4.49) and add up the results. To do so, we'll need to substitute the following values:

$$\varphi = 23.5°, G = 6.67 \times 10^{-11}\ \mathrm{m}^3\ \mathrm{kg}^{-1}\ \mathrm{s}^{-2}, \left[\frac{C-A}{C}\right] \text{of Earth}$$

$$= 0.003273763 \tag{4.50}$$

The mass of Sun $M_s = 1.989 \times 10^{30}$ kg, the mass of Moon $M_m = 7.36 \times 10^{22}$ kg, the average distance between Sun and Earth $R_s = 1.496 \times 10^{11}$ m, and the average distance between Moon anFVVd Earth $R_m = 3.844 \times 10^8$ m. In the end, the period of Earth's axial precession can be calculated to be:

$$2\pi\left[\frac{3G}{2}\left(\frac{C-A}{C}\right)\frac{\cos\varphi}{\omega}\left(\frac{M_s}{R_s^3} + \frac{M_m}{R_m^3}\right)\right]^{-1} \approx 25,500 \text{(years)} \tag{4.51}$$

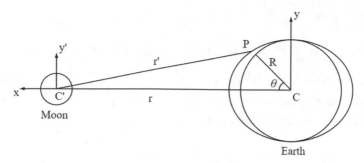

Fig. 4.11 A schematic description of the tide of the Earth-Moon system. Referenced from Fig. 3 of Zhao and Jiao [35]

As the result is only 300 years apart from the actual value, which is 25,800 years, we can safely assume that our derivation process is quite accurate.

Appendix 4.2: Measuring the Influence of Moon's Gravity on Earth Tides with the Law of Universal Gravitation[38]

This proposition demonstrates how we can analyze the influence of lunar gravity on Earth's tidal movements with the law of universal gravity. As shown in Fig. 4.11, in a rotating reference frame where the center of mass of the moon and earth sits at the center, the effective force working on the oceans on Earth is the sum of Moon's gravity and the centrifugal force arising from the inertia of the Moon-Earth co-orbiting. If you look at Fig. 4.11, you can see the distribution of this effective force, which draws the seawater on Earth along the direction of the line joining the two bodies and shapes the Earth into an ellipse. As illustrated below, the highest points of the two tidal bulges both sit on the same line mentioned above, which explains why locations that are 180° apart in longitude would see simultaneous tides and contribute to the phenomenon of double day tide.

In Fig. 4.11, G is the gravitational constant, M_m the mass of Moon, C/C' the center of Moon/Earth, P the height of seawater, whose mass is Δm, $\left| \overrightarrow{r\prime} \right|$ the distance from P to the Moon's barycenter

---

[38]The mathematical derivation and the two figures of this proposition are referred to Zhao and Jiao [35] as well as Chap. 10 of Morin [24].

$\left(=\sqrt{r^2+R^2-2rR\cos\theta}=r\prime\right)$, and $\left|\vec{R}\right|$ the distance from P to the center of Earth $\left(=\left|\overrightarrow{CP}\right|=R\right)$. As for the distance between the centers of Moon and Earth, it can be represented as $\left|\vec{r}\right|=\left|\overrightarrow{CC\prime}\right|=r$. To start the derivation process, let's first use the following equation to represent the lunar gravity exerted on Δm, the seawater on Earth:

$$\vec{f}=\frac{G\Delta m M_m}{r\prime^3}\vec{r}\prime \tag{4.52}$$

On the other hand, the seawater is also acted upon by a centrifugal force that arises from inertia:

$$\vec{f_I}=-\frac{G\Delta m M_m}{r^3}\vec{r} \tag{4.53}$$

Therefore, we know the Moon's tidal drawing force on Earth's oceans should be the sum of (4.52) and (4.53):

$$\vec{f_T}=\vec{f}+\vec{f_I}=G\Delta m M_m\left(\frac{\vec{r}\prime}{r\prime^3}-\frac{\vec{r}}{r^3}\right)=G\Delta m M_m\left(\frac{\vec{r}-\vec{R}}{\left|\vec{r}-\vec{R}\right|^3}-\frac{\vec{r}}{r^3}\right) \tag{4.54}$$

Then, we can derive this equation below based on the geometrical properties observed from Fig. 4.11:

$$\left(\vec{r}-\vec{R}\right)_x=r-R\cos\theta, \tag{4.55a}$$

$$\left(\vec{r}-\vec{R}\right)_y=-R\sin\theta \tag{4.55b}$$

Now, let's represent the horizontal (or x) and vertical (or y) components of the tidal drawing force $\vec{f_T}$ as $\left(\vec{f_T}\right)_x$ and $\left(\vec{f_T}\right)_y$ before simplifying the equations below by only considering the linear term of the lowest order when calculating $\frac{R}{r}$. In the end, we can obtain the approximate values of $\left(\vec{f_T}\right)_x$

and $\left(\vec{f}_T\right)_y$ as:

$$(f_T)_x = \frac{G\Delta m\, M_m}{r^2}\left[\frac{1 - \frac{R}{r}\cos\theta}{\left(1 - \frac{2R}{r}\cos\theta + \frac{R^2}{r^2}\right)^{\frac{3}{2}}} - 1\right] \approx \frac{2G\Delta m\, M_m}{r^3}R\cos\theta$$

(4.56)

$$(f_T)_y = -\frac{G\Delta m\, M_m}{r^3}R\sin\theta \tag{4.57}$$

Judging from (4.56) and (4.57), the tidal drawing force is inversely proportional to r^3. Also, when $\theta = 0$ and $\theta = \pi$, the seawater directly facing and on the opposite side of the Moon would be affected by the largest $\left(\vec{f}_T\right)_x$ and rise to the highest points. On the other hand, when $\theta = \pi/2$, the tides would drop to the lowest level.

As a next step, we will calculate the height difference between high and low tides, also called the tidal range. As illustrated in Fig. 4.12, if we dig two shafts (Shafts 3 and 1) with the cross-section area of ds from the North Pole and the equator along with the directions of x' and y', all the way to the center of Earth, the water levels in them would be h_3 and h_1 assuming both are filled with water constantly (note that r in Fig. 4.12 represents the distance between a given point inside the Earth to its center). Now, we'll conduct a two-part calculation by starting with P_1 and P_3, the in-Earth pressure of the water inside Shafts 1 and 3. When ρ and $g(r)$ represent the density of water (considered a constant in this proposition) and Earth's gravitational acceleration in the range of r, we know that $P_1 = P_3$ when in a state of stability, which we can then use to calculate the tidal range $\Delta h_m = h_1 - h_3$.

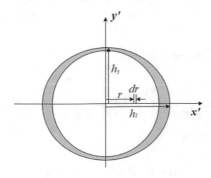

Fig. 4.12 The water level around the Earth's surface. Referenced from Fig. 6 of Zhao and Jiao [35]

Let's now look at Fig. 4.12 and suppose the mass of the water in dr to be dm, which equals $\rho dr ds$. While the amount of gravity exerted on this range of water by Earth is $dm g(r) = \rho g(r) dr ds$, the Moon's tidal drawing force working on it needs to be calculated with (4.56), except that we need to replace R (the radius of Earth) in (4.56) with r and represent the distance between the centers of Moon and Earth with r_m. As such, we can calculate the pressure put on Earth's center by the water in dr to be:

$$dP_1 = \rho \left[g(r) - \frac{2GM_m}{r_m^3} r \right] dr \tag{4.58}$$

By integrating (4.58) over h_1, we can pinpoint the total amount of pressure added by Shaft 1 on the center of Earth:

$$P_1 = \rho \int_0^{h_1} \left[g(r) - \frac{2GM_m}{r_m^3} r \right] dr \tag{4.59}$$

Similarly, we can also calculate the same physical quality of Shaft 3 with h_3:

$$P_3 = \rho \int_0^{h_3} \left[g(r) + \frac{GM_m}{r_m^3} r \right] dr \tag{4.60}$$

As $P_1 = P_3$ in any state of stability, we can obtain the equation below:

$$\int_0^{h_1} \left[g(r) - \frac{2GM_m}{r_m^3} r \right] dr = \int_0^{h_3} \left[g(r) + \frac{GM_m}{r_m^3} r \right] dr \tag{4.61}$$

Which we can further simplify as:

$$\int_{h_3}^{h_1} g(r) dr = \int_0^{h_1} \frac{2GM_m}{r_m^3} r dr + \int_0^{h_3} \frac{GM_m}{r_m^3} r dr \tag{4.62}$$

As the values of h_1 and h_3 are both quite close to R, $g(r)$ can replace the gravitational acceleration on Earth's surface, $g(R) = GM_e/R^2$, where M_e

represents the mass of Earth. Therefore, the integral in (4.62) can be further simplified as:

$$\int_{h_3}^{h_1} g(r)dr = (h_1 - h_3)\frac{GM_e}{R^2} = \frac{GM_e}{R^2}\Delta h_m \qquad (4.63)$$

By applying $h_1 = h_3 = R$ to Eq. (4.62) and combining it with Eq. (4.63), we will come to this result:

$$\frac{GM_e}{R^2}\Delta h_m = \int_0^{R_e} \frac{3GM_m}{r_m^3}rdr = \frac{3GM_m}{2r_m^3}R^2 \qquad (4.64)$$

Which can be simplified again and lead us to Δh_m, the tidal difference that we're seeking to confirm:

$$\Delta h_m = \frac{3M_m}{2M_e}\left(\frac{R_e}{r_m}\right)^3 R_e \qquad (4.65)$$

With the known values of Earth's radius $\left(R = 6.4 \times 10^3\right)$ and mass $\left(M_e = 5.98 \times 10^{24}\right)$, we can use (4.65) to calculate the tide heights triggered by the gravity of Sun and Moon as $\Delta h_s = 0.25$ m and $\Delta h_m = 0.55$ m respectively. When the two bodies exert gravitational forces of the same direction on a specific location, that particular place would see the occurrence of spring tide, whose tidal height is $\Delta h_m + \Delta h_s = 0.80$ m, while the opposite counterpart (when the forces are of opposite directions) would be $\Delta h_m - \Delta h_s = 0.30$ m. Please note that to calculate the tidal range induced by Sun, we need to apply the mass of the Sun $\left(M_s = 1.989 \times 10^{30} \text{ kg}\right)$ and its average distance from the Earth, $\left(r_s = 1.496 \times 10^{11} \text{ m}\right)$, meaning the parameters M_m and r_m in (4.65) have to be replaced with M_s and r_s. When calculating the tidal difference caused by the Moon, on the other hand, it is the mass of the Moon $\left(M_m = 7.36 \times 10^{22} \text{ kg}\right)$ and its average distance from the Earth $\left(r_m = 3.844 \times 10^8 \text{ m}\right)$ that are utilized.[39]

---

[39] According to the calculation of Chap. 10 and Appendix J of Morin [24], the Moon's tidal effect is twice the Sun's.

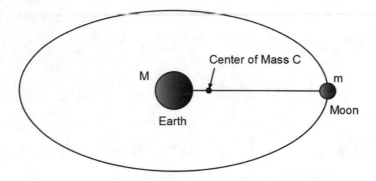

Fig. 4.13 The Earth-Moon system

Appendix 4.3: Exploring the Requirement for Tidal Locking in the Earth-Moon System[40]

It's a widely known phenomenon that the Moon always faces the Earth with the same side, i.e. the period of lunar rotation equals that of the time needed for the Moon to complete a circle around Earth. In this proposition, we will analyze the necessary conditions required for co-orbiting to happen based on the theory that the rotation and revolution of Moon would be synchronized when the total mechanical energy (i.e. momentum plus potential energy) in the Earth-Moon system hits a minimum.

To calculate the total mechanical energy in the Earth-Moon system, we should first safely assume that Earth's rotation speed wouldn't change with the angular velocity of Moon's rotation because the moment of inertia of the former is far larger than that of the latter; in other words, the rotational energy of the Earth can be deemed a fixed value. In addition, considering the angular velocity of the Moon's revolution around Earth changes extremely slowly with time, Kepler's laws of planetary motion still stand as valid. Last but not least, we will also assume the conservation of the overall angular momentum (L) in the Earth-Moon system (i.e. not affected by the gravity of Sun or other planets) and that the Moon's axis of rotation is perpendicular to its plane of revolution.

In Fig. 4.13, which is a representation of the co-orbiting system of Earth and Moon (simply referred to as the Earth-Moon system from this point forward), M is the mass of Earth, m that of Moon, R the distance between them, Ω the angular velocity of Moon's self-rotation, ω that of both bodies' revolution around the barycenter of the Earth-Moon system, l the total

---

[40]The method of solution is partially referred to Taiwan Physics Olympiad TPhO-2016.

orbital angular momentum of their revolution around C, and I the moment of inertia of Moon's self-rotation.

Ever since the formation of the Earth-Moon system, its total amount of energy (potential energy plus the momentum from rotation and revolution) started dwindling and gradually approaching a certain minimum, eventually entering a state of stability due to the cross interaction among the gravity of both bodies, the tidal friction on Earth's surface, and various factors, a process regulated by the four parameters R, Ω, ω, and l, whose mathematical relationship is illustrated in the equations below. First, Eq. (4.66) shows the conservation of the system's angular momentum (L, excluding that of Earth's rotation), which is the sum of the angular momentum of Earth's and Moon's revolution around the barycenter C (l) as well as that of Moon's self-rotation ($I\Omega$). The relationship among the three parameters is illustrated as such:

$$L = l + I\Omega \rightarrow \Omega = \frac{L - l}{I} \tag{4.66}$$

Then, if we consider the balance between the gravitational forces of Earth and Moon, whose masses are M and m respectively, and the centrifugal force of Moon's revolution, we can derive (4.67) below:

$$\frac{GMm}{R^2} = m\left(\frac{M}{M + m}R\right)\omega^2 \rightarrow R^3\omega^2 = G(M + m) \tag{4.67}$$

As for (4.68) below, it represents the orbital angular momentum of the revolution of Earth and Moon around the barycenter C in their shared system:

$$l = m\left(\frac{M}{M + m}R\right)^2 \omega + M\left(\frac{m}{M + m}R\right)^2 \omega \tag{4.68}$$

On the right-hand side of (4.68), the first expression represents the angular momentum of the Moon's revolution around the barycenter C, while the second expression that of Earth. After some simplification, we'll reach (4.69) below:

$$R^2\omega = \frac{M + m}{Mm}l \tag{4.69}$$

Lastly, we know the total mechanical energy in the system is:

$$E = \frac{1}{2}m\left(\frac{M}{M+m}R\omega\right)^2 + \frac{1}{2}M\left(\frac{m}{M+m}R\omega\right)^2 + \frac{1}{2}I\Omega^2 - \frac{GMm}{R}$$

$$= -\frac{G^2M^3m^3}{2(M+m)}\frac{1}{l^2} + \frac{(L-l)^2}{2I} \tag{4.70}$$

Please note, that in the section between the two equal signs in (4.70), the first and second expressions represent the revolutionary momentum of Moon's and Earth's motion around C respectively, the third the rotational momentum of Moon, while the fourth is the combined gravitational energy of both bodies. And with Eq. (4.67), we know that the first, second, and fourth expressions can be combined to become $-\frac{G^2M^3m^3}{2(M+m)}\frac{1}{l^2}$, the first item on the right-hand side of the second equal sign in (4.70); in addition, (4.66) can be rewritten as the second item, i.e. $\frac{(L-l)^2}{2I}$.

As (4.70) tells us the total energy of the Earth-Moon system can be calculated with the function of l, the orbital angular momentum of both bodies' revolution around C, we can differentiate the overall system energy over l, and when the value of differential is zero, we know the energy is at its minimum and can therefore derive the following equation:

$$\frac{dE}{dl} = \frac{G^2M^3m^3}{M+m}\frac{1}{l^3} - \frac{L-l}{I} = \omega - \Omega = 0 \tag{4.71}$$

Under this circumstance, Ω (the angular velocity of Moon's rotation) and ω (that of the revolution around C of Earth and Moon) will be equal values, meaning the overall angular kinetic energy will hit a minimum when Ω equals ω, which proves that when two astronomical bodies revolve around the barycenter of their system (C) due to the influence of universal gravitation, the rotational velocity of the body with a lesser mass would eventually be equalized with the angular velocity of the two bodies' revolution around the center, a phenomenon called tidal locking.

However, such synchronization wouldn't necessarily happen in systems of dual astronomical bodies since the value of dE/dl in (4.71) isn't always zero. In other words, changes in the relationship between E and l won't necessarily generate a minimum. Therefore, for the conditions of tidal locking to be fulfilled, (4.71) has to stand as valid, meaning the equation below must be satisfied as well:

$$(L - l)l^3 = \frac{G^2 M^3 m^3 I}{M + m} \tag{4.72}$$

Due to arithmetic-geometric mean inequality, we know that:

$$\frac{L}{4} = \frac{(L - l) + \frac{l}{3} + \frac{l}{3} + \frac{l}{3}}{4} \geq \sqrt[4]{(L - l)(l/3)^3} \tag{4.73}$$

As (4.73) can be satisfied when $l = \frac{3}{4}L$, if simplified a bit, it can be rewritten as:

$$(L - l)l^3 \leq \left(\frac{L}{4}\right)^4 \times 27 = \frac{27}{256}L^4 \tag{4.74}$$

Which is to say that related values must fit in (4.74) for tidal locking to happen:

$$L \geq \sqrt[4]{\frac{256 G^2 M^3 m^3 I}{27(M + m)}} \tag{4.75}$$

Since (4.75) is the solution to (4.74), it is the eventual requirement for tidal locking.

Since the rotational speeds of both Earth and Moon reduce with tidal changes, based on the conservation law of angular momentum, we can assume the angular velocity of Moon's revolution around Earth to go up. And as the increase in the Moon's tangential speed would strengthen its tendency to deviate from its revolving orbit, it's supposed to move away from Earth bit by bit. In fact, Project Apollo has confirmed that judging from the time needed for laser beams reflected from the mirrors installed by astronauts on the Moon, it's true that the orb is gradually deviating from Earth at the speed of a 3.8 cm annual increase in its revolving radius, which again verifies the validity of Newton's law of universal gravity.

Appendix 4.4: When Will We See Earth's Rotation and Moon's Revolution Synchronized?[41]

It's a known fact that the rotation and revolution of Moon are currently synchronized, yet due to the influence of tidal locking, it's generally predicted that the periods of Earth's rotation and Moon's revolution will need an extremely long time to gradually become the same. In this proposition, therefore, we'd like to make an estimate of how long it'll take for this phenomenon to eventually happen.

For the purpose of creating and deriving our mathematical models, we need to first establish the following conditions:

- The Earth and Moon constitute an isolated system not affected by any external factors
- The rotation axes of both Earth and Moon are perpendicular to the latter's revolution plane
- The distance between Earth and Moon changes so slowly that the orbit of Moon's revolution can be regarded as a circle
- The Moon's self-rotation has been locked and will always be synchronized with its revolution
- The line joining the tidal bulges on Earth forms a fixed angle with that joining Earth and Moon (details will be provided later on).

    In addition, we'll also be using the following values retrieved from related references:
- Mass of Earth: $M_E = 5.97 \times 10^{24} \mathrm{kg}$
- Moment of inertia from Earth's rotation: $I_E = 8.01 \times 10^{37} \mathrm{kg} \cdot \mathrm{m}^2$
- Mass of Moon: $M_M = 7.35 \times 10^{22} \mathrm{kg}$
- Moment of inertia from Moon's rotation: $I_M = 8.77 \times 10^{34} \mathrm{kg} \cdot \mathrm{m}^2$
- Current distance between Moon and Earth: $r_0 = 3.85 \times 10^8 m$, with an annual increase of 3.8 cm
- Current period of Moon's revolution: $T_M = 27.3$ days
- Current period of Earth's rotation: $T_E = 24\,h$
- Gravitational constant: $G = 6.67 \times 10^{-11} N \cdot m^2 / kg^2$.

In the derivation process below, we'll use the below symbols to represent various parameters in physics:

---

[41]The method of solution is partially referred to Asian Physics Olympiad APhO-2001 and International Physics Olympiad IPhO-2009.

- Ω: angular velocity of Earth's rotation (the function of t, its time of rotation)
- ω: angular velocity of Moon's rotation (the function of t, its time of rotation)
- r: distance between Earth and Moon (again the function of t)
- Angular momentum of Earth's rotation: $S_E = I_E \Omega$; angular momentum of Moon's rotation: $S_M = I_M \omega$
- Earth's orbital angular momentum: l_E; Moon's orbital angular momentum: l_M (from the perspective of their revolution around the barycenter of the Earth-Moon system).

With the above conditions and values, we can now start our three-part derivation of what it takes for Earth's rotation to be synchronized with Moon's revolution.

1. Qualitative analysis

As the Earth rotates in the same direction, the Moon revolves around it and has a higher angular velocity, the line between the tidal bulges on Earth would move ahead of and form a small angle θ with the Earth-Moon line due to a friction effect, as shown in Fig. 4.14. For the purpose of visualization and calculation, we'll consider the bulges on both sides of Earth two equivalent masses that're acted upon by Moon's gravity, but as the bulge facing the Moon is under a stronger force, the Earth would be affected by a torque opposite to its direction of rotation and slowly decelerate due to loss of angular momentum.

Meanwhile, as we consider the Earth and Moon an isolated system immune to all external factors in this proposition, the overall angular momentum of this system should be the sum of the angular momentum of Earth's (S_E) and Moon's rotation (S_M) as well as the orbital angular momentum of both bodies (l_E and l_M), which should be a non-changing value. Therefore, if Earth's rotation slows down, the value of $l_E + l_M$ will

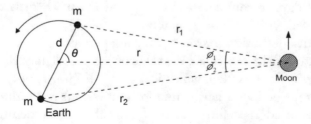

Fig. 4.14 The relative motions and positions of Earth and Moon

have to go up. As for the angular momentum of Moon's rotation S_M, we'll later explain why it's small enough to be disregarded.

$$l_E + l_M = M_E \left(\frac{M_M}{M_E + M_M} r \right)^2 \omega + M_m \left(\frac{M_E}{M_E + M_M} r \right)^2 \omega$$

$$= \frac{M_M M_E}{M_E + M_M} r^2 \omega \qquad (4.76)$$

Following Newton's law of universal gravity, the gravity acting on Earth $\left(\frac{GM_EM_M}{r^2} \right)$ should be the product of M_E and Earth's acceleration, $\left(\frac{M_M}{M_E+M_E} r \right) \omega^2$, which is to say:

$$\frac{GM_EM_M}{r^2} = M_E \left(\frac{M_M}{M_E + M_M} r \right) \omega^2 \qquad (4.77)$$

Equation (4.77) can be simplified as:

$$\omega^2 r^3 = G(M_E + M_M) \qquad (4.78)$$

Which we can then substitute into (4.76) and obtain:

$$l_E + l_M = M_M M_E \left(\frac{G^2}{M_E + M_M} \right)^{1/3} \frac{1}{\omega^{1/3}} \qquad (4.79)$$

According to (4.79), a smaller ω is the only possibility for $l_E + l_M$ to increase, and based on (4.77), a decreased ω means a longer distance between Earth and Moon (r), which scientists working on Project Apollo have confirmed at 3.8 cm per year with laser range-finding results showing the Moon's gradual deviation from Earth. As ω is inversely proportional to the Moon's rotation period, we can predict that it'll take the body more than 27.3 days to complete a circle around Earth when its revolution is synchronized with the latter's rotation.

2. Quantitative derivation based on revolving periods
   In this part, let's first calculate the value of the overall angular momentum in the Earth-Moon system, which is $L = S_E + S_M + l_E + l_M$, and derive therefrom the period of Earth's rotation and Moon's revolution when both are synchronized. As defined previously, the angular momentum of Earth's and Moon's rotation is $S_E = I_E \Omega$ and $S_M = I_M \omega$ respectively. After

substituting the actual values, we will come to the following equations:

$$S_E = I_E \Omega = 8.01 \times 10^{37} \left(\text{kg m}^2\right) \times \frac{2\pi}{23.9 \times 3600(\text{s})}$$

$$= 5.85 \times 10^{33} \left(\text{kg m}^2/\text{s}\right) \tag{4.80}$$

$$S_M = I_M \omega = 8.77 \times 10^{34} \left(\text{kg m}^2\right) \times \frac{2\pi}{27.3 \times 86,400(\text{s})}$$

$$= 2.34 \times 10^{29} \left(\text{kg m}^2/\text{s}\right) \tag{4.81}$$

Again, if we substitute related values into (4.79), the combined orbital angular momentum of Earth's and Moon's revolution around the barycenter of the Earth-Moon system ($l_E + l_M$) can be calculated to be:

$$l_E + l_M = M_M M_E \left(\frac{G^2}{M_E + M_M}\right)^{1/3} \frac{1}{\omega^{1/3}} = 2.86 \times 10^{34} \left(\text{kg m}^2/\text{s}\right) \tag{4.82}$$

Judging from the three equations above, we can tell that the value of S_M is smaller than S_E and $l_E + l_M$ by $10^4 \sim 10^5$ times. Therefore, we'll simply disregard this parameter in the following steps, which leads us to:

$$L = I_E \Omega + M_M M_E \left(\frac{G^2}{M_E + M_M}\right)^{1/3} \frac{1}{\omega^{1/3}} \tag{4.83}$$

Then, we can add up the calculation results above to get the overall angular momentum in the Earth-Moon system:

$$L = 3.44 \times 10^{34} \left(\text{kg m}^2/\text{s}\right) \tag{4.84}$$

When Earth's rotation and Moon's revolution are synchronized, Ω and ω must be the same value, which we represent here as ω_f. Due to the conservation of angular momentum, we can derive the below equation based on (4.84) and (4.85):

$$I_E \omega_f + M_M M_E \left(\frac{G^2}{M_E + M_M}\right)^{1/3} \frac{1}{\omega_f^{1/3}} = 3.44 \times 10^{34} \left(\text{kg m}^2/\text{s}\right) \tag{4.85}$$

Substituting the actual values for the parameters in (4.85) will help us come to (4.86):

$$8.01 \times 10^{37}\omega_f + 3.96 \times 10^{32}\omega_f^{-1/3} = 3.44 \times 10^{34} \qquad (4.86)$$

Whose solution can be calculated as:

$$\omega_f = 1.54 \times 10^{-6}(\text{rad/s}) \qquad (4.87)$$

In other words, when Earth's rotation and Moon's revolution are synchronized, the shared period will become:

$$T_f = \frac{2\pi}{\omega_f} = 4.08 \times 10^6(\text{s}) \qquad (4.88)$$

If we apply the time of a current day, T_f can be calculated at 47.2, meaning the Earth will need 47.2 days to complete a self-rotation.

3. Change of motion patterns in the Earth-Moon system over time
   When addressing this question, we need to first measure the influence of tidal waves with calculation before taking Fig. 4.15 into consideration. As previously mentioned, because the tidal bulges on both sides of the Earth are deemed equivalent masses (m), this value should be proportional to the tidal heights. Just as we've proved in Appendix 4.2, tidal heights are proportional to the cube of the distance between Earth and Moon (r); therefore, we know that $m \propto 1/r^3$. If we represent the distance between the point mass m and the center of Earth as d and assume that the angle

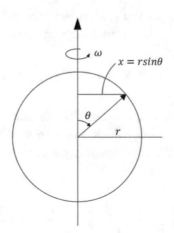

Fig. 4.15 The rotating coordinate system of Earth

θ formed by the line joining two tidal bulges and that between Earth and Moon wouldn't change over time, we can calculate the torque generated by the Moon on Earth with the following steps:

In Fig. 4.15, we can count on the law of cosines to calculate the drawing force exerted on the Moon by m, which is the point mass on Earth closest to it:

$$F_c = \frac{GmM_M}{r_1^2} = \frac{GmM_M}{r^2 + d^2 - 2rd\cos\theta} \tag{4.89}$$

Similarly, the drawing force exerted on the Moon by the point mass on Earth furthest from it (represented as m as well) can also be calculated as:

$$F_f = \frac{GmM_M}{r_2^2} = \frac{GmM_M}{r^2 + d^2 + 2rd\cos\theta} \tag{4.90}$$

Therefore, we know that the torque generated on the Moon by the point mass closest to it (τ_c) is:

$$\tau_c = \left|\vec{r} \times \vec{F_c}\right| = rF_c\sin(\pi - \phi_1) = r\frac{GmM_M}{r_1^2}\sin\phi_1 \tag{4.91}$$

Again, with the law of cosines, (4.16) can be simplified as:

$$\frac{d}{\sin\phi_1} = \frac{r_1}{\sin\theta} \Rightarrow \sin\phi_1 = \frac{d\sin\theta}{r_1} \tag{4.92}$$

Then, by substituting (4.92) into (4.91), we will come to the equation below:

$$\tau_c = r\frac{GmM_M}{r_1^2} \cdot \frac{d\sin\theta}{r_1} = \frac{GmM_Mrd\sin\theta}{\left(r^2 + d^2 - 2rd\cos\theta\right)^{3/2}} \tag{4.93}$$

At this point, we can repeat the previous steps to calculate the torque generated on the Moon by the point mass furthest from it $\left(\tau_f\right)$ to be:

$$\tau_f = \left|\vec{r} \times \vec{F_f}\right| = rF_f\sin(\pi - \phi_2) = r\frac{GmM_M}{r_2^2}\sin\phi_2$$

$$= \frac{GmM_Mrd\sin\theta}{\left(r^2 + d^2 + 2rd\cos\theta\right)^{3/2}} \tag{4.94}$$

Combining (4.93) and (4.94), we now know that the total torque generated by the two point masses on the Moon (τ) is:

$$\begin{aligned}
\tau &= \tau_c - \tau_f \\
&= GmM_M rd\sin\theta\left[\frac{1}{\left(r^2+d^2-2rd\cos\theta\right)^{3/2}} - \frac{1}{\left(r^2+d^2+2rd\cos\theta\right)^{3/2}}\right] \\
&= \frac{GmM_M rd\sin\theta}{r^3}\left[\left(1-\frac{3d^2}{r^2}+\frac{3d\cos\theta}{r}\right) - \left(1-\frac{3d^2}{r^2}-\frac{3d\cos\theta}{r}\right)\right] \\
&= \frac{GmM_M d\sin\theta}{r^2}\cdot\frac{6d\cos\theta}{r} = \frac{6GmM_M d^2\sin\theta\cos\theta}{r^3}
\end{aligned} \tag{4.95}$$

Please note, that we should add a negative sign in (4.95) to signify the direction when considering the overall torque generated by the Moon on Earth's tidal movements, as shown below:

$$\tau = -\frac{6GM_M md^2\sin\theta\cos\theta}{r^3} \tag{4.96}$$

The negative sign in (4.96) means the direction of the torque is opposite from that of Earth's rotation, which will hinder its movement and reduce its speed. Integrating $m \propto 1/r^3$ into (4.96), we can get the result $\tau = dS_E/dt \propto -1/r^6$, which shows that the torque equals the change in angular momentum over time and has an inversely proportional relationship with r to the power of six, and because $L = S_E + S_M + l_E + l_M$ = a fixed value and S_M is small enough to be disregarded, we can further derive:

$$\frac{d(l_E + l_M)}{dt} \propto \frac{1}{r^6} \tag{4.97}$$

In addition, as (4.4) shows $l_E + l_M$ to be proportional to $1/\omega^3$, below equation can therefore be developed:

$$\frac{d\left(\omega^{-1/3}\right)}{dt} = -\frac{1}{3}\omega^{-\frac{4}{3}}\frac{d\omega}{dt} \propto \frac{1}{r^6} \tag{4.98}$$

And with (4.78), which tells us that $\omega^2 r^3$ is a fixed value, we can derive $1/r^6 \propto \omega^4$ therefrom and substitute it into (4.98):

$$\omega^{-\frac{4}{3}}\frac{d\omega}{dt} \propto -\omega^4 \tag{4.99}$$

Which will lead us to:

$$\frac{d\omega}{dt} \propto -\omega^{\frac{16}{3}} \qquad (4.100)$$

If we insert C, a co-efficient of proportionality, into Eq. (4.100), it can be rewritten as:

$$\frac{d\omega}{dt} = -C\omega^{\frac{16}{3}} \qquad (4.101)$$

Then, by integrating the time in (4.101), from when $t = 0$ till $t = t_f$ (when Earth's rotation is synchronized with Moon's revolution), we can generate:

$$t_f = \frac{3}{13C}\left[\omega_f^{-13/3} - \omega_0^{-13/3}\right] \qquad (4.102)$$

In which ω_0 represents the current angular velocity of Moon's revolution. As a result, we simply need to calculate the constant C at this point to obtain the time needed for the synchronization to happen.

To get the value of C, we will make use of the 3.8 cm-result retrieved by researchers of Project Apollo, which is the annual increase in the distance between Earth and Moon. From (4.78), we know that:

$$r = \left(\frac{G(M_E + M_M)}{\omega^2}\right)^{1/3} \qquad (4.103)$$

If we differentiate (4.103) with respect to t, (4.104) will be derived therefrom:

$$\frac{dr}{dt} = -\frac{2}{3}[G(M_E + M_M)]^{\frac{1}{3}}\omega^{-\frac{5}{3}}\frac{d\omega}{dt} \qquad (4.104)$$

Then, from (4.104), we can calculate the current increase rate of the angular velocity:

$$\left(\frac{d\omega}{dt}\right)_{t=0} = -\frac{3\omega_0^{5/3}}{2[G(M_E + M_M)]^{1/3}}\left(\frac{dr}{dt}\right)_{t=0} \qquad (4.105)$$

While a comparison of (4.105) and (4.101) leads us to (4.106):

$$\frac{3\omega_0^{5/3}}{2[G(M_E + M_M)]^{1/3}} \left(\frac{dr}{dt}\right)_{t=0} = C\omega_0^{16/3} \tag{4.106}$$

The constant C can be written as:

$$C = \frac{3\omega_0^{-11/3}}{2[G(M_E + M_M)]^{1/3}} \left(\frac{dr}{dt}\right)_{t=0} \tag{4.107}$$

At this point, we can substitute all the necessary parameters of physics into (4.107) and calculate C to be:

$$C = \frac{3\left(\frac{2\pi}{27.3 \times 86,400}\right)^{-11/3}}{2[6.67 \times 10^{-11}(5.97 \times 10^{24} + 7.35 \times 10^{22})]^{1/3}} \frac{0.038}{365.25 \times 86,400}$$
$$= 6.73 \times 10^6 \left(s^{10/3}\right) \tag{4.108}$$

Now, we can simply substitute the results from (4.108) and (4.87) into (4.102) to come to this final result:

$$t_f = 4.78 \times 10^{17}(s) = 1.52 \times 10^{10} (\text{years}) \tag{4.109}$$

In other words, it'll take 15.2 billion more years for Earth's rotation and Moon's revolution to be synchronized. However, this number is even larger than the universe's age, and nobody knows for sure if the universe can continue to exist for that long. Simply put, it's pretty much impossible for such co-orbiting to take place in the Earth-Moon system.

Last but not least, we'd also like to prove the condition derived from the results by Project Apollo, i.e. when the Moon deviates 3.8 cm from the Earth every year, the latter's rotation will gradually slow down, causing an average of a 1.7 ms-increase in one day on Earth every hundred years. Let's start with Eq. (4.83), where L is a fixed value due to the conservation of angular momentum in the Earth-Moon system, which is why we can differentiate Eq. (4.83) over t and reach the following result:

$$I_E \left(\frac{d\Omega}{dt}\right)_{t=0} = \frac{1}{3} M_M M_E \left(\frac{G^2}{M_E + M_M}\right)^{1/3} \frac{1}{\omega_0^{4/3}} \left(\frac{d\omega}{dt}\right)_{t=0} \tag{4.110}$$

Then, (4.111) can be developed by substituting (4.105) into (4.110):

$$\left(\frac{d\Omega}{dt}\right)_{t=0} = -\frac{\omega_0^{1/3}}{2I_E} M_M M_E \left(\frac{G}{(M_E + M_M)^2}\right)^{1/3} \left(\frac{dr}{dt}\right)_{t=0}$$
$$= 5.59 \times 10^{-22} \left(s^{-2}\right) \tag{4.111}$$

As the definition of the period is known to be $T = 2\pi/\Omega$, we can differentiate T over t and substitute (4.111) to develop:

$$\left(\frac{dT}{dt}\right)_{t=0} = -\frac{2\pi}{\Omega^2}\left(\frac{d\Omega}{dt}\right)_{t=0} = \frac{2\pi \times 5.59 \times 10^{-22}}{\left(\frac{2\pi}{86,400}\right)^2} = 6.64 \times 10^{-13}$$

$$\tag{4.112}$$

As shown in (4.105), each day on Earth would be this much longer every hundred years:

$$6.64 \times 10^{-13} \times 100 \times 365.25 \times 86,400 = 0.002(s) \tag{4.113}$$

Which equals 2 ms after conversion, extremely close to the previously mentioned 1.7 and 2.3 ms.

Appendix 4.5: How the Theory of Pendulum and Its Period Change When in Equatorial and Polar Regions

In 1673, Huygens proposed this theory in his book *Horologium Oscillatorium*: the ratio of a pendulum's period to the time needed for an object to fall from the pendulum's mid-point to its bottom equals the ratio between the perimeter and diameter of a circle whose radius is the length of the pendulum, and this ratio is 3.14159, exactly the π we know today. In the form of an equation, this result of his can be written as $T = 2\pi\sqrt{l/g}$, where T and l represent the period and length of the pendulum respectively, while g is the acceleration of free fall objects (or gravitational acceleration, 9.81 m/s^2). Without the help of calculus or calculation tools, Huygens apparently had to count on his careful observation and precise measurements to obtain a formula with such accuracy[42]. Luckily for us, though, we can simply resort to calculus to derive this pendulum equation:

$$T = 2\pi \sqrt{\frac{l}{g}} \tag{4.114}$$

Amplitude is generally supposed to be very small in the example calculations in common textbooks so that the pendulum movement can be seen as simple harmonic motion, a basis on which the angular frequency of very small vibrations can be easily calculated with Eq. (4.114). However, here we would like to give up this convenient condition by taking the range limit off the amplitude and first calculate the period of a pendulum with normal swing angles before applying the result to the case of minimum angles, whose result will prove the same as in common textbooks.

Now, let's consider a pendulum whose bob has the mass m and the ceiling from which the pendulum hangs to be the zero point of potential energy. Under this circumstance, the sum of the pendulum's momentum and potential energy when it travels furthest from the equilibrium position (swing angle $\theta = \theta_0$, speed $= 0$, potential energy $=$ maximum) should equal that when the bob swings to whichever point on its trajectory ($\theta < \theta_0$), which can lead us to this equation below:

$$\frac{1}{2}m\left(l\dot{\theta}\right)^2 - mgl\cos\theta = 0 - mgl\cos\theta_0 \tag{4.115}$$

Then, (4.115) can be slightly simplified to become:

$$\dot{\theta} = \sqrt{\frac{2g}{l}(\cos\theta_0 - \cos\theta)} \tag{4.116}$$

In the following, firstly, we're going to change the order of integration and calculate the integral value of (4.117). Please note that the range of integral is 0 to θ_0, accounting for 1/4 of the period, so we will need to multiply the

---

[42]Huygens mentioned in his experiment records that the distance traveled by his free falling object in the first second after it dropped was 15.08 "meters" (which equals 16.1ft or about 4.9 m now as the "meter" back then was defined as 1/10,000,000 the length of the meridian). If we apply the value to Galilei's theory for free falling objects that distance is proportional to the square of time, the gravitational acceleration can be calculated to be 32.2 ft/s², which is extremely close to the known value of 32.17 today and proves how accurate Huygens' experiment was.

result by 4, which leads us to this equation below:

$$T = \int dt = 4\sqrt{\frac{l}{2g}} \int_0^{\theta_0} \frac{d\theta}{\sqrt{\cos\theta - \cos\theta_0}} \qquad (4.117)$$

Equation (4.117) represents the period for pendulums with normal swing angles, but as this equation contains an elliptic function and is therefore hard to integrate, we will need to assume the angle to be extremely small, a condition that can help us derive (4.118):

$$\cos\theta \approx 1 - \frac{\theta^2}{2} \qquad (4.118)$$

On the other hand, if we simplify the elliptic function in (4.117), which we can then integrate, the process will yield the below equation, which represents the pendulum period:

$$T = \int dt = 4\sqrt{\frac{l}{g}} \int_0^{\theta_0} \frac{d\theta}{\sqrt{\theta_0^2 - \theta^2}} = 4\sqrt{\frac{l}{g}} \left(\sin^{-1}\frac{\theta}{\theta_0}\right)_0^{\theta_0} = 2\pi\sqrt{\frac{l}{g}}$$

$$(4.119)$$

As Earth's is an oblate spheroid and rotates around its axis once every 24 h, the pendulum period is affected by Earth's shape and rotation when placed at locations of different latitudes. As for the respective influences of both factors, we will conduct an analysis based on Eq. (4.119). Let's start by considering how the period would change when the pendulum is placed in polar and equatorial areas. With the substitution of the average equatorial radius ($R = 6738.14$ km) and the polar radius ($r = 6356.76$ km), the difference can be calculated using (4.120),

$$\Delta T = 2\pi\sqrt{l}\left(\frac{1}{\sqrt{GM/R^2}} - \frac{1}{\sqrt{GM/r^2}}\right) \approx T\left(\frac{R-r}{R}\right) \qquad (4.120)$$

Whose result would be:

$$\frac{\Delta T}{T} \approx 0.057 \qquad (4.121)$$

As a second step, let's look at how the centrifugal force from Earth's rotation affects the period when the pendulum is placed near the equator. Due to the effect of rotation, the gravitational field in the equatorial regions lacks the centrifugal acceleration $\Delta g = R\omega^2 = 0.036$ m/s^2 (ω representing the rotational angular velocity) compared to the fields that are not affected by Earth's rotation. Therefore, the difference caused by the self-rotation of Earth in the pendulum period would be:

$$\Delta T = 2\pi\sqrt{l}\left(\frac{1}{\sqrt{g - \Delta g}} - \frac{1}{\sqrt{g}}\right) \approx T\left(\frac{1}{2}\frac{\Delta g}{g}\right) \qquad (4.122)$$

Which can lead us to this result below:

$$\frac{\Delta T}{T} \approx 0.0018 \qquad (4.123)$$

A comparison of the two calculations above shows the fact that the difference caused by Earth's rotation is almost 30 times smaller than that by varying radiuses in equatorial and polar regions.

Appendix 4.6: Calculating the Difference Between Earth's Equatorial and Polar Radiuses[43]

When calculating the degree of the increase in Earth's equatorial radius (h) under the influence of self-rotation, we need first to suppose that (1) Earth's mass is evenly distributed with the density of ρ, (2) it's only affected by the gravitational force of its own and the centrifugal force from its self-rotation, while the influences of its revolution and the gravity of other astronomical bodies, such as the Sun, should all be disregarded, (3) it rotates at a fixed angular velocity of ω, with all possible movements of its axis disregarded, and (4) changes in the equatorial radius caused by its rotation are by far smaller than its varying radius, i.e. $h \ll r$. With all the conditions specified, let's now try to calculate the value of h with the law of energy conservation.

In the rotating coordinate system illustrated in Fig. 4.15, the Earth has two types of potential energy, one from the universal gravitation, the other from the centrifugal force of its rotation. If Earth's average radius is r (retrieved by calculating the average of radiuses at different locations over the solid angle of the entire Earth), the gravitational energy at any point on Earth can be

---

[43] For details, please see Maxwell [22], pp. 494–496.

represented approximately as mgh since $h \ll r$. Please note that h refers to the difference between r and the actual radius of each location on Earth's surface.

As for the centrifugal potential energy, it can be obtained by integrating the centrifugal force $m\omega^2 x$ over x, which represents the vertical distance between any point on Earth's surface and its axis of rotation. Therefore, the centrifugal potential energy can be derived as:

$$- \int m\omega^2 x dx = -\frac{1}{2}m\omega^2 x^2 \qquad (4.124)$$

As we've assumed the mass of Earth is evenly distributed, the entire spherical surface should be an equipotential plane. Otherwise, the gradient of potential energy would cause a working force that leads to the redistribution of surface mass. As a result, the conservation of Earth's overall potential energy can be written as:

$$mgh - \frac{1}{2}m\omega^2 x^2 = \text{constant} \qquad (4.125)$$

If we substitute $x = r\sin\theta$, which reflects the shape of Earth's surface, into (4.125), we can derive the change function $h(\theta)$, which shows how the degree of Earth's deformation changes with latitude:

$$h(\theta) = \frac{\omega^2 r^2}{2g}\sin^2\theta + C \qquad (4.126)$$

We can then substitute $\theta = \pi/2$ and $\theta = 0$ into the function to calculate the change in equatorial ($h(\pi/2)$) and polar ($h(0)$) radiuses, the difference between the two results being:

$$h\left(\frac{\pi}{2}\right) - h(0) = \frac{\omega^2 r^2}{2g} = \frac{\left(\frac{2\pi}{86,400 \text{ s}}\right)^2 (6380 \text{ km})^2}{2 \times \frac{9.8 \text{ m}}{s^2}} \approx 11 \text{ km} \qquad (4.127)$$

One thing to be noted here is that the difference calculated from (4.127), which is supposed to reflect the difference between equatorial and polar radiuses, is approximately half of the actual value (21.4 km) only. In other words, merely considering the influences of Earth's two types of potential energy is apparently not thorough enough, which is why we need to further conduct an analysis to explore possibly important factors that have been excluded.

Therefore, let's take a look at the previously derived Eq. (4.126) again and try to pinpoint the value of the integral constant *C*. To do this, we'll first need to suppose the overall mass of Earth wouldn't change due to its deformation, which is to say that the mass of the bulges near the equator would equal that of the hollows in polar areas, as shown in Fig. 4.16. Consequently, if we integrate $h(\theta)$, the change function of Earth's deformation over the angles of the entire orb, the result should be zero:

$$\int_0^\pi h(\theta) \cdot 2\pi \sin\theta d\theta = 0 \tag{4.128}$$

We can then substitute (4.126) into (4.128), which we will subseqently integrate and retrieve the value of *C* as:

$$C = -\frac{\omega^2 r^2 \sin^2\theta}{3g} \tag{4.129}$$

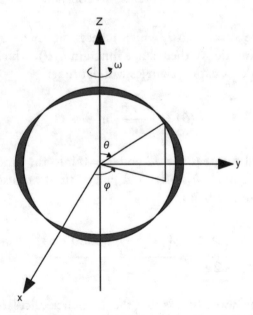

Fig. 4.16 A schematic illustration of the shape-change of Earth's surface due to rotation

Then, the substitution of C back into (4.126) and a bit of simplification will yield:

$$h(\theta) = \frac{\omega^2 r^2}{6g}\left(3\sin^2\theta - 2\right) \tag{4.130}$$

In the derivation process above, we disregarded the tiny influence of Earth's shape (an oblate spheroid, rather than a perfectly round sphere) on the gravitational energy on its surface. In other words, we simply need to take the masses deviating from the spherical frame back into consideration now. As the bulges and hollows are all quite thin (the red and blue sections are negative and positive masses respectively, as shown in Fig. 4.16), we can consider them concentrated on Earth's surface, with the density of $\rho h(\theta)$.

For the purpose of calculation, we'll still suppose $h(\theta)$ to be proportional to $3\sin^2\theta - 2$, but related parameters are slightly different:

$$h(\theta) = f \times \frac{\omega^2 R^2}{6g}(3\sin^2\theta - 2) = \beta(3\sin^2\theta - 2) \tag{4.131}$$

To obtain the value of f, let's consider the potential energy at $(R, 0, 0)$ and $(0, 0, R)$, which represent a point on the equator and the North Pole respectively, to be the same, thereby deriving the below equation:

$$\Rightarrow -2mg\beta - \int_0^{2\pi}\int_0^{\pi} \frac{Gm\rho\beta(3\sin^2\theta - 2)}{R\sqrt{2(1-\cos\theta)}}R^2\sin\theta d\theta d\phi + 0$$

$$= mg\beta - \int_0^{2\pi}\int_0^{\pi} \frac{Gm\rho\beta(3\sin^2\theta - 2)}{R\sqrt{2(1-\sin\theta\cos\phi)}}R^2\sin\theta d\theta d\phi - \frac{m\omega^2 R^2}{2} \tag{4.132}$$

As (4.132) is developed under the assumption that the mass of Earth is evenly distributed, the density ρ can be replaced with $M/\frac{4}{3}\pi r^3$ along with the substitution of $g = GM/r^2$ and yield the following result after integration:

$$1 = f\left\{1 - \int_0^{2\pi}\int_0^{\pi}\left(\frac{1}{\sqrt{1-\sin\theta\cos\phi}} - \frac{1}{\sqrt{1-\cos\theta}}\right)\frac{(3\sin^2\theta - 2)\sin\theta d\theta d\phi}{4\sqrt{2\pi}}\right\}$$

$$\approx f(1 - 0.6) \Rightarrow f \simeq 2.5 \tag{4.133}$$

As the value of integral in (4.133) is approximately 0.6, f can be calculated to be 2.5, which we can then substitute back into Eq. (4.131) and reach the

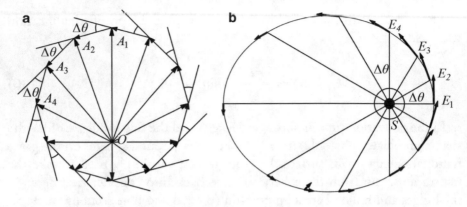

Fig. 4.17 The velocity vectors of Earth's revolution at all moments. **a** A circle; **b** an ellipse

result of the difference in Earth's radiuses at 11 km × 2.5 = 27.5 km, this time larger than the actual value of 22 km. Such deviation is likely to be caused by our excluding the fact that Earth's mass isn't evenly distributed. If we take into consideration that the density near the Earth's surface is lower than that of the interior, the integral of (4.133) should come smaller than 0.6, while f would be lower than 2.5 as well. This way, the final result would also be closer to the actual value.

Appendix 4.7: Feynman's Geometrical Approach of Proving Planetary Orbits to Be Elliptic

In his purely geometrical approach to determining the shape of planetary orbits around Sun, Feynman supposed the Sun to be a fixed point due to its far larger mass than the Earth before seeking to prove the latter's revolving orbit to be elliptic[44]. Below are the three parts of his geometrical derivation.

1. Take the velocity vectors of Earth's revolution at all moments and collect them together so their tails sit at a single point. This way, the tips would actually trace out a perfect circle, as shown in Fig. 4.17. Below is the proving process:

---

[44]The proving process in this proposition is provided by referencing the video *Feynman's Lost Lecture: The Motion of Planets Around the Sun*, which can be found at https://www.youtube.com/watch?v=xdIjYBtnvZU.

Proof As shown in Fig. 4.17b, if we represent the Sun as S and divide the trajectory of Earth into equal parts with $E_1, E_2, E_3 \ldots$ as the points of section, that means $\angle E_1 S E_2 = \angle E_2 S E_3 = \angle E_3 S E_4 = \cdots = \Delta\theta$. We can then draw the velocity vectors of Earth when it's located at $E_1, E_2, E_3 \ldots$ (i.e. $\overrightarrow{OA_1}, \overrightarrow{OA_2}, \overrightarrow{OA_3} \ldots$ in Fig. 4.17a). This way, $\overrightarrow{A_1 A_2}$ can be seen as the change in Earth's velocity when it travels from E_1 to E_2, $\overrightarrow{A_2 A_3}$ that from E_2 to $E_3 \ldots$ and so on. As a next step, we will need to find the limit of $\Delta\theta(\Delta\theta \to 0)$ so that the instantaneous acceleration over Δt can be calculated by dividing $\overrightarrow{A_1 A_2}$ by the time period.

At this point, however, we have to first prove the change in velocity over $\left|\overrightarrow{A_k A_{k+1}}\right|$ to be a fixed number regardless of the value of k. If the distance between E_k and the Sun is r, we can develop the following equation based on Newton's law of universal gravitation:

$$\frac{\left|\overrightarrow{A_k A_{k+1}}\right|}{\Delta t} = \frac{GM}{r^2} \tag{4.134}$$

It's true that Δt and r both change with k, but due to the conservation of angular momentum, we know that:

$$r^2 \frac{\Delta\theta}{\Delta t} = \text{non - changing value} \tag{4.135}$$

Then, the substitution of (4.135) into (4.134) would yield:

$$\left|\overrightarrow{A_k A_{k+1}}\right| \propto \Delta\theta \tag{4.136}$$

As the $\Delta\theta$ corresponding to each $\left|\overrightarrow{A_k A_{k+1}}\right|$ is the same, we know that $\left|\overrightarrow{A_k A_{k+1}}\right|$ is a fixed value. We've also noticed that when $\Delta\theta$ approaches the limit ($\Delta\theta \to 0$), $\overrightarrow{A_k A_{k+1}}$ and $\overrightarrow{E_k O}$ are in the same direction. In addition, the angle of $\Delta\theta$ between $\overrightarrow{E_k O}$ and $\overrightarrow{E_{k+1} O}$ means $\overrightarrow{A_k A_{k+1}}$ and $\overrightarrow{A_{k+1} A_{k+2}}$ form the same angle as well.

Summarizing the derivation above, we can confirm that:

(1) $\left|\overrightarrow{A_k A_{k+1}}\right|$ is a fixed value.

(2) $\overrightarrow{A_k A_{k+1}}$ and $\overrightarrow{A_{k+1} A_{k+2}}$ constantly form the angle $\Delta\theta$.

As such, we can confirm that all the A_k will fall on the circumference of the circle whose center is C. Also, if the angle swept by Earth from E_i to E_j during its revolution around the Sun is θ, $\angle A_i C A_j$ would also equal θ.

2. If we, in the circle shown in Fig. 4.19, choose a point other than the center S and represent this eccentric point as D before choosing another point B on the circumference for the perpendicular bisector of $\overline{BD}$ to intersect with $\overline{BS}$ at E and rotate B for a full circle, the trajectory of E would be an ellipse. Below is the proving process:

Proof As E falls on the perpendicular bisector of BD, we know that $\overline{ES} + \overline{ED} = \overline{ES} + \overline{EB} = \overline{BS}$, regardless of where B or E is. And by simply applying the definition of ellipse, we can prove that the trajectory of E is elliptic.

3. Below is the process of proving the Earth's orbit to be elliptic by combining the results of Part 1 and 2.

Proof If we rotate Fig. 4.18a by 90° clockwise, we'll find it identical with Fig. 4.19, with O, C, $\overrightarrow{OA}$, and the radius $\overline{CA}$ overlapping with D, S, $\overrightarrow{DB}$, and $\overline{SB}$. In addition, the transition from Figs. 4.18a to 4.19 also rotates each velocity vector by 90° clockwise; in other words, as long as we rotate $\overrightarrow{DB}$ using the middle point of $\overline{DB}$ as a center by 90° counter-clockwise, the vector will return to its original direction, same as the tangential direction at point

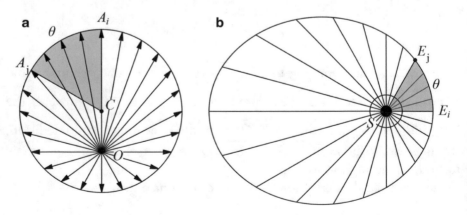

Fig. 4.18 A schematic illustration of the relationship between **a** the circular orbit and **b** the elliptic orbit

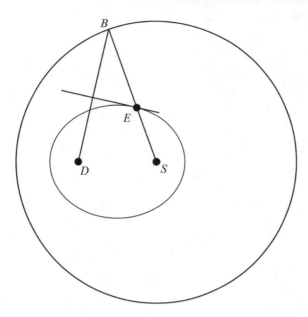

Fig. 4.19 A schematic illustration of the relative positions of the center S, the eccentric point D and a chosen point B

E on the elliptical trajectory in Fig. 4.19. Therefore, all we need to confirm is that this point would overlap with E in Fig. 4.18b. As we already know that the rotating angle of $\overline{ES}$ around S in Fig. 4.19 equals that of $\overline{CA}$ around A in Fig. 4.18a, and the results of Part 1 have also told us that the angle equals that of the rotation of $\overline{ES}$ around S in Fig. 4.18b, we can confirm that the ellipse in Fig. 4.18b and that in Fig. 4.19 actually has an equivalence relation, and that both can represent Earth's revolving orbit. As such, it is proved that the ellipse in Fig. 4.19 is exactly the Earth's revolving trajectory shown in Fig. 4.18b.

5

Special Relativity, Where Time and Space Are Merged

光年之間
所有遠方皆為昨日
空間不想重疊 時間也不能迴復
我們只能漂浮在時空的星海裡
在光年之外的遠方
眺望已過的昨日

In the dimension of light years
All distances were yesterday's news
Space isn't to be overlapped, neither can time recur
All we're left is floating among the stars embedded in time and space
At the distant edge of light years
All looking out at the yesterday that has long passed

Adapted from "In the Dimension of Light Years," a class reflection of Chen YuanYuan (Department of Chinese Literature, National Taiwan University).

It was March 14, 1879. A chubby kid with a big head was born on this chilly spring day into a Jewish family in Ulm, a small town in Southern Germany. He then became the author of five unprecedented scientific publications 26 years later, allowing his talent in the sphere of science to fully shine through and changing everything people knew about the world. This, is the scientific giant, Albert Einstein (1879–1957). Cute as the chubby Einstein was, his parents were slightly worried about the unusually large size of his head and even assumed he was autistic when he still hadn't developed normal linguistic abilities at age three. While Einstein's condition did start improving after he entered elementary school, he also showed symptoms of irritability, which was found to be related to his inability to express himself, and even attacked classmates and teachers from time to time. He disliked groups activities but always achieved top rankings in terms of academic performance. While kids of the same age had their fun playing, he preferred to read or contemplate in his little corner and let his thoughts drift around without bound. Such a daily routine helped cultivate his attentiveness, which was a critical factor that contributed significantly to his success in developing the theory of general relativity as the process involved complex mathematics and the construction of an abstract universe where time and space are established.

So how did he discover relativity? According to Einstein, interest in the abstractness of time and space tends to sprout at an early age. In fact, his slower development was precisely the reason why he was able to think flexibly about time and space and stay motivated in the quest to construct relativity theories even when most people of his age had a fixed mindset already. For many, this also explains why Einstein could stay young at heart his whole life. When he later fell victim to the political persecution of FBI, the child in him became his savior, freeing him from the grip of fear and allowing him to lead a careless life in his imaginary haven. He didn't seem to be affected by the hardships in the real world and even laughed them off with a humorous attitude, candid and straightforward.

5.1   Lonely, Isolated Teenager

In 1895, Einstein dropped out of an elite secondary school in Munich when he just turned 16. Though it took his parents quite some effort to secure him the hard-to-get spot, Einstein himself called the institute an "education machine", not because the graduates were as useful as industrial products, but in the sense that it suffocated students just like the exhaust from factories. After leaving Munich, he transferred to a high school in Switzerland, where he indulged in self-learning for a whole year thanks to the open and lively school atmosphere. Then, strongly encouraged by his math teacher, he signed up for the entrance exam for ETH Zürich (Swiss Federal Institute of Technology), only to be rejected because of his far-from-satisfactory grades in language subjects. Another year of self-learning went by. This time, he was finally able to get the admission and free himself in the infinite universe of physics. As he was exploring the boundless world of knowledge to his heart's content, science saved him from the emptiness and desolation of life, as well as the shackles of ever-changing desire. However, his complete dedication to science also created an invisible wall that stood between him and the real world no matter where he was. Be it Europe or the US, not a single place could cure his loneliness and gave him a sense of belonging, and the lack of emotional support turned out to be the source of his loose, willful behavior later in life. He once wrote to Queen Elisabeth of Belgium to confide how empty and helpless he felt in love life. Besides, he also had affairs with two married women outside of his two broken marriages, one of whom was said to be a sexy spy sent by the Soviet Union's KGB. In a 2003 research, the University of Cambridge in the UK pointed out that judging from Einstein's behavior over the course of his life, it was safe to assume he had the Asperger syndrome, which is typically characterized by developmental delay during infancy and isolation from the outside world in adulthood. Despite such weakness, people with the syndrome tend to possess exceptional IQs and are usually owners of remarkable achievements when it comes to mathematics, physics, arts, and culture. But of course, it wouldn't be fair to celebrate Einstein's success without mentioning his family background. In fact, his love of invention had everything to do with his upbringing.

Einstein was born in a middle-class family, where two other members, his father Hermann Einstein (1847–1902) and uncle Jakob Einstein (1850–1912), were engineers who also had a passion for science. Not only did the two cofound the Einstein electrical company, they also won the sponsorship of the King of Bavaria in 1882, therefore earning a chance to demonstrate at the Munich GlasPalast (Glass Palace) the city lighting system they designed with the carbon filament light bulbs that Thomas Edison (1875–1931) had just invented six years before. Facilities included in the versatile system ranged from generators powered by steam engines as well as transmission and distribution systems to telecommunication systems operated with electricity and music broadcast stations, animating the whole city with the light and rhythm that vibrated with electrical energy. During the seven years between 1882 and 1889, the Einstein brothers' electrical lighting system was widely popularized, setting streets in their hometown, Ulm, ablaze with lights and becoming the star of Oktoberfest. Yet the influence of their invention wasn't just limited to Germany. Even the northern Italian city of Varese jumped on the bandwagon and adopted the system. Surrounded by factories producing electrical facilities, the young Einstein grew up familiar with power and electricity, which was why he was no stranger to the phenomenon of electromagnetic induction, an essential factor in establishing special relativity.

Although Einstein did invent a precision measuring instrument for mild electric currents and design a hearing aid for a deaf singer, he wasn't interested in becoming an engineer and even stopped his own son from pursuing a career in engineering design, which shows the ambivalence in his thinking that bothered him his whole life. As widely known to the public, Einstein had always been a pacifist strongly opposed to war, yet he was also the inventor of the gyrocompass, a highly-responsive navigation instrument independent of Earth's magnetic field that allowed warships to sail safely on the Atlantic Ocean during World War I. Despite winning a patent reward of 20,000 Deutsche Marks with the gyrocompass, Einstein was met with refusal when he tried to transform the planes that the German air force used in World War I with the self-designed cat back-shaped wings because they added difficulty to related operations. After the frustration up in the air, he shifted his focus down under the sea, inventing an electromagnetic ignition fuze for the American navy to use during World War II, but the device was soon deemed a failure as well. Einstein received more than 20 patents throughout his life, but few were actually adopted.

As the 20th century dawned, the Einstein brothers' electrical business faced with fierce competition on the market. As their DC-based facilities were quickly replaced by the AC alternative due to higher transmission loss, the company crashed and finally filed for bankruptcy after years of deficit, forcing the Einstein family to relocate from Munich to Milan. The plan was for Einstein to stay behind and finish secondary school before following their footsteps to Italy, but it turned out he couldn't handle the military-style management of the elite school without family support, so within less than a year, he dropped out and moved to Milan as well.

Einstein had barely worried his parents with rebellious behavior like normal teenagers. Instead, he was absorbed in the abstract world described in a wide range of popular science books. Among them, Aaron Bernstein (1812–1884) put forward a hypothesis in his 20-book series *Popular Books on Natural Science*, like "What if all humans were born without eyes? How would we perceive the world around us?" Such questions gave birth to both micro and macro thought experiments in Einstein's brain, which eventually led him to the speculation that there was an infinite world existing outside the one where humans lived in. In addition, the unexpected influence of the theological and philosophical frameworks discussed in *The Critique of Pure Reason* and *Force and Matter* by Immanuel Kant (1724–1804) and Ludwig Buechner (1824–1899) also helped establish in his experiments a universe that he considered "real". According to his free, unconstrained ideas, the universe was constructed on a simple and widely-applicable principle, making it a natural existence that could fit in anywhere. As well-read as he was, Einstein was most profoundly impacted by two 19th-century scientists who made significant contributions to the field of electromagnetism: Michael Faraday (1791–1867) and James Clerk Maxwell (1831–1879). While Faraday, who started out as a bookbinding apprentice, was recognized with the honor of fellowship from the London Royal Society. Maxwell was hailed by Einstein as the physicist who wielded the most significant influence on the scientific development after Isaac Newton (1643–1727). Why were they held in such high esteem? Well, because they discovered two of the most magical phenomena in the universe: electromagnetic induction and electromagnetic wave. Now, let's first take a look at the lifelong achievements of the two scientific giants in Einstein's eyes.[1]

5.2 Electromagnetic Field: Faraday's Law of Induction

Though legendary in his scientific achievements, Faraday's background wasn't any prominent. Born into a poor blacksmith family south of London in 1791, he was forced to drop out of school at the age of 14 due to the difficult financial situation of his family and sent to a printing factory to work as a bookbinding apprentice, which was actually a bless in disguise because he was exposed to a wide variety of books and therefore able to read about the then advanced scientific disciplines of electrical science, magnetism, chemistry, and so on in the *Encyclopædia Britannica*. Later at the age of 20, a friend gave him the tickets to a series of lectures given by the famous chemist Humphry Davy (1778–1829), which inspired him to apply to work at Davy's lab despite his stuttering problem. Though young and poor, Faraday already knew the drills of leaving good impressions. Therefore, he collected all the notes he made during those lectures and made them a beautifully-crafted book of more than 300 pages using tools in the printing factory. To no surprise, Davy greatly appreciated Faraday's talent as well as determination after receiving the book and hired him as a research assistant. In 1813, as Davy's eyesight was damaged in a lab accident, Faraday was transferred to the Royal Institution of London to assist him with chlorine-related chemical experiments.

Back in those days, Davy's research in electrochemistry was unparalleled in British academia. Although electricity-related topics were highly specialized and advanced in that era, this young scientist was able to demonstrate to

---

[1]The stories of Sect. 5.1 are primarily referenced from Mlodinow [23] and Neffe [26].

members of the Royal Society how to transform electricity into light as early as in 1802, which did make them drop their jaws but still wasn't enough for lighting purposes. The unstable source of electricity in Davy's time was exactly the reason why people had to wait for almost a hundred years till end of the 19th century when Edison invented his incandescent light bulbs. Back then, the most effective way of producing electricity was using lowly-efficient electrolyte batteries,[2] but the method was expensive and far from stable, not to mention the amount generated was very small. As for another type of generator that could provide reliable flows of electricity, it wasn't constructed until Faraday was employed by Davy and discovered the patterns of electromagnetic interaction.

Apart from helping to prepare necessary instruments and materials for experiments, Faraday also conducted experimentation using his off-work time after starting his job at Davy's lab, which led him to discover the patterns of interaction between electricity and magnetic force. In 1821, he read a study by Hans Christian Ørsted (1777–1851), where the Danish scientist noticed a compass needle deflected from magnetic north when placed near an electric cord connected with a battery, showing that the electric current produced a circular magnetic field as it flowed through the wire. Influenced by this precedent, Faraday also devised an experiment,[3] as shown in the Fig. 5.1, by hanging an iron ring (C) above a beaker (A) with a bar magnet secured at the bottom (B) and partially filled with mercury. He then connected this set of apparatuses to the negative electrode of a battery through a wire from

---

[2] In 1938, the German anthropologist Wilhelm Koenig found a 13 cm clay jar containing a copper cylinder that encased an iron rod. The vessel showed signs of corrosion, revealing that an acidic agent, such as vinegar or wine, might have been present. This was likely the very first battery in the human history, used to electroplate gold onto silverware. It's generally agreed that the jar was produced around 225–650 B.C.

[3] See please Cheung [4]. *The Journey of Electricity: Exploring the History of Humans' Utilization of Electricity* (p. 28). Taipei: Commonwealth Publishing Group. The iron ring in Fig. 5.1 was added by Cheung for the convenience of explanation. As for the iron bar E, it was in fact hanged from the ceiling in the original experiment. The Wikipedia page of Faraday (http://en.wikipedia.org/wiki/Michael_Faraday) also illustrates another variant of this experiment, where two glass vessels are both filled with mercury. A wire is extended into one of the vessels from the bottom and connected with the lower end of a bar magnet that can rotate inside the glass, while there is another wire immersed in the mercury whose upper end is secured at a stand above. Meanwhile, the other vessel is shown to have a similar setup, with the sole difference that the bar is secured vertically at the bottom. As such, the instruments contain an induction circuit formed by the wires underneath the vessels, the mercury, the wires on tops of the magnets, and the stand that supports them. When the wires below are charged with currents, the bar in the left vessel and the wire in the other both start showing rotating motion, which is a demonstration of electrically-induced magnetic force. While the bar on the left rotates due to its interaction with the magnetic field generated by the circuit, the wire motions because of the interplay of the bar next to it and the magnetic field around it produced by the current. In other words, the rotation in both vessels are contributed to by the same phenomenon: the interaction between two magnetic fields, one inherent and the other induced by electricity, which proves that electric currents can indeed generate magnetic force.

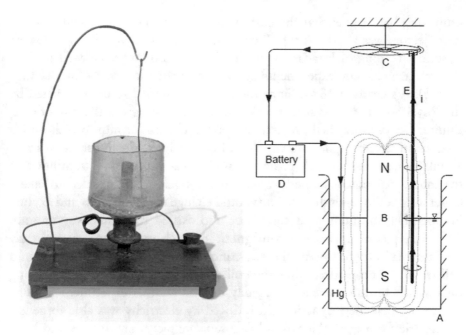

Fig. 5.1 Instruments and illustration of Faraday's motor experiment. *Source* (left): Courtesy of Spark Museum of Electrical Invention https://www.sparkmuseum.org/portfolio-item/faradays-rotating-cup-experiment/

the center of C and hanged on the edge of C a copper bar (E), its other end immersed in the mercury. Meanwhile, a second copper bar F was also placed in the beaker and dipped vertically into the mercury, but was secured on the positive electrode of the battery. As soon as the current passed into the mercury, E started rotating slowly along the ring and moved the other way around when the direction of the current reversed. With the previous knowledge learned from Ørsted's experiment, Faraday attributed the rotation of E to the interaction between the magnetic field of B and that created around E by the current. This was the first-ever demonstration in the human history of transforming electricity into kinetic energy, and the experiment apparatuses also became the earliest prototype of electrical motors.

After inventing the first electric motor in human history, Faraday happily wrote his results into an article, which was published in the *Quarterly Journal of Science* in the same year. However, the publication gave rise to Davy's grudge against this assistant because he wasn't listed as a co-author. Meanwhile, the famous electrochemist William Hyde Wollaston (1766–1828) also accused Faraday of plagiarism, claiming that his research, which was based on a similar experiment and published in the same year, was copied. Yet in fact, his conclusion drawn from the rotation of a charged wire around an unmoving bar magnet that the current in the wire traveled following a spiral trajectory was clearly wrong. However, such accusation still led to Davy's misunderstanding of Faraday, who was therefore no longer allowed to conduct electromagnetic experiments at the lab, a situation that didn't change until Davy's death in 1829. Since then, Faraday started to be recognized by the Royal Society for his studies in chemistry and was given the task of continuing Davy's unfinished project, an optimal chance for him to pick up his research in electromagnetism, which had put on hold for almost ten years.

Taking over the research project, Faraday was immediately reminded of the motor experiment he performed ten years before and decided to take a closer look at the result that currents could change magnetic fields and explore his long-held intuition that the effect also existed the other way around. Before he picked up his electromagnetic research again, there was already a widespread experiment in the scientific circle: an iron bar wrapped in a coil would be magnetized after the coil was charged with electricity. Basing his research on this precedent, Faraday devised another experiment in hope of proving that the magnetic field induced by electricity was able to generate electric force again. He wound two separate pieces of wires (A and B, as shown in Fig. 5.2a) around an iron ring and connected A with a galvanometer, B with a battery. Per his observation, when the electric current passed through B, the pointer of the galvanometer connected with A moved for a brief moment before dropping back to 0, and even when B was supplied with continuous current, there was still no electric force generated in A. In other words, the result was quite the opposite of his previous experiment where

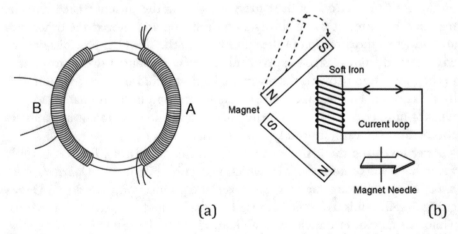

(a)   (b)

Fig. 5.2 Faraday's experiment of magnetic force-induced electricity, inspired by Figs. 5.4 and 5.5 of the website http://highscope.ch.ntu.edu.tw/wordpress/?p=66605

magnetic force was induced by electricity because this time, the magnetic field, despite its stability, proved unable to produce any stable current.

Pondering on the result over and over again, Faraday noticed with his sharp observation that the movement of the pointer was almost synchronized with when the magnetic field was generated, therefore suspecting that perhaps only dynamic magnetic fields could induce electric currents. To prove this hypothesis, he designed the apparatuses shown in Fig. 5.2b, which was consisted of an iron bar and a conductive coil, and under the extension of the coil, a magnetic needle was put in place as its movement would signal the generation of magnetic fields. In order to trigger changes to the magnetic field in the iron, Faraday placed a bar magnet on both ends of it, each closing in on the iron with the N and S poles, and moved the upper magnet up and down while the other remained stationary. While he was doing so, the intensity of the magnetic force in the iron changes proportionally to its distance from the bar magnet, generating dynamic changes in the magnetic field. At the same time, the magnetic needle under the wire also started pointing left and right, signaling the induction of an electrical current in the coil. In addition, the direction of the current varied when the bar magnet was put close or taken away, which proved Faraday's assumption that dynamic magnetic fields could induce electricity. If represented with a modern equation, Faraday's law of electromagnetic induction can be written as $V = -\frac{d\Phi_B}{dt}$, where V is the induced electromotive force (or voltage), and Φ_B the magnetic flux. As implied in this equation, the induced electromotive force in the coil equals the amount of change that the magnetic flux in the iron has undergone.

In this experiment, however, it was hard to observe precisely how the intensity of the magnetic field changed when the bar magnet was being moved back and forth, which prompted Faraday to devise a different experiment that could keep the magnetic field changing, as shown in Fig. 5.3a. In fact, this 1831 experiment was the most famous for him and even evolved into the first generator that Faraday himself invented. The components of the setup included A: a horse-shaped magnet, D: a disk that could rotate, M: an electrical brush touching the edge of the disk, and BB': an external circuit. When the disk turned, the movement induced an electric current that traveled outward from the center toward the rim due to changes in the magnetic flux passing through the dish. Then, the current flowed out through the brush, through the external circuit, and back into the center of the disk through the axle, thereby forming a circuit.

For this result, Faraday came up with the following explanation: when any area of the disk, which could be imagined as a closed circuit, cut through the applied magnetic field, electric currents would be generated in the circuit.

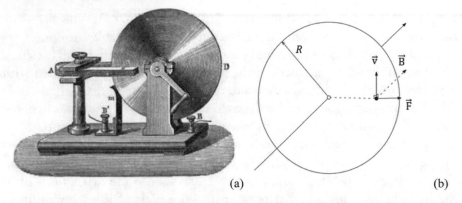

(a)   (b)

Fig. 5.3 Instruments and illustration of Faraday's generator experiment (Émile Alglave & J. Boulard (1884) The Electric Light: Its History, Production, and Applications, translated by T. O'Conor Sloan, D. Appleton & Co., New York, p.224). *Source* (left): https://commons.wikimedia.org/wiki/File:Faraday_disk_generator.jpg Created by E. Alglave in 1884

However, as the closed-circuit didn't really exist, his explanation wasn't complete and could even lead to the Faraday paradox.[4] Yet since the pattern of electron movement has been clarified now, we can calculate the radial voltage in the disk using the concept of Lorentz force. First, as implied in Fig. 5.3b, the Lorentz force equation is[5]:

$$\vec{F} = q\left(\vec{E} + \vec{v} \times \vec{B}\right) \tag{5.1}$$

In this equation, $\vec{F}$ is the Lorentz force, q the electricity carried by electrons, $\vec{E}$ the applied electrical field, $\vec{v}$ the speed of electron movement, and $\vec{B}$ the applied magnetic field. Since there was no electric field applied in Faraday's experiment ($\vec{E} = 0$), the equation becomes $\vec{F} = q\vec{v} \times \vec{B}$, and again, because "electric field" is defined as the Lorentz force working on a charged particle ($E = F/q$) and $F = qvB$, we know that the effect of the Lorentz force equals that of the equivalent electric field $E = vB$. As for the electrical field

---

[4]Although radial currents were indeed observed in Faraday's experiment, the equation $V = -d\Phi_B/dt$ only stands true when the generated voltage is confined to a closed circuit. However, the disk experiment wasn't particularly devised for such a circuit, so what actually happened was a lot more complicated than explained here, which is exactly the reason behind the Faraday paradox, meaning the law of electromagnetic induction isn't applicable to all situations. For details, please refer to https://en.wikipedia.org/wiki/Faraday_paradox.

[5]See Stewart [31] for details of derivation.

gradient, it can be represented as:

$$\Delta V = -\vec{E} \cdot \Delta \vec{r} \tag{5.2}$$

When considering the electric potential difference at the center and the rim of the disk, we need to integrate (5.2) from the center ($r = 0$) to the rim ($r = R$) because of the continual changes in the equivalent electric field:

$$V = \int_0^R vBdr = \int_0^R B\omega rdr = B\omega R^2/2 \tag{5.3}$$

Equation (5.3) represents the radial voltage produced in the disk, but in fact, when electrons were pushed away from the disk center, another Lorentz force was generated and started working along the tangent lines, which was opposite to the direction of the disk movement and therefore slowed its rotation down. The fact that the angular kinetic energy of the disk was transformed into electrical energy indirectly showed the law of energy conservation.

Just like the previously mentioned motor, Faraday's generator was also a first in human history although just a prototype. It was similar to batteries in that both could produce currents, but their mechanisms and effects were entirely different since the electromagnetic generator required no chemical reaction but merely the movement of a charged circuit in a magnetic field to provide stable currents. The mutual induction between electricity and magnetic force inspired later scientists to devise overlapping circuits, which they placed in stationary magnetic fields, charged with electricity, and rotated. Such a setup would trigger regular changes in the magnetic flux passing through the circuits and therefore generated alternating currents whose directions reversed periodically, and this was the basic form of modern generators. By contrast, if we charge wire circuits placed in a stationary magnetic field with electric currents, they will also be showing rotation motion, a phenomenon that helped inventors create motors for power generation. From the 18th century on, electric motors and generators became two of the most effective catalyst for human material civilization.

In fact, however, Faraday's influence on basic research was far more significant than on such scientific applications, which is largely because he introduced a whole new dimension to natural science by coming up with the concept of "field" and wielded a lasting impact on the realm of physics. What inspired this notion of "field" was actually his obsession with the curving patterns formed by iron filings between magnets. As such curves connected the north and south poles of a magnet as if they were the routes taken by the mutual drawing force between the two poles, he decided to call them "field lines." Along these lines, magnetic force transmits in space without any medium, which is essentially the opposite of the general understanding in classic mechanics that no force can be transmitted without physical contact. When Newton developed the law of universal gravitation, he didn't propose any concept related to "force field" or anything of that sort although gravitational field did exist. Today, "field" is understood as a physical quantity that has a value for each and every point in spacetime, forming an orderly pattern. This notion, after established by Faraday, was also extended to represent common quantities like temperature and pressure, making it an indispensable concept in our daily lives.

This said, a mathematical model that could regulate the behavior of electromagnetism was something that Faraday, who didn't have a solid mathematical background after all due to the lack of formal education, struggled to establish despite his more than 600 experiments.[6] Therefore, it wasn't until about half a century later did a young, talented physicist render his law of electromagnetic induction with a seemly-normal but actually extraordinary set of formulas, and this, is the Scottish scientist, James Clerk Maxwell. In a 1931 celebration of Maxwell's 100th birthday, Einstein even hailed this set of equations as the most important influence on the development of physics ever since Newton's time.

After Faraday invented the first prototype generator in the world in 1831, which looked like what's shown in Fig. 5.3, the German engineer Ernst Werner von Siemens (1816–1892) also made the first generator of alternating currents in 1866 by constructing its actuation mechanism. After years of development, multi-phase alternators were finally used widely for power generation from 1891 onward, with the frequencies of the generated currents falling mostly between 16 and 100 Hz. The invention of the generator pushed the human civilization into the electric era, where almost all facilities of human activities are powered by electricity, which is why more than 50% of the global energy consumption nowadays is used for the generation or to support equipment and activities that require electric power.

In addition to the generator, Faraday's earlier invention, i.e. the apparatus in Fig. 5.1 that he used to develop the law of electromagnetic induction in 1821, was also adapted into electric motors, which work in the same way as generators by counting on the movement of charged wire circuits in magnetic fields except that the directions of energy transformation of both instruments

---

[6]Complete records of Faraday's more than 600 experiments, though most of them ended in failure, can be found in the Faraday Museum of the Royal Institution.

are opposite. While the circuits in generators are rotated by applying external forces to produce electric currents, those in motors are charged with electric currents that trigger their motion in magnetic fields. Though developed relatively earlier, the electricity that motors needed to operate had to be supplied with batteries before generators were popularized. Therefore, although the first motor was built as early as in 1740, it wasn't till 1870 when this equipment started to be used for commercial purposes. Among different types of motors, the induction motor of Nikola Tesla (1856–1943), which he made following Faraday's law of electromagnetic induction, was deemed the most successful, and later variants, such as the synchronous motor, stepper motor, ultrasonic motor, etc. have never been lacking, either.

Since the Industrial Revolution started driving the rapid development of human civilization after the 18th century, most activities have been powered by engines and motors. In most industrialized countries in the modern era, the industrial sector accounts for 50% of the nationwide energy consumption, among which 60–70% of the electricity is used for motors. In fact, motors can be found in every corner of our lives because they produce power for air conditioners, refrigerators, fans, microwaves, electric vehicles, elevators, and so on. As such, it goes without saying that the contribution of motors to our everyday activities is beyond reach for other industrial products.

5.3 Maxwell's Theory of Electromagnetic Waves

Maxwell was born in a unique family, where members all held some sort of popular jobs like lawyer, judge, musician, politician, poet, and merchant, although miner was among the list. His birth year 1831 coincided with when Faraday started dedicating himself entirely to the research of electromagnetism. Even better is that he grew up and received education during the years when electromagnetic studies and applications were booming in full swing. For example, the German mathematician and physicist Carl Friedrich Gauss (1777–1855) constructed a signal line roughly a kilometer long in Göttingen with the help of his assistant Wilhelm Weber (1804–1891) to transmit information between an observatory and his lab. The data were sent through the line using alternating currents of opposite directions as binary indicators with every single character encoded based on the binary system. Their encoding technique was later improved by the American inventor Samuel Morse (1791–1872) and his assistant Alfred Vail (1801–1859) to become the famous Morse code, which Morse combined with his self-invented telegraph to dispatch the first-ever telegram in human history in 1838. The next year, the Great Western Railway Company of England quickly took advantage of the invention and sent a telegram out from Paddington for over 21 km to West Drayton.

In 1847, the year when studies on electric communications started growing vibrantly, the 16-year-old Maxwell entered the University of Edinburgh, where he enjoyed chemistry, electrical science, magnetism, optics, mathematics, etc. but particularly loved this last subject. After graduating in 1850, he took a trip to the University of Cambridge to take on the Mathematical Tripos exams, where 16 pages of questions had to be answered within eight days. All the participants had strong references and exceptional college grades to enter the exam, and to be considered competent, what they needed was solid mathematical concepts and outstanding arithmetic skills. Unsurprisingly, Maxwell passed with his incredible talent and knowledge, winning himself the position of math teaching assistant, but in fact, he was just way overqualified for the job. In 1859, he won the Adams Prize, a prestigious astronomy award in that era, with the essay "On the Stability of the Motion of Saturn's Rings". If you think the prize sounds familiar, it was exactly named after the mathematician John Couch Adams, who predicted the existence of Neptune with a complex series of calculations in 1846 (see Sect. 4.12 of Chap. 4).

Identifying the reason behind the stable rotation of Saturn's rings around it had always been a difficult task considered a benchmark challenge in the astronomical circle. Although Saturn's gravity was known to be the major force to keep the rotating rings in stable motion, it was just one of the complex array of working forces. For example, the Cassini Division, a 4,800-km-wide ring was formed due to the gravitational pull from Saturn's moons. After dedicating two years to related mathematical derivation, Maxwell found out that a regular solid ring could not possibly be stable, while a fluid ring tended to break up into blobs due the force of wave action. Therefore, he concluded that the rings must be composed of numerous small particles that he called "brick-bats," each orbiting Saturn independently. In order to understand the derivation process Maxwell went through, we will provide a detailed description in "Appendix 5.1: Small Particles That Form the Rings of Saturn" to calculate the minimum distance from the planet (d) for surrounding substances (moons, ice pieces, fluids, rocks, etc.) to stick together while caught between Saturn's gravity and centrifugal forces. Although adopting a different logical perspective in the Appendix, we'll still cover the essential mathematical skills for addressing this issue.

More than 100 years after he proposed the "brick-bats" structure based on mathematical calculation, it was proved with a photo sent back from Saturn by Voyager 1, a space probe launched by NASA. In addition, his clear, rational derivation that led to the accurate conclusion also served as an endorsement for him to continue his mathematical research at Trinity College. William Thompson (1810–1886), a famous professor of mathematics at the Cambridge University who was also Dean of the Trinity College at that time, thought Maxwell was a rare talent and encouraged him to look into Faraday's electromagnetic experiments, hoping that he could develop a mathematical model for the law of electromagnetic induction. This suggestion was right up Maxwell's alley because describing natural phenomena with mathematical formulae had generally been accepted as a best practice by European physicists. After Newton developed the equation for universal gravitation and succeeded in explaining the motion on Earth and the interaction between astronomical bodies, such a mathematical approach grew all the more popular and was basically elevated to the mainstream status as it was a powerful tool indispensable in analyzing the natural world. Maxwell, among all, was well aware of the potential of the discipline. Therefore, he decided to integrate it with Faraday's concept of "field".

Due to his mathematics-based study of electromagnetism, Maxwell, who was merely 31 years old back then, was entrusted with a research program by members of the Royal Society in the hope that he could collect and systemize all known theories and physical quantities of electromagnetism. Extremely honored by the Society's recognition, he paid an immediate visit from Cambridge to London to see the 70-year-old Faraday, who was thrilled to know that someone was interested in his research and provided details about all the electromagnetic experiments he'd ever performed though afflicted by dementia. With the precious resources provided by Faraday, what first surfaced in Maxwell's mind was the experiment by the Danish scientist Hans Christian Ørsted (1777–1851), where the compass needle was observed to deflect from the magnetic north when placed near an electric cord connected with a battery. After exploring existing literature, Maxwell realized that Ørsted's study didn't just pique the curiosity of Faraday but also the French mathematician André-Marie Ampère, who further designed a follow-up experiment where he proved that two parallel wires charged with same/opposite electric currents would attract/repel each other. As Ampère himself was a mathematician, he later developed a mathematical explanation for phenomenon, i.e. Ampère's circuital law that we know today.[7]

---

[7] See Stewart [31] for details of derivation.

Using Ampère's law as a basis, Maxwell started a well-organized, step-by-step derivation process of all related formulae and physical quantities, as per required by the Royal Society. Eleven years of painfully hard work went by, he finally completed the "Treatise on Electricity and Magnetism", with twelve formulae of electromagnetism listed in more than 400 pages to explain the phenomena observed by Faraday as well as some other scientists in their experiments. In fact, these formulae could even be used to calculate accurate results for many supposed scenarios. Now, let's take a walk through how he derived the "Maxwell's equations", but meanwhile, please note that what we will be focusing on below is the version that was later reformulated by the British physicist Oliver Heaviside (1850–1925).

Back in those days, people tended to compare the drawing force between electricity and magnetic field to universal gravitation and assumed that electromagnetic force was also inversely proportional to the square of distance, which Faraday found unreasonable and hard to accept. Therefore, he came up with the "line of force" notion as an attempt to explain non-contact forces, such as gravity, magnetism, electricity, etc. As for the related "field", his lack of proper training made it impossible for him to develop any mathematical law to regulate the concept, but luckily, Maxwell helped out by publishing the paper "On Faraday's Lines of Force" in 1855, where force lines were integrated with electric and magnetic fields. According to Maxwell's understanding, lines of force could be compared to tubes carrying fluids, while the direction of the force field constituted by the lines was decided by that of the tubes. In addition to reviewing Faraday's research, he also provided an analysis of Gauss's laws of electricity and magnetism in the same paper.

In his law of electricity, Gauss set forth the following definitions: (1) a line of electric force is directed away from a positive charge and toward a negative one, (2) the strength of an electric field is proportional to the number of field lines per unit area, i.e. the electric force per unit positive charge, and (3) the direction of a field is that of the force it'd exert on a positive charge, i.e. the tangential direction of the force line. Based on these definitions, the law indicates that the number of electric force lines (electric flux) originating from per unit positive charge is $1/\varepsilon_0$, while ε_0 refers to the permittivity (also called electric permittivity). As the number of electric lines received by per unit negative charge is $1/\varepsilon_0$ as well, the total lines passing through a closed surface can be calculated by dividing the total number of charges enclosed in that surface by ε_0:

$$\oint \vec{E} \cdot d\vec{a} = \frac{q}{\varepsilon_0} \tag{5.4}$$

In (5.4), $d\vec{a}$ is a small area on the surface, and $\vec{E}$ the intensity of electric field, whose works on the same direction as the outward normal vector of the area. If we multiply $\vec{E}$ by $d\vec{a}$ and retrieve the integral of the entire surface, the number of electric lines passing through can be calculated, and the value has to equal that of q (total number of charges enclosed by the surface) divided by ε_0. Here, however, we'll discuss an exception to explore the physical meaning of this equation.

Consider a spherical shell with the radius r and the electric charge q_1 at its center. If the electric field on the surface (E) is a fixed value, the value of $\oint \vec{E} \cdot d\vec{a}$ will equal that of E multiplied by $4\pi r^2$ (surface area of the shell)

$$E \times 4\pi r^2 = \frac{q_1}{\varepsilon_0} \tag{5.5}$$

Therefore, E can be calculated as:

$$E = \frac{q_1}{4\pi r^2 \varepsilon_0} \tag{5.6}$$

And because the intensity of electric field is defined as the electric force per unit positive charge, if there are two charges q_2 & q_1 whose distance is r, the force exerted on q_2 will be:

$$q_2 E = \frac{q_1 q_2}{4\pi r^2 \varepsilon_0} \tag{5.7}$$

Which shows that the working force on q_2 is inversely proportional to the square of r, the distance between the two charges, and this relation is what we now call "Coulomb's law". It should also note that a $q_1 q_2$ larger than 0 represents a repulsion force, and vice versa. In other words, electric fields tend to result in a repulsive force between particles with charges of the same sign, and an attractive force between charges of opposite signs. As a next step, we're going to rewrite (5.4) as a differential equation in divergence form:

$$\nabla \cdot \vec{E} = \frac{\rho}{\varepsilon_0} \tag{5.8}$$

In (5.8), ρ is the density of electric charges.

Similar to the electric field, the intensity of magnetic field can also be considered the number of magnetic lines passing through a unit area, i.e. magnetic flux. As for the direction of the field, it's tangent to the field line, which is also the direction of the force working on the north pole of a magnetic needle. With these definitions, Gauss's law of magnetism indicates that no magnetic line within any given surface has a starting or ending point because there hasn't been any monopole magnet found in the universe, so no magnetic line can cut through closed surfaces. In other words, the magnetic flux of any closed surface should be 0. Borrowing the expression of electric field above, the law of magnetism can be rendered as this integral equation below:

$$\oint \vec{B} \cdot d\vec{a} = 0 \qquad (5.9)$$

In (5.9), B represents the intensity of magnetic field, which can also be expressed as the differential equation below:

$$\nabla \cdot \vec{B} = 0 \qquad (5.10)$$

With the definitions of both electric and magnetic fields confirmed, Maxwell started exploring the interaction between electricity and magnetism. As he was first reminded of Ørsted's experiment, where an electric current produced a circular magnetic field as it flowed through a wire, he tried to reproduce the phenomenon by assuming the magnetic field to be made of vortices. According to his postulate, electric charges would cause the vortices in a field to swirl without changing their position when passing by because moving charges were supposed to generate magnetic force. Under this situation, the rotation axis of the vortices should be perpendicular to the swirling direction of the magnetic lines. Imagine a revolving door at your local mall. The charges are like customers who do rotate but cannot move the doors when they enter or exit. This should clarify the concept, right? In fact, the equation for calculating the intensity of magnetic fields was first formulated by Ampère, which is why we call it Ampère's circuital law:

$$\oint \vec{B} \cdot d\vec{l} = \mu_0 I. \qquad (5.11)$$

What's the physical meaning of this equation? In (5.11), $\vec{B}$ is the magnetic field. If we divide a closed circuit into small units of length ($d\vec{l}$) and perform a line integral around the entire circuit, the value will equal that of multiplying the circuit's internal current I with magnetic permeability (μ_0). If expressed as a differential equation in curl form, (5.11) will become:

$$\nabla \times \vec{B} = \mu_0 \vec{J}, \qquad (5.12)$$

where $\vec{J}$ is the density of electric currents, representing the number of currents per unit cross-sectional area.

Last but not least, let's take a look at electromagnetic induction. In Faraday's second experiment that we introduced, the magnetic influx in the wire circuit changed when approached by the bar magnet, therefore inducing an electric field where currents were generated. The mathematical equation expressing his law of electromagnetic induction, as previously mentioned, is $V = -\partial\Phi/\partial t$, where $\Phi = \int \vec{B} \cdot d\vec{a} =$ the magnetic flux passing through a certain cross-sectional area, while V is the induced electromotive force (a special name for voltage generated through electromagnetic induction) of the closed circuit binding the area. As V and the induced electric field $\vec{E}$ are the associated physical quantities mutually convertible, $V = -\partial\Phi/\partial t$ can be expressed as this following differential equation:

$$\nabla \times \vec{E} = -\frac{\partial \vec{B}}{\partial t} \tag{5.13}$$

Although (5.13) was sufficient to explain how dynamic magnetic field would generate electric currents, Maxwell didn't stop just here because his belief in Nature's symmetry gave him a disturbing feeling that there had to be a formula that could describe the phenomenon of dynamic electric field producing magnetic force as well. By studying Faraday's experiments and existing theorems of electromagnetism, he concluded that magnetic fields could be generated by (1) electric currents moving through wires, as stated in Ampère's circuital law and represented with $\mu_0\vec{J}$ in (5.12), and (2) changes in the electric field over time. Therefore, he added an expression to (5.12), creating the new equation below:

$$\nabla \times \vec{B} = \mu_0\vec{J} + \mu_0\varepsilon_0\frac{\partial \vec{E}}{\partial t} \tag{5.14}$$

With this genius Maxwell's Addition, which represents the "displacement current", the originally unrelated (5.8), (5.9), (5.13), and (5.14) are now a set of simultaneous differential equations, i.e. the famous "Maxwell's equations."

Since the dynamic magnetic field produces electricity and vice versa, continuous interaction between both factors should eventually generate electromagnetic waves, shouldn't it? To prove this hypothesis, Maxwell reexamined the four equations mentioned in the previous paragraph and succeeded in establishing the electromagnetic wave equation. Since his derivation was quite straightforward, we'll briefly walk through the process below. First, we know that

$\rho = 0$ and $\vec{J} = 0$ are both true in a vacuum. Multiplying (5.13) by ∇, we can obtain:

$$\nabla \times \left(\nabla \times \vec{E}\right) = -\nabla \times \left(\frac{\partial \vec{B}}{\partial t}\right) \qquad (5.15)$$

The expression on the left-hand side of the equal sign can be simplified as (5.16) based on the identity relation below:

$$\nabla \times \left(\nabla \times \vec{E}\right) = \nabla\left(\nabla \cdot \vec{E}\right) - \nabla^2 \vec{E} = -\nabla^2 \vec{E} \qquad (5.16)$$

As for $-\nabla^2 \vec{E}$, it's obtained by substituting $\nabla \cdot \vec{E} = \rho/\varepsilon_0 = 0$ (due to the nature of vacuum) into (5.8).

On the other hand, we can reformulate the expression on the right-hand side of the equal sign in (5.15) as:

$$-\nabla \times \left(\frac{\partial \vec{B}}{\partial t}\right) = -\frac{\partial}{\partial t}\left(\nabla \times \vec{B}\right) = \frac{\partial}{\partial t}\left(\mu_0\varepsilon_0\frac{\partial \vec{E}}{\partial t}\right) = \mu_0\varepsilon_0\frac{\partial^2 \vec{E}}{\partial t^2} \qquad (5.17)$$

In (5.17), the expressions after the second equal sign are also obtained by substituting $J = 0$ into (5.14) based on the nature of vacuum. Now, by substituting (5.16) and (5.17) into (5.15), we'll come to this equation below:

$$\nabla^2 \vec{E} = \mu_0\varepsilon_0\frac{\partial^2 \vec{E}}{\partial t^2} \qquad (5.18)$$

Repeating the same process, (5.19) can also be derived for magnetic field:

$$\nabla^2 \vec{B} = \mu_0\varepsilon_0\frac{\partial^2 \vec{B}}{\partial t^2} \qquad (5.19)$$

Both (5.18) and (5.19) are wave equations that regulate the behavior of electric and magnetic waves respectively. Here's what we can tell from these two equations: the transmission speed of electromagnetic waves is $1/\sqrt{\mu_0\varepsilon_0} = c$. Also, the phase difference between electric and magnetic fields is 90°, while they oscillate at mutually-perpendicular directions and alternate in tandem with the time interval of half a period. Such electromagnetic waves, formed by the mutual induction as well as the coexistence of electricity and magnetism, are known to travel at the speed of c in the vast universe.

Although Maxwell's Addition did bring life to the four simultaneous equations, the nature of such combined waves generated through the interaction between two different types of waves was still perplexing to him. Just when he was bogged down trying to figure things out, he realized that there were two physical constants in his equations, permittivity ε_0 and vacuum permeability μ_0. He went on to multiply the two constants and calculate the square root of the product before obtaining the reciprocal of 299,792,458 m/s. Such a large number was previously unseen in the world of physics, so it naturally aroused Maxwell's curiosity. Judging from the equations, the number seemed to be the speed of the combined waves triggered by electric and magnetic fields. Therefore, it was probably fair to call the waves "electromagnetic." Yet the term was unprecedented in the world of natural science, and nobody had ever even attempted to describe or define such waves.[8] Yet as coincidence would have it, when reviewing existing literature, Maxwell stumbled upon Ole Rømer's observation of Jupiter's moon Io, where the Danish astronomer calculated the speed of light to be 220,000 km/s. This discovery came as a big surprise because other than the reciprocal mentioned before, Rømer's result was the only different physical quantity of such enormous value in the universe. Although both values seemed totally unrelated, Maxwell's intuition somehow suggested that they could actually be the same quantity, which prompted him to propose this revolutionary hypothesis: the combined wave produced by electricity and magnetism is a type of electromagnetic wave traveling at light speed; in other words, light is also an electromagnetic wave.

Such a groundbreaking discovery was detailed by Maxwell in his more-than-400-page-long treatise, and its publication immediately sparked the trend to look for electromagnetic waves, which unfortunately led to no real progress for as long as 15 years. It wasn't until 1888 did the German scientist Heinrich Hertz (1857–1894) finally discover the first-ever electromagnetic wave, a radio wave with a one-meter wavelength, 1,000,000 times that of visible light. After a complete series of tests, Hertz proved the wave to be compliant with Maxwell's equations in terms of physical properties and was able to create the radio wave with his self-invented instruments as well. From then on, scientists were convinced that Maxwell's theory was correct, i.e. electromagnetic waves are produced by the interaction between electric and magnetic fields and travel at the speed of light.

---

[8]During his later years in Switzerland, Faraday did a simple experiment while leading a laid-back life due to poor body conditions, trying to validate his intuition that light was the product of the interaction between electricity and magnetism. Though he generated the magneto-optic effect with a magnet and polarizer, the result received little attention from the scientific community. It wasn't until Maxwell established his equations were people reminded of this experiment.

Quite naturally, such a significant discovery was inundated with skepticism and criticism just like Newton's law of universal gravitation, but this time, it was scholars of the Newtonian school who fired the shots because Newton had always upheld the principle that relative motion had to be considered whenever one tried to address any issue related to speed. In other words, the speed of electromagnetic waves was supposed to depend on that of the observer or properties of the medium through which they were transmitted. However, Maxwell never adopted any coordinate system as a reference frame for the observer or considered the transmission speed in any medium while deriving his equations, implying that the speed of electromagnetic waves wouldn't be affected by the observer or medium, which went completely against the classical rules. Under Newton's then mainstream framework, all mechanical waves, such as water and sound waves, were known to need a medium for transmission whose properties would significantly impact the speed of waves along with the observer's movement. While regarding the lack of medium, one could simply contend that electromagnetic waves were able to travel in vacuum, there was no easy answer for the absence of the observer. Therefore, there was widespread suspicion that Maxwell's equations were, in fact, a tremendous physical paradox that nobody was able to prove for or against, until 1905. Thirty years after presented to the public, his theory was finally confirmed correct and adopted by Einstein as the basis for his studies on relativity.

In 1873, Maxwell proposed the theory that the oscillation of electric and magnetic fields would trigger the generation of electromagnetic waves, which would travel perpendicularly to the field direction at light speed. As assertive as he was, not a single physicist was able to produce electromagnetic waves until 1880 when Hertz finally succeeded. While the wavelength of the radio wave he produced was much shorter than that of visible light, it did satisfy Maxwell's definition of the electromagnetic wave. Despite Hertz's success, a group of physicists, including Oliver Lodge (1851–1940), member of the Royal Institution of London, were quick to express their doubt that his result could be of any practical value, not to mention serving communication purposes. Their ungrounded criticism, nevertheless, was soon questioned by Guglielmo Marconi (1874–1937), an Italian youngster in his early 20s.

Just like Faraday, Marconi also devised his experiment with simple components, using a Leyden jar and a Ruhmkorff coil to spark a signal before receiving it with a coherer, which made a complete set of communications equipment. After experimenting several times, he concluded that the coherer could receive signals sent out by a Leyden jar that was as far as 30 m away and further proved such signals able to travel through physical objects as long as they were propelled with sufficient energy. A few years later, with countless experiments performed and improvements made, Marconi's research team in Newfoundland, Canada, finally succeeded in receiving signals sent out from the Southwest coast of England on a cold winter day in December of 1901, an achievement that should, to a certain degree, be attributed to Karl Ferdinand Braun (1850–1918) because he added to the Hertz oscillator used for the experiment two resonant circuits that were coupled together to reduce the resistance to radio waves and greatly extend the range of transmission. Due to his effort in helping to establish the wireless communications network that encompassed the Atlantic Ocean, Braun shared the Nobel Prize for physics with Marconi 1909.

Although Marconi's coast-to-coast experiment was a big success, he didn't really get to acquire an accurate understanding of how electromagnetic waves traveled in Earth's atmosphere before passing away. For example, while it was in fact the ionosphere that reflected electromagnetic waves to the ground, he thought it was the antennae that made the transmission of signals possible. In addition, he also argued that longwave could travel greater distances than shortwave but was unable to provide any theoretical rationale. Years later, Edward Victor Appleton examined this contention with a series of experiments (1892–1965) and found out that radio waves could be reflected among different layers of the ionosphere with varying angles depending on their frequencies and proved the existence of the Heaviside as well as the Appleton (later named after himself) layers, which won him the 1947 Nobel Prize for Physics. In "Appendix 5.2, Reflecting the Mechanism of Electromagnetic Waves in the Ionosphere", we will take a look at how such reflection works and develop a theoretical equation to prove that an electromagnetic wave must have a wavelength that's over a certain threshold to be reflected by the ionosphere.

The invention of wireless communication generated an immeasurable influence on politics, economy, and military operations. Let's just take the two World Wars for example. Were it not for the instruments of wireless communication, ships and planes would've been clueless and had to navigate without knowledge of their surroundings in the vast oceans and boundless sky. After a few hundred years' developments, the precision and usability of wireless technologies have been elevated to an unprecedented new height. In fact, cellphones, one of the indispensable gadgets in our modern life, are highly precise receivers/senders of wireless signals.

After Maroni's invention of wireless communication was widely acclaimed, related studies sprang up like mushrooms, where radio waves, microwaves, visible light, ultraviolet, X rays, γ rays, and many other electromagnetic waves we're familiar with today were brought into the focus of research. Despite the variety, all members of the electromagnetic family transmit at light speed, from the lowest-frequency radio wave to the highest-frequency radiation ray. Their wavelengths, though, vary a lot from thousands of kilometers down to the Planck length, but what's true of all electromagnetic waves is that they all travel continuously with no limit as to how far they can go. In today's atmosphere, it isn't just nature's electromagnetic waves that 're ubiquitous but the artificial ones as well, such as the radioactive rays still emitted from the waste of the 1945 nuclear explosion as well as the low-frequency waves telecommunication companies fill a certain air-band with due to the use of the Internet, satellite television, cell phone, etc. Just how surrounded we are by electromagnetic waves in the universe is indeed a wonder.

Though revealing one of the most profound secrets of the universe with his great theory that electromagnetic waves of various wavelengths traveled at the speed of light in every corner of the world, Maxwell died fairly young at the age of 48 of stomach cancer. Despite his untimely death, his scientific achievements served as yet another positive example of the power of science after Newton's law of universal gravitation, making scientists confident that accurate mathematical models and calculations could help humans predict Nature's phenomena, visible or invisible alike. It was exactly the momentum of the winning streak that Einstein rode on to develop his theory of relativity and extend classic mechanics, which only addressed absolute time and space (such as the solar system), to the light-speed universe constituted by the warping of space and time.

5.4   The Miracle Year: Five Groundbreaking Papers

With his vision focused on the scientific giants before him, the young Einstein often forgot to keep on good terms with the people around him. When he graduated from college in 1900, his hope of remaining to work as a teaching assistant was shattered because professors were offended by his defiant style of learning. However, the rejection also directed him to the path that took him straight into academia: publishing papers. After researching day and night for several months, he completed his academic debut, two papers on fluid mechanics that he described as valueless but were still published in *Annalen der Physik* (*Annals of Physics*). Unfortunately, the two studies also triggered a bitter disagreement from his college advisor Heinrich Friedrich Weber (1843–1912) and led him into the quagmire of unemployment due to the university's authoritative influence. In fact, his situation became so dire that his father even wrote to Weber as an apology and begged for forgiveness but only to no avail. After struggling for more than a year, Einstein finally landed a job as a private tutor in September of 1901 but was fired only six months later after he described high school education as suffocating kids just like the exhaust from factories right in front of the tutored student. Seeing him losing the teaching gig, his father again jumped into send his papers of fluid mechanics to the Leipzig physicist Friedrich Wilhelm Ostwald (1853–1932) along with a letter but received no response. That said, Ostwald seemed to be appreciative of the papers after all since he was the first person to recommend Einstein to the Nobel Committee for Physics ten years later. Just when things felt like he'd hit rock bottom, Einstein's luck finally started changing. This time, he was introduced to Minister of the Swiss Patent Office because the father of his close college classmate Grossmann Marcell was friends with the official. Thanks to Marcell's help, he finally secured a job as a level-3 assistant examiner in Bern in June of 1902 and acquired Swiss citizenship.

Having gone through the hardship of two and a half years' unemployment, Einstein was determined to take this job seriously like a mature, responsible adult. Reflecting on the five years between 1902 and 1907 that he spent at the Patent Office, he said the experience was just like living in a scientific heaven tucked in the mundane world as he was constantly surrounded by the work of inventors who enjoyed thinking and exploring, and reviewing all their applications was nothing short of a brainstorming session because he had to evaluate all kinds of creative ideas. In addition, he also got to exercise his intuition when judging the novel ideas that were yet to be executed from varying angles, as if he was taking a sneak peek at other scientists' minds, which was simply perfect for someone curious like him. He said in later years that it was precisely in that period when many of his genius physical concepts were formed and went on to be refined and eventually shine the brightest light in 1905. In fact, almost all authors of Einstein's biographies called 1905 his Miracle Year because in the 23 papers that he had in *Annalen der Physik*, which was the most important scientific journal in his era, 21 were completed and published in 1905.[9] Among these papers were five that addressed the contradictions and issues arising from interdisciplinary studies and achieved the integration of various academic fields by breaking the barriers. Below, we will take a walk through the five masterpieces:

---

[9]Referenced from Neffe [26].

5.4.1 Paper 1: A New Determination of Molecular Dimensions

Among the five papers, the first to be published was named "Eine Neue Bestimmung der Moleküldimensionen" ("A New Determination of Molecular Dimensions"). This paper provided proof of the still-questioned existence of atoms and molecules by pointing out that the physical properties of a solution, such as viscosity, could be used for accurate calculations of the size and number of solvent molecules and laid down a solid foundation for colloid chemistry. In fact, it was also the doctoral thesis that he submitted to the University of Zurich, helping him pass the final examination for doctoral candidates. He first calculated the viscosity coefficient between pure water and water with low concentration of dissolved sugar. Then, with the help of the Stokes-Einstein relation, he developed a formula to represent the relationship between the two liquids. As the last step, the Avogadro constant and the radius of table sugar molecules were determined by comparing the equations derived from the two calculations. The paper showcases Einstein's exceptional skills and profound wisdom in the field of physics with extremely complex mathematical derivations as well as strategical hypotheses and simplifications. For easier understanding, though, we'll simplify his research into a proposition containing only the critical steps. For details, please see "Appendix 5.3: Measuring the Radius of Table Sugar Molecules."

5.4.2 Paper 2: Brownian Motion

Nine days after the first paper was presented, Einstein published a second paper about the movement of atoms and molecules, named "On the Motion of Small Particles Suspended in a Stationary Liquid, as Required by the Molecular Kinetic Theory of Heat", this time only focusing on the motion of two types of elementary particles in liquids. This topic had in fact been researched before by the British scientist Robert Brown (1773–1858) in 1827 when he looked through a microscope at the water-immersed pollen of a plant. There, he observed the constant but irregular zigzag motion of small colloidal particles, and this phenomenon was later named after him as the "Brownian motion". In this second paper, Einstein adopted the angle of statistical mechanics and defined particle movement as a kind of Brownian motion which could also be understood as the result of the interaction between the macroscopic concept of temperature and microscopic molecules. Due to the effect of Brownian motion, collisions would take place among particles and water molecules, triggering energy redistribution in the water solution of pollen, with each particle and water molecule allocated a momentum of $(3/2)kT$ at the center. Of course, the average velocity of particles is a lot lower than that of water molecules due to larger mass, but they still move in a continuous manner and keep changing directions because of constant collisions.

After explaining the Brownian motion, Einstein also established a diffusion equation for pollen particles. According to him, when the particles started to diffuse, it could easily be proved through calculations that the density of them would increase along with their distances from the original point. Based on his rationale, the random fluctuation of pollen in liquids was caused by the high-frequency impact of ultrafine particles (water molecules) from all directions, but as the pollen particles were hit with molecules of different frequencies and impact levels, their directions of motion could change at any moment, which contributed to the random nature of Brownian motion. In order to describe the random movement of pollen, Einstein conducted a two-phase derivation to calculate the average distance traveled before each collision, which led him to the motion equation for pollen particles, which is what we call the "diffusion equation" today. He then introduced his self-measured diffusion coefficient and related physical quantities into the equation before calculating the average distance traveled between each collision to be 1.8×10^{-6}m, which perfectly fit the pollen movement observed by Brown in his experiment. For details about the two-phase derivation, please see "Appendix 5.4: Average distance traveled by pollen particles in liquids through random motion."

Both papers mentioned above clearly proved the existence of atoms and that their size could be measured accurately, which elevated the particle from a ghost-like element dismissed by scientists as unreal to the basic unit of careful mathematical calculations. These two articles soon entered the limelight of the European academia after publication and made Ostwald, who'd refused to introduce any job for Einstein and long denied the existence of atoms, change his mind. Even Max Planck (1858–1947) also acclaimed the discoveries of Einstein's research in his speech in Berlin, which was quite rare for an authoritative figure like him. In 1895, Jean Baptiste Perrin (1870–1942), who discovered cathode rays to be streams of tiny particles, devised an experiment to prove Einstein's equation and was unexpectedly awarded the Nobel Prize for Physics in 1926.[10] In addition to the recognition of the prize, this

---

[10]It might strike readers as confusing that it was Perrin, who proved Einstein's theory with an experiment, rather than Einstein himself, who was awarded the Nobel Prize. Yet in fact, such a decision was probably made based on the convention the committee had at the time that the Prize couldn't be given to the same person more than once since Einstein had already won it in 1923 for

experiment has also become a critical procedural basis for contemporary pharmaceutical processing, which can be said to be the most critical contribution of Perrin research.

5.4.3 Paper 3: Special Relativity

Although the first two papers were both important in their own ways, Einstein's most popular and influential work was the third, the one dedicated to special relativity. As we've mentioned before, what made Einstein develop the theory of special relativity was his desire to solve the "paradox" arising from Maxwell's equations: all electromagnetic waves travel at the speed of light regardless of the observer's motion. To prove the seemingly-paradoxical truth backed by Maxwell's well-structured equations, Einstein was sure that the relativity of time and space was a crucial factor that couldn't be overlooked. Since the two elements are related to the observer's motion state at the same time, they should both be considered "relative", correct? This, is the central concept of relativity, which modified the hypothesis of Newtonian mechanics that all motion in the universe was based on an absolute frame of reference constituted of one absolute time and space only.

After establishing his discourse in (1) the relativity of time and space as well as (2) the constancy of light speed, Einstein provided a comprehensive interpretation of classic mechanics, electromagnetics, optics, and several related phenomena, all on the basis of these two concepts while also integrating these fields with the notion of relativity. Then, on June 30 of 1905, he finally unveiled the theory of special relativity with the paper "On the Electrodynamics of Moving Bodies," which detailed his mathematical derivation of the relation of time and space. Basically, his mathematical model was based on Minkowski's space-time system and also made use of the Lorentz transformations to prove the invariance of physical quantities in time and space. As for why the title of a paper on special relativity would contain an unrelated word like "electrodynamics." Well, it was because he originally set out to explore why the equivalent mass of an electron would change along with its direction of motion in electric field. He surprisingly chose to look at the question from the perspective of time and space instead of adopting the conventional approach of electric or magnetic research. This eventually led him to derive the equation of mass-energy equivalence, revealing the profound truth that mass is equivalent to energy—the key concept of special relativity.

---

his research in the photoelectric effect. Marie Curie was the only exception in the early days of the award, but later on, John Bardeen also won the Nobel Prize for Physics twice for his contribution to the field of transistors and superconductivity in 1956 and 1972.

For details regarding his derivation process, please see "Appendix 5.5: How the Equivalent Mass of an Electron Change with Its Direction of Motion in an Electric Field."

5.4.4 Paper 4: Mass-Energy Equivalence

In addition to facilitating the unprecedented integration of light, time, and space, Einstein's discovery of the spacetime universe also led to the immensely influential equation to the human civilization: $E = mc^2$, which signifies that the energy of an object is equivalent to the product of its mass multiplied by the square of light speed. In fact, he only happened to find out about the equation when conducting research into the momentum conservation in Minkowski's spacetime system to establish the rule of special relativity. Yet what came as more of a surprise for him was probably how quick the equation was applied to nuclear fission, where massive energy was released out of decreased mass caused by collisions between neutrons and nuclei. As the derivation process was fairly simple, the fourth paper "Does the Inertia of a Body Depend on its Energy Content?" was merely three pages long. As for the publication, it also happened in the same year, on September 27. For more about his research on momentum conservation, details will be provided in Sect. 5.12.

Just like Newton's equation of universal gravitation, this formula of mass-energy conversion also contains three physical quantities only but has proven more than enough to accurately describe slow motion, light speed, and everything in between. Today, many young scientists who admire Einstein keep the equation $E = mc^2$ by their side in one form or another, such as hanging it on the wall, so they can be reminded of their love of science from time to time. In later sections, we will take a look at Einstein's derivation process and research basis of special relativity.

5.4.5 Paper 5: Photoelectric Effect

In October of 1900, Planck's presentation of the blackbody radiation law at the annual meeting of the Physical Society of Berlin officially opened the era of quantum research, but his studies didn't receive much recognition in the physics community immediately because it was hard for scientists to accept the concept of discontinuity when continuum structures that regulated phenomena such as electric and gravitational fields were the academic mainstream. However, Einstein's curiosity was piqued because Planck's claim that the intensity of light depended on its frequency was in perfect accordance with the photoelectric effect, which had always been his field of interest. As his research showed that the energy level of ejected electrons depended on the frequency, rather than the intensity of the light that hit them, he explained this result with the innovative notion of "light quanta" (later called "photon") by contending that it was photon-constituted wave packets through which light transmitted energy, and that just as Planck had pointed out, the energy released equaled the light frequency multiplied by the Planck constant.

Basing his research on the photoelectric effect, which was discovered by Hertz, Einstein claimed that only a light whose frequency was over a certain threshold had enough energy to eject electrons from their trajectories when it struck a metal surface; therefore, just increasing the intensity of light wouldn't help emit electrons. In fact, the photoelectric effect is what drives a lot of common phenomena in our everyday life, and Einstein's contributions lie in the fact that he used the concept of light quanta in interpreting these phenomena and enhanced people's understanding of Planck's blackbody theory. Summarizing related results, he published the paper "A Heuristic Point of View of the Production and Transformation of Light" in 1905, which helped Planck and himself win the Nobel Prize for Physics in 1918 and 1923 respectively while introducing the brand new field of physics: quantum mechanics.[11]

5.5 Foundation of Relativity: Light Speed Over Ether

Though he did provide a thorough analysis of electromagnetic waves with his equation, Maxwell also left behind two paradoxes: (1) all electromagnetic waves travel at the speed of light, and (2) there's no minimum or maximum wavelength for such waves although the known waves with the shortest and longest wavelengths are γ rays and radio waves. For Einstein, the first claim was apparently paradoxical because if two people moving at different velocities observe the light speed at the same time, the results were supposed to be their respective speed plus that of light based on the principles of Newtonian mechanics, so the two observers shouldn't end up with the same value. In other words, if light speed is a fixed value, that means the Newtonian concept of "speed," which is defined based on absolute time and space, is fallacious somehow. To resolve the paradox, Einstein proposed the theory of

---

[11] See also the story illuatrated in Chap. 10 of Mlodinow [23].

relativity, which he constructed on the foundation of relative time and space. As for Maxwell's second paradox, it was actually what inspired Einstein to initiate the studies on quantum mechanics. To trace his research to the very beginning, though, let's first talk about the fundamental question that he had to answer when trying to resolve the conflict between light speed as a fixed value and the theory of classic mechanics: what is light?

5.5.1 What Is Light?

Light is probably the most dazzling, enchanting, and even enthralling physical quantity in the universe of all time since we can not only see its brightness but feel its warmth as well. For thousands of years, however, nobody could describe for sure what light was. It wasn't until the 17th century did Newton claim that light was generated through the combined movement of numerous particles, while Hooke compared it with waves like sounds and ripples. In the 19th century, Maxwell concluded through a well-organized mathematical derivation that light was a type of electromagnetic wave traveling at the speed of 300,000 km/s (or more precisely, 299,792,458 m/s), whereas after the 20th century, Einstein started contending that light didn't need any medium for transmission and traveled at a speed (a cosmological constant) that no substance could ever exceed. After more than 300 years of discussion and debate, the nature of light has finally been clarified, but in the process of understanding the nature of light, there were individual breakthroughs on some critical issues that contributed to the fruitful results that we enjoy now, and to walk through the history of scientists' exploration of light, we need to go all the way back to Galilei's era.

As early as in 1638, Galilei had tried to measure the speed of light using plane mirrors to reflect rays among towers, but as his experiments were subjected to restrictions of imprecise instruments like water clocks and hourglasses, it was of course quite impossible to measure accurately the extremely short amount of time needed for light to travel among the towers. Nevertheless, as light continued to draw everyday attention due to its ubiquity, the Danish scientist Rømer finally obtained some sort of result when he wasn't really expecting it 67 years after Galilei invented the telescope in 1676. After a long-term observation of Io, the closest moon to Jupiter, he found out that the natural satellite would enter the shadow of Jupiter periodically, but the time between each occurrence was increased by about 22 s incrementally. Therefore, he figured that the increase in time interval was caused by the growing distance traveled by light, which meant the distance between Earth and Io was changing as well. In other words, the claim made by Newton that light speed was infinite was factually wrong. As the law of universal gravitation was still in its heyday and dominated the world of physics, Rømer's discovery was undoubtedly a massive blow to the Newtonian school of thought.

In this paragraph, we're going to look at more details about Rømer's revolutionary experiment. Out of the four moons of Jupiter discovered by Galilei, Io is the closest one to the planet and has a revolving period of 42.26 h, while Jupiter's period of revolution around the sun is 11.86 years. Since almost all the planets in the solar system revolve on the ecliptic plane, there's this one time each year for Earth and Jupiter to be on the same side of the Sun, which is what gave Rømer the chance to conduct his experiment. According to his logs, as Jupiter has a revolving radius 5.2 times that of Earth and needs 11.86 years to complete a circle around Sun, its tangent velocity is about 13.07 km/s. Now, suppose the point of observation is set in a coordinate system rotating along with Jupiter around Sun. Under this circumstance, when Earth approaches Jupiter, we'll see Io disappear behind the planet, and if light speed was infinite, the time interval between two sightings of Io's disappearance would've been 42.26 h. In contrast, if light travels at a fixed speed, then the time interval should be shorter than 42.46 h because the distance between Earth and Jupiter is supposed to decrease by 4,551,944 km during two disappearances, which fits the result of Rømer's experiment. In fact, the time interval he measured was roughly 22 s shorter than the actual value and could be converted to the light speed of 220,000 km/s,[12] 30% deviant from the actual speed 299,792 km/s. Remember though, this was the first experiment in human history where the speed of light was measured using an indirect approach and proved to be a finite number. Though deemed unorthodox by some in that era, this attempt was indeed a big step in the development of physical science. For details of related calculations, please see "Appendix 5.6: Rømer's Light Speed Experiment."

Despite the experiment, which proved that light speed wasn't infinite, it was still difficult for many scientists to accept the result since it went against the then prevalent concept. Therefore, the topic became untouched yet again for roughly half a century and only surfaced by chance when the British astronomer James Bradley (1693–1762) was measuring the distance between Earth and a star while examining if lights from different sources would travel at varying speeds, which coincided with Einstein's field of interest. As far as Bradley was concerned, when observed on Earth using a telescope, the light from a star would deviate to certain degree after entering the tube due to

---

[12] Rømer didn't convert his measurement into light speed right away. Instead, it was later scientists who calculated the value of 220,000 km/s out of his result as well as the actual light speed, which is roughly 300,000 km/s, in 1727.

the motion of Earth. Simply put, the light that enters from the center of the objective lens would leave the eyepiece from a spot that's somehow deviated from its center, which is what we call the aberration of light. To get a clear image of a star, therefore, the telescope used for observation has to be slightly adjusted following the direction of Earth's rotation to offset the blurriness caused by light aberration. In later sections, we will reconstruct Bradley's calculation process. For details, please refer to "Appendix 5.7: Light Aberration Caused by Earth's Revolution."

5.5.2 Does Ether Really Exist?

Since light travels at a finite speed that's unaffected by the motion state of its source, then what is the medium of transmission that gives light such properties? Before Einstein, there were two widely-circulated explanations in academia. Some, such as Newton, considered light to be constituted of particles, while others claimed that light was a type of wave. Though the latter was more generally accepted because of the similarity between light and wave interference and many experiments did show through projected light patterns that it traveled with wavelike properties, nobody was sure what the transmission medium was. Yet as all the known waves back then had been found to have their corresponding media (such as air for acoustic waves and water for ripples), Maxwell and some other scientists agreed that it was through a medium called "ether" that light was traveled, which was why the wave theory of light started to be developed on this basis ever since.

In fact, ether wasn't a new concept but one that had existed since Aristotle's era 2,000 years before. While natural philosophers back then considered ether a ubiquitous substance filling all corners of the universe, it was brought up in the 18th century again to explain the phenomena of universal gravitation, light, electromagnetic waves, etc. while defined as a rigid material with zero mass that was able to penetrate any object in motion without changing its own status. Although scientists were keen on describing the possible properties of it, nobody had ever witnessed its existence, not to mention provide proof of any sort. This said, there was in fact a young couple of researchers who intended to prove that either was a real thing by devising the following set of precision instruments in their Berlin lab. They are Albert Michelson (1852–1931) and Edward Morley (1838–1923). Although their device is hailed to be the first-ever interferometer, the experiment result was far from satisfactory.

The Michelson-Morley experiment was based on the assumption that ether was a substance quietly filling the universe that could be felt as gusts of wind during Earth's self-rotation and revolution around Sun due to the relative motion between the moving planet and the stationary ether. Bearing this hypothesis in mind, the pair designed a set of optical instruments that we call the "interferometer" today, where all the apparatuses were floating in a trough of mercury and free to rotate, as shown in Fig. 5.4. Meanwhile, a light was emitted from the source on the left-hand side and divided into two perpendicular beams A and B after passing through the silvered beam splitter in the middle, whose distance from both reflectors was L. Now, if we suppose the ether wind blows from right to left at the speed of v, A and B would be parallel and perpendicular with the wind respectively. The time for A to travel to the reflector and return can be calculated as such:

$$\frac{L}{c-v} + \frac{L}{c+v} = \frac{2Lc}{c^2 - v^2} = \frac{2L}{c \cdot \left(1 - \frac{v^2}{c^2}\right)} \tag{5.20}$$

As for B, it travels along with a slant line relative to the ether wind at the speed of c. According to the Pythagorean theorem, the time needed for it to travel to and back from the reflector is:

$$\frac{L}{\sqrt{c^2 - v^2}} + \frac{L}{\sqrt{c^2 - v^2}} = \frac{2L}{c\sqrt{1 - \frac{v^2}{c^2}}} \tag{5.21}$$

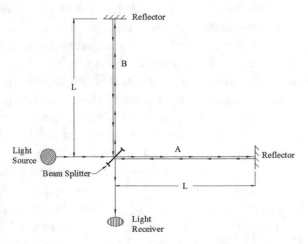

Fig. 5.4 Experiment instruments used by Michelson and Morley to measure ether wind

From these two equations, we can tell that (5.20) is (5.21) multiplied by $\sqrt{1 - \frac{v^2}{c^2}}$. In other words, if ether wind did exist, it'd generate different influences on the speed of A and B; i.e. though traveling equal distances, A and B would reach their destination at different time points due to their directions and the impact from the ether wind. However, despite Michelson and Morley's repeated efforts in changing the position of the instruments and modifying the setup to enhance the accuracy of their measurements, A and B still arrived at exactly the same time without any exception, which left them no choice but to admit the failure to prove the existence of ether. Many researchers also conducted similar experiments later on, but nobody managed to succeed.

Although Michelson and Morley had no luck with their attempt, most scientists in their era still acknowledged the ether theory, including Maxwell, discoverer of electromagnetic waves, and Heinrich Lorentz (1853–1928), whose invention of the inertial frame of reference allowed Einstein to develop the theory of special relativity. Basically, it was a widely accepted concept that light transmitted at a constant speed through ether, which filled the universe in a stationary manner. To understand this theory from a mathematical point of view, researchers considered ether the reference frame of the "absolute stationary" state in the universe, while Fitzgerald and Lorentz even proposed the length contraction theory in 1889 and 1892, trying to explain the failure of the Michelson-Morley experiment. According to them, the length of everything moving in ether would be measured to be shorter than its proper length when in relative motion with the substance. To be more specific, if an object's actual length is L, the measurement would be shortened to $L' = \sqrt{1 - \frac{v^2}{c^2}}L$. In terms of the failed experiment detailed about, the parameter L in (5.20) should be replaced with L':

$$\frac{2L'}{c \cdot \left(1 - \frac{v^2}{c^2}\right)} = \frac{.2L}{c \cdot \sqrt{1 - \frac{v^2}{c^2}}} \tag{5.22}$$

Since (5.22) and (5.21) are essentially the same, the failure of the experiment could now be justified. Under the framework of this explanation, although the measured length of a motioning object was indeed supposed to be shorter, the actual result still remained to be L because the measuring scale would also be shortened by the same proportion and make it nearly impossible for observers to spot the difference between L and L'.

Despite the seemingly plausible justification, the theory of length contraction was in fact quite far-fetched and soon denied by Einstein although he quite admired Lorentz. According to Einstein, ether was nothing but a human-made creation that God had never placed in the universe. Plus, multiple experiments had shown the substance to be unprovable time and again, he boldly announced that the ether theory should be discarded and replaced by the constancy of light speed: i.e. the speed of light was a fixed value that wouldn't change with the observer's motion. To prove this claim, he engaged in several thought experiments and figured that light could only travel at a constant velocity when time wasn't considered an absolute concept independent of space. In later sections, therefore, we will take a look at how he managed to think out of the box of absoluteness and established his groundbreaking theory by discarding all preconceptions and engaging in simple but accurate deduction.

5.6   Integration of Space and Time

Time is a rather abstract notion and one of the first natural phenomena that humans observed as far back as in the prehistoric era through periodical changes such as sunrise and sunset, the Moon's different shapes, the transition of seasons, and tidal movements. Therefore, people started to develop the concept of time based on such recurring events and established calendar systems as a reference for activities in the agricultural society. As we know today, such calendars essentially described the regularity of astronomical phenomena, which made astronomy the first scientific discipline that human beings engage in. The notion of "time" wasn't challenged until the beginning of the 20th century when Einstein had to rethink its definition while attempting to develop the theory of relativity. Since we already know that light travels at a fixed speed of roughly 300,000 km/s, which is a cosmological constant, definitions of distance and time should both be reconsidered bearing in mind that speed is the former divided by the latter.

To continue the following discourse, we need to first establish a spacetime coordinate system to accurately describe events such as sending out a light, catching a ball, starting a stopwatch, detonating a bomb, etc.—in fact, any of our actions and subtle changes in the universe. Regardless of the nature of an event, we will only focus on the "time" and "place" of its occurrence. Therefore, any event can be treated as a point in the spacetime coordinate system and represented as (t, x, y, z), meaning that it took place at the time and place t and (x, y, z). One thing to be noted is that since observers have their own perspectives when looking at any given event, one single event might not have the same coordinates in different systems.

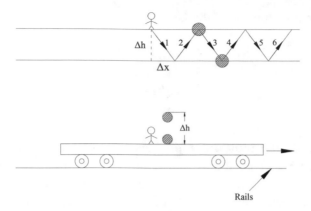

Fig. 5.5 The thought experiment on relative space

Before Einstein developed the theory of relativity, time had always been deemed absolute. In other words, regardless of the observer's perspective, any event was deemed to have one unique time of occurrence (t) only. When it comes to space, though, it was a lot easier for people to detect the relativity of it. Consider this scenario, for example: when a player on a train bounces a basketball to the floor continuously at a constant speed, what he sees is a ball that moves vertically up and down, while what a passenger standing on the platform sees is a ball moving both vertically and horizontally in a zigzag route, as demonstrated in Fig. 5.5. As the distance traveled by the ball observed from the passenger's perspective is a lot longer than that in the player's eyes, the relativity of space can easily be explained.[13]

To describe this example with precise terms under the framework of space and time, we should say that there are two events A and B, happening at the moment when the player lets go of the ball and when it returns into his hand after bouncing off the floor respectively. Suppose the train is moving relative to the platform along the x-axis at a constant speed of v. If A's place of occurrence is observed to be $(0, x_0, y_0, z_0)$ in the spacetime coordinate system by both the player and the passenger under the assumption that time is absolute, then B's time of occurrence should be perceived as the same (t) by both observers as well. As for the position where B takes place, however, it's considered to happen at the same spot as A from the player's perspective, i.e. $(x', y', z') = (x_0, y_0, z_0)$, while for the passenger on the platform, the relation becomes $(x, y, z) = (x_0 + vt, y_0, z_0)$. In fact, the space relativity

---

[13] Also reference to Cox and Forshaw [8].

here can be represented with the Galilean transformation:

$$
\begin{cases}
x' = x - vt \\
y' = y \\
z' = z \\
t' = t
\end{cases}
\tag{5.23}
$$

Which shows the transformation between two inertial frames of reference[14] (S and S', which moves in relation to the former at the speed of v along the x-axis) that overlap with each other when $t = 0$.

In addition to the equations of transformation between two inertial frames, Galilei also proposed the principle of relativity, pointing out that physical laws should remain invariant in all frames alike and that no mechanical experiment could help distinguish between stationary inertial frames and the ones moving at constant speeds. If we differentiate the time parameter in (5.23) twice, we'll be able to tell that the acceleration of any all objects should be the same in both frames. Since acceleration is an invariant in the transformation, plus Newton's law of mechanics dictates that an object will accelerate while entering a different state of motion upon the impact of an external force, it stands to reason that the same mechanical experiment should yield the same result in both reference frames.

So far, what we described above was the transformation of coordinates under the regulation of classic mechanics. From here on, let's take a look at how Einstein approached the same issue. In his theory of relativity, there are two basic postulates:

(1) The laws of physics don't change, even for objects moving in inertial (constant speed) frames of reference.
(2) Light in vacuum regularly travels at the speed of c in any frame of reference.

Please note that any conclusion from the deduction below has to comply with these postulates and cannot be made based on any other hypothesis. In addition, even common phenomena that can easily be seen in everyday life have to be scrutinized with careful examination.

The first principle listed above is an extension of Galilei's principle of relativity and applies to the physical laws in the inertial reference frames in not

---

[14]An inertial frame of reference refers to a coordinate system that moves at a constant speed, i.e. one without acceleration.

just mechanics but electromagnetics as well. In other words, there's no possible way to design an experiment to determine whether a reference frame is stationary or moving at a constant speed. In the discourse of relativity, as a result, there's no such thing as an absolutely stationary frame of reference.

However, the Galilean transformation, which is defined in (5.23), apparently contradicts with the constancy of light speed, which we know by differentiating the equation $x' = x - vt$. As a result $v'_x = v_x - v$ further yields $v'_x = \frac{dx'}{dt'} = \frac{dx'}{dt}$ and $v_x = \frac{dx}{dt}$, we can infer that a light emitted from the S frame along the x−axis will be observed as traveling at the speed of $c - v$ from S', which doesn't comply with Einstein's second postulate because light speed should be a non-changing value. Therefore, we will need to figure out another way of transformation between (t, x, y, z) and $\left(t', x', y', z'\right)$ in order to make sure that the invariance of physical laws is conformed to.

Although it seems difficult to satisfy both postulates at the same time, please take another look at (5.23) before we start to derive a different transforming method. Despite the appearance of t in the transformation of x coordinates, the parameter is in fact an invariant, implying that time plays a different role compared to the place of occurrence and can be considered independently. However, here's another question we need to think about: why did Einstein choose to include "time" in the transformation between reference frames. Was it because absolute time $(t' = t)$ couldn't meet both postulates at the same time? If time isn't absolute anymore, then what does the concept of "time" really stand for, and how do we define it?

For Einstein, all scientific theories had to be verified with experiments, and no physical quantity could be defined without considering how observers measured that specific quantity. Such a principle represents what we call the "operational definition" today. When it comes to the notion of time, it's also supposed to be defined with the method of measurement taken into consideration. Since light speed is a constant value, optic signals seem to be a reasonable choice for measuring unit, but before trying to define what "time" really is, we should first discuss a basic question: is simultaneity consistent across all frames of reference? In other words, if two events happen simultaneously in a specific inertial frame, will they still be perceived as simultaneous in a different frame after the coordinates are transformed between the two systems? To get to the bottom of this question, we need first to determine whether events 1 and 2 in the following scenario happen simultaneously in the S inertial reference frame with an operational definition of time. Let's suppose there are two observers A and B standing respectively at where events A and B are about to happen; then, let's assign another participant C to stand at the middle point between the two observers. If A and B both send out a light

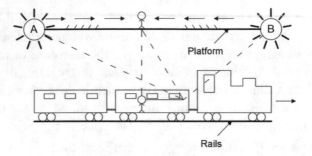

Fig. 5.6 Einstein's thought experiment on simultaneity

towards C when they see 1 and 2 take place, we'll know the two events are simultaneous if C receives both lights at the same time because C's distances to A and B are equal.

After looking briefly at the simple scenario above, let's now proceed to Einstein's train-platform thought experiment. As illustrated in Fig. 5.6, there are two sources of light set up at both ends of the platform, while two observers A and B stand on the left (rear of the train) and right (front of the train) side of the platform respectively. When the train moves at the constant speed of *v* with respect to the platform, the lengths of them are perceived as the same from the reference frame of the platform.[15] Meanwhile, there are two other observers C and D standing on the midpoint of the platform and sitting midway inside the train car. Then, event 1 happens when the light source A is turned on at the moment the train's rear passes by A's standing position, while event 2 takes place when source B is turned on at the moment when the train's front passes by the participant B. As the train and platform are perceived to be of the same length by C, this observer will receive the lights from both sources at the same time, which is why we can infer based on the operational definition of simultaneity that events 1 and 2 are simultaneous in the platform reference frame. What about the reference frame of the train car, then? First, since lights A and B arrive at C simultaneously in the platform frame, we know the two events should have the same spacetime coordinates in the frame of the train; i.e. D will also see the lights arrive at C simultaneously. However, as the lights have not reached C the moment when D rides past C, plus D is moving to the right in relation to C, we can tell that there's no way for C, not to mention D to have received the light from A when light B reaches D. Again, based on the operational definition of spontaneity, we can

---

[15] Precisely speaking, the "reference frame of the platform" refers to the frame of reference where the platform is stationary, but such explanation is too lengthy and therefore simplified for the purpose of easy communication. Similarly, if we mention "the reference frame of something" going forward, we're referring to the frame of reference where that something is in a static state of motion.

infer that event 2 happens before 1 in the frame of reference that motions along with the train.

So what does this thought experiment actually mean? Let's look back at the equations of transformation, where t' is the function of x, y, z, t, expressed as $t' = f_t(x, y, z, t)$. If $\Delta t = 0$ when $\Delta t' \neq 0$, we can be sure that the function $f_t(x, y, z, t)$ is affected by the position coordinates x, y, and z, which is to say that time and space are mutually influential and should there-fore be considered together. In other words, time isn't independent of but intertwined with space. By the point when Einstein proposed his theory, the absolute boundary between time and space, which had long been dictated by classic mechanics, was blurred for the first time ever in the history of scien-tific thought. It was as if the two originally irrelevant concepts joined forces to present a tango performance to the rhythm of the universe.

From the train-platform thought experiment described above, we can also derive the phenomenon of length change, but before explaining the process, let's first recall the operational definition of "length." So, if we're in the inertial frame of reference S, how're we supposed to measure how long a particular object is? As an initial step, we need to prepare a scale and assign two people A and B to the positions that the front and rear ends of that object, which might be moving in this frame, are expected to reach. The moment when the front arrives where A is stationed (event 1), A marks its position on a piece of paper, while the same applies to B and the rear (event 2). If the two events take place simultaneously in S, then the length of the object equals the distance between where events 1 and 2 take place (for the operational definition of "simultaneity," please refer to previous sections). However, here's one more issue to be considered: how should we define a universal "measuring unit" for observers in different frames to compare their measurements? The operational definition of "scale unit" won't be detailed toward the end of this chapter, but here, we can simply use it to examine whether the length of this object would be measured differently in S and S' under the framework of the Galilean transformation. Given that $x' = x - vt$, i.e. $\Delta x' = \Delta x - v\Delta t$, plus A and B have to act simultaneously when measuring the object in S ($\Delta t = 0$), we can derive the equation $\Delta x' = \Delta x$. As for S, simultaneity in this frame requires that $\Delta t' = 0$, and since $t' = t$ implies $\Delta t' = \Delta t$, we can still derive $\Delta x' = \Delta x$. In conclusion, the length of an object wouldn't change even when transformed through the Galilean equations.

What about in the framework of relativity, then? To answer this question, we need to go back to the platform-train thought experiment. In the frame that's stationary with respect to the platform, the rear of the train reaches A (event 1) at the same moment as when the front reaches B (event 2), so in this frame, the lengths of the train and platform are equal. Nevertheless, as event 2 happens prior to 1 in the frame of the train, the train car should be perceived as longer than the platform because its rear has not yet reached A when the front passes by B. Therefore, we know the length of an object also changes with the transformation between reference frames, another property that's different from the Galilean system.

To help readers understand more clearly the mathematical significance of this thought experiment, let's reexamine it from a quantitative point of view. Suppose the train car's length is L when static. Under this circumstance, the distances from D to both the front and rear of the car are $L/2$. As a next step, we'd like to calculate the time interval between the occurrences of events 2 and 1. Please note that it doesn't really matter that we haven't established an operational definition for length because our discussion is limited to the train reference frame and will not involve the transformation of coordinates, not to mention the comparison of lengths or time intervals measured in different frames.

One thing to be emphasized again, though, is that all discussions here are based on D's perspective. To facilitate understanding, we'll also be using the three consecutive graphs in Fig. 5.7 for illustration. In the reference frame of the train, as the platform moves backwards at the speed of v and has a shorter length than the train car, the car's front first reaches B and prompts observer B to send out a light (event 2), as illustrated in Fig. 5.7a. Then, the midpoint of the train (D) speeds past that of the platform (C) at $t' = 0$ in Fig. 5.7b, and lastly, the train's rear passes by A and triggers observer B to send out a light as well (event 1), as shown in Fig. 5.7c. As the distances between B and C as well as A and C are both half the platform's length, the distance between C and D equals $L/2$ minus half of the platform in scenarios (a) and (c). Now, if we suppose the events in (a) and (c) happen at $t' = -\varepsilon$ and $t' = \varepsilon$ respectively, the distances between C and D at these two points of time are both $v\varepsilon$. If lights A and B needs Δt_C and Δt_B to reach C after sent out by observers A and B, we can derive the following equations because the speed of light in this frame of reference remains to be c:

$$c\Delta t_B = \left(\frac{L}{2} - v\varepsilon\right) + v\Delta t_B \qquad (5.24)$$

$$c\Delta t_C + v\Delta t_C = \left(\frac{L}{2} - v\varepsilon\right) \qquad (5.25)$$

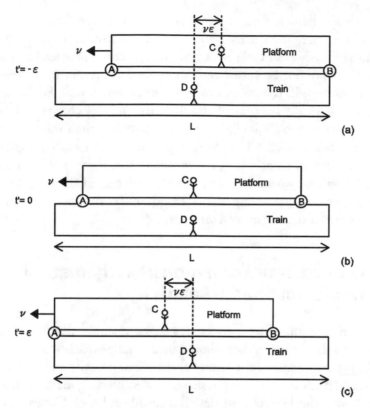

Fig. 5.7 Three-step illustration of Einstein's thought experiment on simultaneity

And because both lights should reach C at the same time regardless of the reference frame, the following equation stands true as well:

$$-\varepsilon + \Delta t_B = \varepsilon + \Delta t_C \qquad (5.26)$$

With (5.24), we can derive the solution of $\varepsilon = Lv/2c^2$, which helps us express the time interval between events 1 and 2 in the train frame as:

$$\Delta t'_{1-2} = 2\varepsilon = \frac{Lv}{c^2} \qquad (5.27)$$

In (5.27), the subscript 1–2 refers to the occurrence time of event 1 minus that of event 2, and because $\Delta t'_{1-2} > 0$, we know that the two events take place at different points of time, which verifies the thought experiment of simultaneity shown in Fig. 5.7.

Toward the ending of this section, let's make a quick summary of what we've discussed up to this point. In both the Galilean transformation and

Newtonian mechanics, time is an invariant in the transformation between frames of reference, and the length of an object is also a non-changing value, while in the framework of relativity, time and space are inseparable; therefore, simultaneity may not be consistent throughout all frames of reference, and length can change in the process of transformation as well. So far, however, we've only got to the concept of simultaneity and the comparison between lengths but haven't really looked at how to calculate time interval or length since we haven't defined a scale for the pair of quantities that applies to all frames of reference. Considering the postulate that light speed is a constant, though, we will only need to define the scale for one of them. As long as we can come up with a measuring unit of length, the corresponding unit of time can be calculated therefrom, and vice versa.

5.7   Lorentz Transformations: Mathematical Model for Special Relativity

Since the integration of time and space is the essential factor in describing all motion in the universe, there should be a strictly-structured mathematical model in place to regulate the seemly abstract but actually real phenomenon. Ever since Kepler's time, many physicists started holding mathematics in awe because it was the best way to describe simple rules of Nature and reveal secrets of the universe hidden deeply under the surface of the world's phenomena, which scientists often failed to realize about until specific laws were discovered. To explore the subtleties of the universe and understand the order of the natural world, therefore, integrating mathematical studies and physical experiments was the optimal approach. In light of the enormous help borrowed from mathematical analyses by physicists in the 300 years before him, Einstein was well aware of how important the integration of the two disciplines was and therefore started following the paths of previous scientists right at the beginning of his development of the relativity theory.

The most important mathematical mechanism behind special relativity is no other than changing the Galilean transformation to a set of equations for transforming between reference frames that satisfy the two basic postulates of special relativity. In the following paragraphs, therefore, we're going to derive these equations, and because they were first developed by Lorentz, the transformations are named after him.[16]

---

[16]The Lorentz transformations were in fact proposed by the scientist in 1904 to justify the results of the Michelson-Morley experiment. As far as Lorentz was concerned, when an observer moved at a certain speed relative to the substance of ether, the length of any object would be contracted along

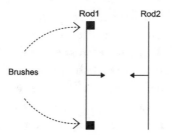

Fig. 5.8 Thought experiment on Lorentz transformations

Before starting the process, let's make sure the issue is solved using accurate terminology. Suppose the coordinate axes of the two inertial frames of reference S and S' are overlapped momentarily when $t = t' = 0$, and S' is moving relative to S at the speed of v along the direction of $+x$. Now, here comes the question: for any particular event, how do we transform its spacetime coordinates (t, x, y, z) in S to that in S'(t', x', y', z')?

One thing to be noted first is that so far, we've only been focusing on the length and time along the x-axis because there is a relative speed on this direction in the frames of both the train and platform that causes a change in length. Yet for y and z, can we be sure that there's no length change on these directions just because the relative speed does not have components along the two axes? Although most people may intuitively perceive the answer to be yes, we still need to conduct the thought experiment below to be on the safe side.

Suppose there are two rods of the same length, which is L when stationary, as shown in Fig. 5.8. Now, if we install a brush coated with paint on the two ends of rod 1 and place both rods in parallel with the y-axis before pushing them at the same constant speed towards each other along x, what will happen when we observe this event from two different frames of reference? Let's start by looking at the two scenarios below:

(1) Scenario A: In the reference frame where rod 1 is stationary, rod 2 is observed to be longer than it is.
(2) Scenario B: In the reference frame where rod 1 is stationary, rod 2 is observed to be shorter than it is.

the direction of the observer's motion, and based on this postulate, he established a complete set of equations for the transformation between reference frames different from that developed by Galilei. In his paper "On the Electrodynamics of Moving Bodies" published one year later in 1905, however, Einstein adopted a different approach to derive the same results with the two basic postulates of special relativity, thereby giving the Lorentz transformations a whole new interpretation.

If scenario A was true, rod 2 should be brushed with paint by rod 1 because of its larger length, but considering the principle of symmetry, rod 1 should also be longer than rod 2 in the reference frame where rod 2 is stationary, which is to say that there's no way the brushes can ever touch rod 2. In other words, the results observed from the two frames are contradictory, which is also true of Scenario B. Since both scenarios do not make sense, there's only one possibility left: in the reference frame where rod 1 is stationary, the observed length of rod 2 is equal to its actual length. Such a result proves that motioning along the x-axis won't cause the length to change on the direction of y or z after the transformation between reference frames, and this physical quantity doesn't change with the time or place of observation. Consequently, we can derive the equations of transformation for the y and z coordinates as:

$$y' = y \tag{5.28}$$

$$z' = z \tag{5.29}$$

Since the y and z coordinates are invariants in the transformation process, an issue left undiscussed before is now resolved. Remember we've previously mentioned the lack of a well-defined scale for time and length that applies to all inertial frames? Now we have two options because either y or z can serve as the measuring unit, which is to say that any scale has to be calibrated to the y- or z-axis before used to determine object length to be of practical value. What about the time scale, then? Well, we can simply draw a line of the length c along the y- or z-axis and easily define one unit of time as that needed for light to travel from one end of the line to the other.

Now that both the equations for transforming the y and z coordinates and the measuring units of time and length have been established, we can shift our focus to the most important part of Lorentz's equations: the transformation of x and t coordinates. So far, we've only confirmed that simultaneity and length both change in the process of such transformation; as for other factors, we don't have any idea of what the equations might look like. Due to such lack of insight, it's difficult to conclude anything straight out of a thought experiment, which is why we need to start with a simple mathematical derivation to explore possible forms of the equations.

First, we know that the most common form of a transforming equation is $x' = f_x(x, y, z, t)$ and $t' = f_t(x, y, z, t)$, but as previously indicated, the y and z coordinates do not change in the process of transformation when a relative speed doesn't have components along the two axes. In other words, even if we randomly move the origin of a reference frame on the yz plane, the final equations will remain the same, which is why we can simplify the functions mentioned above as:

$$x' = f_x(x, t) \tag{5.30}$$

$$t' = f_t(x, t) \tag{5.31}$$

Then, for a more specific illustration, let's imagine there are events 1 and 2, whose coordinates S and S' are (x_1, t_1), (x_1', t_1') and (x_2, t_2), (x_2', t_2') respectively. Based on (5.30) and (5.31), we can derive the following equations:

$$x_1' = f_x(x_1, t_1) \tag{5.32}$$

$$t_1' = f_t(x_1, t_1) \tag{5.33}$$

$$x_2' = f_x(x_2, t_2) \tag{5.34}$$

$$t_2' = f_t(x_2, t_2) \tag{5.35}$$

Considering (5.32)–(5.35), even if we randomly move the x and t axes around, the equations of transformation will still stay the same as long as the two sets of coordinates are overlapped when $t = t' = 0$. Therefore, we can move the origins of both axes to the spacetime coordinates of event 1's place of occurrence, which will make the coordinates of event 2 in S and S' $(x_2 - x_1, t_2 - t_1)$ and $(x_2' - x_1', t_2' - t_1')$ respectively. Next, by applying (5.30) and (5.31) again, we can retrieve the below equations:

$$x_2' - x_1' = f_x(x_2 - x_1, t_2 - t_1) \tag{5.36}$$

$$t_2' - t_1' = f_t(x_2 - x_1, t_2 - t_1) \tag{5.37}$$

Then, substituting (5.32) and (5.34) into (5.36) will lead us to:

$$f_x(x_2, t_2) - f_x(x_1, t_1) = f_x(x_2 - x_1, t_2 - t_1) \tag{5.38}$$

To satisfy (5.38), the function $f_x(x, t)$ has to be of the same linear form as:

$$x' = f_x(x, t) = Ax + Bt \tag{5.39}$$

In (5.39), A and B are both constants[17] that affect the relative speed of the two frames of reference (v). Now that we have the solution of x', let's try to figure t' out as well. First, if we substitute (5.33) and (5.35) into (5.37), we'll be able to derive:

$$f_t(x_2, t_2) - f_t(x_1, t_1) = f_t(x_2 - x_1, t_2 - t_1) \tag{5.40}$$

As a next step, we can express the function $f_t(x, t)$ in this linear form below:

$$t' = f_t(x, t) = Cx + Dt \tag{5.41}$$

In (5.41), C and D are both constants affecting the relative speed of the two frames of reference (v).

Although (5.39) and (5.41) are already established as transformation equations, we still need more thought experiments to figure out how the constants, A, B, C, and D are to be expressed. Yet before proceeding, let's take a looks at how (5.39) and (5.41) embody (1) the inconsistency of simultaneity, and (2) the change of length. Regarding the concept of simultaneity, we can tell from (5.41) that coordinates of the two events satisfy the equation $\Delta t' = C\Delta x + D\Delta t$, which we can apply to the previously-discussed platform-train thought experiment, where events 1 and 2 happen at the same time ($\Delta t = 0$) but at different places ($\Delta x \neq 0$) in the platform system. Therefore, if the constant $C \neq 0$, that means the time interval between the two events observed from the frame of the train ($\Delta t'$) isn't 0, either. In other words, the two events take place at different points of time.

When it comes to length, we can derive from (5.39) the equation $\Delta x' = A\Delta x + B\Delta t$. To measure the length of an object (Δx) in S, the coordinates

---

[17] For readers who would like to know how this conclusion came to be, please refer to the details below; however, we can also rest assured not to go through the entire process because (5.39) has been well examined. In fact, it is no difficult task to verify this equation through mathematical calculation. First, we can partially differentiate (5.38) over x_2 and obtain:

$$\left. \frac{\partial f_x}{\partial x} \right|_{x_2, t_2} = \left. \frac{\partial f_x}{\partial x} \right|_{x_2 - x_1, t_2 - t_1}$$

Then, we can do the same with t_2 and retrieve:

$$\left. \frac{\partial f_x}{\partial t} \right|_{x_2, t_2} = \left. \frac{\partial f_x}{\partial t} \right|_{x_2 - x_1, t_2 - t_1}$$

As both equations have to be true regardless of the values of x_2, t_2 and x_1, t_1, we know the values of $\frac{\partial f_x}{\partial x}$ and $\frac{\partial f_x}{\partial t}$ are both constants at any set of coordinates. Suppose:

$$\frac{\partial f_x}{\partial x} = A; \frac{\partial f_x}{\partial t} = B$$

Then, we can develop the following function:

$$f_x(x, t) = Ax + Bt + C_0$$

But if we substitute $x_2 = x_1$ and $t_2 = t_1$ into (5.38), the result will be yielded as:

$$f_x(0, 0) = 0$$

Which is why the constant $C_0 = 0$.

of both ends of this object have to be measured "simultaneously", implying $\Delta t = 0$. If the object is stationary in S', then $\Delta x'$ represents its length when in a static state of motion, implying that:

$$\Delta x = \frac{\Delta x'}{A} \qquad (5.42)$$

As a result, if $A \neq 1$, measurements of the object done from S and S' will yield different results, but if this object isn't stationary in S', then $\Delta x'$ will not necessarily equal its length in this frame, either. Why? According to the operational definition of "length," the coordinates of both ends of the object have to be recorded simultaneously in S', meaning that $\Delta t' = 0$ has to be true, but in light of $\Delta t' = C\Delta x + D\Delta t$, $\Delta t'$ can't possibly be 0 when $\Delta t = 0$ but $\Delta x \neq 0$. So in (5.42), $\Delta x'$ refers to the coordinates of the measured object's front end at a given moment (t_1) minus that of its rear at t_2 ($t_1 \neq t_2$). For $\Delta x'$ to represent the object length, the coordinates of the object's rear end at t_2 have to be overlapped with that of its front at t_1, and the most simple scenario, is when the object is stationary.

After further clarifying the concepts of spontaneity and length, we can now try to figure out how to express A, B, C, and D with equations by referring back to the Michelson-Morley experiment (see related discussions in Sect. 5.5). Although originally meant to prove the existence of ether and ending as a failure, this experiment can still help us derive part of the Lorentz transformations and understand the phenomena of time dilation and length contraction.

5.8   Time Dilation and Length Contraction

Let's now take a look at the Michelson-Morley experiment in Fig. 5.9a. The instruments are stationary in relation to the S' inertial frame of reference, which moves towards the right (i.e. along the $+x$ direction) relative to S at the constant speed of v. In this section, we will be conducting analyses from the perspectives of both reference frames to derive the equations for transforming coordinates between them. First, light is sent out from the source on the left side and divided into two beams that travel up and right after passing through the splitter. Suppose the axes of S and S' are overlapped at the moment when the light reaches the splitter ($t = t' = 0$), where the origins of both frames are located. Then, when the two divided beams hit the mirrors on the upper and right-hand side, they will both be reflected back to the splitter, so we can represent the upper beam's return to the splitter as event 1, and that of

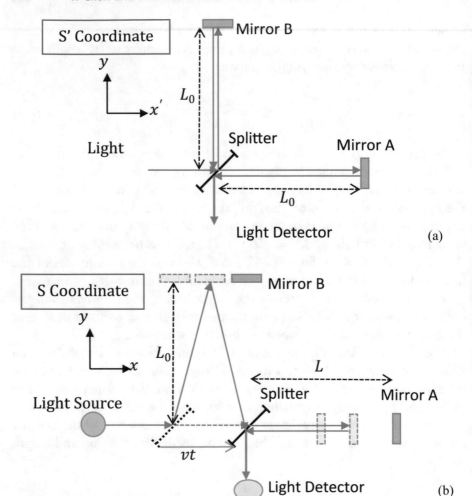

Fig. 5.9 Thought experiment explaining time dilation and length contraction with Michelson-Morley experiment

the other beam event 2. Now that we already know that the distances from the splitter to both mirror A and B are L_0 in S', we can be sure that in this reference frame, events 1 and 2 happen at the same place simultaneously:

$$t_1' = t_2' = \frac{2L_0}{c} \tag{5.43}$$

Simply put, events 1 and 2 have the same spacetime coordinates, meaning that their time and location of occurrence should remain the same across different reference frames.

Next, we will move to S to analyze the same experiment with Fig. 5.9b. Before trying to transform a frame of reference, we need to first confirm the unit of length in it to ensure valid results. Per previous discussions, the length of an object that's parallel to the y-axis doesn't change in the process of transformation, so we know the distance from the splitter to mirror B remains to be L_0. However, the distance to mirror A might change as it sits on the direction of the relative speed v. In order to figure out how the value will change exactly, let's first represent this distance (in S) as L and assume it to have been measured according to our operational definition of length. As the whole set of instruments moves to the right at the speed of v in S, if we observe the experiment from this frame, the route of the light that travels to mirror B before returning to the splitter is composed of two slant lines. If we represent time of occurrence for event 2 (the moment when the reflected light reaches the splitter) in S as t_2, the length of each slant line can be expressed as $ct_2/2$ because light speed is c, and, with the Pythagorean theorem, we can derive the below equation based on the triangular shape in Fig. 5.9b:

$$L_0^2 + \left(\frac{vt_2}{2}\right)^2 = \left(\frac{ct_2}{2}\right)^2 \tag{5.44}$$

From (5.44), the solution of t_2 can be found to be:

$$t_2 = \frac{2L_0}{\sqrt{c^2 - v^2}} = \frac{\frac{2L_0}{c}}{\sqrt{1 - \frac{v^2}{c^2}}} \tag{5.45}$$

Then, the below equation can be derived after comparing (5.43) and (5.45):

$$t_2' = t_2\sqrt{1 - \frac{v^2}{c^2}} \tag{5.46}$$

On the other hand, the time needed for the different light to travel from the splitter to mirror A and back to its original spot in S can be calculated with two components: Δt_+ (from the splitter to mirror A) and Δt_- (from mirror A back to the splitter):

$$c\Delta t_+ = L + v\Delta t_+ \tag{5.47}$$

$$c\Delta t_- = L - v\Delta t_- \tag{5.48}$$

Then, with (5.47) and (5.48), we can find the solutions to be:

$$\Delta t_+ = \frac{L}{c - v} \tag{5.49}$$

$$\Delta t_- = \frac{L}{c + v} \tag{5.50}$$

Since events 1 and 2 take place simultaneously, we know that $t_1 = t_2 = \Delta t_+ + \Delta t_-$, which can help us derive the below equation based on (5.45), (5.49), and (5.50):

$$L = L_0 \sqrt{1 - \frac{v^2}{c^2}} \tag{5.51}$$

Equation (5.51) embodies a phenomenon called "length contraction", which refers to the fact that the length of an object that's in relative motion to an observer at the speed of v will be measured to be $L_0\sqrt{1 - v^2/c^2}$ times shorter than its stationary length by this observer.

With (5.51), we can then develop the equations for the constants A, B, C, and D. Since the experiment instruments in Fig. 5.9a are stationary relative to S', we can combine (5.42) and (5.51) before substituting $\Delta x' = L_0$ and $\Delta x = L$, deriving:

$$A = \frac{1}{\sqrt{1 - \frac{v^2}{c^2}}} \tag{5.52}$$

And because S' is in relative motion to S at the speed of v along the direction of $+x$, we know that the coordinates that satisfy $x = vt$ in S is precisely the origin of S' ($x' = 0$). By applying this relation to (5.39), we can derive:

$$B = -vA \tag{5.53}$$

Then, the substitution of (5.52) and (5.53) back into (5.39) will yield:

$$x' = \frac{x - vt}{\sqrt{1 - \frac{v^2}{c^2}}} \tag{5.54}$$

Now that we've developed (5.54), here's a way to consider the physical meaning it contains: a ruler that's l in length and stationary relative to S' is placed in parallel with the x'-axis, its two ends positioned at $x' = 0$ and $x' = l$ respectively. When observed from S, the perceived length will shorten to $l\sqrt{1 - v^2/c^2}$ due to the contraction effect, so the x coordinates of the ruler's two ends will become $x = vt$ and $x = vt + l\sqrt{1 - v^2/c^2}$. Since any random value of l should be able to satisfy the second equation, we can suppose l equals the x' coordinate that we'd like to transform, thereby rewriting the original equation as $x = vt + x'\sqrt{1 - v^2/c^2}$, essentially the same as (5.54).

Up to this point, we've noticed something important: if S' can move relative to S at the speed of v, we can also see S as moving at $-v$ in relation to S'. Therefore, if we replace v, x, and t in (5.54) with $-v$, x', and t', the result should also be a valid transformation equation:

$$x = \frac{x' + vt'}{\sqrt{1 - \frac{v^2}{c^2}}} \qquad (5.55)$$

Then, the substitution of (5.54) into (5.55) leads us to:

$$x = \frac{\frac{x - vt}{\sqrt{1 - \frac{v^2}{c^2}}} + vt'}{\sqrt{1 - \frac{v^2}{c^2}}} \qquad (5.56)$$

which can be expressed as the below equation after some transposition of terms:

$$t' = \frac{t - \frac{v}{c^2}x}{\sqrt{1 - \frac{v^2}{c^2}}} \qquad (5.57)$$

Since (5.57) and (5.41) are both solutions to t', a simple comparison of them can help confirm the values of C and D. As such, we've got the equations for transforming all four coordinates from the S to S' frame of reference, i.e. (5.28), (5.29), (5.54), and (5.57), and to conduct the transformation the other way around (S' to S), we simply need to follow how we converted (5.54) into (5.55) and replaced v, x, and t with $-v$, x', and t'. Listed below in (5.58) are all the equations that we have derived, which constitute the famous

Lorentz transformations:

$$\begin{cases} x' = \dfrac{x-vt}{\sqrt{1-\frac{v^2}{c^2}}} \\ y' = y \\ z' = z \\ t' = \dfrac{t-\frac{v}{c^2}x}{\sqrt{1-\frac{v^2}{c^2}}} \end{cases} \text{or} \begin{cases} x = \dfrac{x'+vt'}{\sqrt{1-\frac{v^2}{c^2}}} \\ y = y' \\ z = z' \\ t = \dfrac{t'+\frac{v}{c^2}x'}{\sqrt{1-\frac{v^2}{c^2}}} \end{cases} \tag{5.58}$$

For conciseness, we often use the parameters $\beta = \frac{v}{c}$ and $\gamma = \dfrac{1}{\sqrt{1-\beta^2}}$ to simplify these equations as:

$$\begin{cases} x' = \gamma(x - \beta t) \\ y' = y \\ z' = z \\ ct' = \gamma(ct - \beta x) \end{cases} \text{or} \begin{cases} x = \gamma(x' + \beta t') \\ y' = y \\ z' = z \\ ct = \gamma(ct' + \beta x') \end{cases} \tag{5.59}$$

After confirming the values of A, B, C, and D with (5.51), we can now use the Lorentz transformations to examine the thought experiment in Fig. 5.9 and see if the same conclusion will be reached. Let's first look at (5.46), which represents the time of occurrence for event 2 (when the light reflected by mirror B arrives at the splitter), by adopting the perspective of observers in S and S':

(1) The observer in S: In this frame, event 2 happens where $x_2 = vt_2$. Therefore, we can apply this relation to the below equation in the Lorentz transformations:

$$t'_2 = \dfrac{t_2 - \frac{v}{c^2}x_2}{\sqrt{1 - \frac{v^2}{c^2}}} \tag{5.60}$$

And derive the result $t'_2 = \sqrt{1 - v^2/c^2}\,t_2$, which fits (5.46).

(2) The observer in S': In this frame, event 2 happens where $x_2 = 0$. We can again apply this relation to the below equation in the Lorentz transformations:

$$t_2 = \frac{t_2' + \frac{v}{c^2}x_2'}{\sqrt{1 - \frac{v^2}{c^2}}} \tag{5.61}$$

And retrieve this result:

$$t_2 = \frac{t_2'}{\sqrt{1 - \frac{v^2}{c^2}}} \tag{5.62}$$

which also fits (5.46).

In fact, (5.62) represents "time dilation", which refers to the phenomenon where the time interval between two events that happen at the same place but not simultaneously in a specific reference frame measured by an observer who's in relative motion to this frame at the speed of v will dilate from the actual interval $\Delta\tau$ to $\gamma\Delta\tau$. As for $\Delta\tau$, which is the measurement done from the frame where the events take place, it's called the "proper time interval". One thing to be noted here is that (5.60) can only be applied when transforming a proper time interval to its improper counterpart because (5.62) isn't true when $x_2' \neq 0$. Let's consider another typical example often cited in the discussion of relativity: Suppose there's a spacecraft in S' moving relative to the Earth, which sits in S, and sends out a signal at $t' = t_1'$ to inform people on the planet that the current time in the spacecraft is t_1' before doing it again at $t' = t_2'$. If the distance between the spacecraft and Earth is L at the time of occurrence for both signals, the time needed for people on Earth to receive them should both be L/c.[18] Now, if we transform t_1' and t_2' into t_1 and t_2 through the Lorentz transformations, we know that people on Earth will receive signals 1 and 2 at $t = t_1 + L/c$ and $t = t_2 + L/c$ respectively.

According to the equation of time dilation $t_2 - t_1 = \gamma(t_2' - t_1')$, the observers on Earth are supposed to think clocks in the spacecraft move slower than those on the planet. In fact, it's not just clocks but also everything else in the spacecraft. For example, the astronauts' breaths, heartbeats, as well as

---

[18]One thing to be particularly noted here is that time dilation isn't caused by the time interval between two different signals but simply an inevitable result of the Lorentz transformations. To determine the moment when people on Earth actually receive a certain signal, however, the time needed for that signal to travel to Earth is indeed a crucial factor. For example, when we lay our eyes on Vega, which is 25.3 light years far from us, what we see is in fact how the star looked 25.3 years ago.

thinking and aging speeds, will all be perceived as slower by the same percentage[19] due to the effect of time dilation. However, it should again be emphasized that all these events mentioned above only feel slower for the observers on Earth, not for the astronauts themselves. Take heartbeat as an example. If we assume an astronaut has an average resting heart rate of 72 beats a minute and the spacecraft is moving at the speed of $v = 0.8c$ relative to the Earth, then from the perspective of an observer on Earth, the heart rate will become $72\sqrt{1 - 0.8^2} = 43.2$ beat/m, while the astronaut will still perceive the rate as 72 times every minute. On the other hand, since Earth is also in relative motion to the spacecraft with the speed v, people in the spacecraft are also supposed to perceive the time on Earth to be slower by the same rate. In other words, both the astronauts and people on Earth think each other's time proceed more slowly than their own, which contributes to a symmetrical situation that makes it hard to be sure whether it's the spacecraft or Earth that really is moving. Perhaps this is Nature's way of ensuring the first postulate of relativity will always be satisfied, but isn't it contradictory for the two groups of observers to both think their time proceeds faster? To answer this question, we can first try to identify the logical errors in the statements below:

(1) Observers on Earth think only $\Delta\tau/\gamma$ has passed in the spacecraft when the time passed on Earth is $\Delta\tau$.
(2) People in the spacecraft think only $\Delta\tau/\gamma^2$ has passed on Earth when the time passed in the craft is $\Delta\tau/\gamma$.
(3) If (1) and (2) were both true, the value of $\Delta\tau/\gamma^2$ would be smaller than $\Delta\tau$, causing contradiction.

Apparently, the equation of time dilation cannot possibly satisfy both scenarios (1) and (2) because if $\Delta\tau$ isn't 0, either Δx or $\Delta x'$ can't be 0, either, which is to say that either Δt or $\Delta t'$ is not a proper time interval. Therefore, the list above should be rephrased as below for clarification:

(1) Suppose there are two events A and B that happen on Earth with the time interval of $\Delta\tau$ (measured with a clock on Earth). If observers on Earth perceive the events C and D, which happen in the spacecraft, to take place simultaneously with A and B respectively, then the time interval between C and D will be $\Delta\tau/\gamma$ (measured with a clock in the spacecraft).
(2) Suppose there are two events C and D that happen in the spacecraft with the time interval of $\Delta\tau/\gamma$ (measured with a clock in the ship). If

---

[19] If not, astronauts would be able to perceive that it is the spaceship, rather than the Earth, which is moving, causing the situation to obviously go against the first postulate of relativity.

observers in the spacecraft perceive the events E and F, which happen on Earth, to take place simultaneously with C and D respectively, then the time interval between E and F will be $\Delta\tau/\gamma^2$ (measured with a clock on Earth).

(3) Since the definitions of "simultaneity" are different on Earth and in the spacecraft, E and F do not necessarily happen at the same time as A and B even if observers think they do. In other words, there's no point in comparing the time intervals on Earth and in the spacecraft.

The above discussion serves as a reminder that whether it's the equation of length contraction or time dilation, we always need to examine discreetly if they're applicable to specific scenarios, and of course, the easiest way, which is also least prone to fallacies, is to always adopt an operational definition when discussing different events and figure out the answer with the Lorentz transformations.

Next up, we're going to use the Lorentz transformations to verify (5.27), which we derived through the train-platform thought experiment. Suppose the reference frame of the train S' is in relative motion to that of the platform S at the speed of v toward the direction of $+x$. Below are two different approaches to analysis that'll lead us to the same result:

(1) Approach 1: In the platform frame of reference S, events 1 (light source A turns on) and 2 (light source B turns on) happen at the same time, which is why we have $\Delta t_{1-2} = 0$; in the train frame of reference S', the distance between the places of occurrence for events 1 and 2 is $\Delta x'_{1-2} = -L$. With (5.61) in the Lorentz transformations and some subsequent calculations, we can obtain:

$$\Delta t_{1-2} = \frac{\Delta t'_{1-2} + \frac{v}{c^2}\Delta x'_{1-2}}{\sqrt{1 - \frac{v^2}{c^2}}} = 0 \qquad (5.63)$$

As the numerator in (5.63) is 0, we know that:

$$\Delta t'_{1-2} = -\frac{v}{c^2}\Delta x'_{1-2} = \frac{v}{c^2}L \qquad (5.64)$$

Which fits (5.27), hence the verification.

(2) Approach 2: With (5.60) in the Lorentz transformations and some subsequent calculations, we can obtain:

$$\Delta t'_{1-2} = \frac{\Delta t_{1-2} - \frac{v}{c^2}\Delta x_{1-2}}{\sqrt{1 - \frac{v^2}{c^2}}} = \frac{-\frac{v}{c^2}\Delta x_{1-2}}{\sqrt{1 - \frac{v^2}{c^2}}} \tag{5.65}$$

And again, with (5.54) in the Lorentz transformations, we can derive:

$$\Delta x_{1-2} = \sqrt{1 - \frac{v^2}{c^2}}\Delta x'_{1-2} + v\Delta t_{1-2} = -L\sqrt{1 - \frac{v^2}{c^2}} \tag{5.66}$$

Then, the substitution of (5.66) into (5.65) will lead us to (5.27). Judging from the above derivation, we can be sure that no matter what approach or which perspective (frame of reference) we adopt, as long as we can apply the Lorentz transformations correctly, the result will always come out the same regardless of how many equations are used in the process, which clearly shows the consistency and elegant symmetry of such transformation.

Last but not least, we can't call the derivation complete without examining whether the Lorentz transformations comply with the two basic postulates of relativity. Even though we've based our entire derivation solely on both of them, there's no proof that they can be satisfied at the same time. As the last step, therefore, we need to verify if these postulates are generally applicable to all scenarios.

For Postulate 1, the laws of physics don't change, even for objects moving in inertial frames of reference. There's no possible way, in other words, to design an experiment to determine whether a reference frame is stationary or moving at a constant speed. In fact, not even once did we mention that a particular object or reference frame is moving with the speed v in the whole process of deriving the Lorentz transformations. Instead, we always phrase such statements as something like "in relative motion to a certain object or reference frame at the speed of v," which fully embodies the essence of "relativity" and makes sure the first postulate is satisfied. As for how to construct a set of physical laws regarding momentum and energy conservation that remain invariant in all transformation, we'll provide more details in the following section.

As for Postulate 2, the light in vacuum constantly travels at the speed of c in any frame of reference. This (the constancy of light speed) is the most groundbreaking hypothesis in the theory of relativity. To examine if it's true, we need to first derive the transformation equation for speed as a part of the Lorentz transformations. Suppose a certain point mass moves at the speed of

$\vec{u} = (u_x, u_y, u_z)$ in the S frame and $\vec{u}' = \left(u'_x, u'_y, u'_z\right)$ in S'. If we differentiate the Lorentz equation of transformation over time, the expressions for the three components of speed can be derived as:

$$u'_x = \frac{dx'}{dt'} = \frac{\frac{dx-vdt}{\sqrt{1-\frac{v^2}{c^2}}}}{\frac{dt-\frac{v}{c^2}dx}{\sqrt{1-\frac{v^2}{c^2}}}} = \frac{\frac{dx}{dt} - v}{1 - \frac{v}{c^2}\frac{dx}{dt}} = \frac{u_x - v}{1 - \frac{v}{c^2}u_x} \qquad (5.67)$$

$$u'_y = \frac{dy'}{dt'} = \frac{dy}{\frac{dt-\frac{v}{c^2}dx}{\sqrt{1-\frac{v^2}{c^2}}}} = \frac{u_y\sqrt{1-\frac{v^2}{c^2}}}{1 - \frac{v}{c^2}u_x} \qquad (5.68)$$

$$u'_z = \frac{dz'}{dt'} = \frac{dz}{\frac{dt-\frac{v}{c^2}dx}{\sqrt{1-\frac{v^2}{c^2}}}} = \frac{u_z\sqrt{1-\frac{v^2}{c^2}}}{1 - \frac{v}{c^2}u_x} \qquad (5.69)$$

So if a certain mass point moves at light speed in S, i.e. $\left|\vec{u}\right| = \sqrt{u_x^2 + u_y^2 + u_z^2} = c$, its motion speed in S' will be:

$$\left|\vec{u}'\right| = \sqrt{u'^2_x + u'^2_y + u'^2_z}$$

$$= \sqrt{\left(\frac{u_x - v}{1 - \frac{v}{c^2}u_x}\right)^2 + \left(\frac{u_y\sqrt{1-\frac{v^2}{c^2}}}{1 - \frac{v}{c^2}u_x}\right)^2 + \left(\frac{u_z\sqrt{1-\frac{v^2}{c^2}}}{1 - \frac{v}{c^2}u_x}\right)^2}$$

$$= \frac{\sqrt{\left(u_x^2 + u_y^2 + u_z^2\right) + v^2\left(1 - \frac{u_y^2}{c^2} - \frac{u_z^2}{c^2}\right) - 2u_x v}}{1 - \frac{v}{c^2}u_x}$$

$$= \frac{\sqrt{c^2 + v^2\left(\frac{u_x^2}{c^2}\right) - 2u_x v}}{1 - \frac{v}{c^2}u_x} = c \qquad (5.70)$$

With the above derivation, we've successfully proved that the speed of light is indeed constant under the framework of the Lorentz transformations.

As a complementary step, we'd like to provide an additional way of examining the validity of the Lorentz transformations for those who're interested: in the framework of the Galilean transformation, transforming a set of coordinates moving at the speed of v toward the direction of $+x$ for two consecutive times yields the same result as transforming only once the same set of coordinates motioning at the speed of $2v$ along the same direction. So when it comes to the Lorentz transformations, will the result remain the same? Theoretically, it has to because otherwise, such transformation would make no sense whatsoever. The only thing to be noted is that $2v$ has to be replaced with the velocity-addition formula derived from (5.67). What we need to determine here is: if $u'_x = v$ in S', what will be the value of v in S? To answer this question, let's first reverse-transform (5.67) as:

$$u_x = \frac{u'_x + v}{1 + \frac{v}{c^2} u'_x} \tag{5.71}$$

Then, the replacement of u'_x with v in (5.71) will lead us to:

$$v' = \frac{v + v}{1 + \frac{v}{c^2} v} = \frac{2v}{1 + \frac{v^2}{c^2}} \tag{5.72}$$

As a next step, we can try to resolve this question: under the framework of the Lorentz transformations, if we conduct two consecutive transformations to a set of coordinates moving at the relative speed of v toward the direction of $+x$ and only one transformation to the same coordinates motioning at the relative speed of $2v/(1 + v^2/c^2)$ along the same direction, will we obtain the same results? For those who're interested, we encourage you to try determining the answer through self-devised verification steps.

Something worth our attention is that, when the relative speed between two frames of reference is far smaller than light speed, the effect of relativity is extremely hard to observe. For example, since we all know that a train's speed can't possibly be compared with that of light, there's no way for us to see its length change when in motion. As another way to verify the validity of the Lorentz transformations, we can try to determine whether his equations will be degraded to those in the Galilean transformation when $v \ll c$. The answer is positive, but we will leave the examination process to readers who'd like to dig deeper into this topic. In contrast, the effect of relativity becomes very obvious when objects motion at enormous speeds, and the closer the speed is to that of light, the less negligible the effect is. Such scenarios will start to surface in later sections, so it's necessary to have a well-structured system of mathematical language to regulate related discussion.

Such mathematical language is closely related to three properties derived by Einstein when he was developing the theory of relativity: the invariance of distance, the causal relationship, and invariance. The so-called "distance" here doesn't refer to that in three-dimensional space but the distance between two events in four-dimensional spacetime, and according to Einstein, this physical quantity should remain invariant across all inertial frames of reference. Regarding its definition, details will be provided later on. As for the causal relationship, it refers to the notion that the order of all events has to be maintained and can't be reversed under any circumstances. Last but not least, "invariance" addresses the concept that invariant physical quantities (such as momentum and energy, which both satisfy the principle of conservation when required conditions are met) and laws should not change in any way due to the transformation of inertial reference frames.

5.9 Special Relativity Concept 1: Invariance of Spacetime Distance

In everyday language, "distance" is understood as the straight line distance between two points. In the Euclidean space, the distance (Δl) between the two points in a three-dimensional Cartesian coordinate system (x_1, y_1, z_1) and (x_2, y_2, z_2) is defined as:

$$\Delta l = \sqrt{\Delta x^2 + \Delta y^2 + \Delta z^2} \tag{5.73}$$

In this equation, $\Delta x = x_1 - x_2$, $\Delta y = y_1 - y_2$, and $\Delta z = z_1 - z_2$. In the world of Newtonian mechanics, Δl remains invariant in the process of the Galilean transformation even though the coordinates do change if the axes are randomly translated or rotated. Under the framework of special relativity, however, Δl may be different after the Lorentz transformations, which is why we need to figure out a way to redefine "distance" so that Δl stays the same even after the transformation of coordinates and random translation or rotation of frame axes.

Since the major breakthrough of Einstein's theory of relativity lies in his integration of time and space, it stands to reason that to derive an equation for a physical quantity that stays invariant in the transformation of coordinates, we need to take both the elements into consideration. However, since what we aim to discuss here is the concept of distance, to include the factor of "time", we have to first align its fundamental unit with that of "distance" because the latter is the former multiplied by speed. To eliminate an unnecessary degree of freedom, "speed" must be a non-changing value here, which leaves us one choice only: the light speed. Therefore, we can express the spacetime coordinates of any event as (ct, x, y, z), so that the fundamental units of all four axes are measurements of length, and the distance between two events becomes the function of $c\Delta t$, Δx, Δy, and Δz. In this four-dimensional spacetime system, we can rewrite the distance Δs as (5.74) following how we've defined the distance Δl in three-dimensional space with (5.73):

$$\Delta s^2 = (c\Delta t)^2 + \Delta x^2 + \Delta y^2 + \Delta z^2 \qquad (5.74)$$

If we substitute (5.59) in the Lorentz transformations into (5.74), however, we'll see that (5.74) doesn't represent an invariant in the transforming process of coordinates because:

$$\begin{aligned}
(c\Delta t')^2 &+ \Delta x'^2 + \Delta y'^2 + \Delta z'^2 \\
&= \left[\gamma(c\Delta t - \beta\Delta x)\right]^2 + \left[\gamma(\Delta x - \beta\Delta t)\right]^2 + \Delta y^2 + \Delta z^2 \\
&\neq (c\Delta t)^2 + \Delta x^2 + \Delta y^2 + \Delta z^2
\end{aligned} \qquad (5.75)$$

As the cross term containing $\Delta x \Delta t$ in (5.75) cannot be eliminated, Δs is not the same before and after the Lorentz transformations, meaning that (5.74) isn't the correct equation for expressing length as an invariant.

Now that we know the traditional definition of distance doesn't apply, let's switch to a different point of view and approach this topic with a thought experiment. Suppose you're taking a high-speed train traveling from your home-town to the working city, and the time on your watch the moment you get on and off the train in the two cities are 8 AM (event A) and 10 AM (event B) respectively. From the perspective of the Earth reference frame, if the train moves at a constant speed of v from A to B without any change of direction, how can we find an equation to express the spacetime distance Δs between events A and B so it's perceived to be the same by all observers, no matter it's your family at home, your colleagues at work, yourself when sitting in the train, or even aliens flying at a constant speed in the outer space?

Now, let's first look at this scenario from your perspective. As events A and B both happen in the train car where you sit, their places of occurrence are essentially the same for you, which is to say that the two hours that passed by is precisely the proper time interval ($\Delta \tau = 2$). For your family at home, however, the spatial distance between the events is 300 km, while the time interval is supposed to be a bit longer than two hours due to the effect of time dilation. Specifically speaking, they will think the time proceeds slower on the train than at home by a percentage of $\sqrt{1 - v^2/c^2}$. Therefore, the relationship between $\Delta \tau$ and Δt can be expressed as:

$$\Delta \tau = \sqrt{1 - v^2/c^2}\, \Delta t \qquad (5.76)$$

So far, it seems like (5.76) is the only result we can get, and some may find the v in it disturbing because the equation for the spacetime distance Δs should apply equally to all scenarios regardless of the relative speed between different frames. Yet in fact, v can be represented with Δx, Δy, Δz, and Δt. In this case, for example, the speed equals 300 km divided by Δt and can be written as:

$$v = \frac{\sqrt{\Delta x^2 + \Delta y^2 + \Delta z^2}}{\Delta t} \qquad (5.77)$$

We can then substitute (5.77) back into (5.76) and obtain (5.78):

$$c^2 \Delta \tau^2 = c^2 \Delta t^2 - \left(\Delta x^2 + \Delta y^2 + \Delta z^2 \right) \qquad (5.78)$$

Since the expression on the right side of the equal sign in (5.78) applies to all observers who're stationary or in a constant state of motion (such as your family and the aliens in this example), $c^2 \Delta t^2 - (\Delta x^2 + \Delta y^2 + \Delta z^2)$ should be an invariant. Therefore, we can define the opposite number[20] of it as Δs, the spacetime distance between events A and B:

$$\Delta s^2 = -c^2 \Delta \tau^2 = -c^2 \Delta t^2 + \Delta x^2 + \Delta y^2 + \Delta z^2 \qquad (5.79)$$

Please note that in this equation, Δs^2 can be positive or negative. As for its physical meaning, details will be provided later. If we substitute (5.59) in

---

[20] In some publications, Δs is defined as $\Delta s^2 = c^2 \Delta \tau^2 = c^2 \Delta t^2 - \Delta x^2 - \Delta y^2 - \Delta z^2$, which is only different from our definition by one negative sign. However, this doesn't really affect the physical meaning of the equation. More important is that one needs to be clear which definition is used for related calculations.

the Lorentz transformations into (5.79):

$$-\left(c\Delta t'\right)^2 + \Delta x'^2 + \Delta y'^2 + \Delta z'^2$$
$$= -\left[\gamma(c\Delta t - \beta\Delta x)\right]^2 + \left[\gamma(\Delta x - \beta\Delta t)\right]^2 + \Delta y^2 + \Delta z^2$$
$$= -(c\Delta t)^2 + \Delta x^2 + \Delta y^2 + \Delta z^2 \qquad (5.80)$$

This shows that the distance does remain unchanged through the transformation of coordinates, just as we've expected.

As the idea of describing distance in four-dimensional spacetime with (5.79) was proposed by Minkowski,[21] the spacetime geometry expressed through this equation is also called the Minkowski space, which deeply inspired Einstein when he was later developing the theory of general relativity (for details, please see Chap. 6). In addition to constructing such a geometric space, Minkowski also came up with a diagram to illustrate the relationship between space and time, named the Minkowski diagram. If we only consider the relative motion on the x-axis, the y and z coordinates remain the same. Therefore, we can disregard the two dimensions for now and simply draw the ct and x axes in the S inertial frame of reference, which is shown as the two-dimensional diagram shown in Fig. 5.10. Meanwhile, we can also draw the corresponding axes in other reference systems, such as ct' and x' in S', as well as ct'' and x'' in S". In Fig. 5.10, for example, S' and S" are in relative motion to S toward the directions of $+x$ and $-x$ at the speeds of v and $-v$ respectively. One thing to be noted is that all three frames have to be overlapped when $t = t' = t'' = 0$. Otherwise, they won't have the same origin.

To prove that the axes in S and S" shown in Fig. 5.10 aren't orthogonal, let's first go through the below process. Due to (5.58) in the Lorentz transformations, the x'-axis in S' should satisfy $t' = \left(t - \frac{v}{c^2}x\right)/\sqrt{1 - v^2/c^2} = 0$, which leads us to the expression of x' as $ct = vx/c$ in the illustration (the line passing the origin with the slope of β). On the other hand, since the ct'-axis is also supposed to satisfy $x' = (x - vt)/\sqrt{1 - v^2/c^2} = 0$, its expression can be derived to be $x = vt$, making it the line passing the origin with the

---

[21] Minkowski was Einstein's professor of mathematics in the University of Zurich and had in fact called this student a lazy dog due to his indifference in class. However, Minkowski later played an indispensable role in the development of relativity theory by constructing a space tailored to Einstein's postulate regarding the order of events. Unfortunately, he died in 1909 of acute appendicitis at the age of 44, too early for him to take up the challenge of abstruse mathematical calculations required of general relativity.

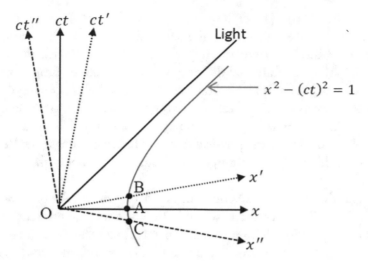

Fig. 5.10 Minkowski's hyperbola and three coordinate systems

slope of $1/\beta$.[22] As for S", we can draw its two axes following this approach, but please note that neither will be orthogonal to its counterpart in the S' frame.

Now that Fig. 5.10 is complete, we can use it to explain the train-platform thought experiment. When the train and platform are stationary relative to S and S' respectively,[23] the two sources of light, which are turned on simultaneously but at different locations in S', can be represented with O and B in Fig. 5.10. Yet when projected onto the ct-axis, the two points correspond to different values of t, meaning the two lights (events O and B) are not sent out at the same time in S. More specifically, the x coordinate of B should be L since L is the distance between the two events in S, the reference frame of the train. Then, as B falls on x', we can calculate its ct coordinate to be Lv/c with the x' slope of v/c, which again leads us to (5.27).

One thing worth our attention is that in the Minkowski diagram, the length between any two points is defined by (5.79), which is different from how the Euclidean space defines it. According to (5.79), any point on the hyperbola $x^2 - (ct)^2 = 1$ in Fig. 5.10 has the same distance with the origin. Now, let's consider A, B, and C in the illustration, whose coordinates

---

[22]Intuitively speaking, even without the Lorentz transformations, we can still be sure that $x' = 0$ will correspond to $x = vt$ based on the definition of relative speed.

[23]Please note that S and S' are opposite to each other in our verification of the Lorentz transformations. Such adjustment is made for clear illustration in the Minkowski diagram. For better understanding, we can imagine that the platform (S') is in relative motion to the train (S) at the speed of v towards the direction of $+x$ (left).

are $(x, ct) = (1, 0)$, $(x', ct') = (1, 0)$, and $(x'', ct'') = (1, 0)$ respectively. Since all three points have the same distance with the origin O, we can define the distance from the hyperbola's cross point with the x-axis of any inertial reference frame to the origin O as the length unit of that axis, which is why the curve is also called the "unit hyperbola." During the transformation between frames of reference, however, such a unit of length will be changed as well, which is why we need to multiply any distance by the percentage of change for the physical quantity to remain the same. Below, we will use time dilation as an example to explain how to apply the Minkowski diagram.

Let's consider the events O and B in Fig. 5.11, which take place at $x' = 0$ in S'. The time difference $\Delta\tau$ between both of them is the proper time interval in this frame. To derive an equation to express the time interval in S (Δt), we need first to assume the angle formed by ct' and ct to be θ (i.e. $\tan\theta = \beta$), rather than simply claim that $\Delta t = \Delta\tau \cos\theta$ just because B can be projected onto the ct-axis as C. As we've mentioned before, the Minkowski diagram and Euclidean space work differently, so any axis has to be calibrated to the length unit of the frame it belongs to before any comparison can be conducted. In Fig. 5.11, for example, the ct'-axis intersects with the hyperbola at A, whose coordinates in S' and S are $(x', ct') = (0, 1)$ and $(x, ct) = (\gamma\beta, \gamma)$ respectively. As a result, the unit of length of ct' is that of ct multiplied by $\sqrt{(\gamma\beta)^2 + \gamma^2} = \sqrt{(1 + \beta^2)/(1 - \beta^2)}$. In other words,

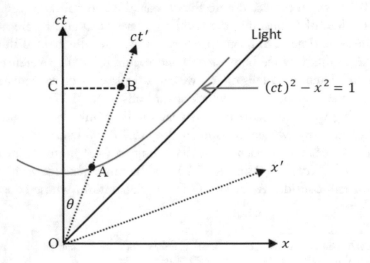

Fig. 5.11 Lorentz transformations in Minkowski diagram

the spacetime relationship between the two frames can be expressed as:

$$\Delta t = \left(\Delta \tau \sqrt{\frac{1 + \beta^2}{1 - \beta^2}} \right) \cos \theta = \frac{\Delta \tau}{\sqrt{1 - \beta^2}} \tag{5.81}$$

As you might have noticed, (5.81) happens to be the equation of time dilation as well. In the example above, the proper time interval in S' was transformed into that in S, but even with the inverse transformation (S to S'), the result will also remain the same.

With a similar approach, we can also derive the equations of length contraction (5.51) and speed transformation (5.67) on the Minkowski diagram, which is basically equivalent to the Lorentz transformation. While Lorentz formulated clear equations for transforming between frames of reference, Minkowski provided a framework for us to understand the actual meaning of such transformation in a visualized way. Therefore, we will also be using his diagram to explain another important principle of special relativity, the invariance of causality.

5.10 Special Relativity Concept 2: Invariance of Causality

Again, let's recall the train-platform thought experiment described in Sect. 5.6. Back in our initial discussion, we took quite a detour in explaining why the observer D sitting midway in the train car sees light B earlier than light A, which might be confusing for some readers because there seems to be a much more intuitive way to approach the order of the events: from the perspective of the platform system, since the train is moving towards B, D is naturally closer to B than A when both lights emit, and since light speed is constant, light B would apparently arrive at D earlier than A, right? Such reasoning is indeed correct in the frame of the platform, but in the system of the train, can we still be sure that light B would hit D first? This question involves a causal relationship,[24] and since we hadn't proved the invariance of causality in Sect. 5.6, a more complicated process of verification was adopted instead.

---

[24]The invariance of causality isn't a basic postulate in the theory of relativity but is always true in any derivation developed based on the two fundamental principles. This is the main point we're trying to make in this section.

Before we dive into the topic of this section, here's a general question for us to consider: if event A happens after O in the S inertial frame of reference, is it possible for us to find a frame where the order of events reverses? This, is the main issue we'd like to address here. For some, it's perhaps tempting to answer this question by directly resorting to the Lorentz transformations. Suppose event A happens at (x, t)/ (x', t') in S/S', while event O takes place at the origin of both frames. With (5.57) in the Lorentz transformations, we're going to examine whether $t > 0$ and $t' < 0$ can coexist. For the occurring order of the two events to be reversed in S and S', the below equation has to be satisfied:

$$t < \frac{v}{c^2}x \qquad (5.82)$$

After some transposition and substitution of $v < c$,[25] we can derive from (5.82) the below equation:

$$\frac{x}{t} > \frac{c^2}{v} > c \qquad (5.83)$$

In terms of physical meaning, (5.83) represents the fact that no substance or information can travel faster than light,[26] so events O and A do not just have no relation of causality but are unrelated whatsoever. As a result, the following conclusion can be reached: if the occurring order of two events reverses in two inertial frames of reference, they must be entirely unrelated. To put it the other way around, the causality between two related events has to be invariant, which we'll be able to see more clearly in the Minkowski diagram.

When using the diagram, it's advisable to move the origin of the reference frame to the place of occurrence for one of the events (such as the point O) so we can better observe the relationship between both events. Let's take Fig. 5.12 as an example. To find out whether a certain event happens before or after O in S', we need to see if its coordinates locate above or below the x'-axis. Following this logic, we can tell from Fig. 5.12 that in the S' frame of reference, B and C take place before and after O respectively. Therefore, before trying to determine the relation of causality between two events, we

---

[25]In the Lorentz transformations, $v \geq c$ has no meaning.

[26]Regarding this statement, we'll provide more details in Sect. 5.11: Invariance of Physical Quantities. As for the speed of information, it's generally accepted to be slower than light speed but has not been proved with any solid approach. To serve our purpose, though, we'll simply assume the hypothesis to be correct for the time being.

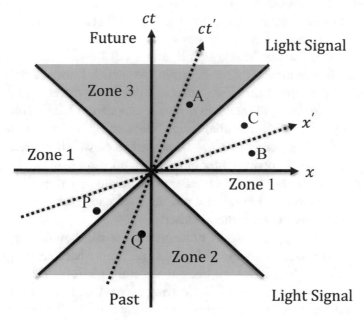

Fig. 5.12 Illustration of event causality

need first to confirm which areas of the S frame in the diagram can possibly be where x' and ct' are located.

In the framework of special relativity, the relative speed v between any two reference frames has to be slower than light speed c. In the Minkowski diagram, therefore, the slope of ct' always has to be greater than 1, meaning the positive side of the axis should fall in Zone 3 in Fig. 5.12. By contrast, since the slope of x' has to be smaller than 1, the positive side of it should locate in Zone 1.

Imagine we're observing from the point O in the spacetime system now. As all substances and information travel at lower speeds than that of light in the framework of special relativity, all events that can affect O should fall in Zone 2 and 3. Due to the definition of (5.79), the spacetime distance between any two events in these zones and O is called the "timelike interval" and has to satisfy $\Delta s^2 < 0$. As for Zone 1 events, since they don't cause any impact on O, there's no causality to be discussed anyway. However, the distance between any two events in Zone 1 and O, which is also called the "spacelike interval", has to satisfy $\Delta s^2 > 0$. What about $\Delta s^2 = 0$? Well, it represents the path of all lights that pass through O. If we draw out the third dimension (i.e. the y- or z-axis) in the Minkowski diagram as well, the collection of all lights that meet the equation $\Delta s^2 = 0$ will form a pair of cones, commonly called the "light cones".

If an event in Zone 2 can travel to O at a speed that's lower than c, then all events in the same zone can cause impact to O. Therefore, we can see Zone 2 as the "influential past" because it'll always fall below the x'-axis, which corresponds to the formula $t' < 0$, no matter which frame of reference we put it in. On the other hand, if an event in Zone 3 can travel to O at a speed lower than c, similarly, all events in this zone can affect O. Therefore, we can consider this zone the "influential future" because it'll always fall above the x'-axis, which corresponds to $t' > 0$, regardless of which reference frame. As for Zone 1, the occurring order of events happening here can be reversed due to change of frames (e.g. B can take place before and after O in S' and S respectively), but since O and B are not involved with each other, there's no relation of causality between them, either.

Up to this point, we've again proved that the Minkowski diagram and Lorentz transformations always yield the same results because they're essentially equivalent. Now, let's look back at the train-platform thought experiment and consider the events of D receiving lights A and B. Since the space-time interval between the two events is obviously timelike (because they happen at the same location in the train system), their order of occurrence must be the same whether observed from the frame of the train or the platform. We already know that from the perspective of the platform reference system, D receives light B before A. But is there any possibility that D might perceive A to arrive first? To answer this question, we can make D raise the right and left hand respectively when light A and B arrives. If A took place prior than B, there would be a brief period of time when D only had the right hand raised, but meanwhile, the observer on the platform would see D raise the right hand immediately after light B arrives, which is contradictory to our original instruction and goes against the theory of special relativity per the discussion in this section. As for the spacetime interval between the events of A and B sending out a light respectively, it is spacelike since they happen simultaneously but at different places in the platform frame. As a result, there's no fixed order of occurrence for the two events, meaning their simultaneity will become irrelevant.

5.11 Special Relativity Concept 3: Invariance of Physical Quantities

In this section, we're going to address the invariance of physical quantities. In the world of nature, there are certain physical quantities that 're invariant, such as momentum and energy. For the invariance of a particular physical quantity to stay consistent across all inertial frames of reference, this quantity has to be an invariant in any frame; to be defined as an invariant, on the other hand, a physical quantity has to satisfy certain criteria. For example, for a system to achieve momentum and energy conservation, the net external force and work exerted on the system has to be zero. Before starting to discuss the concept of invariance, we need to figure out a proper definition for momentum, energy, and working force in the framework of relativity so that these quantities stay invariant across all reference frames while satisfying the rules of Newtonian mechanics when corresponding speeds are far slower than light speed.

5.11.1 Momentum Conservation

Let's first consider a simple two-dimensional collision, as shown in Fig. 5.13a. In this coordinate system, A and B are two identical particles with the same

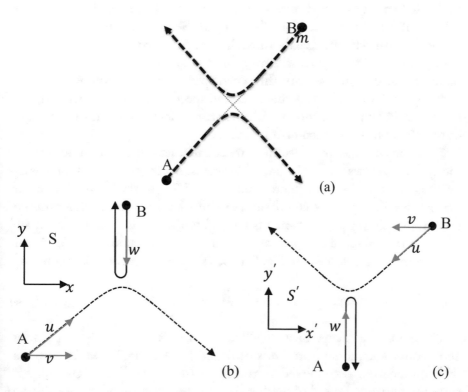

Fig. 5.13 Collision of two identical particles observed from different frames of reference: **a** Origin located outside both particles **b** Origin overlapped with B **c** Overlapped with A

mass of m that collide with each other when motioning at the same speed before parting towards opposite directions, which has to be with the same velocity for compliance with the law of momentum conservation. In addition, the speed has to remain unchanged after the collision to satisfy the law of energy conservation. Please note that the coordinate systems in Fig. 5.13a have already rotated so the horizontal speeds of the two particles are equal.

Now, we're going to observe the same event from two other different inertial frames of reference. From the perspective of S (Fig. 5.13b), which is moving toward the left-hand side with the same horizontal speed as B, the particle first moves downwards in a vertical motion with the speed w and after the collision, returns upwards following the same vertical path with the unchanged speed w. As for A, it's perceived as moving at the speed of u towards the upper ride-hand corner and colliding with B before heading towards the lower right-hand corner at the same speed of u. In this frame, the horizontal speed component of A is represented as v.

As for Fig. 5.13c, it represents the perspective of S', which has the same horizontal speed as A and moves towards the right side relative to S at the velocity of v along the direction of $+x$. In the S' frame of reference, the paths of the two particles are exactly those in Fig. 5.13b rotated by $180°$. While A moves vertically upwards with the speed w and downwards following the same path at the same velocity after participating in the collision, B motions towards the lower left-hand corner at the speed of u, collides with A, and leaves for the upper-left corner with the same velocity of u. In this reference frame, the horizontal speed of B is $-v$.

Based on the rules of Newtonian mechanics, let's first determine whether the S and S' frames satisfy the law of momentum conservation in the scenarios illustrated in the three figures above. In (5.23) of the Galilean transformation, both the y and t coordinates are invariants in the transformation of frames, meaning speed components parallel to the y-axis do not change, and since the vertical speed components of A and B are the same in Fig. 5.13a, the speed components, u, v, w, in Fig. 5.13b, c should satisfy the below equation:

$$\sqrt{u^2 - v^2} = w \qquad (5.84)$$

Therefore, the change of vertical momentum is $2mw$ for both particles, and because the motion directions of A and B are opposite, we 're sure that their momentum is conserved vertically; as for the horizontal counterpart, since the horizontal speed of neither A nor B has changed after the collision, we know the conservation law is satisfied.

Now, let's take a look at the same scenarios from the angle of special relativity. Please note that speed components along the y direction might change after the Lorentz transformations, but since we've derived the equations for speed transformation, we can represent w with u and v, just as in (5.84). In Fig. 5.13c, the $-y'$ speed component of particle B ($\sqrt{u^2 - v^2}$) can be associated with that in Fig. 5.13b (represented as w) using the equation of transformation that we've derived through the Lorentz transformations, i.e. (5.68):

$$\sqrt{u^2 - v^2} = w\sqrt{1 - v^2/c^2} \tag{5.85}$$

Since (5.85) equals (5.84) multiplied by $\sqrt{1 - v^2/c^2}$, based on the definition of momentum in Newtonian mechanics, which is $\vec{p} = m\vec{v}$, momentum conservation isn't satisfied under this situation.

Therefore, we'll need to follow the approach adopted in Sect. 5.9 and determine a new definition for momentum under the framework of relativity so the law of momentum conservation can still be satisfied after the Lorentz transformations. Let's first revisit the actual meaning of momentum and why its conservation has to be fulfilled. In Newton's second law of motion, both momentum and working force are defined according to $\vec{F} = d\vec{p}/d$. In other words, what we need to do here isn't develop an equation between force and momentum but formulate an appropriate set of definitions for these concepts so we can analyze related scenarios on a correct basis and ensure the results satisfy the laws of Newtonian mechanics when corresponding speeds are far lower compared to light speed.

In Newton's equation of momentum $\vec{F} = d\vec{p}/dt$, a force changes the motion state of the object it works on, while the motion state is defined by the object's momentum. By pushing a stationary object, we can make it move towards the direction of our force, which is to say that $\vec{p}$ and the speed $\vec{v}$ share the same direction; if not, the speed direction would change along with that of the force. As the Newtonian equation is linear in nature, both momentum and force can be superposed. Suppose the momentum of the particles 1 and 2 are $\vec{p}_1$ and $\vec{p}_2$, while the working forces acting on them are $\vec{F}_1$ and $\vec{F}_2$. Under this circumstance, we can consider these particles a system where the overall momentum and working force are $\vec{p} = \vec{p}_1 + \vec{p}_2$ and $\vec{F} = \vec{F}_1 + \vec{F}_2$, which satisfies Newton's second law of motion $\vec{F} = d\vec{p}/dt$.

Then, we can divide $\vec{F}_1$ into "the force from particle 2 that works on particle 1 $\vec{f}_{21}$ " as well as an external force exerted on particle 1 from outside the system and do the same with $\vec{F}_2$. As such we can calculate the total force acted upon the system to be:

$$\vec{F} = \vec{F}_1 + \vec{F}_2 = \left(\vec{f}_{21} + \vec{F}_{\text{out},1} \right) + \left(\vec{f}_{12} + \vec{F}_{\text{out},2} \right)$$

$$= \left(\vec{F}_{\text{out},1} + \vec{F}_{\text{out},2} \right) + \left(\vec{f}_{21} + \vec{f}_{12} \right) = \vec{F}_{\text{out}} + \left(\vec{f}_{21} + \vec{f}_{12} \right) \tag{5.86}$$

On the right side of the equal sign in (5.86), $\vec{F}_{\text{out}} = \vec{F}_{\text{out},1} + \vec{F}_{\text{out},2}$ is the total external force working on the system, while $\left(\vec{f}_{21} + \vec{f}_{12} \right)$ represents the internal force working between the particles inside the system. For $\vec{F}_{\text{out}}$, we can obtain $d\vec{p} = \vec{F}dt$ through transposition, but $\vec{F}dt = \vec{F}_{\text{out}}dt + \left(\vec{f}_{21} + \vec{f}_{12} \right)dt = 0$ also has to be met to satisfy the law of momentum conservation. In this equation, $\vec{F}_{\text{out}}dt$ refers to the momentum given from outside the system and obviously has to be zero since the system under discussion here is independent. As for $\vec{f}_{21}$ and $\vec{f}_{12}$, they represent an action and its reaction force, which, according to Newton's third law of motion, are equal but of the opposite directions while generated and disappearing at the same time[27]. Therefore, we know $\left(\vec{f}_{21} + \vec{f}_{12} \right)dt = 0$. Please note that in the law of momentum conservation, "simultaneity" is very important because the amount of momentum change is calculated by integrating force overtime, which is why there would be a short period of time when the law was violated if $\vec{f}_{21}$ and $\vec{f}_{12}$ did not work at the same time.

---

[27] On the basis of relativity, however, simultaneity might cause deviation because signals are transmitted at limited speeds. The most obvious example is force at a distance. Consider a situation where a gravitational force exists between particles A and B. If we move B even just by a little bit, the force exerted on it will still change immediately. Yet according to the theory of general relativity, the transmission speed of gravity equals that of light, which is why A will not feel any immediate change. In fact, this is exactly the reason why there exists no perfectly rigid body in the world of relativity, but with normal collisions, such scenarios are mostly approached with the assumption that the colliding objects only engage in a transient encounter when they're extremely close in distance, which is why the conservation of momentum can still be considered satisfied.

The above analysis can be applied to systems of two or more particles, but please note that the item $\left(\vec{f}_{21} + \vec{f}_{12}\right)$ has to contain the internal forces working between any pair of particles. Summarizing the above discussion, we know that a system whose net external force $\vec{F}_{out}$ is zero satisfies the law of momentum conservation as long as each force works simultaneously, which is to say that $\vec{F}$ is 0 at every moment. One particular situation is where we divide an object moving at a certain speed with the mass m into two components, each with the mass $m/2$. Despite the division, the momentum of the two smaller components combined still equals that of the original object. In other words, when the speed of a specific object remains invariant while its mass is divided in half, its momentum will also decrease to 1/2 of the original value, which tells us that momentum and mass are directly proportional.

Up to this point, we've discussed the definition of momentum as well as the physical meaning of momentum conservation. In addition, we've also found out momentum has the following properties: (1) $\vec{p}$ and $\vec{v}$ share the same direction, and (2) momentum and mass are directly proportional. Based on these findings, we can express momentum ($\vec{p}$) as the general equation below:

$$\vec{p} = f\left(\vec{v}\right)m\vec{v} \tag{5.87}$$

In (5.87), $f\left(\vec{v}\right)$ is an unknown scalar function of the speed $\vec{v}$ and can be derived based on the fact that the overall momentum of a system has to fulfill the conservation law when the net external force working on the system is 0. Another property of $f\left(\vec{v}\right)$ is that it should stay in the same form regardless of the direction of $\vec{v}$ because in an inertial reference system where spacetime is evenly distributed, there exists rotational symmetry. Simply put, $f\left(\vec{v}\right)$ is the function of v even when the speed direction changes, which is why (5.87) can be expressed as:

$$\vec{p} = f(v)m\vec{v} \tag{5.88}$$

Now, if we apply (5.88) to the scenario of a two-dimensional collision, the amount of change in A's momentum in the S frame in Fig. 5.13b should be $2f(u)m\sqrt{u^2 - v^2}$ (direction $-y$), while the change in B's momentum is $2f(w)mw$ towards the opposite direction ($+y$). As the momentum has to remain the same before and after the collision, the change in A's momentum should equal that of B:

$$2f(u)m\sqrt{u^2 - v^2} = 2f(w)mw \qquad (5.89)$$

In this step, we can replace w with u and v by applying (5.85):

$$f(u)\sqrt{1 - v^2/c^2} = f\left(\sqrt{\frac{u^2 - v^2}{1 - v^2/c^2}}\right) \qquad (5.90)$$

As shown above, (5.90) is a functional equation that stands true as long as both u and v are real numbers and $u \geq v$. To examine the physical meaning of (5.90), let's first look at the special scenarios below, starting with $u = v$, which we can substitute into (5.90)[28] and obtain:

$$f(v) = \frac{f(0)}{\sqrt{1 - v^2/c^2}} \qquad (5.91)$$

Then, the substitution of (5.91) into (5.88) yields:

$$\vec{p} = \frac{f(0)m\vec{v}}{\sqrt{1 - v^2/c^2}} \qquad (5.92)$$

For our result under the situation of $v \ll c$ to be consistent with $\vec{p} = m\vec{v}$ in Newtonian mechanics, we'll substitute the function $f(0) = 1$ into (5.91) and derive:

$$f(v) = \frac{1}{\sqrt{1 - v^2/c^2}} \qquad (5.93)$$

By substituting (5.93) into (5.90), we also find out that (5.93) does satisfy (5.90). Based on (5.88), we've established the equation for expressing

---

[28] Strictly speaking, (5.90) can't really be satisfied when $u = v$ because in the derivation of (5.90) from (5.89), we eliminated $\sqrt{u^2 - v^2}$ on both sides of the equal sign. However, we can still consider u to be approaching v. Such derivation isn't very thorough from the perspective of mathematics, but since a solution that fits all related conditions can still be derived at the end, the result is deemed sufficient for the purpose of physical analysis.

momentum under the framework of special relativity[29]:

$$\vec{p} = \frac{m\vec{v}}{\sqrt{1 - v^2/c^2}} = \gamma m\vec{v} \tag{5.94}$$

As for the working force $\vec{F}$, it can be derived by substituting (5.94) into $\vec{F} = d\vec{p}/dt$:

$$\vec{F} = \frac{d}{dt}\left(\gamma m\vec{v}\right) = m\vec{v}\frac{d\gamma}{dt} + \gamma m\frac{d\vec{v}}{dt} \tag{5.95}$$

From which we can observe the fact that $\vec{F}$ isn't proportional to acceleration $(d\vec{v}/dt)$ anymore and that the two parameters might even work in different directions.

5.11.2 Energy Conservation

In this section, let's shift our focus to the definition of energy and the physical meaning of energy conservation, starting with "kinetic energy". In classical mechanics, the total work applied on a certain object by all the forces exerted on it equals the amount of change in its kinetic energy, i.e. $\vec{F} \cdot d\vec{r} = dE_K$, in which $\vec{r}$ and $d\vec{r}$ represent the object's three-dimensional coordinates and displacement respectively. On the left-hand side of the equal sign, $\vec{F} \cdot d\vec{r}$ is defined as the work done by the working force $\vec{F}$, while on the other side of the sign is dE_K, representing the amount of change in kinetic energy. In fact, this equation is quite similar to $\vec{F}dt = d\vec{p}$ because while simultaneity

---

[29] In some publications, the γm parameter in (5.94) is phrased as "dynamic mass", which increases with an object's motion speed and becomes unlimited when the speed approaches that of light, so no working force of a limited value can cause the object to travel faster than light. If we represent the dynamic mass with the speed v as $m_v = \gamma m$, the corresponding momentum can be expressed as $\vec{p} = m_v\vec{v}$. In this book, however, we won't be using the term "dynamic mass", which Einstein himself didn't appreciate, either, because the property "mass" doesn't undergo any change in nature no matter an object is moving or stationary while the word "dynamic" often leads to the misunderstanding that the object would go through certain internal change caused by its motion. Yet in fact, such a term is nothing but an alternative arising from the use of a different spacetime geometry.

is indispensable in the conservation of momentum, what matters for energy conservation is the spatial consistency of the working force $\vec{F}$.

As such, the change in the amount of momentum in the dual-particle system can be expressed as:

$$\vec{F}_1 \cdot d\vec{r}_1 + \vec{F}_2 \cdot d\vec{r}_2 = dE_{K1} + dE_{K2} \tag{5.96}$$

In (5.96), $\vec{F}_1$ and $\vec{F}_2$ are the external forces working on particles 1 and 2 respectively, while $\vec{r}_1$ and $\vec{r}_2$ as well as E_{K1} and E_{K2} are their position vector and kinetic energy. We can then divide $\vec{F}_1$ into "the working force exerted by particle 2 on 1 ($\vec{f}_{21}$)" and the external force from outside the system that acts on particle 1 ($\vec{F}_{\text{out},1}$) while doing the same with $\vec{F}_2$ before substituting the results into (5.96) and retrieving:

$$\vec{F}_{\text{out},1} \cdot d\vec{r}_1 + \vec{F}_{\text{out},2} \cdot d\vec{r}_2 = d(E_{K1} + E_{K2}) - \left(\vec{f}_{21} \cdot d\vec{r}_1 + \vec{f}_{12} \cdot d\vec{r}_2\right) \tag{5.97}$$

On the left-hand side of the equal sign in (5.97) is the total work done by external forces to the system. As for $-\left(\vec{f}_{21} \cdot d\vec{r}_1 + \vec{f}_{12} \cdot d\vec{r}_2\right)$ on the other side of the sign, the expression is a bit more complicated. While $\vec{f}_{21}$ and $\vec{f}_{12}$ are a pair of action and reaction forces, they can be further categorized as conservative and non-conservative. The conservative group, including gravity, electrostatic force, strong interaction force, etc., can be expressed as effective potential energy U_{12}, which is why (5.97) can be rewritten as:

$$\vec{F}_{\text{out},1} \cdot d\vec{r}_1 + \vec{F}_{\text{out},2} \cdot d\vec{r}_2 = d(E_{K1} + E_{K2} + U_{12})$$

$$-\text{work done by non - conservative force} \tag{5.98}$$

As for the non-conservative group (such as friction loss), there's no specific way of calculation. In elastic collisions, normally there's no energy loss caused by non-conservative forces. If no work is done by external forces, either, then the kinetic and potential energy in the system has to be a fixed value so it satisfies the law of energy conservation.

As for the equation of the kinetic energy E_K, we can derive it through $dE_K = \vec{F} \cdot d\vec{r} = (d\vec{p}/dt) \cdot d\vec{r}$. The substitution of (5.94) into this equation will lead us to:

$$dE_K = \frac{d}{dt}\left(\frac{m\vec{v}}{\sqrt{1 - v^2/c^2}}\right) \cdot d\vec{r} \tag{5.99}$$

Simplifying[30] the expression on the right-hand side of the equal sign in (5.99) with $\vec{v} = d\vec{r}/dt$, we will obtain:

$$dE_K = d\left(\frac{mc^2}{\sqrt{1 - v^2/c^2}}\right) \tag{5.100}$$

Since the kinetic energy of a stationary object is defined as 0, we can, with the help of (5.100), derive the expression for this physical quantity to be:

$$E_K = \frac{mc^2}{\sqrt{1 - v^2/c^2}} - mc^2 = \gamma mc^2 - mc^2 \tag{5.101}$$

Denoted by (5.101) is an important limit of the universe: the kinetic energy of an object, which can be calculated from (5.101), will approach infinity if it moves at a speed approaching light speed. Therefore, the object can never be propelled to equal or even exceed light speed when acted upon by finite working forces. On the other hand, when an object moves at a speed v that's far slower than light speed, $\gamma - 1 \approx v^2/2c^2$ is true, we can therefore

---

30

$$\frac{d}{dt}\left(\frac{m\vec{v}}{\sqrt{1 - v^2/c^2}}\right) \cdot d\vec{r}$$

$$= d\left(\frac{m\vec{v}}{\sqrt{1 - v^2/c^2}}\right) \cdot \frac{d\vec{r}}{dt} = \vec{v} \cdot d\left(\frac{m\vec{v}}{\sqrt{1 - v^2/c^2}}\right)$$

$$= m\vec{v} \cdot \left(\frac{\vec{v}(-1/2)(-2v/c^2)dv}{(1 - v^2/c^2)^{3/2}} + \frac{d\vec{v}}{\sqrt{1 - v^2/c^2}}\right)$$

$$= \frac{mvdv}{(1 - v^2/c^2)^{3/2}} = d\left(\frac{mc^2}{\sqrt{1 - v^2/c^2}}\right)$$

substitute this equation into (5.101) before deriving:

$$E_K \approx \frac{v^2}{2c^2}mc^2 = \frac{1}{2}mv^2 \qquad (5.102)$$

Showing that (5.102) satisfies the expression for kinetic energy in Newtonian mechanics.

5.12 The Wonder of Special Relativity: Mass-Energy Formula

What we discussed in the previous section is a perfectly elastic collision where no change is undergone by the particles except for their speeds, which is to say that the combined mass of them is the same before and after the colliding event. In an imperfect elastic collision, however, the scenario is different. In fact, it's precisely this difference that led to a discovery that Einstein himself didn't quite expect, either: the mass-energy formula.

If we consider the scenario in Fig. 5.14, where the S frame of reference contains two objects with the same mass m that collide with each other after traveling at the same speed u along the y-axis and form the new stationary object C together. If we see the objects A and B as a system, the overall kinetic energy it contains before and after the collision is $E_{K,i} = 2mc^2\left(\frac{1}{\sqrt{1-u^2/c^2}} - 1\right)$ and $E_{K,f} = 0$ respectively. Yet according to (5.98), the work done by the external force to the system is 0 because there's simply not any. Therefore, assuming there's no non-conservative force participating in the process of collision, $E_{K1} + E_{K2} + U_{12}$ has to be a fixed value, which is to say that the kinetic energy existing before the collision will be transformed into and stored as a certain form of potential energy between A and B, as illustrated in

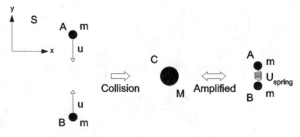

Fig. 5.14 Energy conservation in a system of dual-particle collision, observed from S frame

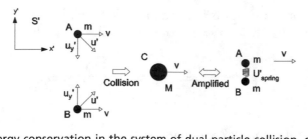

Fig. 5.15 Energy conservation in the system of dual-particle collision, observed from S' frame

Fig. 5.14. Imagine there's a small spring[31] placed between A and B which gets compressed to store the kinetic energy in the form of internal energy when the two particles collide. Based on the principle of energy conservation, the internal energy can be expressed as:

$$U_{spring} = E_K = 2mc^2 \left(\frac{1}{\sqrt{1 - \frac{u^2}{c^2}}} - 1 \right) \tag{5.103}$$

Now, let's consider a different system shown as in Fig. 5.15, where both objects move at the speed of v towards the direction of $+x'$ when observed from the S' frame of reference, which is in relative motion to S with the speed v along the $-x$ direction. After transforming from S to S' frame using the Lorentz equations, A and B's velocity along the y'-axis becomes u'_y, while their speed u' changes to $\sqrt{u'^2_y + v^2}$. Based on (5.68), we can derive:

$$u'_y = \sqrt{1 - \frac{v^2}{c^2}} u \tag{5.104}$$

Then, the integration of (5.104) and $u' = \sqrt{u'^2_y + v^2}$ will lead us to:

$$\sqrt{1 - \frac{u'^2}{c^2}} = \sqrt{1 - \frac{v^2}{c^2}} \sqrt{1 - \frac{u^2}{c^2}} \tag{5.105}$$

In this step, let's examine the preconditions for the conservation of momentum and energy in the S' reference frame. First, the momentum along

---

[31] What we assume here is a spring that's 0 in mass but able to store energy.

the direction of y'-axis is obviously an invariant, while the total momentum of A and B parallel to the x'-axis before the collision can be expressed as:

$$p'_i = \frac{2mv}{\sqrt{1 - \frac{u'^2}{c^2}}} = \frac{2mv}{\sqrt{1 - \frac{v^2}{c^2}}\sqrt{1 - \frac{u^2}{c^2}}} \qquad (5.106)$$

As for the total post-collision momentum:

$$p'_f = \frac{2mv}{\sqrt{1 - \frac{v^2}{c^2}}} \qquad (5.107)$$

From (5.106) and (5.107), we can tell that $p'_i \neq p'_f$, which is to say the momentum of the system is different before and after the collision. For the law of momentum conservation to be satisfied, we need to explain the difference, which is why we should consider the only possibility: the spring also has a momentum that can be calculated as below:

$$p'_{\text{spring}} = p'_i - p'_f = \frac{2mv}{\sqrt{1 - \frac{v^2}{c^2}}}\left(\frac{1}{\sqrt{1 - \frac{u^2}{c^2}}} - 1\right) = \frac{\left(\frac{U_{\text{spring}}}{c^2}\right)v}{\sqrt{1 - \frac{v^2}{c^2}}} \qquad (5.108)$$

As for energy conservation, the total kinetic energy of A and B before the collision is:

$$E'_{K,i} = \frac{2mc^2}{\sqrt{1 - \frac{u'^2}{c^2}}} = \frac{2mc^2}{\sqrt{1 - \frac{v^2}{c^2}}\sqrt{1 - \frac{u^2}{c^2}}} \qquad (5.109)$$

While the total post-collision kinetic energy is:

$$E'_{K,f} = \frac{2mc^2}{\sqrt{1 - \frac{v^2}{c^2}}} \qquad (5.110)$$

Following the law of energy conservation, we can calculate the elastic potential energy stored in the spring in the S reference frame:

$$U'_{\text{spring}} = E'_{K,i} - E'_{K,f} = \frac{2mc^2}{\sqrt{1 - \frac{v^2}{c^2}}}\left(\frac{1}{\sqrt{1 - \frac{u^2}{c^2}}} - 1\right) = \frac{\left(\frac{U_{\text{spring}}}{c^2}\right)c^2}{\sqrt{1 - \frac{v^2}{c^2}}} \qquad (5.111)$$

Up to this point, we've figured out several properties that the spring has in both frames of reference. In S, the spring is stationary with zero momentum and U_{spring} of energy, the value of which can be calculated through (5.103), while in S', it moves at the speed v, with the momentum p'_{spring} and energy U'_{spring}, which are expressed with (5.108) and (5.111) respectively. In other words, the spring has $\left(\frac{U_{spring}}{c^2}\right)c^2\left(\frac{1}{\sqrt{1-v^2/c^2}}-1\right)$ times more energy in S' than in S, its behavior identical to that of a particle with a mass of $m_{eff} = U_{spring}/c^2$. As a result, we can deem the system constituted by A, B, and the spring to be a particle (represented as C) with the mass $M = 2m + U_{spring}/c^2 = 2m + m_{eff}$ that satisfies the laws of momentum and energy conservation in any inertial frame of reference.

In this example, we can convert the energy U_{spring} to the equivalent mass $m_{eff} = U_{spring}/c^2$ so that both the laws of momentum and energy conservation can be fulfilled in any inertial frame of reference. In fact, it was boldly hypothesized by Einstein that mass and energy are mutually convertible. According to him, an object with the mass m can generate the amount of energy calculated through $E = mc^2$, and vice versa. This is the famous mass-energy formula, which also shows that a stationary object with the mass m has the rest energy mc^2. Meanwhile, we know from (5.101) that an object's energy is larger when it moves at the speed of v than when stationary due to the equation $E_K = \gamma mc^2 - mc^2$. Therefore, the total energy possessed by an object can be expressed as:

$$E = \text{Rest Energy} + \text{Kinetic Energy} = mc^2 + E_K = \gamma mc^2 \qquad (5.112)$$

With this equation above, let's examine whether the system of "A + B + Spring" can be used as a generally applicable example in the natural world. First of all, the spring represents the concept of potential energy, i.e. U_{12} in (5.98). In any stably existing object in the natural world, the net potential energy of the interaction between particles (U_{12}) is a negative number. Therefore, the energy contained in two particles when they're combined is lower than when both exist independently, which is why the former situation is more stable than the latter. Let's look at a specific example here: the nucleus of helium consists of two protons plus two neutrons and sees the strong attraction between any pair of particles in it, which is why the nucleus can exist in a stable state thanks to its negative potential energy. Because of the same reason, if we measure the mass of the helium nucleus, the results will come out showing $m\left(^4_2He\right)$ to be smaller than the combined mass of two protons and two neutrons ($2m_n + 2m_p$). However, as long as the nucleus isn't impacted by particles with high energy and the protons and neutrons contained therein aren't separated using other methods, it'll only render the normal behavior of a particle that has the mass of $m\left(^4_2\text{He}\right)$ without showing that a certain part of its mass is actually the potential energy from the interaction between particles. In fact, the enormous amount of energy absorbed or released in the course of nuclear fission is precisely the result of the change in potential energy during the process of the collision and reformation among nuclei. Details regarding this topic will be provided later on.

At this point, we've figured out how to express the momentum and energy contained in a particle and confirmed that they're interconvertible. However, since the above derivation is based entirely on two special cases of collision, you might wonder if our results are applicable to all scenarios alike. Below, we're going to adopt a more precise approach in order to find out the equations for transforming momentum and energy, just as we've derived the transformation equations for time and place in Sect. 5.8.

Before diving into the process, let's first reorganize the expressions for momentum and energy. For the former, i.e. (5.94), we can rewrite it as the below equation by applying (5.76):

$$\vec{p} = \frac{m}{\sqrt{1 - v^2/c^2}} \frac{d\vec{r}}{dt} = m \frac{d\vec{r}}{d\tau} \tag{5.113}$$

In (5.113), $\vec{r} = (x, y, z)$ represents the coordinates of an object in a three-dimensional space, while $d\tau$ is the proper time interval, referring to the interval between two events' time of occurrence measured from the same place. As an object is deemed stationary from its own perspective, we can interpret $d\tau$ as the time perceived to have passed from the perspective of the object's frame. As for dt, it's the time passed measured from the frame that's in relative motion to the object at the speed of v.

On the other hand, the energy of an object originally expressed with (5.112), can be rewritten following the same approach:

$$E = \frac{mc^2}{\sqrt{1 - v^2/c^2}} = mc^2 \frac{dt}{d\tau} \tag{5.114}$$

Please note that $d\tau$ is an invariant in the transformation of frames, just as we've discussed in Sect. 5.10. Therefore, (5.113) and (5.114) can be integrated to become:

$$\left(\frac{E}{c}, p_x, p_y, p_z\right) = m\frac{d}{d\tau}(ct, x, y, z) \tag{5.115}$$

As both the mass m and proper time interval $d\tau$ are invariants, and (ct, x, y, z) satisfies the Lorentz transformations, so should the coordinates $\left(E/c, p_x, p_y, p_z\right)$ whose equations of transformation will appear as follows when written in its entirety:

$$\begin{cases} p'_x = \dfrac{p_x - \frac{v}{c^2}E}{\sqrt{1-\frac{v^2}{c^2}}} \\ p'_y = p_y \\ p'_z = p_z \\ E' = \dfrac{E-vp_x}{\sqrt{1-\frac{v^2}{c^2}}} \end{cases} \text{Or} \begin{cases} p_x = \dfrac{p'_x + \frac{v}{c^2}E'}{\sqrt{1-\frac{v^2}{c^2}}} \\ p_y = p'_y \\ p_z = p'_z \\ E = \dfrac{E'+vp'_x}{\sqrt{1-\frac{v^2}{c^2}}} \end{cases} \tag{5.116}$$

If the momentum and energy defined by (5.113) and (5.114) can both fulfill the conservation laws of the two physical quantities in an inertial frame of reference, the substitution of $\left(\Delta E/c, \Delta p_x, \Delta p_y, \Delta p_z\right) = (0, 0, 0, 0)$ into (5.116) should also yield the result $\left(\Delta E'/c, \Delta p'_x, \Delta p'_y, \Delta p'_z\right) = (0, 0, 0, 0)$, which proves the equation of momentum can still be true even after the Lorentz transformations. As such, the requirement we set out at the beginning is now satisfied.

The above equations can be further simplified. Though we've rendered position and momentum vectors as $\vec{r} = (x, y, z)$ and $\vec{p} = \left(p_x, p_y, p_z\right)$ using the concept of vector in three-dimensional space, what's in question now is four-dimensional spacetime, meaning we need to extend our discussion by one dimension. As a result, the element of vector will also go from three to four-dimensional, normally called the "four-vector." For the purpose of distinction, four-vectors are expressed as the form of x^μ (with a superscript μ, the Greek letter normally used for positive integers including 0), as opposed to $\vec{r}$ (with an arrow on top) or $\mathbf{r}$ (boldface) for its three-dimensional counterpart. For a four-vector, the subscript $\mu = 0, 1, 2, 3$, signifies ordinal, rather than exponential numbers. In addition, not any set of coordinates constituted of four components can be deemed a four-vector. To be qualified, the coordinates have to satisfy the equations of the Lorentz transformations.

Here are some four vectors that are commonly seen and how they're represented: coordinates in four-dimensional spacetime, $x^\mu = \left(x^0, x^1, x^2, x^3\right) = (ct, x, y, z) = \left(ct, \vec{r}\right)$; speed in four-dimensional spacetime (four-speed), $v^\mu = \left(\frac{dx^0}{d\tau}, \frac{dx^1}{d\tau}, \frac{dx^2}{d\tau}, \frac{dx^3}{d\tau}\right) = \left(\gamma c, \gamma\vec{v}\right)$, where $\vec{v}$ is a three-dimensional speed; momentum in four-dimensional spacetime (four-momentum), $p^\mu = mv^\mu = \left(\gamma mc, \gamma m\vec{v}\right) = \left(\frac{E}{c}, \vec{p}\right)$, where $\vec{p}$ and E are expressed with (5.113) and (5.114) in special relativity. As such, the conservation of momentum and energy can be integrated and phrased as conservation of four-momentum (p^μ). In fact, the two quantities are essentially an inseparable pair in the framework of special relativity, just like the concepts of time and space. Please note that in any four-dimensional spacetime, the rate of time change is calculated by dividing vector by $d\tau$ because the latter is an invariant in the transformation of coordinates, same as the reason why we always use dt as the denominator in three-dimensional space since t isn't affected by transformation of frames, either. To be more specific, Newtonian mechanics is built on a system of three-dimensional space where time is an invariant, while special relativity is constructed with four-dimensional spacetime where the invariant is the proper time.

After the brief introduction, let's now talk about how to calculate the length of a four-vector. As pointed out in Sect. 5.10, the square of the distance between x^μ and the origin in the Minkowski space is $-(ct)^2 + x^2 + y^2 + z^2$. Similarly, the square of a four-speed can be calculated through $-(\gamma c)^2 + \gamma\vec{v} \cdot \gamma\vec{v} = -c^2$ to be a constant, which means that all objects travel at the same speed in spacetime under the structure of special relativity. Last but not least, we can obviously tell that the square of four-momentum is that of four-speed multiplied by m^2, i.e. $-\left(\frac{E}{c}\right)^2 + p^2 = -m^2 c^2$, which can also be expressed in a more common form:

$$E^2 = p^2 c^2 + m^2 c^4 \qquad (5.117)$$

Similar to three-vectors, the coordinates of a four-vector may undergo certain changes in the transformation of reference frames, but its length will always stay the same.

As the last part of our discussion, let's consider the physical meaning of (5.117). When $p = 0$, i.e. when an object is stationary, it still has the rest energy $E = mc^2$. When its mass $m = 0$, we will obtain the equation $E =$

pc, meaning that light (or photon) itself doesn't have any mass but possesses energy and momentum, while the ratio of the two is the light speed. We can also observe from either (5.94) or (5.112) that if the energy and momentum of the transmission of a certain light flash are both definite values, the flash must be traveling at light speed. If you still remember, we actually started out by assuming that light speed is a constant c at the very beginning. The result that we've just retrieved after such lengthy, detailed discussion simply proved that our original assumption is correct. Meanwhile, such a conclusion also shows how we've stuck to Einstein's two basic postulates of relativity (see Sect. 5.6) in our derivation process, which is why no conflict can be found in our entire discourse.

Now that we've got all the theoretical concepts of special relativity covered, perhaps some of you have started admiring Einstein's deep thinking power because what we discussed in the past few sections isn't just the relativity theory itself but also includes the logic of his thought experiments. In fact, his 1905 paper "On the Electrodynamics of Moving Bodies" wasn't aimed at proposing the framework of special relativity but addressing a topic regarding the mass of electrons, which had caused quite some controversy among physicists back then because the equivalent mass of an electron was known to be different when it moved horizontally and vertically in the electric field. In trying to derive the expressions for the two different masses, Einstein detailed his results in developing the theory of special relativity in the first half of the paper while dedicating the second half to solving issues of electrodynamics with those results. Back in those days, mainstream physicists were fond of a fundamental worldview based on various electronics theories derived from electromagnetics, which explained why some of them had come up with varying results as to the mass of electrons with such theories. Yet Einstein was the only one who was able to approach the issue from a simple and fundamental perspective by rethinking space and time, two physical quantities that everyone was familiar with, and defining them in a much more accurate way. With a series of well-structured logical derivation, his research yielded some very unexpected conclusions. Perhaps other physicists should've known better not to prioritize complicated theories and calculations over fundamental perspectives since physics is a realm of simplicity filled with the beauty of symmetry after all.

Nevertheless, the discovery that energy was contained in matter stayed silent for quite a while after revealed by Einstein because no scientist, not even the great discoverer himself was able to determine the physical meaning of the mass-energy formula, not to mention study its possible use cases in the field of engineering. Some of you might remember that we've mentioned in the derivation of this chapter that what's contained in atoms is a type of potential energy, but this conclusion actually wasn't reached until a century after special relativity was established. Although the notion is simply an extension of Einstein's discovery, the release of energy from atoms is indeed an important milestone marking the actual application of special relativity and also the only way so far to generate energy with mass, which summarizes the mechanism behind what we now call the "nuclear fission reaction." In addition to generating energy, however, such a mechanism was also used against enemies and caused significant destruction, pushing the human civilization toward annihilation. In the below section, we'll provide an overview of how scientific and technological development in this field has been influencing the world and threatening lives.

5.13 Far-Reaching Influence of Relativity Theory Applied to Engineering: Birth of Nuclear Bombs[32]

At the end of 1938, Lise Meitner (1878–1968) and her nephew Otto Frisch (1904–1979) sat by a leaning trunk on the grass of a park in the freezing city of Stockholm while frantically jotting down the results they had just achieved through the derivation process of how energy could be generated out of nuclear fission chain reaction, a research that they developed on the basis of the experiment conducted two months before by Otto Hahn (1879–1968) and Friz Straussmann (1902–1980) at the Kaiser Wilhelm Institute for Chemistry in Berlin, where large amounts of thermal energy was released out of uranium through impacts on the element. Meitner and Frisch came up with the hypothesis that there was a huge number of neutrons produced in the experiment process, creating enough collisions to maintain the chain reaction. In addition, they also calculated the energy generated from one nuclear fission to be approximately 200 MeV, which they attributed to the decrease of mass equaling that of roughly 0.2 neutron. This conclusion reminded Meitner of Einstein's mass-energy formula $E = mc^2$ and made her wonder if the technique of nuclear fission was exactly how humans could apply the theory of special relativity to the field of engineering. Therefore, she shared her thoughts with Niels Bohr (1885–1962), a close friend in Denmark for him to announce at the annual meeting of the American Physical Society held in

---

[32]Referenced from Reed [29].

Washington D.C. in January, 1939, causing an enormous sensation in the American academia.

In half a year's time, two Jewish physicists Leo Szilard (1898–1964) and Eugen Wigner (1902–1995) visited Einstein in Long Island, New York and explained to him the technique of nuclear fission and how it could be applied to producing nuclear bombs with massive power, which provoked a strong reaction from Einstein as he never expected his own theory to be used for such destructive weapons. Therefore, he wrote to President Roosevelt and prompted him to initiate a nuclear program as soon as possible in case Nazi Germany went one step ahead in producing such bombs. This letter successfully convinced the president to establish the large-scale Manhattan Project, which led to the production of one uranium and four Plutonium nuclear bombs in three and a half years' time. Then, two of them were dropped to Japan's Hiroshima and Nagasaki, putting an end to World War II and stopping the death toll at about 50,000,000. So, how is it that a simple nuclear fission reaction can generate such enormous amount of energy? Below, we will dive into the realm of nuclear physics and necessary parameters in producing nuclear bombs through a three-part discussion.

5.13.1 Chain Reaction of Nuclear Fission

First, let's take a look at why the total mass of nuclear fuel, such as uranium 235 and Plutonium-239, would decrease after nuclear fission. To put it another way, the question can be rephrased as: why is the mass of an atomic nucleus not equal to that of all protons and neutrons in it combined? In fact, the difference is referred to as the "mass defect", which equals the binding energy required for a nucleus to stay in a stable state, i.e. the minimum energy needed to separate the particles inside. The reason behind why an atomic nucleus can exist stably is the interplay between the strong force (attraction) and Coulomb force (repulsion) among the particles inside the nucleus, but as the mass of the nucleus goes up along with its atomic number, eventually the repulsion will outweigh the attracting force. When this happens, we know that the mass threshold has been exceeded, and that the nucleus won't be able to remain stable. Suppose we have an atomic nucleus with the mass of m, the combined mass of all the protons and neutrons inside is m_0, and the binding energy is U. Based on Einstein's mass-energy formula, we have $mc^2 + U = m_0 c^2$, i.e. $m = m_0 - U/c^2$.

From the energy perspective described above, we can create Fig. 5.16 to demonstrate the relationship between the average binding energy in atomic nuclei and their mass numbers based on the equation $m_0 = m + U/c^2$.

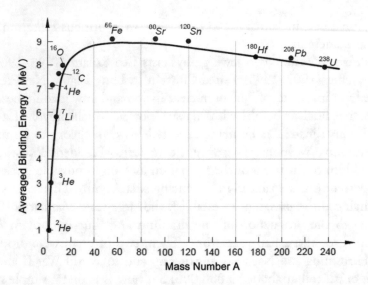

Fig. 5.16 Relationship between average nucleus binding energy and mass number. Based on the information of the website https://en.wikipedia.org/wiki/Nuclear_binding_energy

The average binding energy on the x-axis is calculated by dividing the total binding energy of a nucleus by its mass number (i.e. the numbers of protons and neutrons combined). As we can observe from the curve, when a nucleus has a very small mass number, its average binding energy goes up drastically along with the increase in the number of mass units and reaches the maximum when the unit count is around 60, e.g. in the case of iron (Fe). If the mass number exceeds that threshold, however, the binding energy will start moving inversely, showing that the strong force is more powerful than the Coulomb force in nuclei with relatively lower mass, vice versa, and when the mass number is around 50–60, both forces are roughly the same in small nuclei. On the other hand, as the repulsion in heavy elements are generally stronger, nuclear fission is triggered more easily, producing several lighter elements whose binding energy combined is lower than that of the original heavy counterpart. The difference in the energy level before and after the fission is exactly how much it takes to divide the heavy element, and it is the neutrons that hit the nucleus that provide the necessary energy.

5.13.2 The Critical Mass of Nuclear Fuel

When nuclear fission was found to be able to generate enormous amounts of energy, a perfect mechanism for producing super powerful bombs, the search for proper nuclear fuel naturally took the scientific circle by storm. Meanwhile, another popular research topic was calculating the critical mass of U235, which had already been proved an effective fuel for nuclear explosions. As we've mentioned before, a nuclear chain reaction can only proceed continuously when neutrons are generated faster than consumed; as for the speed of neutron generation, it's proportional to the mass of the element used for a nuclear explosion. Once the mass of an element exceeded a certain threshold, neutrons will be produced at a faster speed than exhausted. This threshold value, which is called the "critical mass," is a critical parameter that has to be confirmed before nuclear bombs can be designed.

As crucial as the parameter was, studies regarding the critical mass of U235 didn't go exactly that smooth, which was in fact a result of the international situation back in those days. As the war had just broken out on the European territory and the Asia-Pacific War was also brewing, it was just quite impractical to sponsor labs for related experiments when most countries were desperately short of war supplies already. Therefore, most scientists could only approach the issue through theoretical calculation, which yielded highly inconsistent results and became a headache for not just physicists but also the officials in charge of making related decisions. An example was Frisch, who derived a theoretical model for nuclear fission in cooperation with Meitner in Sweden. He later calculated the critical mass of U235 to be 500 g, basing his research on the theory by John Archibald Wheeler (1911–2008) and Niels Henrik David Bohr (1885–1962) that slow neutrons could trigger nuclear fission. The French physicist, Francis Perrin (1901–1992), on the other hand, claimed that as much as 40 tons of natural uranium ore had to be processed to extract enough U235 that could reach the mass threshold. As 0.72% of uranium ore is constituted by U235, the critical mass he proposed was 288 kg. Rumor has it that it was exactly the outrageously large value calculated by Werner Heisenberg (1901–1976) that dissuaded Hitler from developing such "Jewish bombs". Even toward the actual completion of nuclear bombs in 1944, such widely varying numbers would still pop out of labs from time to time.

5.13.3 Calculating Critical Mass

Despite the divergent results back in that era, this field has seen solid progress with well-developed calculations over the years, so in below paragraphs, we'll try to calculate the critical mass of a spherical-shaped nuclear bomb with a simple model of neutron diffusion. Suppose the nuclear fuel is a circular ball with the radius of R, while the density of neutrons in it is $n(r, t)$. Under such circumstances, the number of neutrons passing through a unit area per unit time (a.k.a. the neutron flux) can be expressed as:

$$J(r, t) = -D \frac{\partial n(r, t)}{\partial t} \tag{5.118}$$

Let's suppose D, the coefficient of diffusion is a constant and that each collision will produce v neutrons. As such, the change in the number of neutrons in the sphere-shaped bomb can be calculated through the diffusion equation below:

$$\frac{\partial n}{\partial t} = -\frac{1}{r^2} \frac{\partial}{\partial r} \left(r^2 J \right) + \frac{v - 1}{\tau} n \tag{5.119}$$

Here, we can substitute (5.118) into (5.119) and simplify the result as below with the equation $\frac{1}{r^2} \frac{\partial}{\partial r} \left(r^2 n \right) = \frac{1}{r} \frac{\partial^2}{\partial r^2} (nr)$:

$$\frac{\partial n}{\partial t} = \frac{D}{r} \frac{\partial^2}{\partial r^2} (nr) + \frac{v - 1}{\tau} n \tag{5.120}$$

Due to the decay of radioactive particles, we can assume the decay rate of the neutrons to be an exponential function:

$$n(r, t) = n_1(r) e^{\frac{v_1}{\tau} t} \tag{5.121}$$

where $v_1 \geq 0$ has to be true for the chain reaction to be triggered. Then, the substitution of (5.121) into (5.120) will lead us to:

$$\frac{d^2}{dr^2} (rn_1) + \frac{v - v_1 - 1}{D\tau} (rn_1) = 0 \tag{5.122}$$

where we can find the solution to be:

$$n_1 = \frac{1}{r} \left[A \sin \left(\sqrt{\frac{v - v_1 - 1}{D\tau}} r \right) + B \cos \left(\sqrt{\frac{v - v_1 - 1}{D\tau}} r \right) \right] \tag{5.123}$$

In (5.123), A and B are constants of integration to be determined by boundary conditions.

For n_1 not to be divergent when $r = 0$, B has to be 0. Also, the boundary condition $n_1(R) = 0$ has to exist when $r = R$; otherwise, neutrons would be constantly motioning towards outside the spherical bomb, which is why we know $\sin\left(\sqrt{\frac{v-v_1-1}{D\tau}}R\right) = 0$. Observing from this equation, we can tell that

the shortest possible radius R_c is $\sqrt{\frac{v-v_1-1}{D\tau}}R_c = \pi$ while the corresponding v_1 should be 0, thereby deriving:

$$R_c = \pi\sqrt{\frac{D\tau}{v-1}} \qquad (5.124)$$

As (5.124) represents the shortest radius required for the sphere formed by nuclear fuel, the smallest possible volume can be calculated therefrom, which we can then multiply by the density of the fuel and eventually obtain the critical mass.

5.13.4 Ways to Lower Critical Mass

While the above calculation was conducted assuming the bomb was made entirely of a sphere-shaped lump of nuclear fuel, it's not the case with nuclear bombs in the real world, where fuel is generally covered with a layer of neutron reflector. In addition to separating fuel and other devices in bombs, this layer is mainly used to reflect neutrons so those that have not participated in any collision can return to generate further chain reactions, which helps lower the critical mass for successful nuclear fission. Now, let's take a look at how the critical mass might change for a bomb covered with a neutron reflector.

Suppose the density of neutrons in the reflector is $n'(r)$. Though we can follow the previous approach of calculation, the number of neutrons in the bomb isn't supposed to change because the motion of the particles is restricted by the outer layer. Therefore, $n'(r)$ has to satisfy:

$$\frac{D}{r}\frac{d^2}{dr^2}(n'r) = 0 \qquad (5.125)$$

Which is an exception to (5.120). Again, if we represent the external radius of the reflecting shell as R' and adopt $n'(R') = 0$ as a boundary condition, the integration of (5.125) will be:

$$n'(r) = C\left(\frac{1}{r} - \frac{1}{R'}\right) \qquad (5.126)$$

For easier calculation, we can assume R' to be far larger than R, which will lead us to $n'(r) = \frac{C}{r}$ and yield the following equation when considered with the boundary condition:

$$n'(R) = n_1(R), \quad \left.\frac{\partial n'}{\partial r}\right|_{r=R} = \left.\frac{\partial n_1}{\partial r}\right|_{r=R} \tag{5.127}$$

As such, the minimum radius can be calculated to be:

$$R'_c = \frac{\pi}{2}\sqrt{\frac{D\tau}{\nu - 1}} \tag{5.128}$$

Please note that without the regulation of (5.127), neutrons would be accumulated on the interface of $r = R$. Comparing (5.124) and (5.128), we can tell that the minimum radius of a bomb with a reflector (represented by the latter equation) is only half that of one without an external shell (the former). In other words, the critical mass required for a nuclear bomb covered by a reflector is as little as 1/8 of that without one, indeed a dramatically significant difference.

As the last step, let's substitute the actual values regarding U235 into (5.128), i.e. $D = 2.75 \times 10^5 \mathrm{m}^2/\mathrm{s}$, $\tau = 8.10 \times 10^{-9}\mathrm{s}^{-1}$, and $\nu = 2.50$. From this equation, we can calculate the minimum radius to be $R'_c \approx 0.061\mathrm{m}$. Then, if we multiply this value by the density of the nuclear fuel $\rho = 19.07 \times 10^3 \mathrm{kg/m}^3$, the critical mass of U235 bombs can thus be determined as 17.7 kg. However, due to the safety factor added by military research units to ensure a successful explosion, the overall mass of the U235 used in the Little Boy, the uranium bomb eventually produced out of the Manhattan Project, was over 40 kg, more than 2.5 times that calculated above.

5.13.5 Power of Nuclear Bombs[33]

In the early morning of July 16th, 1945, an orange ball blazing fire that seemed brighter than thousands of suns combined suddenly lit up the sky above the vast Death Valley in the state of New Mexico at 5:30 AM, its strong shock waves shattering all the buildings and cars within two kilometers of radius and high-temperature radiation leaving all the sand and gravel within 400 meters of radius burnt to become green-yellowish glass-like crystallized structures. This was where the Manhattan Project's first atomic bomb made of plutonium was tested. In just about a month, another bomb, this time made of uranium, was already on the American air force's B29 bomber before

---

[33] Referenced from Chen [3].

dropped from on top of Hiroshima. After exploding about 600 m above the downtown area, the bomb destroyed all the city's wooden architecture and killed over 70,000 people with the heat that was over 1000 degrees of Celsius. Merely a few days later, a plutonium nuclear bomb was again thrown to Nagasaki, resulting in Japan's unconditional surrender and bringing an end to World War II. In below sections, we're going to use two sample calculations to explain the denotation mechanism of the three bombs mentioned above, whose unleashed power was equal to that of more than 10,000 tons of TNT.

5.13.5.1 Uranium Bomb

First, the chemical reaction of nuclear fission induced by U235 for uranium atomic bombs can be expressed as:

$$^{235}_{92}U + ^{1}_{0}n \rightarrow ^{141}_{56}Ba + ^{92}_{36}Kr + 3^{1}_{0}n + \Delta E \tag{5.129}$$

In (5.129), ΔE is the amount of energy released. In the above chemical reaction, the mass of all participating elements is: $m\left(^{235}_{92}U\right) = 235.044u$, $m\left(^{141}_{56}Ba\right) = 140.914u$, $m\left(^{92}_{36}Kr\right) = 91.926u$, and $m\left(^{1}_{0}n\right) = 1.00867u$, while $1u = 931.502 MeV/c^2$. Therefore, the energy released from each $^{235}_{92}U$ nucleus in the fission process is:

$$\Delta E = \left[m\left(^{235}_{92}U\right) - m\left(^{141}_{56}Ba\right) - m\left(^{92}_{36}Kr\right) - 2m\left(^{1}_{0}n\right)\right]c^2 = 173.9 MeV \tag{5.130}$$

For each gram of U235, as a total of $6.02 \times 10^{23}/235.044 \, ^{235}_{92}U$ atoms are contained therein, the energy generated can be calculated as below if all of them undergo the nuclear fission reaction:

$$\frac{6.02 \times 10^{23}}{235.044} \times 173.9 MeV = 7.13 \times 10^{10} J (joule) \tag{5.131}$$

The Hiroshima bomb dropped by the American in 1945 was 440 kg in total, where the nuclear fuel $^{235}_{92}U$ accounted for 45 kg of it. Based on the above calculation, theoretically the energy produced by this atomic bomb should be:

$$7.13 \times 10^{10} (J/g) \times 45 \times 10^3 (g) = 3.2 \times 10^{15} J \tag{5.132}$$

However, the actual energy released after the denotation was mere 6.7×10^{13}J, which is to say that the explosion efficiency of the uranium bomb was only 2.1%, while the energy was released in a variety of forms, such as through the emitted photons as well as colliding particles, whose momentum was impulsively converted into thermal energy in the process of collision.

5.13.5.2   Plutonium Bomb

First, the chemical equation of nuclear fission induced by Pu239, the nuclear fuel for Plutonium atomic bombs:

$$^{239}_{94}\text{Pu} + ^{1}_{0}\text{n} \rightarrow ^{134}_{54}\text{Xe} + ^{103}_{40}\text{Zr} + 3^{1}_{0}\text{n} \tag{5.133}$$

In (5.133), ΔE is the amount of energy released. In the above chemical reaction, the mass of all participating elements is: $m\left(^{239}_{94}\text{Pu}\right) = 239.0521634\text{u}$, $m\left(^{134}_{54}\text{Xe}\right) = 133.90539\text{u}$, $m\left(^{103}_{40}\text{Zr}\right) = 102.92660\text{u}$, and $m\left(^{1}_{0}\text{n}\right) = 1.00867\text{u}$, while $1\text{u} = 931.502\text{MeV}/c^2$. Therefore, the energy released from each $^{239}_{94}\text{Pu}$ nucleus in the fission process is:

$$\Delta E \approx \left[m\left(^{239}_{94}\text{Pu}\right) - m\left(^{134}_{54}\text{Xe}\right) - m\left(^{103}_{40}\text{Zr}\right) - 2m\left(^{1}_{0}\text{n}\right) \right]c^2 = 188.9\text{MeV} \tag{5.134}$$

For each gram of Pu239, as a total of $6.02 \times 10^{23}/239.05216$ $^{239}_{94}\text{Pu}$ atoms are contained therein, the energy generated can be calculated as below if all of them undergo the nuclear fission reaction:

$$\frac{6.02 \times 10^{23}}{239.05216} \times 188.9\text{MeV} = 7.62 \times 10^{10}\text{J(joule)} \tag{5.135}$$

The Nagasaki bomb dropped by the American on August 9th of in 1945, i.e. the Fat Man, was 4,545 kg in total, where the nuclear fuel Pu239 only accounted for 6.2 kg of it. Based on the above calculation, theoretically the energy produced by this atomic bomb should be:

$$7.62 \times 10^{10}\,(\text{J/g}) \times 6.2 \times 10^{3}\,(\text{g}) = 4.7 \times 10^{14}\text{J} \tag{5.136}$$

Although the actual amount of energy released was just 8.4×10^{13}J, merely 17.8% of the theoretically value, the explosion of this second bomb was still a lot more efficient than that of the one thrown to Hiroshima (2.1%).

As the nuclear bombs were detonated one by one, governments across the world started recognizing the extreme power that could be garnered with nuclear weapons, which was why so many countries hopped onto the band wagon of nuclear research, only to threaten the development of human civilization. Therefore, the Chicago University, which was well-aware of the possible destruction that could be triggered by such weapons, proposed the concept of the "Doomsday Clock" in its publication the *Bulletin of the Atomic Scientists* as a way to represent the likelihood of a man-made global catastrophe. The Clock's original setting in 1947, when the Cold War was just about to start, was seven minutes to midnight, a symbol of a complete breakout of nuclear wars across the globe that would eventually cause the hypothetical annihilation of the human race. As time proceeded, however, the *Bulletin*'s Science and Security Board also adjusted the clock to reflect the international situation. While the Cuban Missile Crisis in 1962 didn't result in any adjustment of the clock because it was resolved in a short period of time, the clock was set backward by seven minutes when the United States and Soviet Union signed the first Strategic Arms Reduction Treaty (START I) in 1991, leaving mankind two minutes away from midnight. Then in 2007, the Board included the global climate change as a potential threat, which set the clock forward by two minutes in 2015, and recent years have also seen the advancement of the clock due to North Korea's test missiles and the unrest in Ukraine. As the clock has been adjusted more and more often, it isn't hard to see the possibility of an all-out nuclear war has largely been increased. At the moment, the minute hand is merely two minutes away from midnight.

5.14 Internal Energy of Matter: More Powerful Within Than Outside

In light of the enormous explosive power of nuclear bombs, it's not hard to tell that the "mass" in Einstein's mass-energy formula $E = \gamma mc^2 \approx mc^2 + (1/2)mv^2$ doesn't just refer to a substance itself but also the energy it contains, which can roughly be divided into mass energy and kinetic energy. Yet how is the conversion of mass and energy embodied in our everyday life? Well et al. (2009) had a great way of illustrating the phenomenon: imagine a bucket of armed mousetraps whose springs were all wound-up and released in a very short amount of time, which triggered no other change than the traps' drastic jumps up in the air. When this happens, the potential energy of the springs would be released as the kinetic energy of the jumping motion, which is $(1/2)mv^2$. In other words, the mousetraps, now snapped shut and no longer useful, would become lesser in mass, except that the decrease is so tiny that we can't really detect it. Again, imagine a box full of hot gas whose temperature gradually goes down as the heat leaked out of the walls. Similar to the mousetraps, the box has more mass when hot than cold. As for how subtle the difference is, we can provide an example with coal, which becomes lesser in mass after burning. Suppose we have a coal burning boiler with an energy releasing power of 1,000 W. After burning for eight hours, the overall energy released therefrom is roughly 30,000,000 J, which, if divided by the square of light speed ($m = E/c^2$), would yield the mass of about 10^{-6}g, indeed an extremely small amount.

Although the internal energy of most substances in our daily lives is too little to be detected, not to mention applied to any use case, the phenomenon of mass-energy conversion can in fact be observed in the production process of fossil energy, one of the energy sources most used by human beings. When fossil fuels are burned, the hydrocarbons contained inside would actually undergo a series of chemical changes, where the bonds between atoms are broken before reconnected. While the former reaction requires energy input, the decrease in the mass of atoms after they're reconnected would lead to energy output. Again, let's consider coal as an example. During the coal-burning process, the bonds in big molecules consisting of carbon and hydrogen atoms are broken down before smaller molecules are formed by particles released therefrom, which causes a decrease in the total mass and hence the generation of energy. To determine how the re-bonding among molecules results in the change of mass, we can take a look at hydrogen, the lightest of all atoms. While a hydrogen atom is formed by an electron surrounding a proton, its mass is 2×10^{-35}g lesser than that of an independent electron and proton combined, which is equivalent to approximately $13.6 \text{eV}/c^2$ (eV: electron volt, the amount of energy gained by the charge of a single electron moved across an electric potential difference of one volt). As for how the bonds are reconstructed, electromagnetism is the driving force because mutual attraction exists between electrons and protons, which carry negative and positive electric charges respectively.

In addition to electromagnetic force, there also exists a really strong nuclear force in atomic elements that aggregates protons carrying the same electric charges inside the nuclei, which, judging from its effect, is way more powerful than electromagnetic force. As much as we'd like to observe just how robust this strong force is, it's barely possible to reproduce it in the real word. If we consider two protons with positive electric charges, the amount of repulsion force between them is inversely proportional to their distance, while the strong nuclear force would draw them towards each other. When their distance is shortened to a certain degree, one of the protons would release a positron, which is essentially a particle with a positive electric charge, before becoming a neutron and emitting a neutrino, a process called the β decay. While a proton is $938.3 \, \text{MeV}/c^2 \approx 1.673 \times 10^{-30} \text{g}$ in mass, the same physical quantity of Deuterium, which is made of one proton and one neutron (two protons bonded), becomes $1875.6 \, \text{MeV}/c^2$, lesser than that of two independent protons by $1 \, \text{MeV}/c^2$. In other words, about 1 MeV's energy is taken away with the released positron and neutrino. Half of this energy was converted into the mass of the positron, while the other half (0.5 MeV) is 40,000 times that of the energy emitted when a proton is bonded with an electron (13.6 eV) and 100,000 times that of the 6 eV released when hydrogen and oxygen form water through combustion. As a result, we can tell the energy emitted from nuclei is simply too much larger than that of burning reaction, which also explains why nuclear bombs are tens of thousands of times more powerful than TNT.

Appendix 5.1: Small Particles That Form the Rings of Saturn

In this proposition, we're going to prove that small particles (moons, pieces of ice, fluids, rocks, etc.) around Saturn have to be apart from the planet by a minimum distance of d for them to stick together under the effects of gravitational and centrifugal forces. For easy calculation, let's suppose Saturn and the small celestial body that we're going to use as an example here both are spherical in shape and have evenly distributed density. The radius and density of Saturn and the celestial body are represented as R_S, ρ_S and r, ρ. Figure 5.17 is an illustration of the rings of Saturn.

Suppose the celestial body is x away from Saturn in distance. If we consider a point mass (m) on the body's surface, it's under the impact of three different forces: (1) the gravitational force of Saturn, (2) the centrifugal force from its revolution around Saturn, and (3) the gravitational force from itself. Among them, (1) and (2) combined will result in a tidal force working on the body's closest point to and furthest point from the planet, the values of which can be calculated as:

$$\frac{2G\left(\rho_S \frac{4}{3}\pi R_S^3\right)m}{x^3} r. \qquad (5.137)$$

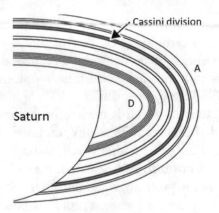

Fig. 5.17 The illustration of the rings of Saturn

$$\frac{G\left(\rho\frac{4}{3}\pi r^3\right)m}{r^2},\tag{5.138}$$

While the tidal force points outward from the celestial body's barycenter, its gravitational force can be calculated with this equationwhich points inward on the radial direction. For the celestial body to not be pulled apart, the inward force has to be stronger than the outward counterpart, which is why the following inequality holds correct:

$$\frac{G\left(\rho\frac{4}{3}\pi r^3\right)m}{r^2} \geq \frac{2G\left(\rho s\frac{4}{3}\pi R_S^3\right)m}{x^3}r.\tag{5.139}$$

After some simplification, we can retrieve the below equation:

$$x \geq R_S\left(\frac{2\rho s}{\rho}\right)^{\frac{1}{3}}.\tag{5.140}$$

This is the condition that has to be satisfied for the celestial body to revolve around Saturn in a stable manner. Namely, its radius has to be larger than the value calculated from (5.140).

In the above process, however, we didn't consider the possible self-rotation of the body, which can also generate a centrifugal force. Therefore, if we represent the angular velocity of the rotation and the resulting centrifugal force as ω and $mr\,\omega^2$, the above equation should be changed to:

$$\frac{G\left(\rho\frac{4}{3}\pi r^3\right)m}{r^2} \geq \frac{2G\left(\rho s\frac{4}{3}\pi R_S^3\right)m}{x^3}r + mr\,\omega^2.\tag{5.141}$$

If this celestial body has been tidally locked, then ω equals the orbital angular velocity (Ω) of its revolution around Saturn, which can be calculated as:

$$\frac{G\left(\rho_S \frac{4}{3}\pi R_S^3\right)m}{x^2} = mx\Omega^2, \tag{5.142}$$

By substituting what's on the right side of the equal sign in (5.141) into (5.142), where the centrifugal force is considered, we'll be able to obtain (5.143),

$$mr\Omega^2 = \frac{G\left(\rho_S \frac{4}{3}\pi R_S^3\right)m}{x^3} r. \tag{5.143}$$

$$\frac{G\left(\rho \frac{4}{3}\pi r^3\right)m}{r^2} \geq \frac{3G\left(\rho_S \frac{4}{3}\pi R_S^3\right)m}{x^3} r. \tag{5.144}$$

Which can be further simplified as:

$$x \geq R_S \left(\frac{3\rho_S}{\rho}\right)^{\frac{1}{3}}. \tag{5.145}$$

The result of (5.145) is 1.5 times that where the centrifugal force is disregarded. If we suppose the celestial body to be a block of rocky debris with the same density as Saturn ($\rho_S = \rho$), we can derive the following conclusions: (1) $d_1 \approx 1.26R_S$: centrifugal force disregarded, the distance between the debris and Saturn (d_1) is 1.26 times that the planet's radius (2) $d_2 \approx 1.44R_S$: centrifugal force considered, the distance between the debris and Saturn (d_2) is 1.44 times the planet's radius.

Yet the truth is, $\rho_S \neq \rho$, which is why we have to substitute the following parameters into (9): the mass of Saturn $M_S = 5.68 \times 10^{26}$kg, the radius of Saturn $R_S = 60,720$km, the density of Saturn $\rho_S \approx 700$kg/m³, and the density of the small particles in the rings of Saturn (mostly ice pieces)$\rho \approx 900$kg/m³, which will lead us to two different results: (1) $d_1 \approx 1.16R_S$, centrifugal force disregarded, the distance between the debris and Saturn (d_1) is 1.16 times that the planet's radius, and (2) $d_2 \approx 1.32R_S$, centrifugal force considered, the distance between the debris and Saturn (d_2) is 1.32 times that the planet's radius. As shown in Fig. 5.17, the rings of Saturn are broadbands, while the radius of the inner- (D) and outer-most (A) rings are $1.11R_S$ and $2.27R_S$ respectively, showing that the results we derived are quite close to the actual radius of the D ring. As for the minimum distance that has to exist between the small body and Saturn to prevent the former's disintegration, it's what we commonly called the Roche limit.[34]

Please note that the above derivation is based on the assumption that the small celestial body is a rigid object, which leads us to the conclusion that rocky debris around Saturn would gather near its innermost ring. If the small particles in the rings are fluids, the minimum distance will become the following[35] considering the centrifugal force as:

$$d = 2.44 R_S \left(\frac{\rho_S}{\rho}\right)^{\frac{1}{3}}. \tag{5.146}$$

If we supposed the density of the fluid to be $\rho \approx 1000 \, \text{kg/m}^3$ and substitute this value into the above equation, the result will be $d \approx 2.17 R_S$, meaning the small body in our example would still be located inside the broad rings of Saturn but closer to the outermost one.

With all these said, the working forces affecting the rings of Saturn are a lot more complicated than what we've considered here. For example, the Cassini Division, which is as wide as 4,800 km (the dark ring in Fig. 5.18), was the result of the pulling forces from many of Saturn's moons. As indicated by the conclusion of this proposition, however, the gravitational force of Saturn is still the main factor behind the stable revolution of the rings around it; as for the forces from the planet's moons, they only cause slight impacts on the bands.

Appendix 5.2: Reflecting Mechanism of Electromagnetic Waves in the Ionosphere

For the purpose of this proposition, let's first define the ionosphere as the region from 50 to 500 km above the Earth's surface. Due to the radiation of ultraviolet rays, a component of sunlight, a certain part of the gas molecules in Earth's atmosphere will be ionized. When the chemical combination and decomposition reach a balance, the atmosphere will see a steady, continuous distribution of electrons and positive ions. As the mass of positive ions is way

---

[34]The Roche limit refers to the distance between an astronomical body and another one when its own gravitational force equals the tidal force exerted by the counterpart. If the distances between both bodies is smaller than the Roche limit, one of them will start to disintegrate and eventually become part of a ring that revolves around the other body.

[35]The derivation process of this equation is extremely complicated and can't yet be explained with any simple mathematical approach.

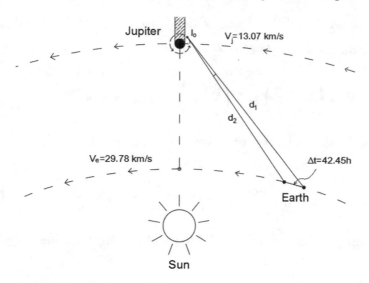

Fig. 5.18 The positional relationship between Earth and Jupiter

larger than that of electrons, only the latter will oscillate when triggered by electric fields.

Now, we can represent the density of electrons in the ionosphere as N, and the electric charge/mass of an electron as $-e/m$. Suppose we're sending an electromagnetic wave with the angular frequency of ω into the ionosphere. Under the circumstance where the amplitude of the electric field only changes with the z coordinate, the field can be expressed as a function of height $\vec{E}(z)$, and this is where Maxwell equations come in:

$$\nabla \times \vec{E} = -\frac{\partial \vec{B}}{\partial t} \tag{5.147}$$

$$\nabla \times \vec{B} = \mu_0 \vec{J} + \mu_0 \varepsilon_0 \frac{\partial \vec{E}}{\partial t} \tag{5.148}$$

The curl of (5.147) can be retrieved as:

$$\nabla \times \left(\nabla \times \vec{E} \right) = -\frac{\partial}{\partial t} \left(\nabla \times \vec{B} \right) \tag{5.149}$$

Under the assumption that the density of electric charges in the ionosphere is still small enough to be disregarded even when the disturbance from electromagnetic waves cause the electrons to be redistributed, we can rewrite the expression on the left-hand side of the equal sign in (5.148) as (5.149) by applying the equation $\nabla \cdot \vec{E} = \frac{\rho}{\varepsilon_0} = 0$:

$$\nabla \times \left(\nabla \times \vec{E} \right) = \nabla \left(\nabla \cdot \vec{E} \right) - \nabla^2 \vec{E} = -\nabla^2 \vec{E} \tag{5.150}$$

Then, the differentiation of (5.149) over time will lead us to:

$$-\frac{\partial}{\partial t} \left(\nabla \times \vec{B} \right) = -\mu_0 \frac{\partial \vec{J}}{\partial t} - \mu_0 \varepsilon_0 \frac{\partial^2 \vec{E}}{\partial t^2} \tag{5.151}$$

And lastly, we can substitute (5.149) and (5.150) back into (5.148) and obtain:

$$\nabla^2 \vec{E} = \mu_0 \frac{\partial \vec{J}}{\partial t} + \mu_0 \varepsilon_0 \frac{\partial^2 \vec{E}}{\partial t^2} \tag{5.152}$$

Please note that the electric current density $\vec{J}$ can be calculated from the speed of electrons, which is why we need to consider their motion equation:

$$m \frac{d\vec{v}}{dt} = -e\vec{E} \tag{5.153}$$

If we differentiate $\vec{J} = -Ne\vec{v}$ over time and substitute (5.152) into the result, the below equation can be derived:

$$\frac{\partial \vec{J}}{\partial t} \approx -Ne \frac{d\vec{v}}{dt} = \frac{Ne^2}{m} \vec{E} \tag{5.154}$$

As a next step, we can substitute (5.154) back into (5.153) and obtain the equation for the electric field in the ionosphere:

$$\nabla^2 \vec{E} = \mu_0 \frac{Ne^2}{m} \vec{E} + \mu_0 \varepsilon_0 \frac{\partial^2 \vec{E}}{\partial t^2} \tag{5.155}$$

As the electric field $\vec{E}(z)$ in (5.155) is the function of z, the Laplace operator ∇^2 in the same equation can be replaced with $\frac{d^2}{dz^2}$. On the other hand, as the oscillation frequency of the electric field is ω, the operator $\frac{\partial^2}{\partial t^2}$ in (5.155) can again be replaced with ω because the sine wave of this angular frequency, if differentiated twice over time, equals the product of it multiplied by $-\omega^2$, which is why (5.155) can eventually be expressed as (5.156) after the two substitutions:

$$\frac{d^2 E(z)}{dz^2} + \mu_0 \varepsilon_0 \left(\omega^2 - \frac{Ne^2}{\varepsilon_0 m} \right) E(z) = 0 \qquad (5.156)$$

As the next step, let's define plasma frequency as the below expression for convenience:

$$\omega_p = \sqrt{\frac{Ne^2}{\varepsilon_0 m}} \qquad (5.157)$$

As such, (5.157) can be simplified as:

$$\frac{d^2 E(z)}{dz^2} + \mu_0 \varepsilon_0 \left(\omega^2 - \omega_p^2 \right) E(z) = 0 \qquad (5.158)$$

Then, we're going to consider the situation where an electromagnetic wave is sent into the ionosphere from below at a certain angle. As the ionosphere is extended along the $+z$ direction, we'd like to know whether this wave can be transmitted after its emission into this layer of the atmosphere. If so, part of its energy will be dissipated through the ionosphere, and if not, all the wave energy will be reflected. To determine whether transmission is possible in the layer, we should first find out the solution of (5.159) by considering two different situations:

$$\omega < \omega_p$$

Under this situation, the solution of $E(z)$ will decay exponentially: $E(z) \sim e^{-\sqrt{\mu_0 \varepsilon_0 \left(\omega_p^2 - \omega^2 \right)} z}$, showing the impossibility of transmission in the ionosphere and hence the complete reflection of the electromagnetic wave.

$$\omega > \omega_p$$

Under this situation, the solution of $E(z)$ is a sine wave: $E(z)$ $\sim$ $\sin\left(\sqrt{\mu_0\varepsilon_0\left(\omega^2 - \omega_p^2\right)}z\right)$, showing that electromagnetic waves can be transmitted in the ionosphere.

As such, we can conclude that only electromagnetic waves with frequencies lower than that of plasma (ω_p), i.e. with wavelengths longer than a specific value, can be reflected in their entirety.

Appendix 5.3: Measuring the Radius of Table Sugar Molecules

This proposition describes Einstein's two-part derivation of the Avogadro constant and the radius of table sugar molecules:

(1) Part 1

Let's consider a low-density water-sugar solution and represent the viscosity coefficients of pure water and the solution as k and k^* respectively. Then, a series of calculations will yield the ratio of the two values as:

$$\frac{k^*}{k} = 1 + 2.5\varphi \tag{5.159}$$

In the above equation, φ is the volume ratio of table sugar molecules in the solution. In other words, if a sugar molecule is a sphere with the radius of P, and a unit volume of the solution contains n molecules, we can derive the equation $\varphi = n\frac{4}{3}\pi P^3$ and resolve n as $N_A\frac{\rho}{m}$ by representing the weight of the sugar molecules in a unit volume, the molecule mass, and the Avogadro constant as ρ, m, and N_A. Combining the two equations mentioned in this paragraph, the below can be obtained:

$$N_A P^3 = \frac{3}{10\pi}\frac{m}{\rho}\left(\frac{k^*}{k} - 1\right) \tag{5.160}$$

Now, let's substitute the parameters regarding the low-density sugar-water solution (with a weight percentage concentration of 1%) into this equation: $\frac{k^*}{k} = 1.0245$, $m = 0.342\frac{kg}{mole}$, and solution density $= 1003.88\,kg/m^3$, which leads to $\rho = 10.0388\,kg/m^3$ and the below equation:

$$N_A P^3 = 7.97 \times 10^{-5}m^3 \tag{5.161}$$

(2) Part 2

This part of calculation is a reference to Einstein's development of the Stokes-Einstein relation in his paper:

$$D = \frac{RT}{6\pi k N_A P} \tag{5.162}$$

In (5.161), D, R, T, and k represent the diffusion coefficient of sugar molecules, the ideal gas constant, the absolute temperature, and the viscosity coefficient of the sugar solution respectively. Once simplified, the equation can be rewritten as (5.162):

$$N_A P = \frac{RT}{6\pi k D} \tag{5.163}$$

Already known under the temperature of 9.5 °C is $R = 8.31 \frac{J}{mole \cdot K}$, the diffusion coefficient of the sugar solution $D = 3.84 \times 10^{-10} \frac{m^2}{s}$, and the viscosity coefficient of water $k = 1.35 \times 10^{-3} Pa \cdot s$. The substitution of these values into (5) will lead to:

$$N_A P = 2.40 \times 10^{14} m \tag{5.164}$$

Combining the results of (5.163) and (5.164), we can calculate the Avogadro constant (N_A) and radius of table sugar molecules (P) to be:

$$N_A = 4.2 \times 10^{23} \tag{5.165}$$

$$P = 5.8 \times 10^{-10} m \tag{5.166}$$

Please note that the Avogadro constant calculated from (7) is deviant from the actual value (6.02×10^{23}) to a considerable degree, which is probably the result of the simplified model that we adopted. Therefore, the radius of sugar molecules might be inaccurate as well.

Appendix 5.4: Average Distance Traveled by Pollen Particles in Liquids through Random Motion

In this two-part proposition, we're going to simulate how Einstein calculated the path and distance traveled randomly by pollen particles in water solution.

The first part consists of the formulation of a diffusion equation for Brownian particles, in which the diffusion coefficient is related to the mean squared displacement of a Brownian particle. The second part focuses on relating the diffusion coefficient to measurable physical quantities.

(1) Part 1

The first part of Einstein's argument is to determine how far a Brownian particle travels in a given time interval. Classical mechanics is unable to pinpoint this distance because of the enormous number of bombardments a Brownian particle undergoes (roughly 10^{14} per second). Thus, Einstein was led to look at the collective motion of Brownian particles. In this part, we're only going to consider the motion of these particles along the x-axis. Suppose their density at the location x and time point t is $\rho(x, t)$. With a very short time τ passed (which can be regarded as the time interval between two subsequent collisions), the particles will be redistributed and cause a change in their density. If we only focus on one of the many particles and represent the distance it's traveled during τ as Δ, all the possibilities of Δ can be covered by the probability density function $\varphi(\Delta)$, meaning the likelihood of the particle moving from Δ to $\Delta + d\Delta$ within τ is $\varphi(\Delta)d\Delta$. Then, since the sum of all probabilities is 1, we know that:

$$\int_{-\infty}^{\infty} \varphi(\Delta)d\Delta = 1 \tag{5.167}$$

If the particles move in random directions, $\varphi(\Delta)$ should be an even function, i.e.

$$\int_{-\infty}^{\infty} \Delta\varphi(\Delta)d\Delta = 0 \tag{5.168}$$

Now that the particles are redistributed, their density will also change from $\rho(x, t)$ to $\rho(x, t + \tau)$, and since τ is extremely small, the relationship between the two functions can be expressed as:

$$\rho(x, t + \tau) \approx \rho(x, t) + \tau \frac{\partial \rho}{\partial t} \tag{5.169}$$

In fact, $\rho(x, t + \tau)$ can also be represented using $\rho(x, t)$ in combination with $\varphi(\Delta)$. As the particle travels from x to $x - \Delta$ at the start and end time points of t and $t + \tau$, we can multiply $\rho(x - \Delta, t)$ by the probability $\varphi(\Delta)d\Delta$ and differentiate each Δ to derive $\rho(x, t + \tau)$, yielding:

$$\rho(x, t + \tau) = \int_{-\infty}^{\infty} \rho(x - \Delta, t)\varphi(\Delta)d\Delta \approx \int_{-\infty}^{\infty} \left[\rho(x, t) - \Delta\frac{\partial\rho}{\partial x} + \frac{\Delta^2}{2}\frac{\partial^2\rho}{\partial x^2}\right]\varphi(\Delta)d\Delta$$

$$= \rho(x, t) \int_{-\infty}^{\infty} \varphi(\Delta)d\Delta - \frac{\partial\rho}{\partial x} \int_{-\infty}^{\infty} \Delta\varphi(\Delta)d\Delta + \frac{\partial^2\rho}{\partial x^2} \int_{-\infty}^{\infty} \frac{\Delta^2}{2}\varphi(\Delta)d\Delta$$

$$= \rho(x, t) + \frac{\partial^2\rho}{\partial x^2} \int_{-\infty}^{\infty} \frac{\Delta^2}{2}\varphi(\Delta)d\Delta \qquad (5.170)$$

Please notice that the last equation in (5.170) is derived by applying (5.167) and (5.168). Then, the comparison of (5.168) and (5.170) will lead us to:

$$\frac{\partial\rho}{\partial t} = \left(\int_{-\infty}^{\infty} \frac{\Delta^2}{2\tau}\varphi(\Delta)d\Delta\right)\frac{\partial^2\rho}{\partial x^2} \qquad (5.171)$$

The expression on the right-hand side of the equal sign in (5.171) is defined by Einstein as the diffusion coefficient D, i.e.

$$D = \int_{-\infty}^{\infty} \frac{\Delta^2}{2\tau}\varphi(\Delta)d\Delta \qquad (5.172)$$

Subsequently, the substitution of (5.168) back into (5.167) will yield (5.166):

$$\frac{\partial\rho}{\partial t} = D\frac{\partial^2\rho}{\partial x^2} \qquad (5.173)$$

And this is the famous diffusion equation.

As denoted by this equation, the diffusion of Brownian particles can be observed when a lot of them co-exist in a solution with an uneven density, and such diffusion is essentially the accumulated displacement of the many particles that follow the Brownian motion at the same time. If we consider a situation where all particles start diffusing from a set origin at the start point of time $t = 0$, the solution of (5.167) can be calculated as:

$$\rho(x, t) = \frac{\rho_0}{\sqrt{4\pi Dt}}e^{-\frac{x^2}{4Dt}} \qquad (5.174)$$

With this solution, changes in the distribution and position of Brownian particles in a solution over time can thus be calculated. In particular, we can use this equation to determine the square-mean displacement (x^2) in a one-dimensional space:

$$x^2 = \frac{\int_{-\infty}^{\infty} x^2 \rho(x, t)dx}{\int_{-\infty}^{\infty} \rho(x, t)dx} = 2Dt \tag{5.175}$$

(2) Part 2

This part seeks to represent the results from Part 1 with measurable physical quantities in order to verify the concluded theory. Let's first consider Stokes' Law and a spherical particle whose mass, radius, and endured viscosity in a specific solution are m, a, and $-(6\pi\eta a)\vec{v}$, with η representing the viscosity coefficient. Hence, the motion equation of this particle along the direction of the x-axis is:

$$m\ddot{x} = -(6\pi\eta a)\dot{x} \tag{5.176}$$

Now, if we multiply the expressions on both sides of the equal sign in (5.176) by x and retrieve the average value of the entire equation:

$$m\frac{d}{dt}x\dot{x} - 2\frac{1}{2}m\dot{x}^2 = -(6\pi\eta a)x\dot{x} \tag{5.177}$$

Please note that (5.177) was derived with the application of $x\ddot{x} = \frac{d}{dt}(x\dot{x}) - \dot{x}^2$. Assuming the Equipartition Theorem is true (i.e. the solution is in a state of thermal equilibrium), the energy contained in it would be evenly distributed to each particle, which leads us to $\frac{1}{2}m\dot{x}^2 = \frac{1}{2}kT$, where k is the Boltzmann constant $(1.38065 \times 10^{-23} J/K)$ and T the temperature K. Substituting this equation into (5.177), (5.178) can be obtained therefrom:

$$m\frac{d}{dt}x\dot{x} = -(6\pi\eta a)\left(x\dot{x} - \frac{kT}{6\pi\eta a}\right) \tag{5.178}$$

If the distribution of the particles in the overall solution has entered an equilibrium state, $\frac{d}{dt}x\dot{x} \approx 0$ should be satisfied, which can help us simplify (5.178) as:

$$x\dot{x} = \frac{1}{2}\frac{dx^2}{dt} = \frac{kT}{6\pi\eta a} \tag{5.179}$$

According to (5.179), the square mean displacement of any particle after each time interval of t is:

$$x^2 = \frac{kT}{3\pi\eta a}t \tag{5.180}$$

Then, a comparison of (9) and (14) will lead us to:

$$D = \frac{kT}{6\pi\eta a} \tag{5.181}$$

And this is the famous Stokes-Einstein relation, whose equation can be used to predict the radius of table sugar molecules and the Avogadro constant. For details, please refer back to "Appendix 5.3: Measuring the Radius of Table Sugar Molecules."

Appendix 5.5: How the Equivalent Mass of an Electron Changes with Its Direction of Motion in an Electric Field

According to the theory of special relativity, when acted upon by a working force perpendicular and parallel to its direction of motion, a particle carrying electric charges exhibits different equivalent masses. For the purpose of this proposition, let's look at the motion behavior of an electron carrying the electric charge of q traveling with an instantaneous speed of $\vec{v}$ in an evenly distributed electric field $\vec{E}$. First, the definition of momentum under the structure of special relativity tells us that:

$$\frac{d}{dt}\left(\frac{m\vec{v}}{\sqrt{1 - v^2/c^2}}\right) = q\vec{E} \tag{5.182}$$

By representing $\vec{v}$ and $\vec{E}$ with their components, we can express the $\hat{x}$ component of what's on the left-hand side of the equal sign in (1) as:

$$\frac{d}{dt}\left(\frac{mv_x}{\sqrt{1 - \frac{v_x^2+v_y^2+v_z^2}{c^2}}}\right) = m\frac{\left(1 - \frac{v_y^2+v_z^2}{c^2}\right)\frac{dv_x}{dt} + \frac{v_xv_y}{c^2}\frac{dv_y}{dt} + \frac{v_xv_z}{c^2}\frac{dv_z}{dt}}{\left(1 - \frac{v_x^2+v_y^2+v_z^2}{c^2}\right)^{3/2}} \tag{5.183}$$

Now, here are the two different scenarios we need to consider:
(A) Electrons moving in parallel with the electric field

Suppose the applied electric field $\vec{E}$ works along the direction of x with an intensity of E, while the electron moves at the instantaneous speed of v toward the same direction. As (17) still equals qE under this circumstance, we can introduce the speed state of the electron into the equation:

$$\frac{m}{\left(1 - v^2/c^2\right)^{3/2}} \frac{dv_x}{dt} = qE \tag{5.184}$$

Comparing (3) with $F = m_{||}a$ from Newtonian mechanics, we can calculate the equivalent mass of the electron as:

$$m_{||} = \frac{m}{\left(1 - v^2/c^2\right)^{3/2}} = \gamma^3 m \tag{5.185}$$

Represented by (4) is the horizontal mass.
(B) Electrons moving perpendicularly to the electric field

Here let's suppose the applied electric field $\vec{E}$ works along the direction of x with an intensity of E, while the electron moves at the instantaneous speed of v toward the perpendicular direction (y). As (17) still equals qE under this circumstance, we can also introduce the speed state of the electron and obtain:

$$\frac{m}{\sqrt{1 - v^2/c^2}} \frac{dv_x}{dt} = qE \tag{5.186}$$

Comparing (5) with $F = m_{\perp}a$ from Newtonian mechanics, we can calculate the equivalent mass of the electron as:

$$m_{\perp} = \frac{m}{\sqrt{1 - v^2/c^2}} = \gamma m \tag{5.187}$$

And this is the vertical mass.

With (19) and (21), we can tell that $m_{||}$ and $m_{\perp}$ are different values. In addition, the difference also grows larger along with the increase in speed.

Appendix 5.6: Rømer's Light Speed Experiment

Suppose Earth and Jupiter revolve around Sun following circular orbits whose radiuses are r and R at the angular velocities of ω_E and ω_J, respectively. As the time needed by Jupiter and Earth to complete a circle around Sun is approximately 11.86 years ($V_j = 13.07$km/s) and 1 year ($V_e = 29.78$ km/s), we know that $\omega_E > \omega_J$. If observed from the rotating reference frame where Jupiter is stationary and one of the axes passes through the center of Sun and forms a perpendicular angle with both planets' rotational plane, the angular velocity of Earth's revolution would be $\omega = \omega_E - \omega_J$. If we illustrate the positional relationship between Earth and Jupiter based on the actual proportion of their radiuses, as shown in Fig. 5.18, you can see that when they're on the same side of Sun, Io, the closest moon to Jupiter, is about to enter the shadow of it, explaining why the moon becomes invisible to observers on Earth. And when Io enters the shadow the next time, it is after 42.46 h, exactly the revolving period of this moon. As the revolving radius of Io around Jupiter is quite small, we can consider Earth to have the same distance from Jupiter and Io before representing as d, which can be calculated as below based on geometrical relations:

$$d = \sqrt{R^2 + r^2 - 2Rr \cos \omega t} \approx R - r \cos \omega t \tag{5.188}$$

Please note that as $r \ll R$ in (5.187), we only retained the linear term of r/R when expanding the square root with the Taylor series.

If we represent the time points of Io's two consecutive entries into the shadow of Jupiter as t_1 and t_2, the light reflected off this moon at the two moments will arrive at Earth at $T_1 = t_1 + d_1/c$ and $T_2 = t_2 + d_2/c$, the interval between both being:

$$T_2 - T_1 = t_2 - t_1 + \frac{d_2 - d_1}{c} = t_2 - t_1 - \frac{r}{c}[\cos \omega T_2 - \cos \omega T_1] \tag{5.189}$$

Please note that the difference between ωT_2 and ωT_1 is extremely small, and therefore so is the angle swept by Earth on its orbit of revolution (roughly 42.5 h/365 days). As such, (5.189) can be further simplified as:

$$\Delta T = \Delta t + \frac{r \omega}{c} \sin \omega T \Delta T \tag{5.190}$$

Then, ΔT can be resolved as:

$$\Delta T = \frac{\Delta t}{1 - \frac{r\omega}{c}\sin\omega T} \approx \left(1 + \frac{r\omega}{c}\sin\omega T\right)\Delta t \tag{5.191}$$

In other words, based on the results retrieved from the perspective of Earth, which is in relative motion to Jupiter, there is roughly a $(r\omega/c)\Delta t$ difference between the largest and average measurements (i.e. the actual period), which is perfectly in line with the equation of the Doppler effect:

$$\Delta T = \sqrt{\frac{c+v}{c-v}}\Delta t \tag{5.192}$$

where v is the relative speed of Earth and Jupiter (note that linear approximation is adopted for $v \ll c$ in the equation). Based on Rømer's observation, the actual period is 15 s longer than the shorted he'd measured, which is why:

$$\frac{r\omega}{c}\Delta t = 15(s) \tag{5.193}$$

In other words:

$$c = \frac{r\Delta t}{15(s)} = \frac{1.49 \times 10^{11}(m) \times \left(\frac{2\pi}{365\,\text{days}} - \frac{2\pi}{365\,\text{days}\times 11.9}\right) \times 42.46\,\text{h}}{15(s)}$$

$$= 2.77 \times 10^8 \left(\frac{m}{s}\right) \tag{5.194}$$

We note that the light speed Rømer calculated according to his measurements was 220,000 km/s, while our result 277,000 km/s is closer to the actual value of 300,000 km/s. The still-present deviation can be explained with the fact that we postulated both Earth and Jupiter to be revolving on circular orbits, which contributed to the deviation in their periods and the length of their trajectories.

Appendix 5.7: Light Aberration Caused by Earth's Revolution

When the British astronomer James Bradley pursued the measurement of the distance between Earth and stars, he concluded that, when observed on Earth using a telescope, the light from these stars would deviate to some degrees

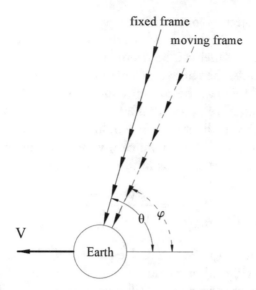

Fig.. 5.19 Illustration of the measurement of the distance between Earth and stars

after entering the tube of telescope due to the motion of Earth. Simply put, light that enters from the center of the objective lens would leave the eyepiece from a spot deviating from the center to a certain degree, a phenomenon called the "light aberration." To get a clear image of a star, therefore, the telescope used for observation has to be slightly rotated following the direction of Earth's self-rotation to offset the blurriness caused by such aberration. In below sections, we're going to prove that the light aberration caused by Earth revolution around Sun, also known as the "annual aberration," is approximately 20.5 ArcSec.

As shown in the figure below, the angle formed by the light reflected off a distant star and the line representing Earth's motion direction is φ from the perspective of an observer standing on the moving Earth. From the viewpoint of a stationary observer on Earth (imagine a world where the Earth didn't revolve around Sun), by contrast, the angle would become θ. The difference between φ and θ is exactly what we call the light aberration, the values of which changes constantly with the moving speed and position of Earth. In fact, the rotation of Earth also causes the diurnal aberration of less than 1 ArcSec,[36] which is small enough to be disregarded compared to the 20.5-ArcSec annual counterpart. Therefore, only the latter is considered in this proposition (Fig. 5.19).

---

[36] Reference: https://zh.wikipedia.org/wiki/%E5%85%89%E8%A1%8C%E5%B7%AE.

As Bradley's object of observation was a distant star and Earth's revolution speed (about 30 km/s) is merely 1/10,000 that of light (roughly 300,000 km/s), the planet can be deemed in uniform motion in relation to light. In other words, the reference frame of Earth is an inertial one where it moves in a straight line at the constant speed of v toward the $-x$ direction. On the other hand, if we postulate the Earth to be a stationary body, any ray of light traveling from the star to it would project a component of $c \cos \theta$ toward the $-x$ direction, which we can, with the Lorentz transformations, move to the frame where Earth is in uniform motion:

$$c' = \frac{c \cos \theta - v}{1 - \frac{c \cos \theta \, \Delta v}{c^2}} \tag{5.195}$$

As the solution of (5.194) should equal $cos\varphi$, the component of light speed c on the $-x$ direction measured from the reference frame where Earth is moving at a constant speed, the following equation can be derived:

$$\cos \varphi = \frac{v'}{c} = \frac{\cos \theta - \frac{v}{c}}{1 - \frac{v}{c} \cos \theta} \tag{5.196}$$

And this is the equation for calculating light aberration.

As the direction and speed of Earth's revolution vary over the course of a year, the value of annual aberration is constantly changing as well, yet the maximum angle can be effectively calculated because light aberration is at its maximum when Earth's revolving direction is perpendicular to the light from distant stars. Therefore, we can substitute the equation $\theta = 90°$ into (5.196) and derive $\varphi = \cos^{-1} \frac{v}{c}$, which is quite close to $\theta = 90°$ because $v/c \approx 10^{-4}$ yet still contains a slight difference, as shown in:

$$\theta - \varphi = 90° - \cos^{-1} \frac{v}{c} \approx 0.00569° = 20.5'' \tag{5.197}$$

As you can tell, the solution of (5.197) is 20.5 ArcSec, exactly the light aberration observed from Earth in the reference frame where it revolves around Sun. In light of the result, we can conclude that the speed of Earth's revolution is so small on the scale of the entire universe that the aberration caused by such relative motion to Sun or any distant star is almost the same.

6

General Relativity, Where Time and Space Are Curved

天地間充滿引力,分佈像黎曼流形
源頭是地球諸星
浩於人心,沛乎蒼冥

Pervading Heaven and Earth is the gravitational force.
It assumes the form of Riemannian manifold,
Originating from stars and Earth,
Sweeping through humans' hearts and filling up the living space.

Adapted from Song of Righteousness, *by Wen Tianxiang (1236–1283), Chinese poet and politician*

In 1905, Einstein published the five papers listed in Sect. 5.4 consecutively, all of which generated a significant influence on the development of physics. Among these papers, what was introduced in the third and fourth were considered the theory of "Special Relativity" (SR) by later scientists, which differentiates it from "General Relativity" (GR), where Einstein extended Newton's law of universal gravitation from Earth's surface to the entire universe. The theory of special relativity has two major flaws as it's only applicable to (1) "inertial" frames of reference (2) where no gravitational force is exerted. A so-called inertial reference frame refers to one where a body with zero net force acting on it does not accelerate, which in fact, represents the majority of the universe because the average density of cosmic matter is mere

9.47×10^{-34}g/cm^3. The mass of these substances per cubic meter is equivalent to that of about five protons, making space a near-vacuum environment. In addition, almost all visible cosmic matter is gathered in giant nebulae, which take up 1% of space only, so the theory of special relativity can actually cover 99% of the universe.

Without the small amounts of astronomical objects and substances, though, the universe would've been a lifeless, stagnant environment. Besides, interstellar phenomena can hardly be disregarded in any attempt to develop a credible astrophysical theory because astronomical bodies are so closely related to human existence. Yet once taken into consideration, the gravitational force of cosmic bodies immediately exposes the two major flaws of special relativity, denying the validity of the entire theory. This had to be a serious headache for Einstein because the theory was, in fact, meant to identify the source of universal gravitation at first. As a result, it became an inevitable task for him to integrate the gravitational force of astronomical bodies into the realm of special relativity. Starting from 1907, he spent eight years to fully construct the gravity-infused framework of general relativity, consummate with the set of "Einstein field equations", which regulates a variety of cosmic phenomena and wields a long-lasting impact on the human civilization. In the below sections, we're going to reexamine how he established the field equations from a modern perspective and try to reconstruct his thought process by clarifying related physical concepts, building mathematical models, and verifying the equations.

6.1   Concept Clarification: How Time and Space Work Under Gravitational Impact

Before actually starting with the theory of general relativity, Einstein first sought to clarify all related physical concepts through thought experiments in order to draw a clear boundary between his theory and Newtonian mechanics. He began by redefining the physical meaning of time and space through the observation of the two elements from the perspective of a non-inertial frame of reference under the effect of gravity. For this purpose, he created the equivalence principle, which he then coupled with an advanced tensor-based mathematical model as well as the curved pattern of time and space to derive the field equations under the structure of the inertial reference frame. Believe it or not, the process took him eight entire years. In the following sections, we'll dive into the thought transformation on both the physical and mathematical levels experienced by Einstein in his journey of establishing the principle of equivalence and the geometry of tensors, two major foundations of general relativity.

6.1.1 Vast Range of Relative Time and Space

Although it's true that the theory of special relativity doesn't apply to non-inertial frames of reference where gravity is present, one would be completely mistaken to think its range of application only covers places far from Earth because the slowing of time caused by motion exists in our everyday life. If not considering this effect, related calculations and analyses will yield highly erroneous results. Using the muon (μ) as an example, let's take a look at how the theory of relativity helped prove the fact that cosmic rays contribute to the generation of muons when entering Earth's atmosphere by colliding with atoms and molecules in the air. As muons are extremely small in mass, they travel close to the speed of light after being generated but only have a lifespan of roughly 2.2 μs (2.2×10^{-6}s). The distance they're able to travel during such a short period of time is merely 660 m, the product of their lifespan and light speed. As a result, there would've been no way for scientists to prove the existence of muons without considering the effect of relativity because the distance traveled before their disappearance is simply too short.

By contrast, if we take time dilation into consideration by applying (5.62) and assume the speed of muons to be 99.99% of light speed, we know through calculations that the lifespan of and distance traveled by this elementary particle are 220 μs and 66 km from the perspective of observers on Earth, while the latter roughly equals the thickness of the planet's atmosphere. In other words, the muons generated as a result of cosmic rays are capable of penetrating the atmosphere and reaching Earth's surface before disappearing for good. On the other hand, if we shift to the perspective of a certain muon, i.e. observe from the inertial frame of reference where it sits on the origin, the atmosphere should become thinner due to length contraction. If we assume the atmosphere is originally 60 km thick, (5.51) can help us calculate its relative thickness to muons, which are moving near the speed of light, to be 600 m. As the distance is shorter than the 660 m they're capable of traveling before their lifespan ends, this result explains why scientists on Earth could detect the existence of these particles. In fact, the European Organization for Nuclear Research (CERN) in Switzerland has used purpose-built particle accelerators to prove through high-speed particle experiments that the lifespan of muons is extended to 63.8 μs when they move at 99.94% the speed of light, 29 times longer than when in the stationary state.

If we ever want to plan for space travel in the future, time dilation and length contraction must be taken into consideration as well. The importance of these two factors can be explained with a thought experiment. Suppose a spaceship powered by nuclear fusion is now departing from Earth for a beautiful planet on the other side of the Milky Way by accelerating for the first ten years of its journey and decelerating during the second decade. Then, it'll return to Earth following the same motion pattern. To simulate an environment with universal gravitation inside the vessel, we shall assume the spaceship to accelerate and decelerate both at the constant speed of 1 g ($9.8 \, m/s^2$). Through a series of calculations factoring in the effect of relativity, we know that 40 years in the reference frame of the moving ship is supposed to be perceived as 58,194 years by observers in the static frame of Earth. As for the destination planet, it's 29,095 light-years from Earth.[1] So if the spaceship accelerates to a higher speed, the time required to travel the distance between both planets should be shorter as well. Judging from this thought experiment, space trips depicted in sci-fi movies are surely not impossible in the future.

Due, however, to the fact that continued acceleration will eventually push the spaceship to travel near the speed of light, the postulate of this thought experiment is hardly achievable for modern scientific technology. Take Voyager 1, launched by NASA in the 1970s, as an example. This space probe has been traveling in the Solar System for more than 40 years now, yet its relative speed to Sun has just reached 50 times the sonic speed ($\gamma \approx 1 + 10^{-9}$) recently, still a long way to go before catching up with light. Even with Juno, the fastest spacecraft manufactured by human beings so far, the speed had also just reached 250 times the sonic speed when the probe arrived at Jupiter in 2016 six years after launched by NASA in 2010, still an extremely long stretch from light speed.[2]

As can be told from both examples, the relativity of time and space has to be considered when trying to understand the vast universe. Otherwise, we'd be encountered with not just theoretical dilemma but technical infeasibility as well. Luckily though, Einstein had long before conducted all the necessary research to integrate the impact of universal gravitation into the framework

---

[1] See Cox and Forshaw [8], the second last paragraph of Chap. 4 for more details. The calculation process will be detailed in later discussion, where $a = g = 9.8 \, m/s^2$ and $c = 3 \times 10^8 \, m/s$ will be substituted into (6.285) and yield the result that ten years in the spaceship ($t' = 10$) is equivalent to 14,549 years on Earth ($t = 14549$). As each lag of the return trip consists of acceleration and deceleration, the time passed on Earth should be four times the said result, i.e. 58,194 years. As for the distance traveled by the spaceship, $t' = 10$ can also be substituted into (14) in Appendix 6.1 and lead us to the resolution of $x = 14547$ (light years). Yet due to the motion pattern of accelerating for the first decade and decelerating during the second, the total distance traveled by the vessel from Earth to the destination planet should be doubled to become $x = 29,095$ light years.

[2] See Lee [20], pp. 40–42 for calculation details.

of relativity. His first step was establishing the principle of equivalence, the most crucial concept in the theory of general relativity.

6.1.2 Gravity and Weak Equivalence Principle

Although gravity is known to come from matter, nobody has really been able to elaborate on the mechanism of its generation. In the 16th century, Galilei conducted a free-fall experiment and found out that without the effect of air friction, a piece of iron and a feather would fall to the ground simultaneously when dropped from the same height. Looking back at this experiment, Newton proposed that it was due to the same acceleration generated by Earth's gravitational force towards both objects that they hit the ground at the same time, an explanation based on his double definition of "mass". Under Newton's framework, mass comes in two types: (1) the inertial mass (i.e. m_I in $\vec{F} = m_I \vec{a}$), which measures an object's resistance to being accelerated by a force, and (2) the gravitational mass, the mass of a body as measured by its gravitational attraction for other bodies (m_G in $\vec{W} = m_G \vec{g}$[3]); the bigger the mass, the stronger the attraction. In Newtonian mechanics, inertial mass equals gravitational mass ($m_I = m_G$), which is also called the "weak equivalence principle". In other words, objects located in the same gravitational field would always accelerate with the same speed ($\vec{a} = \vec{g}$)) in free falls regardless of their mass. In the several hundred years after the law of universal gravitation was established, many scientists strived to prove the weak equivalence principle, and in fact, such efforts have extended to the contemporary era as well. For example, the Apollo 15 astronaut David Scott (1932–) dropped a hammer and a feather at the same time in front of a camera on his moon landing mission in 1971. When both objects fell to the ground simultaneously, he added a humorous comment, saying "Well how about that. Mr. Galilei was correct in his findings." In 2014, BBC also produced a high-cost physics show where Professor Brian Cox (1968–) conducted the same experiment in a giant vacuum chamber containing little air. Originally suspended at 100 m high, an iron ball and a feather were released at the same time and, with the help of slow-motion shooting, could be seen to hit the ground simultaneously.[4]

---

[3]In this equation, $\vec{W}$ is the gravitational force exerted on an object, while $\vec{g}$ is the gravitational acceleration.

[4]See https://www.youtube.com/watch?v=E43-CfukEgs for more.

Despite the seemingly abundant proof of the weak equivalence principle in the modern era of physics, Einstein didn't seem to be satisfied or consider any of the corresponding approaches prudent enough (just imagine if he saw the light-hearted comment made by Scott). In fact, he didn't even agree with Newton's interpretation of the gravitational field due to its incompatibility with the framework of relativity, which can be easily observed through basic concepts regarding the transformation of reference frames. In Newtonian mechanics, the gravitational field can be understood as the distribution of free-fall acceleration as long as the weak equivalence principle stands. Yet in the theory of relativity, we know from the Lorentz transformations that when observed from two inertial frames with varying speeds, the acceleration of the same object is to be perceived as different.[5] Regarding this point, detailed elaboration is provided below.

Suppose there are two inertial frames S and S′ that're completely overlapped when $t = t' = 0$ but start motioning at different speeds afterwards. If we have a static object r in S, the presence of it should come with a gravitational field $\vec{g}\left(\vec{r}\right)$ (a position function) according to the law of universal gravitation. On the other hand, as there's a non-zero relative speed between S′ and S, we know that due to the transformation of acceleration between the two frames, the gravitational field of r should be perceived as $\vec{g}'\left(\vec{r}\right)$ rather than $\vec{g}\left(\vec{r}\right)$ when observed from S′. At this point, though, a contradiction arises because, how can the field not be the same when the mass in both frames is identical when $t = t' = 0$? To this question, the only possible answer is that "mass" isn't the only determining factor of how the gravitational field is distributed. As the stationary object r in S would be perceived as moving at a constant speed from S′, Einstein proposed that a moving object could also change the gravitational field around it. In other words, mass and momentum both play a determining role.

In fact, this wasn't a new concept for Einstein but one that had been in place when he was constructing the theory of special relativity. If you still remember, the mass-energy Eq. (5.112) has already illustrated how the two

---

[5]The acceleration can be calculated by differentiating the object's position over the corresponding time twice in a row, and since both the spatial and temporal coordinates change in the process of frame transformation, the result will surely be different as well. This said, the value of acceleration does remain unchanged through the Galilean transformation, which we've already discussed before. Readers who're interested can use the Lorentz equations to determine how the acceleration changes when the reference frames are transformed. In fact, the derivation process consists of the same steps we've previously adopted for developing the speed transformation Eqs. (5.67)–(5.69).

elements are closely intertwined, just as time and space can never be considered separately. As for how mass and energy should actually be integrated, though, Einstein knew he had some physical concepts to clarify and mathematical tools to develop first.

6.1.3 Physical Phenomena in Reference Frame with Constant Proper Acceleration

Before diving into Einstein's research, we need to first understand what a non-inertial frame of reference is. Following the same approach of dealing with special relativity, here we're also going to conduct some basic thought experiments in order to confirm certain properties of a non-inertial frame before proceeding to mathematical derivation and calculation. The first experiment we're going to engage in has to do with length contraction: suppose we have a car whose length is d when stationary. Based on (5.51), when the car is moving at the speed of v in relation to an observer standing on the ground, its length should be perceived as $d\sqrt{1 - v^2/c^2}$, where c represents light speed. Yet since it's the properties of a non-inertial frame that we're trying to determine here, what we need to focus on is the continuous process of the car's acceleration from the static state to v, its relative speed to the ground. So, imagine we're observing the process from two different points of view:

(1) From the perspective of the frame co-moving with the car, the length should still be perceived as d when the speed reaches v because otherwise, the vehicle would've been disintegrated. As such, we can assume that the effect of length contraction would cause the observer standing on the ground to think the car is shortened to $d\sqrt{1 - v^2/c^2}$.
(2) Alternatively, suppose every single point mass in the car is moving in a straight line with the uniform acceleration of a (the so-called "proper acceleration") when the car is accelerating. Yet once the speed of the vehicle reaches v, each point mass immediately switches to uniform linear motion. In this scenario, the observer standing on the ground would think the length of the car remains to be d because all the point masses share the same pattern of motion.

As you've probably noticed already, although observed from two different frames of reference, the perceived length of the car turned out to be the same, so it's quite obvious that something must've gone wrong in our discussion. Therefore, we're going to explain in below paragraphs the fallacy of this postulate: every single point mass in the car is moving in a straight line with uniform acceleration when the car is accelerating.

For the purpose of accuracy, let's start by clarifying a few concepts that're essential to later elaboration. First, if we divide the acceleration process into extremely short periods of time, the car can be considered to be moving with constant speed during each period. Hence, an observer who's located inside the car during any of these periods is actually observing from an instantaneous inertial frame in uniform linear motion at the speed of the specific period. Simply put, every single moment in the acceleration process corresponds to a different instantaneous inertial frame that moves at a constant speed, and the car's acceleration at each of these moments can be regarded as its relative acceleration to the corresponding instantaneous inertial frame a, also called the "proper acceleration."[6]

Now, let's take a closer look at the assumption of "every single point mass in the car moving in a straight line with uniform acceleration when the car is accelerating" and find out what result this postulate will lead us to. For this purpose, we can simplify the scenario as an inertial frame S fixed on the ground where two objects A and B (representing the front and rear of the car) motion along the x axis. Before any motion takes place, both objects are stationary with the distance of d in between (length of the car), while the coordinates of A and B are $x = d$ and $x = 0$ respectively. When the vehicle starts moving at the instant of $t = 0$, A and B also begin motioning with the same proper acceleration of a. From the perspective of S, therefore, the motion behaviors of both objects are completely identical with their distance remaining as d the entire process of the car's acceleration.

To visualize this thought experiment, we can illustrate the position changes of A and B over time in the Minkowski diagram. In Fig. 6.1, x and ct are two axes in the S frame. As for the curves of two different hyperbolas,[7] they represent the world lines[8] of objects A and B. In the fixed-to-ground frame S, the distance between both objects is always d, and hence, so will that between the intersecting points of the two curves and any straight line parallel with x. Nevertheless, when it comes to S′, a certain instantaneous inertial frame

---

[6]The time period between the occurrences of two different events inside the car measured from the frame of the vehicle is called the "proper time interval" because in this particular frame, the place of occurrence for any event is the same (i.e. in the car itself). The time displayed on any clock in the car is considered the "proper time", while the car's relative speed to itself is the proper speed (but of course, the value is always 0). Similarly, the vehicle's relative acceleration to itself is the "proper acceleration" as well.

[7]Mathematically speaking, the curves do fit the strict definition of hyperbola. As for how to prove this point, more details will be provided in the derivation of (6.255).

[8]The world line of a certain object refers to its motion trail in a spacetime coordinate system.

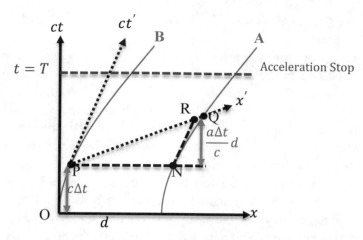

Fig. 6.1 Motion trails of A and B in the Minkowski diagram

co-moving with B, things are a bit different. When B reaches the point P, for example, S′ will be composed of x' and ct' as two of it axes and P as the origin. Per previous discussion, the slope of ct' is c/v', while v' is the instantaneous speed of B in relation to S. Therefore, we know that ct' is tangent to the world line of B. In addition, the slope of x' is v'/c, reciprocal of the slope of ct'.

As a next step, let's attach a clock to A as well as B to determine whether the clocks would move at varying speeds when observed from different reference frames. While both clocks apparently move at equal speeds in S the inertial frame, what will happen when we switch to S′? To answer this question, we can again observe Fig. 6.1 for easier understanding. At the instant when the objects are about to start accelerating but haven't actually moved, they can be deemed static. When $t = 0$, an observer in the frame of B would think the clock on A also shows $t = 0$. By contrast, when both objects are already motioning after an extremely short amount of time Δt, the instantaneous speed of B becomes $a\Delta t$ and the slope of the x' axis $a \cdot t/c$. As Δt is a very small value, the world line of A is virtually vertical. Hence, the value of its intersecting point with x' will be $c\Delta t + (a\Delta t/c) \times d = \left(1 + ad/c^2\right)c\Delta t$ on the ct axis, which is also the value that's supposed to be displayed on the A clock from the perspective of B. One thing to be noted is that, as the speeds of A and B are way slower compared to the light speed c, the time displayed on the clocks attached to both objects should be nearly identical to that in

S.[9] Therefore, we know that in S', the ratio of the speeds of clocks A and B is $(1 + ad/c^2)$.

Up to this point, the question that we set out to answer has been resolved. Once the observer in the frame of B starts to think the clock on A moves faster than that on B, the same observer will also see A as gradually moving away from B. Such a phenomenon will eventually cause the distance between A and B to become larger than d, which clearly implies the disintegration of the car. As a result, what we postulated above can't be the actual pattern of how the vehicle accelerates.

So how can we find out the real pattern of the acceleration? From the perspective of S', every point mass in the car has to be stationary for the entire vehicle to stay together. Now, suppose the origin of S' ($x' = 0$) has the proper acceleration of a. If we fix a clock at $x' = d$, its the proper acceleration should be not just smaller than a but in fact equal to $a/(1 + ad/c^2)$. The reason is that in the S' frame, time proceeds faster at $x' = d$ than at $x' = 0$, so the corresponding proper acceleration should also be slower by the same percentage of $a/(1 + ad/c^2)$. Now, if we look back at the original scenario, the proper acceleration at any position where the distance from the rear end of the car is $x' = d$ has to be exactly $a/(1 + ad/c^2)$ for the vehicle to remain in its original shape without being disintegrated, stretched out, contracted, or twisted. Such a result is actually quite intuitive and fits any general rule of thumb.

From the above discussion regarding the basic properties of non-inertial frames of reference, we can reach two important conclusions:

(1) In a reference frame motioning with proper acceleration, the further a point is located along the direction of acceleration, the faster time proceeds.

(2) In a reference frame motioning with proper acceleration, the value of proper acceleration of any given point mass in an object located in this frame has to be precisely $a/(1 + ax/c^2)$ to prevent the object from being disintegrated ($x =$ the distance between a specific point and the rear end of the object).

---

[9]In fact, if considering the effect of time dilation, the speed of A and B should be fine-tuned to $a\Delta t$, giving us the correction factor of $\sqrt{1 - (a\Delta t/c)^2}$. Then, expanding this expression, we'll find out that the lowest term of Δt is second-order, which is why the parameter can be reasonably disregarded.

6.1.4 Deeper Dive Into Principle of Equivalence

After concluding the properties of accelerating frames of reference, let's continue by looking at the impact that'll be exerted when the gravitational field is introduced into such a frame. In this part of discussion, we're going to apply the weak equivalence principle introduced before, where m_G (gravitational mass) equals m_I (inertial mass). After Galilei and Newton published their findings about this principle, the phenomenon of $m_G = m_I$ had always been considered a coincidence without much effort in trying to determine what it actually implied, until Einstein presented the theory of relativity. It was Einstein who first thought of the weak equivalence principle as the key to integrating the law of universal gravitation into the world of relativity. To understand his approach, we need to conduct a thorough series of thought experiments based on the principle of weak equivalence.

Imagine you're in a spaceship holding a lead ball, which contributes to the feeling of weight in your hand. Meanwhile, the normal force between your feet and the floor also allows you to sense the weight of your own body. As far as Einstein was concerned, such a scenario can possibly happen under two circumstances only. The first is when the ship motions in the gravity-free space with the acceleration of $\vec{a}$, causing all objects inside the vessel to feel the effect of an inertial force $(-m_I\vec{a})$ that points toward the opposite direction of the acceleration. Therefore, both you and the lead ball should have a tendency to be pulled towards the direction of this force, which is mainly generated by the inertial mass m_I. The other possibility is when the spaceship sits on a certain planet. Under this situation, the force from the planetary gravitational field $(m_G\vec{g})$ works on both you and the ball, hence leading to the feeling of weight. Compared to the first circumstance, this alternative is a result of the gravitational mass m_G. According to Einstein, as long as $m_I = m_G$, both scenarios should create the same outcome in the spaceship, so there's no way for anyone inside to tell whether the ship is in the gravitational field of $\vec{g}$ or motioning with the acceleration of $\vec{a} = -\vec{g}$ through mechanical experiments. Therefore, he also went on to propose that if no mechanical or any other type of experiment can help determine which of the possible circumstances is the actual situation in the ship, then the weak equivalence principle would be a crucial factor to consider.[10]

---

[10]As long as we restrict ourselves to purely mechanical processes in the realm where Newton's mechanics holds sway, we are certain of the equivalence of the systems K and K'. *But this view of ours will not have any deeper significance unless the systems K and K' are equivalent with respect to all physical processes, that is, unless the laws of nature with respect to K are in entire agreement with those*

Foɪ easier understanding, let's now move this experiment to Earth's surface, where we know of a uniform gravitational field $\vec{g}$. Due to the effect of the weak equivalence principle, people on Earth should feel the same way as those in the spaceship, which motions with the acceleration of $-\vec{g}$. Meanwhile, we also know from the discussion in Sect. 6.1.3 that time would proceed faster at the front end of the ship than at the rear, so the higher up a spot is on Earth, the faster the time there goes. Precisely speaking, the speed of time procession at the elevation of H is faster by the percentage of $\left(1 + gH/c^2\right)$ compared to a spot at the sea level. In other words, if we send out a photon with the frequency f from H towards below, the frequency of the photon will become roughly $\left(1 + gH/c^2\right)f$ when received on the ground, a phenomenon called "gravitational redshift".[11] Physicists in the past designed quite some experiments to verify the value, but as it turned out, the theory was just extremely hard to prove because even at the elevation of 2,000 m ($H = 2000$), the value of gH/c^2 is still merely 2×10^{-13}, showing very little effect of gravitational redshift. As a result, experiments of this kind have to be extremely precise, which is why many scientists failed in their efforts.

Despite so, some still wouldn't give up this pursuit. In 1959, Robert Pound (1919–2010) and Glen Rebka (1931–2015) at Harvard University decided to base their research on the Mössbauer effect (i.e. recoilless nuclear resonance fluorescence) to measure the frequency transfer of ^{57}Fe 14.4 keV with a spectroscopy of transition. The experimental mechanism lies in the fact that when returning to the ground state after excitation, an atom will emit a photon that can be absorbed by an atom of the same type that's also in its ground state. Yet when an atom emits or absorbs a photon, atomic recoil can result from the conservation of momentum, which is why the atom will carry a certain amount of kinetic energy and cause unpredictable frequency shifts in the photon that's emitted or absorbed. To eliminate such unpredictability, Pound and Rebka bound the ^{57}Fe atom used in the experiment in a giant lattice placed 22.5 m above ground level, so a photon could be emitted towards below for another ^{57}Fe on the ground in the excited state to absorb. Due, however, to the effect of gravitational redshift, the frequency

---

with respect to K. By assuming this to be so, we arrive at a principle which, if it is really true, has great heuristic importance. For by theoretical consideration of processes which take place relatively to a system of reference with uniform acceleration, we obtain information as to the career of processes in a homogeneous gravitational field.— *Einstein, 1911*.

[11] Gravitational redshift refers to the shift of wavelength of a photon to longer wavelength (the red side in an optical spectrum) when observed from a point in a lower gravitational field, which results in a lower frequency and a longer wavelength. For more details, please refer to https://en.wikipedia.org/wiki/Gravitational_redshift.

of the photon would've been increased when it hit the bottom, and because atoms could only absorb photons with frequencies in a fixed range, the ^{57}Fe had to move away from the emitted photon at a certain speed due to the Doppler shift before becoming able to absorb it. Therefore, by adjusting the atom's speed of motioning away from the photon, the researchers should be able to measure the frequency transfer in the photon caused by the gravitational redshift, which in turn could help verify the redshift equation. As easily feasible as this might sound, the speed of ^{57}Fe atoms are extremely slow at $(2 \times 10^{-15}) \times (3 \times 10^8) = 6 \times 10^{-7}$ m/s, so it was with great precision that the experiment was carried out in order for countable results.

In fact, the experiment described above can also be used to prove the weak equivalence principle compatible with the theory of relativity and even quantum mechanics. For this purpose, here's another thought experiment to consider: if we send out a photon with the frequency f from the elevation H when the photon hits the ground, its frequency will have changed to f'. According to Planck's theory, the photon carries the amount of energy hf at the height of H, where h is the Planck constant. Thus, we can convert the energy contained in the photon to the equivalent mass based on the mass-energy formula $E = mc^2$.[12] One thing to be noted is that, when we were deriving this formula in Chap. 5, all discussion was conducted in the realm of kinematics with no involvement of gravitational force, so for the sake of accuracy, $E = mc^2$ should be rewritten as $E = m_{\mathrm{I}}c^2$ because m_{I} is the inertial, rather than the gravitational mass. As such, the equivalent inertial mass

---

[12]One thing to be clarified here is that in real-world experiments, there's no way for us to directly convert the photon's energy into an equivalent mass, yet the same purpose can be served by using an atom as a medium. For example, an atom with the exact difference of hf between its ground and excited states would be a feasible option. Through below steps, we also provide an actual experimental process for reference:

(Set up two systems, S1 and S2, on the ground and at the elevation of H where energy measuring instruments are installed and radiation can be emitted/absorbed.)

(1) S2 sends towards the ground the energy E in the form of radiation, which S1 received as $E(1 + gH/c^2)$
(2) Move an object with the gravitational mass of m_{G} from H to the ground to release the gravitational energy of $m_{\mathrm{G}}gH$ (work done by the object).
(3) When S1 sends the energy E to the object, its gravitational mass will increase to m'_{G}.
(4) Lift the object back to the height of H by applying an $m'_{\mathrm{G}}gH$ amount of work.
(5) The object then sends the energy E back to S2.

From the law of conservation, we know that $m'_{\mathrm{G}} - m_{\mathrm{G}} = E/c^2$, which is caused by the energy E. Source of methodolofy: "On the Influence of Gravitation on the Propagation of Light", By A. Einstein. Annalen der Physik, 35, pp. 898–908, 1911.

of the photon can be expressed as:

$$m_I = hf/c^2 \tag{6.1}$$

While falling from the height of H, this amount of mass will do the work of $m_G g H$ due to the effect of universal gravitation, so when the photon hits the ground, the total energy it carries will change to $hf + m_G g H$. According to the law of conservation, we should be able to convert all this energy into a photon with the frequency f':

$$hf + m_G g H = hf' \tag{6.2}$$

As it's already known that $f' = \left(1 + gH/c^2\right) f$, we can substitute this equation into (6.2) along with (6.1) and retrieve $m_I = m_G$ after some simplification. Hence, the weak equivalence principle is again proved.

Now that we're positive that the photon can be converted into an equivalent gravitational mass, here comes another question: is this photon subject to the deviation caused by the gravitational field? To determine the answer, we need to return to the spaceship and assume it's moving forward with the proper acceleration of 1 g to ensure the equivalence between the gravitational environment inside the ship and Earth's gravitational field. Now, if we send out a laser ray from the right-side wall of the spaceship, this ray will be perceived as traveling in a straight line from the perspective of the inertial frame outside the vessel. Yet as the ship is accelerating, when the light reaches the left-side, this other wall will have already moved forward by a certain distance. In other words, the arrival point of the laser ray will be located behind the projected point of its origin by this certain distance. On the other hand, if we observe this same event from the inside, we'll perceive the laser light as falling towards the rear end of the ship with acceleration as if worked upon by Earth's gravitational force. Hence, we can come to the conclusion that light does get deflected by the gravitational field.

The bending of light caused by the gravitational field was actually described in great detail in Einstein's 1991 paper "On the Influence of Gravitation on the Propagation of Light". Towards the end of this publication, he also calculated the deflection angle of light when traveling close by Sun. Regarding this topic, we will provide more details later on, but to facilitate subsequent discussions, let's first look at some simple concepts here. First, the warping of light under the influence of solar gravity involves the effect of a large-scale gravitational field whose strength varies with its location. To properly handle such changes, we need to divide the gravitational environment around Sun into extremely small areas, enough so that the gravitational field within all of them can be expressed as the previously mentioned position function $g(x) = a/(1 + ax/c^2)$.

Such an approach is in fact similar to how modern computers store images, where they are divided by software programs into many squares (pixels) of different color shades represented by varying gray values from the range of 0–255. Therefore, as long as an image is divided into a sufficient number of small-enough pixels, it can be rendered just as real as how it'd appear in front of your eyes through computer graphics. Now, let's imagine we're running a program to divide a three-dimensional gravitational field into many small areas, each with a nearly uniform function of gravity. This way, any of these areas can be seen as a spaceship moving in a gravity-free space with an acceleration equaling the value of the gravitational force contained in that specific area. Then, we can simply merge all these areas to construct the overall gravitational field, and this is the strong equivalence principle proposed by Einstein. If you're wondering whether this principle can be derived from the weak counterpart, though, the answer is no because the latter is in fact just an exception of the former.

Although the principle of strong equivalence is extremely important, Einstein's research was flawed due to the insufficient precision in his division of the gravitational field, which caused the discontinuity of physical quantities in neighboring areas. In fact, the deflection angle he calculated in the 1911 paper was hence merely half the actual value (1.75ArcSec). If interested, readers are encouraged to try calculating the warping of light when traveling close to Sun by combining areas with a uniform gravitational field with the previously mentioned methods regarding the equivalent mass of photons and the accelerating spaceship. As for the result, it's expected to come out half the actual value, just as Einstein had retrieved.

6.1.5 Mathematical Model and Spacetime Geometry Describing Reference Frame with Constant Proper Acceleration

In the previous section, we discussed certain properties of reference frames moving with constant proper acceleration, but only on a conceptual level without much mathematical calculation. From here onward, we're going to derive the equations for transforming spacetime coordinates between inertial frames and frames with constant proper acceleration,[13] just as how we

---

[13] A friendly reminder of the definition of proper acceleration: in relativity theory, the proper acceleration an object is the acceleration measured from the instantaneous inertial frame of the object itself.

introduced the Lorentz transformations in Chap. 5. First, let's start with a crucial question: suppose there are an inertial frame S and a frame with constant proper acceleration S' that are relatively static and overlapped when $t = t' = 0$. At the moment when t exceeds 0, S' also starts to move with the constant proper acceleration of a toward the direction of $+x'$. Now, if there's a certain event whose spacetime positions are (ct, x, y, z) and (ct', x', y', z') in S and S' respectively, how do we transform between the two sets of coordinates?

Before jumping into the derivation process, there's one thing that needs to be clarified: as the speed of time procession changes with x' coordinates in S', the time interval between two events will also change with their places of occurrence. To facilitate our discussion, let's first agree on the rule here that when we say the spacetime coordinates of a certain event are (ct', x', y', z') in S', we mean this event happens at (ct', x', y', z') from the perspective of an observer who's located at $x' = 0$. Then, with a series of calculations in "Appendix 6.1: Transformation between Reference Frame with Constant Proper Acceleration and Inertial Frame," we'll come to the below result:

$$\begin{cases} ct = \left(x' + \frac{c^2}{a}\right) \sinh \frac{at'}{c} \\ x = \left(x' + \frac{c^2}{a}\right) \cosh \frac{at'}{c} - \frac{c^2}{a} \\ y = y' \\ z = z' \end{cases} \quad (6.3)$$

Last but not least, let's take a look at the spacetime events perceived from the perspective of an observer in a frame moving with constant proper acceleration since the focus of this section is the exploration of spacetime geometry. If you still remember, it was the Minkowski diagram that we used for illustration when discussing the theory of special relativity in Chap. 5. Back then, we defined the spacetime interval ds between two events as having to satisfy (5.79):

$$ds^2 = -c^2 d\tau^2 = -c^2 dt^2 + dx^2 + dy^2 + dz^2 \quad (6.4)$$

As the value of ds doesn't change with the Lorentz transformations, the interval between any pair of events can be calculated from (6.4) to be the same in all inertial frames. To extend the discussion into proper-acceleration

---

As for the so-called "constant proper acceleration", it refers to the fact that the instantaneous inertial frame moves with the same acceleration at all time, i.e. when observed from the perspective of the object in question, the value of its acceleration is equal at every moment.

frames, we'll need to use (6.3) to derive an equation for calculating the space-time interval between two given events measured by an observer in such a reference frame. For this purpose, we can first differentiate the equation for transforming the x coordinate:

$$dx = \cosh\frac{at'}{c}dx' + c\left(1 + \frac{ax'}{c^2}\right)\sinh\frac{at'}{c}dt' \qquad (6.5)$$

Subsequently, we can again differentiate the transformation equation for t:

$$cdt = \sinh\frac{at'}{c}dx' + c\left(1 + \frac{ax'}{c^2}\right)\cosh\frac{at'}{c}dt' \qquad (6.6)$$

As for the y and z coordinates, since they don't change in the process of transformation, we know that:

$$dy = dy' \qquad (6.7)$$

$$dz = dz' \qquad (6.8)$$

Now, if we substitute (6.5)–(6.8) back into (6.4), a bit of simplification will lead us to:

$$ds^2 = -c^2d\tau^2 = -\left(1 + \frac{ax'}{c^2}\right)^2 c^2dt'^2 + dx'^2 + dy'^2 + dz'^2 \qquad (6.9)$$

As such, we've come to the result that we set out to look for: (6.9) is the equation for calculating the spacetime interval between two events from the perspective of an observer in a frame with proper acceleration,[14] which is only different from (6.4) in that it contains the parameter of time interval dt'. In addition, (6.9) also shows the non-uniform pattern of time changing with the x' coordinate.

To look further into the significance of (6.9), let's return to the principle of equivalence, which tells us that the accelerating frame S' is equivalent to a gravitational field with the acceleration of $g(x) = -a/(1 + ax/c^2)$. If we fix a clock on a free-motioning object in this field, the time displayed on the clock (represented as τ, the proper time for the object) can be calculated with

---

[14]In fact, (ct', x', y', z') is also called the "Kottler-Møller coordinates", one of the many expressions for the famous "Rindler coordinates". Readers who're interested are encouraged to look for more information online with these keywords.

(6.9):

$$dτ = \sqrt{\left(1 + \frac{ax'}{c^2}\right)^2 - \frac{1}{c^2}\left(\frac{dx'^2 + dy'^2 + dz'^2}{dt'^2}\right)}\,dt'$$

$$= \sqrt{\left(1 + \frac{ax'}{c^2}\right)^2 - \frac{v'^2}{c^2}}\,dt' \tag{6.10}$$

In the above equation, $v' = \sqrt{dx'^2 + dy'^2 + dz'^2}/dt$ is the speed of the object in S'. If we only consider the scenario of low-speed ($v' \ll c$) weak gravitational field ($ax' \ll c^2$), (6.10) can be approximated as:

$$dτ \approx \sqrt{1 - \frac{v'^2}{c^2}}\left(1 + \frac{ax'}{c^2}\right)dt' \tag{6.11}$$

Upon observation, we can tell that (6.11) reflects two phenomena: (1) time dilation in the framework of special relativity ($\sqrt{1 - v'^2/c^2}$), i.e. clocks in motion tick slower, and (2) gravitational time dilation under the effect of general relativity, i.e. clocks with higher gravitational potential energy tick faster. If we regard both types of dilation as separate factors under the conditions of weak gravitational field ($ax' \ll c^2$) and low speed ($v' \ll c$), the multiplication of their respective expressions will also yield (6.11).

6.2   Riemannian Manifold: Geometry of Curved Time and Space

After publishing the paper "On the Electrodynamics of Moving Bodies" in 1905, Einstein spent two more years contemplating the role played by the central force in Newtonian mechanics—gravity, in the world of relativity. Eventually, he managed to figure out the equivalence principle through a thought experiment, thereby putting an equal sign between the acceleration of a frame in a gravity-free environment and the static local gravitational field where the frame is located. This equivalent relationship extended the theory of special relativity, which was only applicable to inertial frames, to local gravitational fields. Needless to say, this was a glorious triumph for Einstein.

Yet the initial victory didn't guarantee immediate success. Several years after starting to engage in the research of gravity, Einstein was still stuck in the unknown. Although he did predict the degree of light bending caused by solar gravity to be 0.85ArcSec in 1911 (which he later modified as the actual value 1.75ArcSec in 1915 with his field equations) and proved the phenomena of time dilation and gravitational shift through the equivalence principle, all these were imprecise early-stage results in lack of a strictly-organized mathematical model. Meanwhile, he was also aware that once the gravitational field was introduced, the originally flat Minkowski spacetime would be curved and twisted by gravitational force into a complex geometric space. Just when it seemed like he'd come to his wits' end, he took up a teaching position at his alma mater University of Zurich in July of 1912, where his old friend and classmate Grossmann happened to be Head of the Math Department at that time. After Einstein described his difficulties, Grossmann told him that the Riemannian geometry, which Bernard Riemann[15] (1826–1866) had established in 1853, might be exactly what he needed.

Einstein had always trusted Grossmann on scientific insights, so after obtaining the advice, he immediately started looking into the mathematical realm although he wasn't familiar with it at all. It didn't take him long to find out that the Riemann geometry hadn't been adopted by many physicists despite having existed for more than half a century, which was probably a result of the extremely complex mathematical calculations involved in the model. Although scientists before him tended to avoid the Riemann geometry unless absolutely necessary, Einstein decided to take a leap of faith as he'd wasted so much time searching for a mathematical tool that could actually help with his research. Therefore, despite knowing he was walking into the storm, he still decided to fully dedicate himself to the studies of the Riemann manifold with firm resolve and determination.

In order to facilitate readers' understanding of the not-so-friendly Riemann geometry, we'll be providing a simplified introduction below. The following section will start with the definition of curved space and time before moving forward to mathematical tools like metric tensor, parallel transfer, the Levi-Civita connection, and the Christoffel symbols. Lastly, we'll wrap up the model establishment with the Riemann curvature tensor, which describes the curvature of space and time.

---

[15]A student of the famous mathematician Gauss, Riemann had actually completed "On the Hypotheses Which Lie at the Bases of Geometry", where he established the Riemannian manifold, by the early age of 27, but the paper didn't receive much attention because the mathematical calculations were way too complicated and geometrical models too abstruse. Therefore, the research wasn't published until 1868 after he died.

6.2.1 Definition of Curved Time and Space

To determine the definition of curved space and time, let's first examine the physical meaning of (6.9) before starting to derive the geometric properties of spacetime curvature. Represented by this equation is the interval between a pair of events perceived from the perspective of an observer in a four-dimensional proper-acceleration frame. Quite intuitively, a four-dimensional space is supposed to be constructed with four separate axes, but the actual lengths represented by the scale unit of each axis might not be the same at different spots in the space. For example, the unit length of the temporal axis at $x' = x_1$ is $(1 + ax_1/c^2)$ times that at $x' = 0$. To be clear, though, we can't attribute this phenomenon to the dilation caused by curved space and time because it's an inevitable result of observing the same spacetime from different frames. In other words, we'll need to understand the basic properties of coordinate transformation and spacetime curvature in order to avoid such confusion.

For the transformation of coordinates, let's start with a two-dimensional space, which, in terms of its definition, can be as simple as a curved surface. So, let's just suppose there's an ant that lives on a curved surface. In the world of this ant, which knows no concept of height (the third dimension), it's simply possible to describe all directions with expressions of left, right, front, and back, while the location of any point can also be determined with merely two parameters. By contrast, if we look at a two-dimensional "flat" surface, it can be treated as a Cartesian coordinate system with the position of any spot therein represented with the parameters x and y. As for the distance between any two points (ds) in this system, it can be calculated with the Euclidean distance formula:

$$ds^2 = dx^2 + dy^2 \qquad (6.12)$$

In fact, apart from the Cartesian coordinate system, there are many other choices that we can use to describe the position of the point (x, y). In a polar coordinate system, for example, the same point can be expressed as r and θ, with the former representing the distance between this point and the origin, the latter its polar angle.

Since both systems can be used to describe the same space, they should also be mutually transformable in one way or another. In our example, the equations for transformation are:

$$\begin{cases} x = r \cos \theta \\ y = r \sin \theta \end{cases} \qquad (6.13)$$

Next up, it is how to calculate the distance between any pair of points in the two types of coordinate systems that we want to figure out. Although the process of calculation might be different, the end results should be the same. For the distance between two points in a polar coordinate system, we can first differentiate (6.13):

$$\begin{cases} dx = \cos\theta\, dr - r\sin\theta\, d\theta \\ dy = \sin\theta\, dr + r\cos\theta\, d\theta \end{cases} \tag{6.14}$$

Then, the substitution of (6.14) into (6.12) will lead us to the equation for distance:

$$ds^2 = (\cos\theta\, dr - r\sin\theta\, d\theta)^2 + (\sin\theta\, dr + r\cos\theta\, d\theta)^2 = dr^2 + r^2 d\theta^2 \tag{6.15}$$

As can be observed from (6.12) and (6.15), although the distance equations in Cartesian and polar systems appear in different forms, they can still describe the same space. One thing worth special attention, though, is that this space has to be a flat, gravity-free environment because the presence of gravitational force will lead to curved space and time. It should also be noted that the difference between the two equations is caused by the fact that the same distance is described from two different coordinate systems, not by spacetime curvature.

What, then, is curved space and time, and what are some of the typical properties? As it might be overwhelming to jump right into four dimensions, let's first think back to the ant that lives on the curved surface and ask this question: does the ant have any possible way of knowing that the surface is curved? In fact, this is somehow similar to how ancient people were able to find out that Earth was not flat but round. However, while human beings could count on the positions of distant stars, shadows from sunlight, as well as solar and lunar eclipses as evidence, the ant has no access to anything outside the surface. Hence, it can only resort to certain geometric properties by drawing lines, measuring distances, and determining angles. But how does the ant draw a straight line on a curved surface in the first place? Below we propose two different methods:

(1) After drawing all possible routes between two points, choose the shortest one and it should be a straight line.
(2) Start from any chosen point and draw a short line to reach a second point, where another short line is drawn in parallel with the first one.

Keep repeating the same steps. In the end, the ant should be able to retrieve a straight line by connecting all the short lines.

Through mathematical calculation, the two different approaches can be proved to yield the same result. As for the length of the final straight line, it's basically all the short lines combined. When it comes to the measurement of angles, as the Riemannian geometric space that Einstein chose for his research is a smooth continuum, we can simply focus on a small-enough area on the curved surface and consider it flat. In other words, the measuring of angles can be done on this local flat surface.

If we draw a triangle following the approaches described in the line-drawing section above, the results will come out as shown in Fig. 6.2. On a flat surface, the triangle will appear as in Fig. 6.2a, where the three interior angles total 180°. When it comes to a spherical surface (which we may loosely think of as Earth), the change of shape opens up a lot more possibilities, one of which consists of a line overlapped with the equator and the other two connecting two different points on the divider with the North Pole, as illustrated in Fig. 6.2b. Under this situation, the total interior angles will become $90° + 90° + \theta > 180°$, where θ represents the angle formed by the lines intersecting at the Pole. If we move the triangle to an irregular curved surface, however, the sum of the three angles in Fig. 6.2c will grow much more complicated than can be explained in simple terms.

Yet the complexity of geometric properties on curved surfaces doesn't stop here. As our ant is capable of drawing lines and measuring lengths, we know it can also connect all the points with the same distance to a certain point O and form a circle centered on this point. If this circle is formed on a flat surface, the ratio of its perimeter and diameter should be π. When moved to a spherical surface, the ratio should be slightly smaller than π. However,

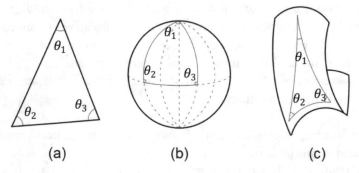

Fig. 6.2 Illustration of different geometries and interior angles: **a** Flat **b** Spherical **c** Riemannian

if put on a complicated curved surface, the value will be quite difficult to calculate.

Apart from the ratio, there's yet another peculiar property about curved surfaces that's indispensable to our following discussion. This time, we're going to make the ant carry a small arrow while walking on the triangle sides on the spherical surface. When taking every single step, this little ant also has to make sure the direction of the arrow is parallel with that at the previous moment. On a flat surface, it's obvious that the arrow will parallel its original direction after the ant walks the entire triangle, but the scenario is quite different on a spherical surface. As you can see below, the direction change of the arrow along the way is illustrated in Fig. 6.3. When the ant completes the entire route, the arrow is 90° apart from its original direction. But in fact, the degree of deviation depends on the size of the area enclosed by the route and the curvature of the surface in that particular area. Such a property will be mentioned again in Sect. 6.2.4 to define the Riemann curvature.

From the ant experiment, something important for us to learn is: to determine whether a 2-space is curved, there's no need to involve the third dimension. After all, all the ant did to decide whether its surface was curved or flat was draw a triangle, form a circle, and walk the triangular route carrying an arrow. More specifically, if a discrepancy is ever found out between the experimental results and the properties of Euclidean geometry, the ant can be sure that it lives in a twisted spacetime, and this is exactly the definition of curved space and time. Please note that the "twisting" in this context refers to the "intrinsic" curvature of the space. We can't just curl a piece of paper into a

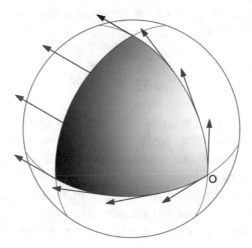

Fig. 6.3 Illustration of the vector in parallel transfer along triangle sides on a curved surface

cylinder and called it a twisted spacetime, a statement solely based on our visual perception from an external point of view. If we place the ant on the paper, the intrinsic geometric properties it observes will still comply with the Euclidean geometry, which is why the cylinder isn't actually a curved space.

Through the thought experiment above, we can tell it isn't very difficult to determine if a 2-space is curved. When the number of dimensions goes up to three, however, things will get a bit trickier and also negate the possibility of dealing with a four-dimensional spacetime first before applying the results back to three-dimensional spaces. As a result, we'll need to adopt a method similar to the ant approach to examine whether the properties of 3-space can fit in the framework of Euclidean geometry. For example, we can draw a circle and measure the ratio of the perimeter and diameter. Even when measured from the same perspective, though, the direction of the plane (xy, yz, and zx) where the circle is located might change the ratio. The same is true for the sum of the interior angles in a triangle, as well as the result of walking the triangular route carrying an arrow.

As a result, more parameters are needed in a 3-space than on a surface to accurately determine the curvature of a certain spot. Some might wonder why we don't just construct a sphere and calculate the ratio of the spherical surface A and the square of the radius r^2. As easy as it might sound, the ratio obtained from this approach actually represents the average curvature of varying directions, which doesn't illustrate the property in twisted spacetime that curved degree changes with position. To clarify this concept, we'll use an exception that won't be derived through field equations till Sect. 6.3.9.[16] Suppose there's a small-enough ball with evenly-distributed density in a space that contains matter, the ratio of the spherical surface and the square of its radius can be expressed as:

$$r - \sqrt{\frac{A}{4\pi}} = \frac{G}{3c^2} M \qquad (6.16)$$

This shows that if there's a mass M existing inside the ball, the three-dimensional space will be curved, causing the ratio to be smaller than expected. Yet if the ball contains no mass at all ($M = 0$), then the spherical surface area will again become $A = 4\pi r^2$, just as in Euclidean geometry.

Now that a space of mere three dimensions is already this complicated, how we're going to tackle the 4-space, the actual focus of general relativity? Well, the only thing we can do for now is to determine the difference between

---

[16]For details, please see Feynman [10].

the spacetime we live in and one without gravity, so let's return to the discussion of the gravitational field. Suppose we have a skyscraper with an observer A on the bottom floor, where the gravitational acceleration is g. Based on the equivalence principle, any physical phenomenon observed by A should be equivalent to what's perceived by a different observer moving upwards with the constant proper acceleration of g. Now, let's send another observer B to the top floor. As the distance between A and B is basically the height of the building h, which doesn't change with time, an issue in our assumption of equivalence will arise.

If the observations of A and B really are equivalent, B's proper acceleration has to be $g/(1 + gh/c^2)$. Yet in fact, it's impossible for the gravitational intensity at B's position to exactly equal this value because the gravitational field of Earth wouldn't auto-adjust to accord with the function $g(h) = -g/(1 + gh/c^2)$. The reason behind our incorrect assumption is the fact that the principle of equivalence doesn't factor in the geometric changes in the spacetime of the gravitational field. Rather, it only allows us to describe a flat, gravity-free space from the perspective of an accelerating observer. Therefore, the principle cannot explain the scenario of the skyscraper experiment, where the 4-space is curved by gravity. When gravitational force comes into the picture, the spacetime on Earth can become totally different from a flat space with no gravity in terms of geometric properties. Since we're unable to transform between the two types of systems, we'll resort to other approaches in the following sections to tackle the complicated curvature of time and space.

6.2.2 Basis Vector and Metric

In this section, we're going to take a detailed look at the interval Eqs. (6.4), (6.9), (6.12), and (6.15). While (6.4) only applies to the Minkowski spacetime, (6.9) represents the interval perceived by any observer with a constant acceleration in a flat spacetime. On the other hand, (6.12) and (6.15) apply to the Cartesian and polar coordinate systems under the framework of flat Euclidean geometry respectively. Of course, we can also describe the geometric properties of a two-dimensional sphere of the radius R with the parameters (θ, φ) in a spherical coordinate system (see Fig. 6.4). However, please note that $r = R$ is a non-changing attribute on the spherical surface. Under this situation, the equation for spacetime interval can be expressed as:

$$ds^2 = R^2 d\theta^2 + R^2 \sin^2 \theta d\varphi^2 \tag{6.17}$$

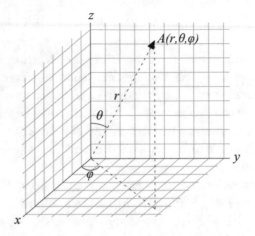

Fig. 6.4 Illustration of parameters in the spherical coordinate system

Among the five different equations for an interval, only (6.17) describes curved space, while all the others are only applicable to the flat counterpart. Such a difference actually lies at the center of our goal for this chapter: after discussing the Riemannian geometry, we should be able to tell whether it's a flat or curved spacetime that a given interval equation describes and quantify the curved degree, which Einstein called the "Riemann curvature". To measure such a curvature, one of the methods is sending our ant to walk a circle carrying an arrow (i.e. a vector) while ensuring the arrow stays in parallel directions the entire time. After the ant returns to the origin, we can measure the amount of change in the direction of this arrow to figure out the curvature. To not get ahead of ourselves, though, let's start by defining the concept of "vector" in curved spacetime and learning how to convey the parallel transport of such vectors using a mathematical approach.

A vector is an object that has both a magnitude and a direction. In a flat space regulated by Euclidean geometry, we can see a vector as an arrow that originates from a certain point in a coordinate system and ends at another. Even after parallel-transported following any random path, the arrow will still remain the same as neither its direction or length will change. Nevertheless, this is not the case in a curved space because the direction and even length of the vector can vary after the parallel transport. Therefore, it's best that we limit our vector to a local spacetime that's small enough. More specifically, the way of describing a certain vector only applies to its local area unless the vector is placed somewhere else by way of parallel transported. This said, varying routes of parallel transport can also lead to different end vectors.

So how do we define a local vector that's restricted to a fixed area? First, the local area for this vector has to be extremely small. Just as we've mentioned

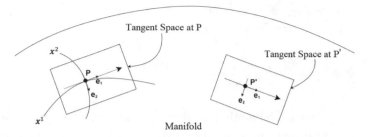

Fig. 6.5 Tangent plane at point *P* for an arbitrary 2D manifold (left square), which is embedded in the 3D manifold on which another tangent plane at *P′* is shown

before about the gravitational field, an area can be considered flat as long as it's small enough. In addition, this vector can't travel too long a distance at a time to avoid any obvious property change. Therefore, the local area of a vector should be a plane that's tangent to its originating point, and in the case of three or four dimensions, a tangent space. Illustrated in Fig. 6.5 are the planes tangent to P and P' on the spherical surface as well as the local vectors in these areas.[17]

With the definition of local vector established, our next task is learning to describe one using a mathematical approach. The most intuitive and convenient way, of course, is expressing a vector with the components along the axes of its local coordinate system. In the two-dimensional space in the left square of Fig. 6.5, for example, we can first represent the basis vectors that run along the directions of x^1 and x^2 as e_1 and e_2. This way, any vector V that originates from P can be expressed as:

$$V = V^1 e_1 + V^2 e_2 \qquad (6.18)$$

In which V^1 and V^2 represent the vector's components on x^1 and x^2 respectively. Please note that although e_1 and e_2 are basis vectors, their lengths are not necessarily 1 (unit vector). In addition, their directions and lengths might vary from one position to another.

In a 4-space, we can also follow the above method in describing local vectors. If we express 0 (which usually represents the temporal axis), 1, 2, 3

---

[17]The illustration of Fig. 6.5 as well as the derivation of this sector are referenced from Chap. 3 of Guidry [12].

(the other three axes) as Greek letters μ, ν, α, β, γ, σ, ρ, κ, etc., a four-dimensional vector can be written as:

$$V = \sum_{\mu=0,1,2,3} V^\mu \mathbf{e}_\mu = V^\mu \mathbf{e}_\mu \qquad (6.19)$$

One thing worth elaboration is that the equation in (6.19) is established with the Einstein summation convention, a notational convention that implies summation over a set of indexed terms in a formula, thus achieving notational brevity. Based on his rule, when a term is indexed with both a subscript and superscript twice or more, all possible values of the term has to be summed to avoid the hassle of writing the sigma sign over and over again. For example, the term $V^\mu \mathbf{e}_\mu$ in (6.19) comes with upper and lower indices, so we need to add up its values when μ is replaced with 0, 1, 2, and 3. In other words, we'll have to take special notice of such terms and determine if the Einstein notation should be followed from now on.

With the convention explained, let's move forward to define the inner product of a vector. If we consider two vectors originating from the same point, $A = A^\mu \mathbf{e}_\mu$ and $B = B^\nu \mathbf{e}_\nu$, their inner product can be calculated on the plane tangent to this particular point:

$$A \cdot B = (A^\mu \mathbf{e}_\mu) \cdot (B^\nu \mathbf{e}_\nu) = (\mathbf{e}_\mu \cdot \mathbf{e}_\nu) A^\mu B^\nu \qquad (6.20)$$

Please note that as the superscript and subscript indices μ and ν appear in pairs, $A \cdot B$ will have 16 possible values ($4 \times 4 = 16$), which we all need to add up. In an orthogonal coordinate system, by contrast, $\mathbf{e}_\mu \cdot \mathbf{e}_\nu$ has to be 0 when $\mu \neq \nu$. Hence, (6.20) will only be composed of four different values. This said, basic vectors aren't orthogonal in all generalized coordinate systems we choose for discussing curved spacetime, so $\mu \neq \nu$ doesn't necessarily mean that $\mathbf{e}_\mu \cdot \mathbf{e}_\nu$ equals 0. To facilitate subsequent calculation, we can define a new tensor below:

$$g_{\mu\nu} \equiv \mathbf{e}_\mu \cdot \mathbf{e}_\nu \qquad (6.21)$$

This way, (6.20) can be rewritten as:

$$A \cdot B = g_{\mu\nu} A^\mu B^\nu \qquad (6.22)$$

As $g_{\mu\nu}$ has $4 \times 4 = 16$ components, the term can be represented as a 4×4 matrix, called a "rank-2 tensor". As $\mathbf{e}_\mu \cdot \mathbf{e}_\nu = \mathbf{e}_\nu \cdot \mathbf{e}_\mu$ implies $g_{\mu\nu} = g_{\nu\mu}$, we know that $g_{\mu\nu}$ is a symmetric matrix. Among its 16 components, as a result, only 10 of them are independent. Something worth noting is that the inner product of any given vector with itself is the square of its length. So for the vector V, the value will be:

$$|V|^2 = V \cdot V = g_{\mu\nu}V^\mu V^\nu \tag{6.23}$$

From the above equation, we can tell that $\sqrt{g_{11}}$ is the unit length of the e_1 axis, while $\sqrt{g_{\mu\mu}}$ represents that of all the other axes. Due to this property, $g_{\mu\nu}$ is also called the "metric tensor" as it shows various unit lengths and the angles formed by any two axes. In fact, $g_{\mu\nu}$ is the function of the x location[18] because it changes with the basis vector in each local area. In other words, as long as we know the value of $g_{\mu\nu}(x)$ at every position in a given space, we can completely understand the space by describing its geometric properties with the metric tensor, the major mathematical tool used by Einstein in constructing the field equations.

Next up, we're going to establish a geometric formula for four-dimensional spacetime using the above-mentioned metric tensor. First, the displacement vector dx for two very close points can be expressed as:

$$dx = dx^0 e_0 + dx^1 e_1 + dx^2 e_2 + dx^3 e_3 = dx^\mu e_\mu \tag{6.24}$$

In which e_0 is the basis vector of the temporal axis, while e_1, e_2, and e_3 represent that of the three spatial axes. Based on (6.23), we can rewrite the square of the interval ds as:

$$ds^2 = g_{\mu\nu}dx^\mu dx^\nu \tag{6.25}$$

Then, applying (6.25) to the Minkowski spacetime, we can derive:

$$ds^2 = -\left(dx^0\right)^2 + \left(dx^1\right)^2 + \left(dx^2\right)^2 + \left(dx^3\right)^2 \tag{6.26}$$

In a Cartesian coordinate system, we know that: $dx^0 = cdt$, $dx^1 = dx$, $dx^2 = dy$, and $dx^3 = dz$. As a result, the metric tensor of the Minkowski

---

[18]Going forward, we'll use x as the uniform representation of a certain "location" (i.e. a certain point of coordinates) in a spacetime.

spacetime can be derived as (6.27) by comparing (6.25) and (6.26):

$$g_{\mu\nu} = \eta_{\mu\nu} = \begin{pmatrix} -1 & 0 & 0 & 0 \\ 0 & 1 & 0 & 0 \\ 0 & 0 & 1 & 0 \\ 0 & 0 & 0 & 1 \end{pmatrix} \tag{6.27}$$

Please note that the Minkowski spacetime is represented with the special sign $\eta_{\mu\nu}$ here to differentiate it from other counterparts. For spacetime in frames with constant proper acceleration, we know from (6.9) that:

$$ds^2 = -\left(1 + \frac{ax^1}{c^2}\right)^2 c^2 \left(dx^0\right)^2 + \left(dx^1\right)^2 + \left(dx^2\right)^2 + \left(dx^3\right)^2 \tag{6.28}$$

Similarly, if we compare (6.28) and (6.25), we can derive:

$$g_{\mu\nu} = \begin{pmatrix} -\left(1 + \frac{ax^1}{c^2}\right)^2 & 0 & 0 & 0 \\ 0 & 1 & 0 & 0 \\ 0 & 0 & 1 & 0 \\ 0 & 0 & 0 & 1 \end{pmatrix} \tag{6.29}$$

As can be told from (6.29), $g_{\mu\nu}$ is the function of the coordinate x^1, meaning the unit length of the temporal axis varies with x^1. While the physical meaning of such variation has been discussed in detail before, we still need to emphasize that although the metric tensors in (6.27) and (6.29) appear in different forms, they actually represent the same spacetime geometry. The only reason why they're not of an identical form is the fact that we adopted two different coordinate systems in describing one single spacetime. For example, while the metric tensor of a two-dimensional Cartesian coordinate system is $\begin{pmatrix} 1 & 0 \\ 0 & 1 \end{pmatrix}$ (see (6.12)), it'll become $\begin{pmatrix} 1 & 0 \\ 0 & r^2 \end{pmatrix}$ when switched to a flat polar system (see (6.15)). Despite the varying forms, it's one flat surface with no curvature that the two metric tensors both describe.

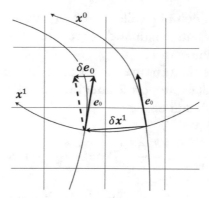

Fig. 6.6 Illustration of basis vector in parallel transport in curved spacetime, inspired by Fig. 2.5 of Hamilton [13]

6.2.3 Parallel Transport of Vector

In this section,[19] we're going to discuss how to parallel-transport a vector, just as in Fig. 6.3. Suppose there's a vector A on a plane tangent to x that we need to move to a very close plane that's tangent to $x + dx$. As dx is an extremely small value, the entire transport can be regarded as done on the same plane. If the vector stays in parallel directions through the entire process with its length unchanged, it'll still be the same vector when the move is completed. However, as the basis vectors at $x + dx$ and x might be different, the component of A along each axis can change as well. As you can see in Fig. 6.6, if we move the basis vector $\mathbf{e}_0(x)$ at x to $x + \delta x^1 \mathbf{e}_1$ along the x^1 axis, the end vector will become $\mathbf{e}_0(x + \delta x^1 \mathbf{e}_1)$ due to the new component on x^1. In the figure below, the dotted arrow represents the theoretical position of the parallel-transported basis vector, while the solid one on its left side is the actual result of the transport.

As a next step, we're going to describe the transport process with a mathematical expression. First, let's express the vector A on the tangent plane at x as $A(x)$ and use $A(x + dx)$ to represent its counterpart that's parallel-transported to $x + dx$. As dx is very small, we can develop the below equation:

$$A(x + dx) = A(x) + \frac{\partial A}{\partial x^\nu} dx^\nu \tag{6.30}$$

---

[19]The derivation of this section is referenced from Chap. 2 of Hamilton [13].

Based on the definition of parallel transport, any vector should remain in the same direction with its length unchanged after moved by a very short distance. Therefore, we know that $A(x + dx) = A(x)$. Meanwhile, the fact that dx^ν is a random value allows us to choose a path that satisfies $\partial A / \partial x^\nu = 0$ as the route for our parallel transport. Under these conditions, we can substitute $A = A^\mu \mathbf{e}_\mu$ into (6.30) and obtain:

$$0 = \frac{\partial A}{\partial x^\nu} = \frac{\partial (A^\mu \mathbf{e}_\mu)}{\partial x^\nu} = \frac{\partial A^\mu}{\partial x^\nu} \mathbf{e}_\mu + A^\mu \frac{\partial \mathbf{e}_\mu}{\partial x^\nu} \qquad (6.31)$$

Please note that $\mathbf{e}_\mu$ itself is the function of a coordinate, so the value of $\partial \mathbf{e}_\mu / \partial x^\nu$ won't necessarily be 0, just as illustrated in Fig. 6.6. In addition, as the components of $\mathbf{e}_\mu$ on all directions can vary after it's moved to a different location, we need to express it as a linear combination of the basis vector $\mathbf{e}_\kappa$:

$$\frac{\partial \mathbf{e}_\mu}{\partial x^\nu} = \Gamma^\kappa_{\mu\nu} \mathbf{e}_\kappa \qquad (6.32)$$

In (6.32), $\Gamma^\kappa_{\mu\nu}$ is the coefficient of the linear combination. As κ is both the upper and lower index for $\Gamma^\kappa_{\mu\nu}$ and $\mathbf{e}_\kappa$, all values of both terms when $\kappa = 0, 1, 2, 3$ have to be added up. Meanwhile, some might have noticed that ν and μ are both subscripted in $\Gamma^\kappa_{\mu\nu}$. This is actually a purposeful arrangement since it's best for all terms in an equation to have the same numbers of subscripts and superscripts. When an index appears in both upper and lower positions, the indices can be offset. As for when a parameter is divided by a one with an upper or lower index, it can be regarded as multiplied by the same parameter, but with the index, position reversed. If we take (6.32) as an example, $\frac{\partial \mathbf{e}_\mu}{\partial x^\nu}$ can be considered equivalent to having two lower indices as $\mathbf{e}_\mu$ has a lower index and x^ν an upper one. On the other side of the equal sign, we can offset the κ in both parameters, leaving $\Gamma^\kappa_{\mu\nu} \mathbf{e}_\kappa$ with two subscripts, same as in the simplified result of $\frac{\partial \mathbf{e}_\mu}{\partial x^\nu}$. The above explanation summarizes the rules to be followed going forward. For all future calculations, we must pay special attention to the number of sub- and superscripts as a way of identifying incorrect steps in our derivation.

Looking back at the coefficient $\Gamma^\kappa_{\mu\nu}$ in (6.32), it actually has a proper name—the Christoffel symbols, also called the coefficient of connection. When we go on to compare vectors on tangent planes, which may vary with their positions in terms of geometric properties, the complicated Christoffel symbols will come in handy as it makes the connection possible between two

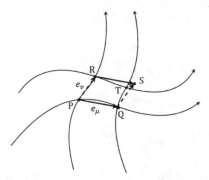

Fig. 6.7 Illustration of how basis vector changes when parallel-transported in curved spacetime

neighboring planes. To be more specific, $\Gamma^\kappa_{\mu\nu}$ can be understood as the projection (along $\mathbf{e}_\kappa$) of the difference between the original basis vector $\mathbf{e}_\mu$ and $\mathbf{e}_\mu$ moved by a unit length along $\mathbf{e}_\nu$.

The Christoffel symbols have another important property: $\Gamma^\kappa_{\mu\nu} = \Gamma^\kappa_{\nu\mu}$. Suppose there are two neighboring axes in a small enough area, as shown in Fig. 6.7. If we represent the basis vectors $\mathbf{e}_\mu$ and $\mathbf{e}_\nu$ as PQ and PR before parallel-transporting $\mathbf{e}_\mu$ from P along $\mathbf{e}_\nu$ by a unit length to R, $\mathbf{e}_\mu$ will then become RS while PRSQ forms a parallelogram. However, the basis vector that shares the direction of $\mathbf{e}_\mu$ at R is RT, so we know the component of ST along $\mathbf{e}_\kappa$ is $\Gamma^\kappa_{\mu\nu}$ due to the definition of Christoffel symbols. Next, we can again parallel-transport $\mathbf{e}_\nu$ from P along $\mathbf{e}_\mu$ by a unit length to Q, which we know will lead $\mathbf{e}_\nu$ to become QS because PRSQ is a parallelogram. As the basis vector sharing the direction of $\mathbf{e}_\nu$ at Q is QT, we can again determine the component of ST along $\mathbf{e}_\kappa$ to be $\Gamma^\kappa_{\nu\mu}$. Comparing the results[20] of the two parallel transports described above, the important property of $\Gamma^\kappa_{\mu\nu} = \Gamma^\kappa_{\nu\mu}$ is thus verified.

Then, the substitution of (6.32) back into (6.31) will lead us to:

$$0 = \frac{\partial A^\mu}{\partial x^\nu}\mathbf{e}_\mu + A^\mu \Gamma^\kappa_{\mu\nu}\mathbf{e}_\kappa = \left(\frac{\partial A^\mu}{\partial x^\nu} + \Gamma^\mu_{\kappa\nu}A^\kappa \right)\mathbf{e}_\mu \qquad (6.33)$$

Please note that the second equal sign in (6.33) is established by swapping κ and μ in $A^\mu \Gamma^\kappa_{\mu\nu}\mathbf{e}_\kappa$. The exchange doesn't affect the mathematical meaning of the term since both indices have to be replaced with 0, 1, 2, and 3. From

---

[20]In fact, the verification process and its result only apply to torsion free spaces, a postulate on which the theory of general relativity is based. Therefore, we won't discuss torsion any further here.

(6.33), we can subsequently derive:

$$\frac{\partial A^{\mu}}{\partial x^{\nu}} = -\Gamma^{\mu}_{\kappa\nu}A^{\kappa} \tag{6.34}$$

Expanding (6.34), we'll obtain:

$$A^{\mu}(x + dx) = A^{\mu}(x) - \Gamma^{\mu}_{\kappa\nu}A^{\kappa}dx^{\nu} \tag{6.35}$$

Which means if we parallel-transport the vector A from x to $x + dx$, the length of the projection A^{μ} along $\mathbf{e}_{\mu}$ will equal the difference between $A^{\mu}(x)$ and $A^{\mu}(x + dx)$, which can be expressed as $-\Gamma^{\mu}_{\kappa\nu}A^{\kappa}dx^{\nu}$.

Up to this point, we've already learned from (6.35) the changing pattern of each component of a vector when it's parallel-transported, but one question remains: how do we determine the value of $\Gamma^{\kappa}_{\mu\nu}$? As some might still remember, we've mentioned before that $g_{\mu\nu}$ can be used to describe all geometric properties in a space, so we'll simply figure out a way of expressing $\Gamma^{\kappa}_{\mu\nu}$ with $g_{\mu\nu}$. Since $\Gamma^{\kappa}_{\mu\nu}$ is defined through the differential of a basis vector, we can differentiate $g_{\mu\nu} = \mathbf{e}_{\mu} \cdot \mathbf{e}_{\nu}$ and retrieve:

$$\begin{aligned}
\frac{\partial g_{\mu\nu}}{\partial x^{\lambda}} &= \frac{\partial(\mathbf{e}_{\mu} \cdot \mathbf{e}_{\nu})}{\partial x^{\lambda}} = \frac{\partial \mathbf{e}_{\mu}}{\partial x^{\lambda}} \cdot \mathbf{e}_{\nu} + \mathbf{e}_{\mu} \cdot \frac{\partial \mathbf{e}_{\nu}}{\partial x^{\lambda}} \\
&= \Gamma^{\kappa}_{\mu\lambda}\mathbf{e}_{\kappa} \cdot \mathbf{e}_{\nu} + \mathbf{e}_{\mu} \cdot \Gamma^{\kappa}_{\nu\lambda}\mathbf{e}_{\kappa} = g_{\nu\kappa}\Gamma^{\kappa}_{\lambda\mu} + g_{\mu\kappa}\Gamma^{\kappa}_{\nu\lambda}
\end{aligned} \tag{6.36}$$

In deriving the terms on the right side of the third equal sign in (6.36), we've actually used the definition of connection coefficient and the symmetrical property of $\Gamma^{\kappa}_{\mu\lambda} = \Gamma^{\kappa}_{\lambda\mu}$ already. As such, we can rotate the μ, ν, and λ in the above equation and obtain the following two:

$$\frac{\partial g_{\nu\lambda}}{\partial x^{\mu}} = g_{\lambda\kappa}\Gamma^{\kappa}_{\mu\nu} + g_{\nu\kappa}\Gamma^{\kappa}_{\lambda\mu} \tag{6.37}$$

$$\frac{\partial g_{\lambda\mu}}{\partial x^{\nu}} = g_{\mu\kappa}\Gamma^{\kappa}_{\nu\lambda} + g_{\lambda\kappa}\Gamma^{\kappa}_{\mu\nu} \tag{6.38}$$

Last but not least, we're going to add (6.37)–(6.38) and subtract (6.36) before dividing the result by 2:

$$g_{\lambda\kappa}\Gamma^{\kappa}_{\mu\nu} = \frac{1}{2}\left(\frac{\partial g_{\lambda\mu}}{\partial x^{\nu}} + \frac{\partial g_{\nu\lambda}}{\partial x^{\mu}} - \frac{\partial g_{\mu\nu}}{\partial x^{\lambda}}\right) \tag{6.39}$$

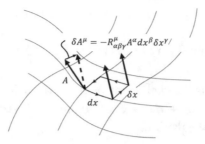

Fig. 6.8 Illustration of vector A parallel-transported 360° in curved spacetime, inspired by the figures shown in https://spaces.ac.cn/archives/4014

Then, if we define $g^{\alpha\lambda}$ as the inverse matrix of $g_{\lambda\kappa}$ (i.e. $g^{\alpha\lambda}g_{\lambda\kappa} = \delta^{\alpha}_{\kappa}$), we can multiply the terms on both sides of the equal sign by $g^{\alpha\lambda}$ and obtain the final result:[21]

$$\Gamma^{\alpha}_{\mu\nu} = \frac{1}{2}g^{\alpha\lambda}\left(\frac{\partial g_{\lambda\mu}}{\partial x^{\nu}} + \frac{\partial g_{\nu\lambda}}{\partial x^{\mu}} - \frac{\partial g_{\mu\nu}}{\partial x^{\lambda}}\right) \qquad (6.40)$$

6.2.4 Riemann Curvature Tensor

Having analyzed the mechanism of parallel-transport, we can now move a vector following a closed route and determine if space is curved through the approach in Fig. 6.3. Meanwhile, we can also calculate the curvature along different directions in the space. To do so, let's suppose we're parallel-transporting the vector A along the path described in (6.41):

$$x \to x + dx \to x + dx + \delta x \to x + \delta x \to x \qquad (6.41)$$

And eventually taking it back to the origin. The amount of change ΔA^{μ} in the component A^{μ} of A along $\mathbf{e}_{\mu}$ can be found illustrated in Fig. 6.8. If ΔA^{μ} is a non-zero value (i.e. if the component has experienced any change), then we know the space is curved.

As a next step, let's express the parallel transport process in Fig. 6.8 as mathematical equations by first considering the vector's move from x to $x + dx$. According to (6.35), we know the relationship between $A^{\mu}(x)$ (the value

---

[21]In the discussion here, δ^{α}_{κ} is defined as $\delta^{\alpha}_{\kappa}\delta^{\alpha}_{\kappa} = \begin{cases} 1, if\ \alpha = \kappa \\ 0,\ else \end{cases}$.

of A^μ at x) and $A^\mu(x \mid dx)$ (when transported to $x + dx$) is:

$$A^\mu(x + dx) = A^\mu(x) - \Gamma^\mu_{\alpha\beta}(x)A^\alpha dx^\beta \tag{6.42}$$

In (6.42), $\Gamma^\mu_{\alpha\beta}(x)$ is the coefficient of connection at x. Similarly, when the transport $x + dx \rightarrow x + dx + \delta x$ is completed, A^μ can be transformed through the following equation:

$$\begin{aligned}
A^\mu(x + dx + \delta x) &= A^\mu(x + dx) - \Gamma^\mu_{\nu\gamma}(x + dx)A^\nu(x + dx)\delta x^\gamma \\
&= A^\mu(x) - \Gamma^\mu_{\alpha\beta}(x)A^\alpha dx^\beta \\
&\quad - \left(\Gamma^\mu_{\nu\gamma}(x) + \frac{\partial \Gamma^\mu_{\nu\gamma}(x)}{\partial x^\beta}dx^\beta\right)\left(A^\nu(x) - \Gamma^\nu_{\alpha\beta}(x)A^\alpha dx^\beta\right)\delta x^\gamma
\end{aligned} \tag{6.43}$$

In (6.43), $\Gamma^\mu_{\nu\gamma}(x + dx)$ is the coefficient of connection at $x + dx$. Please note that the expression between the two equal signs is derived by expanding dx to the first order before replacing $A^\mu(x + dx)$ with (6.42).

Then, if we consider the process of $x \rightarrow x + \delta x \rightarrow x + \delta x + dx$, which is the reverse transport of $x + dx + \delta x \rightarrow x + \delta x \rightarrow x$, the final expression for $A^\mu(x + \delta x + dx)$ can be derived as (6.44), similar to (6.43):

$$\begin{aligned}
A^\mu(x + \delta x + dx) &= A^\mu(x) - \Gamma^\mu_{\nu\gamma}(x)A^\nu \delta x^\gamma \\
&\quad - \left(\Gamma^\mu_{\alpha\beta}(x) + \frac{\partial \Gamma^\mu_{\alpha\beta}(x)}{\partial x^\gamma}\delta x^\gamma\right)\left(A^\alpha(x) - \Gamma^\alpha_{\nu\gamma}(x)A^\nu \delta x^\gamma\right)dx^\beta
\end{aligned} \tag{6.44}$$

Next up, we can subtract (6.43) from (6.44) to obtain the solution of δA^μ, which represents the amount of change in A^μ when A is parallel-transported following the 360° route of $x \rightarrow x + dx \rightarrow x + dx + \delta x \rightarrow x + \delta x \rightarrow x$:

$$\begin{aligned}
\delta A^\mu &= A^\mu(x + dx + \delta x) - A^\mu(x + \delta x + dx) \\
&= -\left(\frac{\partial \Gamma^\mu_{\nu\gamma}}{\partial x^\beta}A^\nu - \frac{\partial \Gamma^\mu_{\alpha\beta}}{\partial x^\gamma}A^\alpha + \Gamma^\mu_{\alpha\beta}\Gamma^\alpha_{\nu\gamma}A^\nu - \Gamma^\mu_{\nu\gamma}\Gamma^\nu_{\alpha\beta}A^\alpha\right)dx^\beta \delta x^\gamma \\
&= -\left(\frac{\partial \Gamma^\mu_{\alpha\gamma}}{\partial x^\beta} - \frac{\partial \Gamma^\mu_{\alpha\beta}}{\partial x^\gamma} + \Gamma^\mu_{\nu\beta}\Gamma^\nu_{\alpha\gamma} - \Gamma^\mu_{\nu\gamma}\Gamma^\nu_{\alpha\beta}\right)A^\alpha dx^\beta \delta x^\gamma \\
&= -R^\mu_{\alpha\beta\gamma}A^\alpha dx^\beta \delta x^\gamma
\end{aligned} \tag{6.45}$$

Please note that δx and dx are only accurate to the second order. In addition, we've also applied the new definition (6.46) in deriving the rightmost term in (6.45):

$$R^{\mu}_{\alpha\beta\gamma} \equiv \frac{\partial \Gamma^{\mu}_{\alpha\gamma}}{\partial x^{\beta}} - \frac{\partial \Gamma^{\mu}_{\alpha\beta}}{\partial x^{\gamma}} + \Gamma^{\mu}_{\nu\beta}\Gamma^{\nu}_{\alpha\gamma} - \Gamma^{\mu}_{\nu\gamma}\Gamma^{\nu}_{\alpha\beta} \qquad (6.46)$$

In fact, it's exactly the Riemann curvature tensor that (6.46) defines, and the defining equation also satisfies the Einstein summation convention that each term should have the same number of valid sub- and superscripts.

As for the mathematical significance of the Riemann curvature tensor $R^{\mu}_{\alpha\beta\gamma}$, (6.72) can give us a clear answer: if we move the vector A along a parallelogram with the length/width of $dx^{\beta}\mathbf{e}_{\beta}/\delta x^{\gamma}\mathbf{e}_{\gamma}$, the component A^{μ} along $\mathbf{e}_{\mu}$ will decrease by $R^{\mu}_{\alpha\beta\gamma}A^{\alpha}dx^{\beta}\delta x^{\gamma}$ when A is returned to the origin. Please note that each side of the parallelogram has a different direction, and each direction can cause changes in the four components of A. Hence, the Riemann curvature tensor has $4^4 = 256$ components, and we can be sure that a space is curved as long as any of the 256 values isn't 0. For example, if we substitute (6.27) or (6.29) into (6.46), a complicated series of calculations will eventually lead us to the result that all components of $R^{\mu}_{\alpha\beta\gamma}$ are 0. In other words, the corresponding space must be flat whether it's a Minkowski spacetime or a coordinate system with a constant proper acceleration. However, if we substitute (6.17) instead, the calculation will result in several non-zero components, showing that the spherical surface is curved. Last but not least, it should be brought to attention that all coefficients of connection Γ can be expressed with metric tensors, which we know from (6.40). Meanwhile, (6.46) also shows that $R^{\mu}_{\alpha\beta\gamma}$ can be represented with Γ. Combining the two relationships, we know the Riemann curvature tensor can fully depend on the metric tensor $g_{\mu\nu}$.

6.3 Einstein's Field Equations

In Sects. 6.1 and 6.2, we've clarified the physical significance of four-dimensional spacetime, established necessary mathematical tools while introducing the Riemannian manifold, and therefore determined how to describe the distribution of curvature in a given space. Yet as Einstein's ultimate goal was to modify Newton's theory in a way for the law of universal gravitation to fit in the framework of relativity, it wasn't just mass (i.e. energy) but the factor of momentum as well that he had to consider in order to describe the curvature of spacetime comprehensively. Essentially, he had to derive a gravitational field equation to regulate the geometric relationship between mass/energy/momentum and spacetime. Once such an equation was established, he could combine it with the distribution functions of the three factors to derive the metric tensor $g_{\mu\nu}$, which could then help determine the motion behavior of objects in spacetime. However, the process of developing this hypothesized field equation was easier said than done, trapping Einstein in a quagmire as if he was trudging through a muddy patch because a 4-space under the effect of gravity was just too complicated. Therefore, we will adopt a simplified, step-by-step approach in this section to recounting how he constructed the enormously influential equation with the concepts of symmetry and invariance.

Before diving into the derivation process, here's something we need to emphasize: when Einstein was developing his field equations, he often resorted to unproved but very precise hypotheses, which were compatible with certain known phenomena and even served as predictions of related ones as well. In fact, such a drill had been practiced by many physicists after Galilei who attempted to explain a wide range of natural events with simple theories. Although their claims weren't necessarily backed by solid evidence at the beginning, oftentimes they could manage to come up with experiments that verified such theories. This is exactly why physics is usually deemed an experimental science. When Newton first created the law of universal gravitation, he didn't explain why the equation consisted of mass and distance, while Einstein didn't elaborate on why matter would cause curvature in space, either. Nevertheless, they both counted on prudent predictions and derivations to eventually achieve the end results—mathematical equations glowing with the beauty of simplicity and symmetry. Even better, these equations can explain numerous physical phenomena. In fact, it's exactly such logical reasoning and deduction that allowed their genius and insight as scientists to shine through. Their intuitive but discreet methodology is also the focus of our following discussion.

6.3.1 Review and Integration—How Newton's Law of Universal Gravitation and Curved Spacetime are Related

For this section, let's first recall the basics of Newton's law of universal gravitation, which is epitomized in the below equation:

$$\vec{F} = -\frac{GMm}{r^2}\hat{r} \tag{6.47}$$

In (6.47), $\vec{F}$ is the gravitational force exerted by the point mass M on m, G the gravitational constant, and $\hat{r}$ the unit vector that shares the direction of the line joining M and m. Generally speaking, $\vec{F}$ can be expressed as $\vec{F} = m\vec{g}$, where $\vec{g}$ represents the gravitational acceleration generated by M at the location of m. As such, (6.47) can be rewritten as:

$$\vec{g} = -\frac{GM}{r^2} \tag{6.48}$$

Although we're generally quite familiar with (6.47) and (6.48), they can only be applied to very limited scenarios, i.e. the gravitational field produced by a point mass or object with spherical symmetry. For a wider range of application, the law of universal gravitation needs to be rewritten by adding the factor of divergence:

$$\nabla \cdot \vec{g} = -4\pi G\rho \tag{6.49}$$

Where ρ is mass density, and $\nabla \equiv \hat{x}\frac{\partial}{\partial x} + \hat{y}\frac{\partial}{\partial y} + \hat{z}\frac{\partial}{\partial z}$. Therefore:

$$\nabla \cdot \vec{g} = \frac{\partial g_x}{\partial x} + \frac{\partial g_y}{\partial y} + \frac{\partial g_z}{\partial z} \tag{6.50}$$

In (6.50), g_x, g_y, and g_z are the strengths of the gravitational field along the directions of x, y, and z. As such, we can further define the gravitational potential energy ϕ as having to satisfy:

$$\vec{g} = -\nabla\phi \qquad (6.51)$$

Then, we can substitute (6.51) back into (6.49) and derive the equation for gravitational potential energy:

$$\nabla^2\phi = 4\pi G\rho \qquad (6.52)$$

Above we summarized the theory of universal gravitation, which Einstein considered incomplete because Newton only described the effect of gravity with the factor of acceleration. In addition, just a transformation of the reference frame (i.e. switch of perspective) can drastically change how a gravitational field is distributed, so Einstein was hoping to formulate a theory that could still apply even after the transformation of coordinates. As we know a flat space with no mass-energy distribution is always flat despite any frame transformation, "curvature" seems to be a good choice in describing the effect of gravitational force, at least for a flat space as it remains 0 anyway. In fact, it was exactly the Riemann curvature that Einstein used to establish his theory on gravity.

6.3.2 Geodesic: Spacetime Path of Freely Moving Object

After deciding to adopt the Riemann curvature, Einstein knew he had to determine the relationship between spacetime geometry and gravitational acceleration in order to figure out the motion behavior of objects or point masses. In Newtonian mechanics, the motion of an object is regulated by $\vec{F} = m\vec{g}$, but in general relativity, a space under the force of gravity is a curved spacetime with a non-zero Riemann curvature, rather than an acceleration field. Hence, there are different rules of motion that objects abide by in flat and curved spaces. In the latter environment, a freely moving object with a given point/time of departure/arrival will always move in a pattern that maximizes the proper time interval.

To verify the above-mentioned rule, we'll start with an inertial frame of reference. Due to the strong principle of equivalence, any type of spacetime can be found to have a local inertial reference frame where the departure and arrival point of a moving object lies at the same spot, and this is the frame from which we're going to observe the motion of the object. It's generally known that an object at rest in an inertial frame stays at rest. If unconstrained by any other limits than the point/time of departure/arrival, such an object can have an infinite number of paths to take. Among all the possible motions patterns, though, the one where the object simply stays at its original location has the longest proper time interval.

As we've discussed before, the proper time interval can be understood as the time passed on the clock attached to a certain object. Once this object goes into motion, we'd think the clock starts to tick slower due to the effect of time dilation. Therefore, the proper time interval is the longest when the object simply stays where it is without moving at all. This rule applies to all types of reference frames because proper time interval remains invariant through the transformation of coordinates.

To express motion rules followed by objects in curved spaces with a mathematical equation, let's consider the most common scenario, where the interval between any two points is $ds^2 = g_{\mu\nu}dx^\mu dx^\nu$. Substituted with a proper time interval, the equation can also become:

$$d\tau^2 = -\frac{1}{c^2}g_{\mu\nu}dx^\mu dx^\nu \tag{6.53}$$

As such, we can prove that an object has to satisfy the below equation to maximize its proper time interval:

$$\frac{d^2x^\kappa}{d\tau^2} + \Gamma^\kappa_{\mu\alpha}\frac{dx^\mu}{d\tau}\frac{dx^\alpha}{d\tau} = 0 \tag{6.54}$$

In fact, (6.54) is the general form of motion equation for objects in curved spacetime, commonly known as the "geodesic equation". For details, please see "Appendix 6.2: Derivation of the Geodesic Equation".

Up till this point, we've covered all the required tools for constructing the relation between gravitational acceleration and spacetime curvature. As the equation of universal gravitation must be compatible with the theory of general relativity in a low-speed weak field, we're going to approximate (6.54) with $d\tau \approx dt$, which leads us to $\frac{dx^0}{d\tau} = \frac{cdt}{d\tau} \approx c$. Meanwhile, $i = 1, 2, 3$ also shows us that $\frac{dx^i}{d\tau} \ll c$, so all values of $\frac{dx^i}{d\tau}\frac{dx^j}{d\tau}$ can be ignored when compared to $\frac{dx^0}{d\tau}\frac{dx^0}{d\tau}$. As such, the result of approximation will be:

$$\frac{d^2x^\kappa}{dt^2} = -\Gamma^\kappa_{00}c^2 \tag{6.55}$$

While (6.40) also tells us that:

$$\Gamma^\kappa_{00} = \frac{1}{2}g^{\kappa\nu}\left(2\frac{\partial g_{0\nu}}{\partial x^0} - \frac{\partial g_{00}}{\partial x^\nu}\right) \tag{6.56}$$

In the framework of Newtonian mechanics, all gravitational fields are considered static (or, to be more precise, the rate of change for gravitational acceleration over time is extremely slow). Therefore, the time rate of change for all metric tensors in such fields can be regarded as 0 ($\frac{\partial g_{\mu\nu}}{\partial x^0} \approx 0$), meaning (6.56) can be simplified as:

$$\Gamma^\kappa_{00} = -\frac{1}{2}g^{\kappa\nu}\frac{\partial g_{00}}{\partial x^\nu} \tag{6.57}$$

Next, let's consider the approximate solution under a weak field, where the metric tensor $g_{\mu\nu}$ nearly equals the Minkowski counterpart, i.e. $g_{\mu\nu} \approx$

$\eta_{\mu\nu} = \begin{pmatrix} -1 & 0 & 0 & 0 \\ 0 & 1 & 0 & 0 \\ 0 & 0 & 1 & 0 \\ 0 & 0 & 0 & 1 \end{pmatrix}$. As for its inverse matrix, we have $g^{\kappa\nu} \approx \eta^{\kappa\nu} =$

$\begin{pmatrix} -1 & 0 & 0 & 0 \\ 0 & 1 & 0 & 0 \\ 0 & 0 & 1 & 0 \\ 0 & 0 & 0 & 1 \end{pmatrix}$. Observing (6.57), we can tell that the equal sign will only

be followed by non-zero elements when $\nu = \kappa$. In addition, $\kappa = \nu = 0$ will lead to the term $\frac{\partial g_{00}}{\partial x^0}$, which can also be approximated to 0 in a static field. As a result, the only non-zero term we're left with is:

$$\Gamma^\kappa_{00} = -\frac{1}{2}\frac{\partial g_{00}}{\partial x^\kappa} \tag{6.58}$$

Then, the substitution of (6.58) back into (6.55) will yield:

$$\frac{d^2 x^\kappa}{dt^2} = \frac{c^2}{2}\frac{\partial g_{00}}{\partial x^\kappa} \tag{6.59}$$

Subsequently, we can compare (6.59) and (6.51) to obtain:

$$\frac{c^2}{2}\frac{\partial g_{00}}{\partial x^\kappa} = -\frac{\partial \phi}{\partial x^\kappa} \tag{6.60}$$

Which is equivalent to:

$$g_{00} = -\frac{2\phi}{c^2} + \text{constant} \tag{6.61}$$

Where we know the constant is -1 as the approximation method is adopted for a weak field. Hence:

$$g_{00} = -\left(1 + \frac{2\phi}{c^2}\right) \tag{6.62}$$

With the Newtonian approximation, a four-dimensional spacetime under a low-speed, weak static field can be expressed as:

$$ds^2 = -\left(1 + \frac{2\phi}{c^2}\right)c^2 dt^2 + dx^2 + dy^2 + dz^2 \tag{6.63}$$

In (6.63), ϕ represents gravitational potential energy. As for the physical significance of the equation, it implies that a clock ticks faster where gravitational potential energy is larger. With this result, we can also come to the important conclusion that the object acceleration calculated simultaneously from the metric tensor Eq. (6.63) and the geodesic Eq. (6.54) is exactly the same as the result under the rules of Newtonian mechanics.

Last but not least, we need to emphasize again that the acceleration of an object isn't necessarily induced by the gravitational field but can be a result of the observer's acceleration as well (for example, when the observer is in a frame with constant proper acceleration). Based on the principle of equivalence, we can't tell which is the actual case in a local environment, but in a larger area, the observer should be able to distinguish the driving force of such acceleration because the gravitational field generated by mass/energy or momentum will produce spacetime curvature. By contrast, a spacetime where no gravity is present will still be perceived as flat even when the observer is moving with acceleration. Therefore, Einstein claimed that the Riemann curvature was an ideal choice for describing the impact of an object on the spacetime where it's located. To further illustrate his point from a mathematical perspective, the term $\nabla \cdot \vec{g}$ in (6.49) is the first-order differential of the gravitational acceleration, which again is the first-order differential of the metric tensor. Therefore, we know $\nabla \cdot \vec{g}$ to be the second-order differential of the metric tensor. When it comes to (6.46), the Riemann curvature contains the first-order differential of the Christoffel symbol Γ, which is also the first-order differential of the metric tensor. In other words, the curvature can also be understood as the second-order differential of the metric tensor. From the explanation above, it shouldn't be too hard to tell the similarity between the Riemann curvature and $\nabla \cdot \vec{g}$, and this is exactly why Einstein decided to replace the latter with a parameter related to the curvature.

6.3.3 Energy-Momentum Tensor

Before getting into details about how Einstein pulled off the replacement, let's first go back to (6.49), where $\nabla \cdot \vec{g}$ regulates the effect of $\vec{g}$ while $-4\pi G\rho$ describes the source of gravity. With the goal of transforming this equation into a form that could fit into the structure of general relativity, Einstein decided to replace the term $\nabla \cdot \vec{g}$ with parameters related to the Riemann curvature. As for what parameters exactly, it'll be the focus of this section. Just like we've mentioned before, the factors leading to the generation of the gravitational field aren't confined to mass but include energy and momentum as well. Since $-4\pi G\rho$ already covers the density of mass in a space, the term should become related to the density of energy and momentum after some kind of derivation. In addition, as we know that $\nabla \cdot \vec{g}$ is embedded in the geometry of spacetime and that all spacetime properties are regulated by the metric tensor $g_{\mu\nu}$, which is a 4×4 matrix, Einstein's new parameters for substitution should consist of a similar matrix as well.

Basing his research on the above reasoning, Einstein started looking for a tensor suitable for representing energy and momentum, which eventually led him to the "energy-momentum tensor". This tensor was no small deal as it was discovered by Minkowski when he was modifying the theories of Maxwell and Lorentz on electrodynamics following the principle of special relativity from 1907 to 1908. In addition, the tensor also plays an important role in the 1911 publication *Das Relativitätsprinzip* (*The Principle of Relativity*) of Max von Laue (1879–1960). Since the energy-momentum tensor was first created on the basis of special relativity, let's introduce some of its properties under the structure of this discipline through Lorentz transformations. Then, we'll extend our discussion to include other forms of the tensor that fit in the context of general relativity in later sections.

The energy-momentum tensor is generally expressed as $T^{\mu\nu}$, with the following definitions: (the range of i and j is 1–3)

1. T^{00}: density of mass. Equivalent to energy density because mass and energy are mutually convertible.
2. T^{0j}: energy flow along the direction of j (often divided by c for unit consistency)
3. T^{i0}: momentum density along the direction of i (often multiplied by c for unit consistency)
4. T^{ij}: density of the momentum influx toward the direction of j, also called the "momentum flow".

Judging from the above points, we can tell that $T^{\mu\nu}$ represents energy flow when μ is 0 and the momentum density along different directions when μ is replaced with 1–3; as for the second upper index, what the term $T^{\mu\nu}$ represents are the density of momentum and momentum flow along different directions when ν is replaced with 0 and 1–3 respectively.[22]

Now, we're going to express all the above-mentioned components and their corresponding physical quantities in proper forms to further determine their properties. In a pressure-free environment (postulated as such because pressure can impact the flow of momentum), energy and momentum can be represented with (5.113) and (5.114). Yet as what we're discussing here is the density of both quantities, the element "volume" has to be taken into consideration as well. Suppose we're looking at a small-enough local area containing mass. When observed from an inertial frame that's static in relation to this mass, the volume and mass density of this area can be represented with dV and ρ. If we switch to an inertial frame moving at the speed of v in relation to the mass, however, the volume will become $\sqrt{1 - v^2/c^2}dV$. This is because the length parallel to the direction of motion is shortened by $\sqrt{1 - v^2/c^2}$ times due to the contraction effect while the length perpendicular to the motion stays unchanged. As for the density, it now becomes $\rho' = \rho/\sqrt{1 - v^2/c^2}$ because the total mass in the area remains the same. Based on the above discussion and (5.113)–(5.114), when the mass in a certain area is perceived to be moving at the speed of v, the energy density T^{00} in this area and the momentum density T^{i0} along the direction of i can be expressed as:

$$T^{00} = \frac{\rho'c^2}{\sqrt{1 - \frac{v^2}{c^2}}} = \frac{\rho c^2}{1 - \frac{v^2}{c^2}} = \rho u^0 u^0 \tag{6.64}$$

---

[22]I.e. the number of a certain element passing through a unit area in a unit time. For example, an energy flow is defined by the amount of energy passing through a unit cross-sectional area in a unit time.

$$T^{i0} = \frac{\rho' v^i}{\sqrt{1 - \frac{v^2}{c^2}}} c = \frac{\rho v^i}{1 - \frac{v^2}{c^2}} c = \rho u^i u^0 \qquad (6.65)$$

The parameter ρ in (6.64) and (6.65) represents the density of mass measured from a frame that's static relative to the mass in the particular area. The 4-speed $u^\mu \equiv \left(c\frac{dt}{d\tau}, \frac{dx}{d\tau}, \frac{dy}{d\tau}, \frac{dz}{d\tau}\right)$ introduced in Sect. 5.11 was also used for simplifying purposes in the derivation of the above equations.

Next, let's move on to the concept of "flow". To determine the momentum flow along the direction of x, we need to know how much momentum passes through a unit area on the yz plane in a unit time. If the moving speed and density of matter along the direction of x are represented with v^x and ρ' from the perspective of an inertial frame, the momentum passing through the area of A on the yz plane during the time period Δt can be expressed as $\gamma \rho' \vec{v} A v^x \Delta t$. As such, we can tell from the existing definition that the momentum flow is $\gamma \rho' \vec{v} v^x$, where $\vec{v}$ is the three-speed vector of the substances in the unit area. In normal scenarios, the components of energy and the i-direction momentum along the direction of j are $\gamma \rho' c^2 v^j$ and $\gamma \rho' v^i v^j$ respectively, which is why we can derive:

$$T^{0j} = \frac{\rho' c v^j}{\sqrt{1 - \frac{v^2}{c^2}}} = \frac{\rho c v^j}{1 - \frac{v^2}{c^2}} = \rho u^0 u^j \qquad (6.66)$$

$$T^{ij} = \frac{\rho' v^i v^j}{\sqrt{1 - \frac{v^2}{c^2}}} = \frac{\rho v^i v^j}{1 - \frac{v^2}{c^2}} = \rho u^i u^j \qquad (6.67)$$

Observing (6.64)–(6.67), we can see that all four of them come in the same form and can therefore be integrated to become a tensor equation:

$$T^{\mu\nu} = \rho u^\mu u^\nu \qquad (6.68)$$

Again, please note that ρ is the density of mass measured from an inertial frame that's stationary relative to the mass in the local area, while u^μ represents the 4-speed. In addition, (6.68) is only true in pressure-free environments. Otherwise, it has to be rewritten as:

$$T^{\mu\nu} = \left(\rho + \frac{P}{c^2}\right) u^\mu u^\nu - P\eta^{\mu\nu} \qquad (6.69)$$

Where $\eta^{\mu\nu}$ and P represent the inverse matrix of the Minkowski metric tensor and pressure respectively. Although (6.69) is applicable to perfect fluids, we won't further discuss the equation as it isn't directly related to our later subjects. However, please note that (6.68) and (6.69) both show the term $T^{\mu\nu}$ to be a symmetrical matrix, i.e. $T^{\mu\nu} = T^{\nu\mu}$.

The establishment of the matrix $T^{\mu\nu}$ can largely facilitate the discussion about invariant physical quantities such as mass, energy, and momentum. This is because the invariance of any of these elements refers to the fact that the difference between its inflow into and outflow from a given area should equal the amount of change happening inside this area. To render this concept in a mathematical form, let's consider the density of a certain physical quantity $T^{\mu 0}$ and its density of flow $T^{\mu i}$ along the direction of i. If we look at a small area $\left[x^1, x^1 + dx^1\right] \times \left[x^2, x^2 + dx^2\right] \times \left[x^3, x^3 + dx^3\right]$ with the volume of $dx^1 dx^2 dx^3$ in the Cartesian coordinate system ($x^1 = x, x^2 = y, x^3 = z$), the total physical quantities contained in this area can be calculated through the multiplication of its density by volume, i.e. $T^{\mu 0} dx^1 dx^2 dx^3$. As for their time rate of change in the same area, it can be expressed as $\frac{T^{\mu 0}(t+dt) - T^{\mu 0}(t)}{dt} dx^1 dx^2 dx^3 = \frac{\partial T^{\mu 0}}{\partial t} dx^1 dx^2 dx^3$, which equals the inflow into x^i minus the outflow from $x^i + dx^i$:

$$\left[T^{\mu i}\left(x^i\right) - T^{\mu i}\left(x^i + dx^i\right)\right]$$
$$\times \text{cross section perpendicular to the direction of } i$$
$$= -\frac{\partial T^{\mu i}}{\partial x^i} \times \left(dx^i \times \text{cross section perpendicular to the direction of } i\right)$$
$$= -\frac{\partial T^{\mu i}}{\partial x^i} \times dx^1 dx^2 dx^3$$

$$(6.70)$$

For a flow along the direction of 1, for example, the cross-sectional area perpendicular to 1 is $dx^2 dx^3$, so the product of dx^1 and $dx^2 dx^3$ is the volume of the area. Combining the flows from all directions, we know the net inflow into the area is:

$$-\sum_{i=1}^{3} \frac{\partial T^{\mu i}}{\partial x^i} \times dx^1 dx^2 dx^3 \qquad (6.71)$$

Which should equal $\frac{\partial}{\partial t}\left(T^{\mu 0}dx^1 dx^2 dx^3\right)$. Hence, we can derive:

$$\frac{\partial T^{\mu 0}}{\partial t} = -\sum_{i=1}^{3} \frac{\partial T^{\mu i}}{\partial x^i} \tag{6.72}$$

With the Einstein summation convention, we can simplify (6.72) into:

$$\frac{\partial T^{\mu v}}{\partial x^v} = 0 \tag{6.73}$$

And of course, (6.73) can also be expressed as the below due to its symmetrical property:

$$\frac{\partial T^{\mu v}}{\partial x^\mu} = 0 \tag{6.74}$$

As $T^{\mu v} = T^{v\mu}$, (6.73) are (6.74) equivalent as well. Please note that when calculating the values of both equations, μ has to be replaced with $0, 1, 2, 3$ because the upper index μ appears as part of both the denominator and numerator and has to contribute to a valid value when $v = 0, 1, 2, 3$.

Last but not least, we should emphasize again that the derivation so far is only applicable to the context of special relativity and will lose validity when moved to curved spacetime. To extend the range of application to general relativity, more mathematical skills will be required, which is why we won't dive into that part of the discussion until related models are completely established.

6.3.4 Ricci Tensor

Based on the derivation above, we already have a general idea that Einstein's field equation should look something like this:

$$\text{Parameter related to Riemann curvature} = \text{Constant} \times T^{\mu v} \tag{6.75}$$

Just as mentioned before, Einstein established his field equation by proposing numerous hypotheses and eliminating the unreasonable ones. It was through a long process of reasoning and deduction that he finally retrieved a set of equations fitting all related requirements. The first challenge he was faced with was the fact that the Riemann curvature comes with one upper and three lower indices, while the term on the other side of (6.75) only has two, causing an asymmetry. To solve this issue, the simplest way is to transform the terms on both sides of the equal sign into 4×4 matrixes, which is why we first need to reduce the number of components in the Riemann curvature from 4^4 to 4×4. Back in Einstein's time, mathematicians hadn't come up with many curvature-related second-order tensors. Among the available ones, he chose the Ricci tensor, which is defined as:

$$R_{\mu\nu} = R^{\sigma}_{\mu\sigma\nu} \qquad (6.76)$$

In fact, (6.76) simply aligns the second lower index in the Riemann curvature with its superscript, so the equation is essentially equivalent to $R_{\mu\nu} = \sum_{\sigma=0,1,2,3} R^{\sigma}_{\mu\sigma\nu}$.

As for why Einstein chose the Ricci tensor, we can identify the reason from the motion behavior of objects in a static, low-speed weak field. If we go back to (6.46), the defining equation for the Riemann curvature, and substitute the definition of the Ricci tensor in:

$$R_{\mu\nu} = R^{\sigma}_{\mu\sigma\nu} = \frac{\partial \Gamma^{\sigma}_{\mu\nu}}{\partial x^{\sigma}} - \frac{\partial \Gamma^{\sigma}_{\mu\sigma}}{\partial x^{\nu}} + \Gamma^{\sigma}_{\kappa\sigma}\Gamma^{\kappa}_{\mu\nu} - \Gamma^{\sigma}_{\kappa\nu}\Gamma^{\kappa}_{\mu\sigma} \qquad (6.77)$$

As T^{00} is the most prevalent variant of $T^{\mu\nu}$ in low-speed environments, we can reasonably predict the same for R_{00} when it comes to $R_{\mu\nu}$. And since $g_{\mu\nu} \approx \eta_{\mu\nu}$ is true in weak fields, the last two terms in (6.77) can be disregarded, leading us to:

$$R_{00} \approx \frac{\partial \Gamma^{\sigma}_{00}}{\partial x^{\sigma}} - \frac{\partial \Gamma^{\sigma}_{0\sigma}}{\partial x^{0}} \qquad (6.78)$$

In addition, as the static status means the differential of time ($\frac{\partial \Gamma^{\sigma}_{0\sigma}}{\partial x^{0}}$) can be omitted, plus the Γ^{σ}_{00} in the above formula can be approximated to (6.58), this equation can be substituted into (6.78) lead us to:

$$R_{00} \approx \frac{\partial \Gamma^{\sigma}_{00}}{\partial x^{\sigma}} = -\frac{1}{2} \sum_{\sigma=0,1,2,3} \frac{\partial^2 g_{00}}{\partial x^{\sigma 2}} \approx -\frac{1}{2} \nabla^2 g_{00} \qquad (6.79)$$

When dealing with (6.79), we need to summate all the values when $\sigma = 0, 1, 2, 3$ because in $\frac{\partial \Gamma^{\sigma}_{00}}{\partial x^{\sigma}}$, the superscript σ appears in both the numerator and denominator. By contrast, (6.58) doesn't contain the same number of upper and lower indices because the element $g^{\kappa\kappa}$ in the inverse matrix of the metric tensor was replaced with 1 in (6.57). Meanwhile, as (6.79) is essentially equivalent to the second-order differential of time when $\sigma = 0$, we actually only have to take into consideration the possibilities of $\sigma = 1, 2, 3$ when adopting the approximation method in a static field. As such, what we need is the sum of the second-order differentials along all three spatial axes, which can be calculated with the Laplace operator ∇^2.

Last but not least, we can substitute (6.62) into (6.79) and retrieve:

$$R_{00} \approx \frac{1}{c^2}\nabla^2\phi = \frac{1}{c^2}\nabla \cdot \vec{g} \tag{6.80}$$

Which is exactly the term to the left of the equal sign in the equation of universal gravitation multiplied by c^2 (see (6.49) or (6.52)). In addition, T^{00} and the density of mass, i.e. the term following the equal sign in (6.52), are also c^2 times different. Therefore, we can represent (6.52) with R_{00} and T^{00} as:

$$R_{00} = \frac{4\pi G}{c^4}T^{00} \tag{6.81}$$

Judging from the form of (6.81), it seems like this equation can be rewritten as (6.82) to apply more generally to a wider range of spacetime scenarios:

$$R_{\mu\nu} = \frac{4\pi G}{c^4}T^{\mu\nu} \tag{6.82}$$

The reason behind this prediction is the perfect compatibility of (6.82) with the law of universal gravitation in a static, low-speed weak field (i.e. Newtonian approximation). However, this seemingly perfect equation has a major flaw as it loses validity once put through the process of transformation. In order to fix this problem, Einstein racked his brain to figure out a solution but made no progress for years, bogged down in a prolonged struggle. At one point, he even considered giving up all his results and switching to a whole new mathematical model. So what exactly is the flaw in (6.82) that sent him into such desperation? Details will be recounted in the following section.

6.3.5 Frame Transformation and Tensor Definition

In fact, the defect of (6.82) shouldn't be too hard to spot: the two lower indices μ and ν in $R_{\mu\nu}$ are in the upper position in $T^{\mu\nu}$, so it doesn't seem very safe to just put an equal sign between two asymmetrical terms. Specifically speaking, there should be some kind of transforming equation that regulates terms with indices in different positions. After our explanation below, you'll soon realize the decision of establishing a transformation rule with "index" as the main focus is an extremely wise one.

To understand what it means for a term to be indexed with sub- or super-scripts, we first need to determine the common properties shared by physical quantities without any index. For example, vector (V, which is never indexed although its component V^μ might be), spacetime interval (ds), proper time interval ($d\tau$), light speed (c), and gravitational constant (G) can all be considered. And yes, you might have noticed already. All the elements listed here are invariants in the transformation of coordinates. The invariance of space-time interval, proper time interval and light speed have been explained in Chap. 5. As for vector, the physical quantity itself doesn't vary with the transformation of reference frames although its components along the axes in the Cartesian and spherical coordinate systems might be different.

Now, let's start with the characteristics of physical quantities with one upper or lower index only. While a basis vector $\mathbf{e}_\mu$ has one subscript, the projection of a vector on a given axis V^μ is a component with an upper index. As for independent variables like x, y, z, t, these should be expressed as x^μ, meaning they're components of the displacement vector $x^\mu \mathbf{e}_\mu$. Suppose there are two frames of reference S and S', where the functions $x^\mu(x'^\alpha)$ and $x'^\alpha(x^\mu)$ represent the transformation between x^μ and x'^μ. Similarly, to derive the equation for transforming between V^μ and $\mathbf{e}_\mu$,[23] we can assume there to be a very small displacement vector dx on the tangent plane in Fig. 6.8 and write it as:

$$dx = dx^\mu \mathbf{e}_\mu = dx'^\alpha \mathbf{e}'_\alpha \tag{6.83}$$

In (6.83), dx is represented with the basis vectors of S and S'. As x^μ is the function of x'^α, dx^μ will satisfy the below equation.[24]

$$dx^\mu = \frac{\partial x^\mu}{\partial x'^\alpha} dx'^\alpha \tag{6.84}$$

---

[23]The function $x^\mu(x'^\alpha)$ can be elaborated as $x^\mu(x'^0, x'^1, x'^2, x'^3)$. In other words, the component of a vector along each axis in S (x^μ) is the function of the component along the corresponding axis in S'.

[24]This is the basic chain rule for finding the derivative of multiple variables, which can be expanded to become:

$$dx^\mu = \frac{\partial x^\mu}{\partial x'^0} dx'^0 + \frac{\partial x^\mu}{\partial x'^1} dx'^1 + \frac{\partial x^\mu}{\partial x'^2} dx'^2 + \frac{\partial x^\mu}{\partial x'^3} dx'^3$$

Then, the substitution of (6.84) back into (6.83) will lead us to:

$$dx = \frac{\partial x^\mu}{\partial x'^\alpha} dx'^\alpha \mathbf{e}_\mu = dx'^\alpha \mathbf{e}'_\alpha \tag{6.85}$$

Comparing (6.83) and (6.85), we can retrieve the equation for transforming the basis vector between S and S':

$$\mathbf{e}'_\alpha = \frac{\partial x^\mu}{\partial x'^\alpha} \mathbf{e}_\mu \tag{6.86}$$

As for the equation of transformation for the component V^μ, let's suppose V has two basis vectors $\mathbf{e}_\mu$ and $\mathbf{e}'_\alpha$ on its tangent plane. As the vector itself doesn't undergo any change in the process of transformation, we know that:

$$V = V^\mu \mathbf{e}_\mu = V'^\alpha \mathbf{e}'_\alpha \tag{6.87}$$

Substituting (6.86) into (6.87), we can then obtain:

$$V'^\alpha = \frac{\partial x'^\alpha}{\partial x^\mu} V^\mu \tag{6.88}$$

As some might have noticed, it's actually the same kind of transformation that (6.88) and (6.84) describe.

Next up, we're going to derive the equation that physical quantities with two upper or lower indices have to satisfy during the transformation of coordinates, starting with the metric tensor $g_{\mu\nu}$. As spacetime interval ds remains invariant when frames are transformed, we know that:

$$ds^2 = g_{\mu\nu} dx^\mu dx^\nu = g'_{\alpha\beta} dx'^\alpha dx'^\beta \tag{6.89}$$

Then, we can substitute into (6.89) the equation for transforming the components of dx, (6.84):

$$ds^2 = g_{\mu\nu} \left(\frac{\partial x^\mu}{\partial x'^\alpha} dx'^\alpha \right) \left(\frac{\partial x^\nu}{\partial x'^\beta} dx'^\beta \right) = g'_{\alpha\beta} dx'^\alpha dx'^\beta \tag{6.90}$$

Hence, the transformation equation for $g_{\mu\nu}$ is:

$$g'_{\alpha\beta} = \frac{\partial x^\mu}{\partial x'^\alpha} \frac{\partial x^\nu}{\partial x'^\beta} g_{\mu\nu} \tag{6.91}$$

Also coming with two indices is the energy-momentum tensor $T^{\mu\nu}$, which satisfies $T^{\mu\nu} = \rho u^{\mu} u^{\nu}$. In this equation, u^{μ} is a 4-speed defined as $u^{\mu} = dx^{\mu}/d\tau$. For any metric tensor $g_{\mu\nu}$ in the Minkowski spacetime, we already know that $d\tau = \frac{1}{c}\sqrt{-g_{\mu\nu}dx^{\mu}dx^{\nu}}$ is true. In order to make sure that $T^{\mu\nu}$ resembles something like this equation,[25] we can leave its definition as $\rho u^{\mu} u^{\nu}$. Yet please note that ρ is an invariant through the transformation of frames, while the 4-speed u is a vector whose components satisfy the regulation of (6.88), i.e.:

$$u'^{\alpha} = \frac{\partial x'^{\alpha}}{\partial x^{\mu}} u^{\mu} \tag{6.92}$$

After substituting (6.92) into the defining equation for $T^{\mu\nu}$, we can obtain the equation of transformation for the energy-momentum tensor as:

$$T'^{\alpha\beta} = \frac{\partial x'^{\alpha}}{\partial x^{\mu}} \frac{\partial x'^{\beta}}{\partial x^{\nu}} T^{\mu\nu} \tag{6.93}$$

Observing (6.86), (6.88), (6.91), and (6.93), we can tell these equations share something in common: in the transformation from S to S', the upper index μ in S has to be multiplied by $\frac{\partial x'^{\alpha}}{\partial x^{\mu}}$ to become the α in S'; similarly, the lower index ν in S needs to be multiplied by $\frac{\partial x'^{\alpha}}{\partial x^{\mu}}$ to become the β in S'. As these rules display a consistent pattern, they can be regarded as actual characteristics of transformation.

Last but not least, let's go a step further to discuss physical quantities with three indices (upper and lower combined, such as $\Gamma^{\kappa}_{\mu\nu}$) and derive the transformation equation for such elements. Starting with the definition of $\Gamma^{\kappa}_{\mu\nu}$ that $\frac{\partial e_{\mu}}{\partial x^{\nu}} = \Gamma^{\kappa}_{\mu\nu} e_{\kappa}$, we can rewrite this equation as the below based on (6.86), which regulates how basis vectors are transformed:

$$\frac{\partial x'^{\alpha}}{\partial x^{\mu}} \frac{\partial}{\partial x'^{\alpha}} \left(\frac{\partial x'^{\beta}}{\partial x^{\nu}} e'_{\beta} \right) = \Gamma^{\kappa}_{\mu\nu} \frac{\partial x'^{\gamma}}{\partial x^{\kappa}} e'_{\gamma} \tag{6.94}$$

The expansion of the term before the equal sign will then lead us to:

$$\frac{\partial^2 x'^{\gamma}}{\partial x^{\mu} \partial x^{\nu}} e'_{\gamma} + \frac{\partial x'^{\alpha}}{\partial x^{\mu}} \frac{\partial x'^{\beta}}{\partial x^{\nu}} \frac{\partial e'_{\beta}}{\partial x'^{\alpha}} = \Gamma^{\kappa}_{\mu\nu} \frac{\partial x'^{\gamma}}{\partial x^{\kappa}} e'_{\gamma} \tag{6.95}$$

---

[25]However, (6.69) has to be rewritten as: $T^{\mu\nu} = \left(\rho + \frac{P}{c^2}\right) u^{\mu} u^{\nu} - P g^{\mu\nu}$.

Next, we can substitute $\frac{\partial e'_\beta}{\partial x'^\alpha} = \Gamma'^\gamma_{\alpha\beta} c'_\gamma$ into (6.95) and retrieve the below equation after some simplification:

$$\Gamma^\kappa_{\mu\nu} = \frac{\partial x'^\alpha}{\partial x^\mu} \frac{\partial x'^\beta}{\partial x^\nu} \frac{\partial x^\kappa}{\partial x'^\gamma} \Gamma'^\gamma_{\alpha\beta} + \frac{\partial^2 x'^\gamma}{\partial x^\mu \partial x^\nu} \frac{\partial x^\kappa}{\partial x'^\gamma} \tag{6.96}$$

And (6.96), is exactly the final equation that regulates the transformation of $\Gamma^\kappa_{\mu\nu}$.

Nevertheless, (6.96) isn't symmetrical since the equal sign is followed by two terms while there's only one on the other side. Hence, $\Gamma^\kappa_{\mu\nu}$ doesn't qualify as a "coordinate tensor" because a tensor has to follow a consistent pattern of transformation. Generally speaking, a tensor indexed with n superscripts and m subscripts is of type $\begin{pmatrix} n \\ m \end{pmatrix}$ and order $n + m$. For example, the component V^μ of a certain vector V is a first-order tensor of type $\begin{pmatrix} 1 \\ 0 \end{pmatrix}$. As for the metric tensor $g_{\mu\nu}$, it's second-order and of type $\begin{pmatrix} 0 \\ 2 \end{pmatrix}$.

With the definition of tensor established, let's now take a look back at (6.82). As the term $T^{\mu\nu}$ after the equal sign follows a consistent pattern of transformation, we can safely assume the same holds true for what's on the other side. In other words, we need to examine whether the Riemann curvature $R_{\mu\nu}$ is a tensor or not. As we know from (6.45), the ΔA^μ before the first equal sign and the $A^\alpha, dx^\beta, \delta x^\gamma$ following the last one are all first-order tensors of the $\begin{pmatrix} 1 \\ 0 \end{pmatrix}$ type, so $R^\mu_{\alpha\beta\gamma}$ apparently has to be a tensor as well. More specifically, it's fourth-order, type $\begin{pmatrix} 1 \\ 3 \end{pmatrix}$, and follows the below equation of transformation[26]:

$$R'^\kappa_{\mu\nu\sigma} = \frac{\partial x'^\kappa}{\partial x^\lambda} \frac{\partial x^\alpha}{\partial x'^\mu} \frac{\partial x^\beta}{\partial x'^\nu} \frac{\partial x^\gamma}{\partial x'^\sigma} R^\lambda_{\alpha\beta\gamma} \tag{6.97}$$

---

[26] Based on the definition that a tensor has to follow a consistent pattern of transformation, if A and B are both tensors, then AB (or expressed more precisely as the tensor product $A \otimes B$) will also qualify as one, which embodies the closure property of tensor multiplication.

As such, we can also confirm the Ricci tensor to be an actual tensor as it's generated through the contraction of the Riemann curvature:[27]

$$R'_{\mu\nu} = \frac{\partial x^\alpha}{\partial x'^\mu} \frac{\partial x^\beta}{\partial x'^\nu} R_{\alpha\beta} \tag{6.98}$$

6.3.6 Covariant, Contravariant & Mixed Tensors

In Section 6.3.5, we established several equations for transforming tensors and explored the significance of their upper and lower indices in such transformation. However, our fundamental issue still hasn't been solved yet: the tensors before and after the equal sign in (6.82) are types $\begin{pmatrix} 0 \\ 2 \end{pmatrix}$ and $\begin{pmatrix} 2 \\ 0 \end{pmatrix}$ respectively, implying two different ways of transformation. Therefore, (6.82) won't necessarily be valid after coordinates are transformed, which must've been a disappointment for Einstein as he definitely wouldn't want to see his field equation only applicable to a certain type of spacetime from the perspective of an observer in a certain status of motion. Even more disappointing was the fact that the subsequent years of attempts with varying methods still failed to repair the defect for him. In a 1913 speech, therefore, he made "The Hole Argument" as a defense case for himself, claiming that no tensor equation could possibly be derived to regulate the transformation of any random frames of reference.[28]

Despite all the setbacks in his research of general relativity, he was progressing quite well in his academic career. In April of 1914, he left the University of Zürich upon the invitation from the Humboldt University in Berlin. Motivated by the higher salary and less teaching burden promised by the new position, Einstein didn't give a second thought about relocating to the German capital, where he soon found the company of his cousin Elsa Löwenthal (1876-1936). In addition to leaving his wife and two kids behind in Switzerland, this move also meant he had to part ways with Grossmann, who'd spared no effort in providing precious advice over the years. Yet inspired perhaps by the fresh, sweet love life with the cousin, the obstacle on his way to constructing a comprehensive theory of general relativity was finally removed. As

---

[27] Although we did use the word "tensor" in previous sections before providing the definition (e.g. when defining metric tensor, the Riemann curvature tensor, and the Ricci tensor), we've actually considered the use of terminology quite carefully. Now that the term is properly defined, you must've figured out why we never called the Christoffel symbols of connection a "tensor".

[28] For more details, readers are referenced to Chap. 2 of Hamilton [13]

if experiencing an epiphany, he proposed to transform the $\begin{pmatrix} 2 \\ 0 \end{pmatrix}$ tensor into the $\begin{pmatrix} 0 \\ 2 \end{pmatrix}$ counterpart and vice versa. As such, the problem with (6.82) was solved.

But why does the exchange of upper and lower indices solve the problem? For tensors with subscripts, transformation works similarly to that of the basis vector $\mathbf{e}_\mu$. Therefore, they're given the name "covariant tensor". As for those with upper notations, the pattern of transformation is quite the opposite compared to the basis vector. Therefore, they're called the contravariant tensor. When it comes to tensors with both upper and lower indices (e.g. the Riemann curvature tensor), these are "mixed tensors". To figure out how to transform between covariant and contravariant tensors, we can start with a first-order tensor of the simplest form, i.e. a vector.

So how do we transform the vector component V^μ into V_μ and make sure the latter shares the transformation pattern of the basis vector $\mathbf{e}_\mu$? This question might seem a bit strange at first because to express the vector component as V_μ, we first need a different basis vector $\mathbf{e}^\mu$ that follows a transforming pattern opposite to that of the original basis vector so that the vector $V = V_\mu \mathbf{e}^\mu$ remains unchanged even after frame transformation. However, such a basis vector doesn't exist yet, so we need to construct a $\mathbf{e}^\mu$ that satisfies the below equation:

$$\mathbf{e}^\mu \cdot \mathbf{e}_\nu = \delta^\mu_\nu = \begin{cases} 1, \; if \; \mu = \nu \\ 0, \; if \; \mu \neq \nu \end{cases} \tag{6.99}$$

As δ^μ_ν is a constant identity matrix that doesn't change through transformation, the transforming patterns of $\mathbf{e}^\mu$ and $\mathbf{e}_\nu$ are exactly the opposite. Theoretically, there must be a way to construct a basis vector $\mathbf{e}^\mu$ that satisfies (6.99). In three-dimensional space, for example, we can assume the vector to be:

$$\mathbf{e}^1 = \frac{\mathbf{e}_2 \times \mathbf{e}_3}{\mathbf{e}_1 \cdot (\mathbf{e}_2 \times \mathbf{e}_3)}$$

$$\mathbf{e}^2 = \frac{\mathbf{e}_3 \times \mathbf{e}_1}{\mathbf{e}_2 \cdot (\mathbf{e}_3 \times \mathbf{e}_1)}$$

$$\mathbf{e}^3 = \frac{\mathbf{e}_1 \times \mathbf{e}_2}{\mathbf{e}_3 \cdot (\mathbf{e}_1 \times \mathbf{e}_2)} \tag{6.100}$$

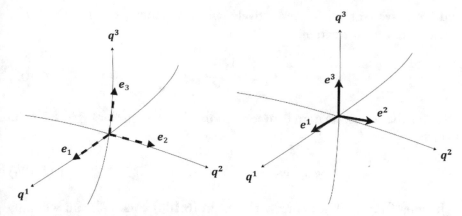

Fig. 6.9 Illustration of two different sets of basis vectors, they will coincide only when the bases are orthogonal

The relationships between $\mathbf{e}^1$, $\mathbf{e}^2$, $\mathbf{e}^3$ and $\mathbf{e}_1$, $\mathbf{e}_2$, $\mathbf{e}_3$ are illustrated in Fig. 6.9 below.

When looking at Fig. 6.9, please note that $\mathbf{e}^1$, $\mathbf{e}^2$, $\mathbf{e}^3$ are perpendicular to the planes formed by $\mathbf{e}_2$ and $\mathbf{e}_3$, $\mathbf{e}_3$ and $\mathbf{e}_1$, $\mathbf{e}_1$ and $\mathbf{e}_2$ respectively. Similarly, we can also construct a basis vector $\mathbf{e}^\mu$ in a 4-space for $\mathbf{e}^0$ to be orthogonal to the space formed by $\mathbf{e}_1$, $\mathbf{e}_2$, and $\mathbf{e}_3$. In addition, the length of $\mathbf{e}^0$ also needs to be adjusted to satisfy the requirement that the inner product of $\mathbf{e}^0$ and $\mathbf{e}_0$ has to be 1. Following this logic, the basis vectors $\mathbf{e}^1$, $\mathbf{e}^2$, and $\mathbf{e}^3$ all have to be orthogonal to the space constructed by the other three vectors, which is exactly what (6.99) means geometrically.

At this point, we already have two basis vectors $\mathbf{e}_\mu$ and $\mathbf{e}^\mu$. Their relationship is regulated by (6.99), while the transformation between them follows the equations below:

$$\mathbf{e}'_\alpha = \frac{\partial x^\mu}{\partial x'^\alpha}\mathbf{e}_\mu \ \& \ \mathbf{e}'^\alpha = \frac{\partial x'^\alpha}{\partial x^\mu}\mathbf{e}^\mu \tag{6.101}$$

In mathematical terminology, $\mathbf{e}_\mu$ is called the "coordinate basis" with its direction parallel to a certain axis. As for $\mathbf{e}^\mu$, it's a "dual basis" whose direction is illustrated in Fig. 6.9. In other words, now we already have two different ways of describing the vector V:

$$V = V^\mu \mathbf{e}_\mu = V_\mu \mathbf{e}^\mu \tag{6.102}$$

Where V^μ and V_μ are the projected components of V along the directions of $\mathbf{e}_\mu$ and $\mathbf{e}^\mu$. These two components are called the "covariant vector" and "contravariant vector" respectively, with the latter following the below equation in transformation:

$$V'_\alpha = \frac{\partial x^\mu}{\partial x'^\alpha} V_\mu \qquad (6.103)$$

As for the transformation between V_μ and V^μ, we can multiply $V^\mu \mathbf{e}_\mu = V_\nu \mathbf{e}^\nu$ by $\mathbf{e}_\alpha$:

$$V^\mu \mathbf{e}_\mu \cdot \mathbf{e}_\alpha = V_\nu \mathbf{e}^\nu K \cdot \mathbf{e}_\alpha \qquad (6.104)$$

Judging from (6.99), the term $\mathbf{e}^\nu \cdot \mathbf{e}_\alpha$ in (6.104) equals δ^ν_α and will only yield a non-zero value when $\nu = \alpha$ (in fact, the only possible result is 1). With the relationship that $\mathbf{e}_\mu \cdot \mathbf{e}_\alpha$ equals $g_{\alpha\mu}$, we can derive the following:

$$g_{\alpha\mu} V^\mu = V_\alpha \qquad (6.105)$$

Similarly, the equation of the inverse transform between V_μ and V^μ can be expressed as:

$$V^\nu = g^{\nu\alpha} V_\alpha \qquad (6.106)$$

Where $g^{\nu\alpha} \equiv \mathbf{e}^\nu \cdot \mathbf{e}^\alpha$ and happens to be the inverse matrix of $g_{\alpha\mu}$.

So far, we've established the pattern of transformation between covariant and contravariant tensors of the first order (vectors). As a next step, we're going to extend this mathematical pattern to generally include all tensors, which can be represented as $T^{\alpha\beta\gamma\cdots}_{\mu\nu\sigma\cdots}$. As the qualification of "tensor" depends on whether or not an element follows a consistent transforming rule, if we define a new physical quantity as:

$$\boldsymbol{T} \equiv T^{\alpha\beta\gamma\cdots}_{\mu\nu\sigma\cdots} \left(\mathbf{e}_\alpha \otimes \mathbf{e}_\beta \otimes \mathbf{e}_\gamma \ldots \otimes \mathbf{e}^\mu \otimes \mathbf{e}^\nu \otimes \mathbf{e}^\sigma \ldots \right) \qquad (6.107)$$

Then $\boldsymbol{T}$ will be an invariant through the transformation of frames. Please note that the symbol $\otimes$ in (6.107) denotes the tensor product, which we can simply regard as normal multiplication. Hence, the $T^{\alpha\beta\gamma\cdots}_{\mu\nu\sigma\cdots}$ in (6.107) can be understood as the component of T along the direction of the basis vector $\left(\mathbf{e}_\alpha \otimes \mathbf{e}_\beta \otimes \mathbf{e}_\gamma \ldots \otimes \mathbf{e}^\mu \otimes \mathbf{e}^\nu \otimes \mathbf{e}^\sigma \ldots \right)$. Similarly, we can rewrite T as the following just by switching the sub- and superscripts in the basis vectors, such as changing $\mathbf{e}_\alpha$ to $\mathbf{e}^\alpha$:

$$\boldsymbol{T} \equiv T^{\beta\gamma\cdots}_{\alpha}{}_{\mu\nu\sigma\cdots} \left(\mathbf{e}^\alpha \otimes \mathbf{e}_\beta \otimes \mathbf{e}_\gamma \ldots \otimes \mathbf{e}^\mu \otimes \mathbf{e}^\nu \otimes \mathbf{e}^\sigma \ldots \right) \qquad (6.108)$$

As for the relationship between $T^{\alpha\beta\gamma\cdots}_{\mu\nu\sigma\cdots}$ and $T^{\beta\gamma\cdots}_{\alpha\mu\nu\sigma\cdots}$, we can apparently express it as the below using (6.105) as a reference:

$$T^{\beta\gamma\cdots}_{\kappa\mu\nu\sigma\cdots} = g_{\kappa\alpha}T^{\alpha\beta\gamma\cdots}_{\mu\nu\sigma\cdots} \tag{6.109}$$

Hence, we know that to move the upper index in a tensor to the lower position, the tensor simply needs to be multiplied by $g_{\mu\nu}$, and for the other way around, the multiplier becomes $g^{\mu\nu}$.

In addition, we can also express the energy-momentum tensor as:

$$T = T^{\mu\nu}\left(\mathbf{e}_\mu \otimes \mathbf{e}_\nu\right) = T_{\mu\nu}\left(\mathbf{e}^\mu \otimes \mathbf{e}^\nu\right) \tag{6.110}$$

Where[29]:

$$T_{\mu\nu} = g_{\mu\alpha}g_{\nu\beta}T^{\alpha\beta} \tag{6.111}$$

At this point, we just need to replace the $T^{\mu\nu}$ in (6.82) with $T_{\mu\nu}$ to retrieve an equation that makes more sense:

$$R_{\mu\nu} = \frac{4\pi G}{c^4}T_{\mu\nu} \tag{6.112}$$

Based on the above deduction, we've confirmed that as long as a reference frame satisfies (6.112), the equation will always be true no matter what type of transformation the frame goes through. Please note that $T_{\mu\nu}$ and $T^{\mu\nu}$ share the same behavior although the indices appear in the opposite positions because both terms are components of the vector T, except that they correspond to different basis vectors. Therefore, (6.112) can also be written as:

$$R^{\mu\nu} = \frac{4\pi G}{c^4}T^{\mu\nu} \tag{6.113}$$

Where $R^{\mu\nu} = g^{\mu\alpha}g^{\nu\beta}R_{\alpha\beta}$. In a weak field, on the other hand, we know that:

$$g_{\mu\nu} \approx \eta_{\mu\nu} = \begin{pmatrix} -1 & 0 & 0 & 0 \\ 0 & 1 & 0 & 0 \\ 0 & 0 & 1 & 0 \\ 0 & 0 & 0 & 1 \end{pmatrix} \tag{6.114}$$

---

[29] Detailed definition of $T^{\mu\nu}$ is in fact $T^{\mu\nu} = (\rho + P/c^2)u^\mu u^\nu - Pg^{\mu\nu}$, which implies $T_{\mu\nu} = (\rho + P/c^2)u_\mu u_\nu - Pg_{\mu\nu}$, where $u_\mu = g_{\mu\alpha}u^\alpha$.

So if we multiply the $g_{\mu\nu}$ in (6.111) by itself, the result will come out as a unit matrix. Simply put, $T_{\mu\nu}$ and $T^{\mu\nu}$ can be considered equivalent under this situation. This result shows we can safely infer that (6.112) is an invariant in the transformation of coordinates which will satisfy the equation of universal gravitation in a static, low-speed weak field.

Although we've turned $T^{\mu\nu}$ into a form that's compatible with general relativity, we still can't be sure whether the equation of $\partial T^{\mu\nu}/\partial x^{\mu} = 0$ in (6.74), which represents the conservation of momentum and energy, will always be true after random frame transformation. This is because $\partial T^{\mu\nu}/\partial x^{\sigma}$ doesn't qualify as a tensor that follows a consistent transforming pattern. As for the reason, please see how is derived below:

$$
\begin{aligned}
\frac{\partial T'^{\alpha\beta}}{\partial x'^{\gamma}} &= \frac{\partial x^{\sigma}}{\partial x'^{\gamma}} \frac{\partial}{\partial x^{\sigma}} \left(\frac{\partial x'^{\alpha}}{\partial x^{\mu}} \frac{\partial x'^{\beta}}{\partial x^{\nu}} T^{\mu\nu} \right) \\
&= \frac{\partial x^{\sigma}}{\partial x'^{\gamma}} \frac{\partial x'^{\alpha}}{\partial x^{\mu}} \frac{\partial x'^{\beta}}{\partial x^{\nu}} \frac{\partial T^{\mu\nu}}{\partial x^{\sigma}} + \frac{\partial}{\partial x'^{\gamma}} \left(\frac{\partial x'^{\alpha}}{\partial x^{\mu}} \frac{\partial x'^{\beta}}{\partial x^{\nu}} \right) T^{\mu\nu} \\
&\neq \frac{\partial x^{\sigma}}{\partial x'^{\gamma}} \frac{\partial x'^{\alpha}}{\partial x^{\mu}} \frac{\partial x'^{\beta}}{\partial x^{\nu}} \frac{\partial T^{\mu\nu}}{\partial x^{\sigma}}
\end{aligned}
\tag{6.115}
$$

As you can see from the above derivation, (6.74) isn't valid anymore after the transformation of coordinates and needs to be modified for a wider range of applicability. However, such modification isn't that easy to make because, in any non-Minkowski spacetime, basis vectors vary with position. In other words, the significance of "along the direction of i" isn't the same at x^{i} and $x^{i} + dx^{i}$; even the value of μ and the size of cross-sectional area change as well. Consequently, to measure the change of flow along i, we can't simply subtract one flow from another, such as $T^{\mu i}\left(x^{i}\right) - T^{\mu i}\left(x^{i} + dx^{i}\right)$, but have to adopt the two-step approach below:

(1) Start by proving that at any position in a 4-space regulated by the Riemann curvature tensor, there exists a local inertial frame with the Minkowski metric tensor where all values of the Christoffel symbol of connection $\Gamma^{\kappa}_{\mu\nu}$ are 0. In other words, change in the local basis vector doesn't need to be considered in deriving the law of conservation. So for this particular frame of reference, it's exactly the energy-momentum conservation that (6.74) represents.

(2) Second, transform the $\partial T^{\mu\nu}/\partial x^{\sigma}$ in (6.74) into a tensor that follows a consistent pattern of transformation, such as $DT^{\mu\nu}/Dx^{\sigma}$ (details to be provided later on). Meanwhile, make sure the term will degrade to $\partial T^{\mu\nu}/\partial x^{\sigma}$ under the Minkowski metric tensor.

If both steps can be fulfilled, $\partial T^{\mu\nu}/\partial x^\mu = 0$ will be true from the perspective of any observer in the local inertial frame of any specific spacetime position and can be deemed equivalent to $DT^{\mu\nu}/Dx^\mu = 0$. In addition, as $DT^{\mu\nu}/Dx^\mu$ is a tensor that follows a fixed transforming pattern, it'll always be 0 no matter what type of transformation is gone through. As such, we know that $DT^{\mu\nu}/Dx^\mu = 0$ is exactly the equation for energy-momentum conservation that we'd hoped to establish. As for the two steps above, we'll provide more concrete details in the following sections.

6.3.7 Local Inertial Frame and Strong Equivalence Principle

To turn step 1 from theory into practice, we first need to validate the principle of strong equivalence, which we've described in Sect. 6.1.4 but haven't proved yet. Before diving into the verification, we should make it clear that mathematically speaking, only a frame that satisfies both the following requirements qualify as a "local inertial frame of reference": (1) having the Minkowski metric tensor as the local tensor, i.e. $g_{\mu\nu} = \eta_{\mu\nu}$, and (2) whose change in basis vector cannot be detected from the local area, i.e. $\Gamma^\kappa_{\mu\nu} = 0$. While the strong equivalence principle dictates that any spot in a four-dimensional spacetime comes with its own local inertial frame, not all the first-order derivatives of $\Gamma^\kappa_{\mu\nu}$ are 0 although the term itself is. This is because the derivatives are directly dependent on the Riemann curvature, a tensor that follows a fixed pattern of transformation. Hence, if any local frame is found to have a non-zero curvature, there will be no way for us to find one where the curvature equals 0 along all directions. This rule will be proved in the following paragraphs.

Suppose there's a reference frame where spacetime geometry is regulated by the metric tensor function $g_{\mu\nu}(x)$. For the purpose of simplification, let's set the point we want to discuss at the origin of this coordinate system ($x = 0$) and find a local inertial reference frame nearby. Specifically, what we need to do is identify a pair of coordinates with the transforming relationship of $x \rightarrow \xi(x)$ that can ensure the transformation of the metric tensor $g_{\mu\nu}(x) \rightarrow g'_{\alpha\beta}(\xi)$ satisfies $g'_{\alpha\beta}(0) = \eta_{\alpha\beta}(0)$. Plus, the value of $\Gamma'^\gamma_{\alpha\beta}(\xi)$ calculated from $g'_{\alpha\beta}$ also has to fulfill $\Gamma'^\gamma_{\alpha\beta}(0) = 0$. To figure out a transformation equation that allows $g'_{\alpha\beta}(0) = \eta_{\alpha\beta}(0)$, we need to go through the following steps.

First, judging from the transforming equation for the metric tensor, we know that:

$$g_{\mu\nu}(x) = \frac{\partial \xi^\alpha}{\partial x^\mu} \frac{\partial \xi^\beta}{\partial x^\nu} g'_{\alpha\beta}(\xi) \qquad (6.116)$$

Then, the substitution of the local property $x = \xi = 0$ into the above equation will lead us to:

$$g_{\mu\nu}(0) = \frac{\partial \xi^\alpha}{\partial x^\mu} \frac{\partial \xi^\beta}{\partial x^\nu} \eta_{\alpha\beta}(0) \qquad (6.117)$$

If we consider this linear mapping below, where A^α_μ is a to-be-determined constant coefficient:

$$\xi^\alpha(x) = A^\alpha_\mu x^\mu \qquad (6.118)$$

Then (6.117) will become:

$$g_{\mu\nu}(0) = A^\alpha_\mu A^\beta_\nu \eta_{\alpha\beta}(0) \qquad (6.119)$$

Please note that (6.119) is a simultaneous set containing 16 equations, so 16 solutions of the variable A^α_μ can be resolved therefrom. With these solutions, we must be able to find a pair of coordinates $\xi(x)$ that satisfies $g'_{\alpha\beta}(0) = \eta_{\alpha\beta}(0)$ but not necessarily $\Gamma'^\gamma_{\alpha\beta}(0) = 0$. At this point, we can first represent ξ with x and implement another transformation $x \to \xi(x)$, which should cause no change in the metric tensor at the origin (otherwise, $g'_{\alpha\beta} = \eta_{\alpha\beta}$ won't be true) but has to ensure that $\Gamma'^\gamma_{\alpha\beta}(0) = 0$:

$$\xi^\kappa(x) = x^\kappa + \frac{1}{2} \Gamma^\kappa_{\alpha\beta}(0) x^\alpha x^\beta \qquad (6.120)$$

As a next step, let's examine whether the transformation described in (6.120) satisfies the first requirement:

$$\left. \frac{\partial \xi^\alpha}{\partial x^\mu} \right|_{x=\xi=0} = \delta^\alpha_\mu \qquad (6.121)$$

Since (6.121) is true regardless of the values of α and μ, we can be sure that the transformation in (6.120) doesn't change the metric tensor at the origin. Then, moving on to the transforming equation for $\Gamma^{\kappa}_{\mu\nu}$, we know by observing (6.96) that:

$$\Gamma^{\kappa}_{\mu\nu}(x) = \frac{\partial \xi^{\alpha}}{\partial x^{\mu}} \frac{\partial \xi^{\beta}}{\partial x^{\nu}} \frac{\partial x^{\kappa}}{\partial \xi^{\gamma}} \Gamma'^{\gamma}_{\alpha\beta}(\xi) + \frac{\partial^2 \xi^{\gamma}}{\partial x^{\mu} \partial x^{\nu}} \frac{\partial x^{\kappa}}{\partial \xi^{\gamma}} \tag{6.122}$$

Substituting the local property $x = \xi = 0$ while considering the require-ment of $\Gamma'^{\gamma}_{\alpha\beta}(0) = 0$, we can derive:

$$\Gamma^{\kappa}_{\mu\nu}(0) = \frac{\partial^2 \xi^{\gamma}}{\partial x^{\mu} \partial x^{\nu}} \frac{\partial x^{\kappa}}{\partial \xi^{\gamma}}\bigg|_{x=\xi=0} \tag{6.123}$$

Then, replacing the term after the equal sign in (6.123) with (6.120), we'll get the below result:

$$\frac{\partial^2 \xi^{\gamma}}{\partial x^{\mu} \partial x^{\nu}} \frac{\partial x^{\kappa}}{\partial \xi^{\gamma}}\bigg|_{x=\xi=0} = \frac{\partial^2 \xi^{\gamma}}{\partial x^{\mu} \partial x^{\nu}} \delta^{\kappa}_{\gamma}\bigg|_{x=\xi=0} = \frac{\partial^2 \xi^{\kappa}}{\partial x^{\mu} \partial x^{\nu}}\bigg|_{x=\xi=0}$$

$$= \frac{\partial^2}{\partial x^{\mu} \partial x^{\nu}} \left(x^{\kappa} + \frac{1}{2} \Gamma^{\kappa}_{\alpha\beta}(0) x^{\alpha} x^{\beta} \right)\bigg|_{x=\xi=0}$$

$$= \frac{1}{2} \Gamma^{\kappa}_{\alpha\beta}(0) \frac{\partial^2 (x^{\alpha} x^{\beta})}{\partial x^{\mu} \partial x^{\nu}}$$

$$= \frac{1}{2} \left(\Gamma^{\kappa}_{\mu\nu}(0) + \Gamma^{\kappa}_{\nu\mu}(0) \right) = \Gamma^{\kappa}_{\mu\nu}(0) \tag{6.124}$$

As proved in the above process, (6.120) does satisfy both the requirements that we set out at the beginning of this section. In other words, it's true that we can find a local inertial frame at any position in a 4-space with the trans-formations regulated by (6.118) and (6.120) executed. Up to this point, we've come up with solid proof through a mathematical approach that as long as a four-dimensional spacetime can be described with a Riemann manifold,[30] then the principle of strong equivalence must be valid as well.

6.3.8 Covariant Derivative

In this section, we'll address the second step listed toward the end of Sect. 6.3.6: transform the $\partial T^{\mu\nu}/\partial x^{\sigma}$ in (6.74) into a tensor that follows a

---

[30]Technically speaking, a torsion-free Riemann manifold.

consistent pattern of transformation, such as $DT^{\mu\nu}/Dx^{\sigma}$. In fact, this process will generate a new operator of differentiation, the covariant derivative. Generally speaking, a tensor will not remain to be one after processed even with the simplest rules of partial differentiation. As shown in (6.115), such calculation always produces an unexpected term in the result. To ensure that a tensor maintains its status, we'll need to address the following question: how do we define a new rule of partial differentiation that, after applied to a general tensor, doesn't disqualify it as one? To solve this issue, we can start with the overall property of a tensor, i.e. calculating the first-order derivative of (6.107) through the following steps:

$$
\frac{\partial T}{\partial x^{\lambda}} = \frac{\partial T^{\alpha\cdots}_{\mu\cdots}}{\partial x^{\lambda}} \left(\mathbf{e}_{\alpha} \otimes \cdots \otimes \mathbf{e}^{\mu} \cdots \right)
$$
$$
+ T^{\alpha\cdots}_{\mu\cdots} \left(\frac{\partial \mathbf{e}_{\alpha}}{\partial x^{\lambda}} \otimes \cdots \otimes \mathbf{e}^{\mu} \cdots \right) + \cdots + T^{\alpha\cdots}_{\mu\cdots} \left(\mathbf{e}_{\alpha} \otimes \cdots \otimes \frac{\partial \mathbf{e}^{\mu}}{\partial x^{\lambda}} \cdots \right) + \cdots
$$
$$
\tag{6.125}
$$

Although we know the differential of a basis vector is $\frac{\partial \mathbf{e}_{\alpha}}{\partial x^{\lambda}} = \Gamma^{\kappa}_{\alpha\lambda} \mathbf{e}_{\kappa}$, we aren't sure yet about how to express that of a dual basis vector.

Hence, let's first retrieve the first-order differential of the identity $\mathbf{e}^{\mu} \cdot \mathbf{e}_{\mu} = 1$ as:

$$
0 = \frac{\partial \left(\mathbf{e}^{\mu} \cdot \mathbf{e}_{\mu} \right)}{\partial x^{\lambda}} = \frac{\partial \mathbf{e}^{\mu}}{\partial x^{\lambda}} \cdot \mathbf{e}_{\mu} + \mathbf{e}^{\mu} \cdot \frac{\partial \mathbf{e}_{\mu}}{\partial x^{\lambda}}
$$
$$
= \frac{\partial \mathbf{e}^{\mu}}{\partial x^{\lambda}} \cdot \mathbf{e}_{\mu} + \Gamma^{\kappa}_{\mu\lambda} \mathbf{e}_{\kappa} \cdot \mathbf{e}^{\mu} = \left(\frac{\partial \mathbf{e}^{\mu}}{\partial x^{\lambda}} + \Gamma^{\mu}_{\kappa\lambda} \mathbf{e}^{\kappa} \right) \cdot \mathbf{e}_{\mu}
$$
$$
\tag{6.126}
$$

Please note that for what's after the last equal sign in (6.126), we swapped the indices μ and κ of the second term in the parentheses. From the above equation, we know that:

$$
\frac{\partial \mathbf{e}^{\mu}}{\partial x^{\lambda}} = -\Gamma^{\mu}_{\kappa\lambda} \mathbf{e}^{\kappa}
$$
$$
\tag{6.127}
$$

Substituting (6.127) back into (6.125) and then reorganizing the sub- and superscripts, we can obtain:

$$
\frac{\partial T}{\partial x^{\lambda}} = \frac{\partial T^{\alpha\cdots}_{\mu\cdots}}{\partial x^{\lambda}} \left(\mathbf{e}_{\alpha} \otimes \cdots \otimes \mathbf{e}^{\mu} \cdots \right) + \Gamma^{\kappa}_{\alpha\lambda} T^{\alpha\cdots}_{\mu\cdots} \left(\mathbf{e}_{\kappa} \otimes \cdots \otimes \mathbf{e}^{\mu} \cdots \right)
$$
$$
+ \cdots - \Gamma^{\mu}_{\kappa\lambda} T^{\alpha\cdots}_{\mu\cdots} \left(\mathbf{e}_{\alpha} \otimes \cdots \otimes \mathbf{e}^{\kappa} \cdots \right) - \cdots
$$

$$= \left(\frac{\partial T^{\alpha \cdots}_{\mu \cdots}}{\partial x^\lambda} + \Gamma^\alpha_{\kappa \lambda} T^{\kappa \cdots}_{\mu \cdots} + \cdots - \Gamma^\kappa_{\mu \lambda} T^{\alpha \cdots}_{\kappa \cdots} \right) (\mathbf{e}_\alpha \otimes \cdots \otimes \mathbf{e}^\mu \cdots)$$

$$(6.128)$$

Where $\frac{\partial T}{\partial x^\lambda}$ is known to be a tensor because T remains invariant through the transformation of coordinates and satisfies the following equation:

$$\frac{\partial T}{\partial x'^\delta} = \frac{\partial x^\lambda}{\partial x'^\delta} \left(\frac{\partial T}{\partial x^\lambda} \right)$$

$$(6.129)$$

As shown in (6.129), $\frac{\partial T}{\partial x^\lambda}$ follows a fixed transforming pattern, and so does $(\mathbf{e}_\alpha \otimes \cdots \otimes \mathbf{e}^\mu \cdots)$. Hence, judging from (6.128), $\left(\frac{\partial T^{\alpha \cdots}_{\mu \cdots}}{\partial x^\lambda} + \Gamma^\alpha_{\kappa \lambda} T^{\kappa \cdots}_{\mu \cdots} + \cdots - \Gamma^\kappa_{\mu \lambda} T^{\alpha \cdots}_{\kappa \cdots} \right)$ should be a tensor as well.

With the above reasoning, we can define D/Dx^λ to be the new way of differentiation, which applies to tensors following the below equation:

$$\frac{DT^{\alpha \cdots}_{\mu \cdots}}{Dx^\lambda} = \frac{\partial T^{\alpha \cdots}_{\mu \cdots}}{\partial x^\lambda} + \Gamma^\alpha_{\kappa \lambda} T^{\kappa \cdots}_{\mu \cdots} + \cdots - \Gamma^\kappa_{\mu \lambda} T^{\alpha \cdots}_{\kappa \cdots}$$

$$(6.130)$$

The special operation denoted by D/Dx^λ is called the "covariant derivative", which will yield the below if applied to the energy-momentum tensor $T^{\mu\nu}$:

$$\frac{DT^{\mu\nu}}{Dx^\lambda} = \frac{\partial T^{\mu\nu}}{\partial x^\lambda} + \Gamma^\mu_{\kappa \lambda} T^{\kappa\nu} + \Gamma^\nu_{\kappa \lambda} T^{\mu\kappa}$$

$$(6.131)$$

From the perspective of a local inertial frame, all the Christoffel symbols of connection are 0, which will lead us to $\frac{DT^{\mu\nu}}{Dx^\lambda} = \frac{\partial T^{\mu\nu}}{\partial x^\lambda}$, and this result does satisfy our requirement. In other words, it's safe to conclude that whether a four-dimensional spacetime is flat or curved, the conservation of energy and momentum can both be expressed through the below equation:

$$\frac{DT^{\mu\nu}}{Dx^\mu} = 0$$

$$(6.132)$$

Last but not least, (6.132) is true in any frame of reference because the covariant derivative of $T^{\mu\nu}$ is still a tensor.

6.3.9   November of 1915—Birth of Field Equation

When Einstein finally managed to understand most tensor properties and related calculations that we've described, he started giving speeches at major academic institutes all over Europe from the beginning of 1915 and spared no effort in explaining the mathematical and physical significance of general relativity. Among these speeches, the most crucial one took place in November of 1915, when he shared his research results with fellows of the Prussian Academy in Germany through four sessions. In fact, what he shared during these meetings was so important that it was later published as proceedings.

The first speech was held on November 4th and saw him give a comprehensive walkthrough of the mathematical operations regarding covariant, contravariant, and mixed tensors. At this point, he'd already broken free from the Hole Argument and mastered the nature of all three types of tensors.

The second speech followed on November 11th, when he introduced the field equation that he constructed based on the Ricci covariant tensor, i.e. (6.113), and explained that $T^{\mu\nu}$ had to satisfy the conservation law of energy and momentum regulated by (6.132). We can tell from his sharing that he was already able to summarize his theory into three simple equations: the field Eq. (6.113), the geodesic Eq. (6.54), and the energy-momentum conservation Eq. (6.132). All three of them can fit in the framework of Newtonian mechanics in low-speed scenarios while satisfying the requirements of general covariance at the same time.[31]

Although the structure of general relativity seemed to be rounded out by now, Einstein still wasn't quite happy about (6.113) because he figured it should only require two equations to describe the motion in four-dimensional spacetime: (1) the field equation to explain how matter curves spacetime with energy and momentum, and (2) the geodesic equation to regulate the motion of objects in curved spacetime. Just as John Wheeler (1911–2008) later commented, general relativity basically consists of matter and spacetime. While the former decides how the latter is curved, the latter dictates the motion pattern of the former. Similarly, Einstein was hoping to describe object motion with a simple and symmetrical theory. Yet as we've mentioned in Sect. 6.3.8, the energy-momentum Eq. (6.132) just seemed to be an indispensable part

---

[31] General covariance refers to the situation where the terms on both sides of the equal sign in (6.113) are second-order covariant tensors. In other words, as long as the equation is true in any given reference frame, it should still be valid even after random transformation of coordinates.

of his theory. This kept bothering him somehow as he believed there must be some kind of mechanism in spacetime geometry that could naturally pull off the conservation of energy and momentum. In his reasoning, this unknown mechanism shouldn't need an extra equation to regulate. One other thing that bugged him was that he still couldn't find out the reason why $DR^{\mu\nu}/Dx^\mu$ had to be 0 unless he applied (6.132) as a restriction. Therefore, he was convinced that the term about spacetime geometry before the equal sign in (6.113) had to be modified. While he was giving the speeches, he was also looking for a form of expression that could ensure that this term was 0 after covariant differentiation was applied over either of its indices. With this goal in mind, he wrote to the mathematician David Hilbert (1862–1943) asking for help. Much to his surprise, Hilbert responded saying he already knew the fix. On November 16th, he sent a brief letter to Einstein about the modified field equation, while the latter also wrote back, confirming they'd come to the same conclusion.[32]

On November 18th, Einstein gave the third speech. In this session, he announced his modified research on the perihelion advance of Mercury and impressed the participants with the accurate result of 43ArcSec per hundred years, which exactly matched the measurement from astronomical observations. Of course, Einstein himself was just as excited about the success as the audience, so he revisited the study about the bending of light in the gravitational field of Sun as well and came up with the result of 1.75 ArcSec. This time, however, he had to wait till May 29th of 1919 for the British physicist Arthur S. Eddington (1882–1944) to prove the value correct by observing a full solar eclipse. Details regarding Eddington's work and sample calculations will be provided in Sect. 6.4.

The last, but also the most important speech took place on November 25th. On this day, he introduced the final version of his set of field equations, which had not just removed the limitation of (6.132) but integrated the equation into the theory of general relativity as well. Now, let's take a look at how he finally came to this satisfactory result.

---

[32]While Einstein was fully dedicated to developing the theory of general relativity, Hilbert also spent quite a lot of effort researching this topic and even invited Einstein to the University of Göttingen to give a speech that lasted for a week in June of 1915. Ever since that, the two of them started exchanging thoughts and sharing results through letters. Although the modified field equation might have been derived by Hilbert first, it's generally agreed that Einstein's original research purpose was the driving force behind the birth of it, while Hilbert's contribution was largely in the field of mathematical construction. Hence, Einstein was credited with the establishment of the field equations.

6.3.10 Field Equation Regulating Energy-Momentum Conservation

Einstein's ultimate goal in perfecting his theory was to express the term $R^{\mu\nu}$ in (6.113) as a covariant tensor that could ensure the equation remained true after covariant differentiation was applied to either of the term's indices. As this hypothesized tensor should still be related to the Riemann curvature, let's first recall some of its properties. As mentioned in Sect. 6.2.4, the Riemann curvature can be fully represented with metric tensor ($g_{\mu\nu}$). Therefore, as long as we familiarize ourselves with the covariant derivative of $g_{\mu\nu}$, we'll be able to understand the Riemann curvature. To retrieve the covariant derivative of $g_{\mu\nu}$, we can use (6.130) as a reference:

$$
\begin{aligned}
\frac{Dg_{\mu\nu}}{Dx^\lambda} &= \frac{\partial g_{\mu\nu}}{\partial x^\lambda} - \Gamma^\kappa_{\mu\lambda} g_{\kappa\nu} - \Gamma^\kappa_{\nu\lambda} g_{\mu\kappa} = \frac{\partial (\mathbf{e}_\mu \cdot \mathbf{e}_\nu)}{\partial x^\lambda} - \Gamma^\kappa_{\mu\lambda} g_{\kappa\nu} - \Gamma^\kappa_{\nu\lambda} g_{\mu\kappa} \\
&= \frac{\partial \mathbf{e}_\mu}{\partial x^\lambda} \cdot \mathbf{e}_\nu + \mathbf{e}_\mu \cdot \frac{\partial \mathbf{e}_\nu}{\partial x^\lambda} - \Gamma^\kappa_{\mu\lambda} g_{\kappa\nu} - \Gamma^\kappa_{\nu\lambda} g_{\mu\kappa} \\
&= \Gamma^\kappa_{\mu\lambda} g_{\kappa\nu} + \Gamma^\kappa_{\nu\lambda} g_{\mu\kappa} - \Gamma^\kappa_{\mu\lambda} g_{\kappa\nu} - \Gamma^\kappa_{\nu\lambda} g_{\mu\kappa} = 0
\end{aligned}
\tag{6.133}
$$

As shown in the derivation above, the covariant derivative of $g_{\mu\nu}$ is constantly 0, a property that we should consider when determining the limitation to be imposed on the Riemann curvature. As a first step, we should follow the definition of the Christoffel symbols of connection and the Riemann curvature to rewrite the latter as the below function of metric tensor:

$$
R^\mu_{\alpha\beta\gamma} = \partial_\beta \Gamma^\mu_{\alpha\gamma} - \partial_\gamma \Gamma^\mu_{\alpha\beta} + \Gamma^\mu_{\nu\beta} \Gamma^\nu_{\alpha\gamma} - \Gamma^\mu_{\nu\gamma} \Gamma^\nu_{\alpha\beta}
\tag{6.134}
$$

$$
\Gamma^\alpha_{\mu\nu} = \frac{1}{2} g^{\alpha\lambda} \left(\partial_\nu g_{\lambda\mu} + \partial_\mu g_{\nu\lambda} - \partial_\lambda g_{\mu\nu} \right)
\tag{6.135}
$$

In (6.134) and (6.135), $\partial/\partial x^\beta$ and D/Dx^β are simplified as ∂_β and D_β for the purpose of concision. As for the next step, we're going to verify the four equations[33] below, which regulate the properties of the Riemann curvature:

$$R_{\mu\alpha\beta\gamma} = -R_{\alpha\mu\beta\gamma} = -R_{\mu\alpha\gamma\beta} \tag{6.136}$$

$$R_{\mu\alpha\beta\gamma} = R_{\beta\gamma\mu\alpha} \tag{6.137}$$

$$R_{\mu\alpha\beta\gamma} + R_{\mu\beta\gamma\alpha} + R_{\mu\gamma\alpha\beta} = 0 \tag{6.138}$$

$$D_\sigma R_{\mu\alpha\beta\gamma} + D_\beta R_{\mu\alpha\gamma\sigma} + D_\gamma R_{\mu\alpha\sigma\beta} = 0 \tag{6.139}$$

To start with, we know that $R_{\mu\alpha\beta\gamma} = g_{\mu\kappa} R^\kappa_{\alpha\beta\gamma}$ judging from (6.110). Then, although the last two terms in (6.134) seem to be extremely complicated, they can be easily eliminated because the coefficient of connection Γ is always 0 from the perspective of local inertial frames. Last but not least, since the Riemann curvature is a tensor, so are all the terms in (6.136)–(6.139), meaning all four equations can apply to any frame as long as they're true in a certain local inertial frame. Hence, we must return to the scenario of local inertial frames, where the Riemann curvature can be expressed as:

$$
\begin{aligned}
R^\mu_{\alpha\beta\gamma} &= \partial_\beta \Gamma^\mu_{\alpha\gamma} - \partial_\gamma \Gamma^\mu_{\alpha\beta} \\
&= \frac{1}{2} g^{\mu\lambda} \left[\partial_\beta \left(\partial_\gamma g_{\lambda\alpha} + \partial_\alpha g_{\gamma\lambda} - \partial_\lambda g_{\alpha\gamma} \right) - \partial_\gamma \left(\partial_\beta g_{\lambda\alpha} + \partial_\alpha g_{\beta\lambda} - \partial_\lambda g_{\alpha\beta} \right) \right] \\
&= \frac{1}{2} g^{\mu\lambda} \left(\partial_\alpha \partial_\beta g_{\gamma\lambda} + \partial_\gamma \partial_\lambda g_{\alpha\beta} - \partial_\beta \partial_\lambda g_{\alpha\gamma} - \partial_\alpha \partial_\gamma g_{\beta\lambda} \right)
\end{aligned}
\tag{6.140}
$$

Simplifying the above equation, we can tell that (6.140) is equivalent to[34]:

$$R_{\mu\alpha\beta\gamma} = \frac{1}{2} \left(\partial_\alpha \partial_\beta g_{\mu\gamma} + \partial_\mu \partial_\gamma g_{\alpha\beta} - \partial_\mu \partial_\beta g_{\alpha\gamma} - \partial_\alpha \partial_\gamma g_{\mu\beta} \right) \tag{6.141}$$

As can be observed from (6.133), the first-order differential ∂g of all metric tensors g in the above equation is 0, but it's not necessarily the case for the

---

[33] Judging from (6.136) to (6.138), $R_{\mu\alpha\beta\gamma}$ only has 20 degrees of freedom though containing 256 possibilities. Similarly, Einstein's set of field equations consists of 16 simultaneous equations, but only 10 are actually different because of symmetry. As 10 equations aren't enough for regulating a spacetime with 20 degrees of freedom, different metric tensors can in fact coexist in the same distribution of energy and momentum. In a gravity-free environment, for example, the metric tensors in inertial frames and the ones with constant proper acceleration are not the same, but what they both describe is the $T^{\mu\nu} = 0$ spacetime.

[34] The derivation is done by multiplying the terms on both sides of the equal sign by $g_{\kappa\mu}$, substituting in the equation of $g_{\kappa\mu} g^{\mu\lambda} = \delta^\lambda_\kappa$, and lastly, replacing κ with μ.

second order counterpart $\partial\partial g$. By simply substituting (6.141) into (6.136), (6.137), (6.138), and (6.139), we can prove all four of them correct.

In fact, Eqs. (6.136)–(6.139) make a comprehensive representation of the properties of the Riemann curvature, which we will use later on, but we still need to find out the reason why $D_\mu R^{\mu\nu}$ has to be 0. Specifically speaking, we have to establish a tensor that will be 0 after covariant differentiation is applied over any of its indices. For this purpose, we should reduce the number of notations in (6.139) to merely two by multiplying the Riemann curvature by $g^{\mu\nu}$, hence moving the first lower index of the curvature to the upper position:

$$D_\sigma R^\nu_{\alpha\beta\gamma} + D_\beta R^\nu_{\alpha\gamma\sigma} + D_\gamma R^\nu_{\alpha\sigma\beta} = 0 \tag{6.142}$$

Then, the replacement of the σ in (6.142) with ν will lead us to:

$$D_\nu R^\nu_{\alpha\beta\gamma} + D_\beta R^\nu_{\alpha\gamma\nu} + D_\gamma R^\nu_{\alpha\nu\beta} = 0 \tag{6.143}$$

As we know that $R^\nu_{\alpha\nu\beta} = R_{\alpha\beta}$ and $R^\nu_{\alpha\gamma\nu} = g^{\nu\mu}R_{\mu\alpha\gamma\nu} = g^{\nu\mu}R_{\mu\alpha\nu\gamma} = -R^\nu_{\alpha\nu\gamma} = -R_{\alpha\gamma}$ from the middle term of (6.136), we can substitute these two equations into (6.143) and obtain:

$$D_\nu R^\nu_{\alpha\beta\gamma} - D_\beta R_{\alpha\gamma} + D_\gamma R_{\alpha\beta} = 0 \tag{6.144}$$

As a next step, we'll define the scalar curvature R as:

$$R \equiv g^{\alpha\beta} R_{\alpha\beta} \tag{6.145}$$

Subsequently, if we multiply (6.144) by $g^{\alpha\beta}$:

$$D_\nu\left(g^{\alpha\beta} R^\nu_{\alpha\beta\gamma}\right) - D_\beta R^\beta_\gamma + D_\gamma R = 0 \tag{6.146}$$

Where the complicated first term can be simplified as:

$$g^{\alpha\beta} R^\nu_{\alpha\beta\gamma} = g^{\alpha\beta} g^{\nu\mu} R_{\mu\alpha\beta\gamma} = -g^{\nu\mu} g^{\alpha\beta} R_{\alpha\mu\beta\gamma}$$
$$= -g^{\nu\mu} R^\beta_{\mu\beta\gamma} = -g^{\nu\mu} R_{\mu\gamma} = -R^\nu_\gamma \tag{6.147}$$

Then, we can substitute (6.147) back into (6.146) and retrieve:

$$-D_\nu R^\nu_\gamma - D_\beta R^\beta_\gamma + D_\gamma R = 0 \tag{6.148}$$

The first two terms in (6.148) are actually the same as they're both summation of the results when v and β are replaced with all possible values. Hence, the equation can be expressed as:

$$D_v R^v_\gamma - \frac{1}{2} D_\gamma R = 0 \qquad (6.149)$$

Which we'll multiply by $g^{\mu\gamma}$:

$$D_v R^{v\mu} - \frac{1}{2} D_\gamma \left(g^{\gamma\mu} R \right) = 0 \qquad (6.150)$$

After some reorganization of the indices, the above equation can be rewritten as:

$$D_\mu \left(R^{\mu v} - \frac{1}{2} g^{\mu v} R \right) = 0 \qquad (6.151)$$

Up to this point, we've accomplished the goal of finding a tensor that satisfies our requirement: $R^{\mu v} - \frac{1}{2} g^{\mu v} R$, a tensor that equals 0 when covariant differentiation is applied over either of its two indices. One important point that needs to be emphasized again, though, is that $D_\mu R^{\mu v}$ isn't necessarily 0, but $D_\mu \left(R^{\mu v} - \frac{1}{2} g^{\mu v} R \right)$ has to be.

As such, we can start rewriting (6.113) by replacing the term before the equal sign with $R^{\mu v} - \frac{1}{2} g^{\mu v} R$ and substituting a constant for $4\pi G/c^4$:

$$R^{\mu v} - \frac{1}{2} g^{\mu v} R = \text{New Constant} \times T^{\mu v} \qquad (6.152)$$

Currently, (6.152) is represented as a contravariant tensor, but of course, it can also be expressed in the form of a covariant tensor:

$$R_{\mu v} - \frac{1}{2} g_{\mu v} R = \text{New Constant} \times T_{\mu v} \qquad (6.153)$$

As the term $4\pi G/c^4$ was established with the Newtonian approximation, we need to reexamine how the equation will turn out after simplified in a low-speed, static weak field. After a series of calculations, we can still retrieve Einstein's field equation:

$$R^{\mu v} - \frac{1}{2} g^{\mu v} R = \frac{8\pi G}{c^4} T^{\mu v} \qquad (6.154)$$

And with the derivation of this final equation, Einstein's theory is now fully validated. Looking back at his research journey, there were two major obstacles that he had to overcome before eventually completing the framework of general relativity. First, he came up with a field equation that satisfied the rules of general covariance and contravariance to ensure the equation would remain true in any random frame of reference. Second, he made sure that this equation complied with the conservation law of energy and momentum because both the expression before the equal sign in (6.154) and $D_\mu T^{\mu\nu}$ turned out to be 0 when D_μ was applied. Solving the above challenges, Einstein reshaped the world of physics with his set of field equations that regulates the motion behavior of all objects in the entire universe.

6.4   Verification and Application of General Relativity

In the previous sections, we started with Newton's law of universal gravitation in search of an equation that can fit in the framework of relativity on the mathematical level but can also regulate the static, low-speed weak fields in classical mechanics. Eventually, we managed to derive the Einstein field equations, which integrate the mass, energy, and momentum of matter with spacetime geometry through complicated tensor calculations. After Einstein published his results, it dawned on many scientists that the motion of the universe was so surprisingly simple and harmonious that it only took one symmetrical, elegant set of equations to regulate. Just like any other great theorem, however, the theory of general relativity had to go through a careful examination to be verified. Although the experimental techniques back in the early 20th century were hardly enough to duplicate the phenomena described by the field equations, a lot of physicists, including Einstein himself, still attempted at such examinations with voluminous calculations and came up with fruitful results. In this chapter, we'll dive into certain phenomena that can be predicted with the field equations by way of mathematical derivation and provide details on how to conduct related experiments for verification purposes. We're also going to discuss a few cases of application in technology for readers to understand that the theory of general relativity isn't just a piece of art in the realms of physics and mathematics but actually wields a huge influence on our everyday life as well.

6.4.1 Schwarzschild Metric

Several months after Einstein introduced his field equations, the German astrophysicist Karl Schwarzschild (1873–1916) proposed a particular solution, which regulated the spacetime geometry near a planet with spherical symmetry. This solution,[35] which can help simulate the external gravitational field of an astronomical body without self-rotation, is as follows:

$$c^2 d\tau^2 = c^2 \left(1 - \frac{2GM}{c^2 r}\right) dt^2 - \frac{dr^2}{1 - \frac{2GM}{c^2 r}} - r^2 \left(d\theta^2 + \sin^2\theta d\varphi^2\right)$$

$$(6.155)$$

Schwarzschild chose to illustrate his solution in a spherical coordinate system (as in Fig. 6.7, center set on the origin with M representing the sphere's mass) because it applied to planets with spherical symmetry. Meanwhile, since it's the external spacetime geometry of planets that this solution regulates, the metric tensor described in (6.231) should satisfy the field equation for vacuum environments ($T_{\mu\nu} = 0$),[36] i.e. $R_{\mu\nu} - \frac{1}{2}g_{\mu\nu}R = 0$. In fact, however, it's only in the area where $r > r_s = 2GM/c^2$ that the Schwarzschild solution applies.

As for the reason, let's first examine the physical meaning of (6.155) and whether it's compatible with the law of universal gravitation. When it comes to the motion pattern of a point mass in spacetime geometry, all discussion starts with the geodesic equation. As we're only going to cover the point masses motioning on the $\theta = \pi/2$ plane, the last term of (6.155) can be simplified as $r^2 d\varphi^2$. After some variational calculation, we can obtain the three forms of the geodesic equation below:

$$\frac{d}{d\tau}\left(\frac{1}{1 - \frac{2GM}{c^2 r}}\frac{dr}{d\tau}\right) + \frac{GM}{r^2}\left(\frac{dt}{d\tau}\right)^2$$

$$+ \frac{\frac{GM}{c^2 r^2}}{\left(1 - \frac{2GM}{c^2 r}\right)^2}\left(\frac{dr}{d\tau}\right)^2 - r\left(\frac{d\varphi}{d\tau}\right)^2 = 0 \quad (6.156)$$

---

[35] In fact, the Schwarzschild solution will overlap with the Minkowski metric tensor when $r \to \infty$, meaning matter that is infinitely far from a given astronomical body doesn't really get affected by its gravity. As for the coordinates t, r, θ, and φ, these are physical quantities measured from the perspective of observers infinitely far from the celestial body.

[36] Details aren't provided here because the calculation process is fairly complicated without much particularly worth mentioning. If interested, however, readers are encouraged to substitute the Schwarzschild solution back into the field equations for verification.

$$\frac{d}{d\tau}\left[\left(1 - \frac{2GM}{c^2 r}\right)\frac{dt}{d\tau}\right] = 0 \tag{6.157}$$

$$\frac{d}{d\tau}\left(r^2 \frac{d\varphi}{d\tau}\right) = 0 \tag{6.158}$$

Where (6.156) can be further simplified as:

$$\frac{d^2 r}{d\tau^2} + \frac{GM}{r^2}\left(1 - \frac{2GM}{c^2 r}\right)\left(\frac{dt}{d\tau}\right)^2 - \frac{\frac{GM}{c^2 r^2}}{1 - \frac{2GM}{c^2 r}}\left(\frac{dr}{d\tau}\right)^2$$
$$- r\left(1 - \frac{2GM}{c^2 r}\right)\left(\frac{d\varphi}{d\tau}\right)^2 = 0 \tag{6.159}$$

As a next step, we're going to examine the physical significance of these equations. When the point mass in question here is stationary,[37] i.e. when $dr/d\tau = d\varphi/d\tau = 0$, we know from (6.155) that:

$$\left(\frac{dt}{d\tau}\right)^2 = \frac{1}{1 - \frac{2GM}{c^2 r}} \tag{6.160}$$

Substituting (6.160) and $dr/d\tau = d\varphi/d\tau = 0$ into (6.159), we'll obtain:

$$\frac{d^2 r}{d\tau^2} = -\frac{GM}{r^2} \tag{6.161}$$

In which $d^2 r/dt^2$ represents the acceleration of the point mass. Therefore, we can be sure that (6.161) is indeed compatible with rules in Newtonian mechanics. Then, assuming the point mass is now in a uniform circular motion, we can derive $dr/d\tau = d^2 r/d\tau^2 = 0$. Substituting it into (6.159), we'll come to the below equation after some simplification:

$$\frac{d\varphi}{dt} = \sqrt{\frac{GM}{r^3}} \tag{6.162}$$

---

[37] In all scenarios where the speed is far slower than light speed, to be more precise.

Which again complies with Newton's law of universal gravitation.[38]

Then, let's move on to examine (6.157) and (6.158), which show the invariance of both $(1 - 2GM/c^2r)dt/d\tau$ and $r^2d\varphi/d\tau$ in a motion process. If we imagine a scenario where a point mass is moving at a low speed in a weak field $(d\varphi/d\tau \approx d\varphi/dt)$ and multiply $r^2d\varphi/d\tau$ by the mass of the point m, the product will turn out equal to the angular momentum $mr^2d\varphi/dt$ of the point mass. Hence, "angular momentum" can be defined with the below equation in the Schwarzschild geometry:

$$L \equiv mr^2\frac{d\varphi}{d\tau} \tag{6.163}$$

While L remains invariant in the moving process, $(1 - 2GM/c^2r)dt/d\tau$ is a bit more complicated. First, we can rewrite (6.155) as:

$$\frac{dt}{d\tau} = \sqrt{\frac{c^2 + \frac{1}{1-\frac{2GM}{c^2r}}\left(\frac{dr}{d\tau}\right)^2 + r^2\left(\frac{d\varphi}{d\tau}\right)^2}{c^2\left(1 - \frac{2GM}{c^2r}\right)}} \tag{6.164}$$

Then, the multiplication of the invariant $(1 - 2GM/c^2r)dt/d\tau$ by mc^2 and the substitution of (6.164) will lead us to:

$$mc^2\left(1 - \frac{2GM}{c^2r}\right)\frac{dt}{d\tau} = mc^2\sqrt{\left(1 - \frac{2GM}{c^2r}\right)\left[1 + \frac{1}{c^2}\left(\frac{1}{1-\frac{2GM}{c^2r}}\left(\frac{dr}{d\tau}\right)^2 + r^2\left(\frac{d\varphi}{d\tau}\right)^2\right)\right]}$$

$$\approx mc^2\sqrt{\left(1 - \frac{2GM}{c^2r}\right)\left(1 + \frac{v^2}{c^2}\right)}$$

$$\approx mc^2\left(1 - \frac{GM}{c^2r} + \frac{v^2}{2c^2}\right) = mc^2 + \frac{1}{2}mv^2 - \frac{GMm}{r} \tag{6.165}$$

Note that the speed of the point mass $v = \sqrt{\left(\frac{dr}{dt}\right)^2 + r^2\left(\frac{d\varphi}{dt}\right)^2}$ was used to derive the second equation above. Finally, as we can observe from (6.165), the expression after the second equal sign represents the sum of the point mass's stationery (mc^2), kinetic, and gravitational energy, which corresponds

---

[38]In Newtonian mechanics, the centripetal force to maintain a point mass in uniform circular motion is deemed to be provided by gravity. Therefore, the centripetal acceleration $r\left(\frac{d\varphi}{dt}\right)^2$ has to equal gravitational acceleration $\frac{GM}{r^2}$, which is why we know that $\frac{d\varphi}{dt} = \sqrt{\frac{GM}{r^3}}$.

to the law of energy conservation in Newtonian mechanics. Hence, "energy" can be defined as follows in the Schwarzschild geometry:

$$E \equiv mc^2 \left(1 - \frac{2GM}{c^2 r} \right) \frac{dt}{d\tau} \tag{6.166}$$

Which is an invariant.

Going through the multi-step verification above, we've confirmed that under the circumstance of a low-speed, weak field, the combination of the Schwarzschild metric tensor (6.155) and the geodesic equation can perfectly fit the theory of general relativity into the framework of Newtonian mechanics. However, what happens if the speed is increased to near-light speed? This question can only be answered with experiments as proof. In the below section, we're going to discuss several phenomena that can only be explained with general relativity and conduct actual calculations to see if the results are compatible with related experimental data.

6.4.2   Light Bending in Gravitational Field

As indicated in Einstein's theory of relativity, all forms of energy have equivalent mass. Even light is no exception. So as you can probably imagine, the equivalent mass of the scorching hot Sun must have an extremely large mass since it accounts for 99.8% of the total mass of the Solar System. It also wouldn't be too surprising if the path of light gets deviated by solar gravity, which results from the Sun's enormous mass. In fact, the equivalence principle and field equations both predicted the degree of such deviation, but the value calculated with the latter was twice that from the former. Therefore, many scientists in Einstein's time started examining his results because everything associated with his groundbreaking research was a guarantee of money and recognition. They didn't care too much whether it was evidence in favor of or against the theory of relativity that they'd find. All they knew was they couldn't miss out on such a terrific chance of achieving overnight fame and raking it in. And just like that, looking for phenomena that could be regulated by the field equations or proving the already proposed ones became one of the most pursued studies in the world of physics. Among the countless aspiring scientists, there was one from the Cambridge University—Arthur S. Eddington (1882–1944).

In 1915, Europe was caught in the middle of World War I, inflicted with warfare and bloodshed that ended up killing almost 100 million people. To make things worse, England and Germany happened to be leaders of the two opposing alliances, so Einstein's research in field equations was actually smuggled by the Dutch astronomer William de Sitter (1872–1934) to the British Islands from the European continent. After reading the paper by the "German Jew", [39] Eddington thought it must be extremely hard to observe the phenomenon of gravitational bending because the light of stars was just too weak compared to the glare of Sun, making it as hard to detect as fireflies in broad sunlight. Therefore, he decided that such observation had to be done during the solar eclipse. But while the Sun is eclipsed once every few years, only people in the shadow of the Moon, when it passes between Sun and Earth, can see the phenomenon take place. Therefore, the eclipse cycle for a certain location on Earth, e.g. Cambridge is about 350 years on average. Of course, Eddington couldn't wait that long, so he immediately built a team and started researching where the next eclipse would be. Despite the ongoing war, he still somehow managed to raise enough funding for his trip to the West African island of Príncipe on May 29th of 1919, which happened to be the day marking the German surrender and end of WWI. [40]

Eddington's team was in fact composed of two people only, himself and his assistant. To make things even harder, the sky above Príncipe Island was obscured by thick clouds that nearly blocked their entire view of the Sun. Besides that, tropical rain also doused the island, almost killing the hope of the two-person team. Despite the desperate situation, Eddington didn't give up. With the help of his assistant, he took 16 pictures intermittently with the camera focused on the Sun's halo when the clouds were dissipated momentarily. While six of them captured the clear sky, four of them were unfortunately damaged due to over-exposure caused by the very high temperature on the island. Out of the remaining two, only one clearly showed the halo surrounded by stars. [41] So several days later, he returned to Cambridge with this

---

[39] Einstein's German citizenship and Jewish origin got him several names from people with different political views, which is epitomized in what the German government called him over the course of time. Before WWI broke out, he was the "Berlin-based scientist", but when WWII was about to erupt, he'd already become a "Swedish Jew" due to the Nazi Party's crackdown on Jewish people.

[40] Eddington wasn't the first scientist to see solar eclipse as a possible way of verifying the theory of relativity. In fact, the eclipse in 1914 also drew the attention of the German astronomer Erwin Finlay-Freundlich (1885–1964), who departed from Berlin on July 19th with a telescope camera for the Crimean Peninsula, which was cast in the shadow of Moon and therefore promised a clear view of Sun's halo. When he arrived on August 24th, though, WWI had already exploded due to the assassination of Austrian Archduke Franz Ferdinand in Sarajevo by the local revolusionist Gavrilo Princip, which caused England and Russia to both declare war on Germany. Therefore, once his large research team set foot on Crimea, all the members got imprisoned by the Russian military with all instruments confiscated. A month later, they were finally released in an exchange of hostages between Germany and Russia but had to endure a long trudge back to Berlin.

[41] Rumor has it that Eddington actually had multiple successful pictures, but he only published one after analyzing all of them comprehensively because the data captured in this particular photo was in perfect accordance with the predictions of field equations.

precious picture and contrasted it against photos taken by the Royal Observatory in Greenwich. As it turned out, certain astronomical bodies seemed to be deviated by 1.6 ± 0.3 ArcSec in position due to the effect of Sun, almost in perfect agreement with the prediction of the field equations.[42] The result put him on cloud nine because this discovery basically marked a 40-year-old German Jew's rise over the giant Newton, who'd enjoyed an unparalleled status in the scientific community over the past 250 years, and who knew he'd be the one to witness this firsthand?

Soon on November 7th of the same year, his discovery was hailed by The Times as initiating a "scientific revolution" that unveiled a great universal law. In fact, the media showed a high level of science literacy because related reports were generally truthful and credible. It seemed like the law of universal gravitation and quantum mechanics had wielded such a significant influence on the world that the media also dedicated considerable effort to understanding and reporting related topics.

Decades later, a more accurate version of Eddington's experiment was conducted by NASA with the probe, Cassini-Huygens, which was launched to Saturn on a space-research mission in 1997. On the way to the planet, the probe started sending occasional signals back to Earth, allowing NASA scientists to confirm that the gravitational redshift did deflect the signals when they passed close to Sun. In addition, the degree of deflection was precisely the same as predicted by the theory of general relativity.

To explain how exactly the predictions are made, we'll calculate the degree of light bending caused by the solar gravitational field based on the equivalence principle and field equations.

(1) Equivalence Principle and Newtonian Approximation [43]

In Sect. 6.1.4, we once postulated that the energy of photons could be converted to an equivalent mass with the mass-energy equation. Now, it's exactly this hypothesis that we'll follow to calculate the degree of light bending. Let's suppose there's a photon with the energy of E traveling from afar to brush over the surface of an astronomical body with the mass of M, which causes a tiny deflection degree of $\Delta\varphi$. As the deflection is extremely small, the path of

---

[42] Such a small number shows how difficult the comparison was for Eddington because one ArcSec of deviation in the sky actually translates to merely 6.25×10^{-5} m when observed from Earth's surface (or, in this case, in the pictures captured from Earth).

[43] In fact, this wasn't the approach adopted by Einstein. In his 1911 paper "On the Influence of Gravitation on the Propagation of Light", he first proves that light speed varies in different parts of a gravitational field through the equivalence principle before combining this finding with the undulatory nature of light (the Huygens–Fresnel principle). However, our results will be the same.

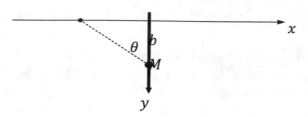

Fig. 6.10 Illustration of how light is deflected by the gravitational force from point mass M

the photon is basically a straight line, while the energy contained in it can be converted into the equivalent mass of m through $E = mc^2$. As such, we can represent the perpendicular distance between the path and the center of the fixed astronomical body as b. As a next step, we need to calculate the impulse J exerted by the body (M in Fig. 6.10) on the photon when it travels from left to right. Please note that the impact must work along the direction of y, just as shown below:

Considering that the angle formed by the line joining the photon/M and the axis of y lies between θ and $\theta + d\theta$ ($x = b \tan \theta$ $x = b \tan \theta$), the impulse generated by the gravitational force along the y direction should be: (distance: $r = b \sec \theta$; time passed: $dt = dx/c$)

$$dJ = \frac{GMm}{r^2} dt \times \cos \theta = \frac{GMm}{b^2 \sec^2 \theta} \frac{d(b \tan \theta)}{c} \times \cos \theta = \frac{GMm}{bc} \cos \theta d\theta \tag{6.167}$$

Therefore, the total impulse is:

$$J = \frac{GMm}{bc} \int_{-\pi/2}^{\pi/2} \cos \theta d\theta = \frac{2GMm}{bc} \tag{6.168}$$

As we know the relationship between the momentum and energy of the photon is $E = pc$, its momentum can be derived as $p = mc$ by combining the formula $E = mc^2$. As for the degree of deflection, it's approximately the ratio of the momentum along the y direction and that parallel with x:

$$\Delta \varphi \approx \frac{J}{mc} = \frac{2GM}{bc^2} \tag{6.169}$$

(2) Exact Method with Field Equations (Schwarzschild Metric)

After the approximation method, we're now going for the exact solution using the Schwarzschild metric. Below, we've listed (6.163) and (6.166), equations that'll be of use in later steps:

$$L = mr^2\frac{d\varphi}{d\tau}, \ E = m\left(1 - \frac{r_s}{r}\right)c^2\frac{dt}{d\tau} \qquad (6.170)$$

While (6.170) contains $r_s = 2GM/c^2$, the limit of $m \to 0$ has to be adopted because light has no mass. It might be tempting to think that both the energy E and angular momentum L will be 0 under this situation, but the truth is the path of light always satisfies $d\tau = 0$,[44] so both parameters can be non-zero values. When $d\tau = 0$, the equation of the Schwarzschild metric (6.155) can help us derive:

$$0 = \left(1 - \frac{r_s}{r}\right)c^2 dt^2 - \frac{1}{1 - \frac{r_s}{r}}dr^2 - r^2 d\varphi^2 \qquad (6.171)$$

The term 0/0 in (6.170) might indeed be an issue in certain situations, but luckily, what we need here is simply the ratio of E and L. If we divide L by E, both m and $d\tau$ can be eliminated:

$$\frac{L}{E} = \frac{r^2}{\left(1 - \frac{r_s}{r}\right)c^2}\frac{d\varphi}{dt} \qquad (6.172)$$

Now that (6.172) is derived, the dt in (6.171) can be represented with $d\varphi$. With a bit of simplification, the equation becomes:

$$\varphi = \int \frac{dr}{r^2\sqrt{\frac{E^2}{L^2c^2} - \frac{1}{r^2}\left(1 - \frac{r_s}{r}\right)}} \qquad (6.173)$$

Since E and L are both invariants, so is the term L/E. Therefore, as long as we can figure out this ratio at a specific moment in the photon's motion, the value of L/E at all times can be confirmed as well. To calculate L/E, the simplest way is to consider the moment when the photon's distance to the astronomical body M is the shortest in its journey (when $r = b$). At this

---

[44]In the Minkowski metric, light speed is c when $d\tau = 0$, which isn't always the case for other metrics under normal circumstances (e.g. in the Schwarzschild metric). For the energy to be a non-zero value, however, $d\tau = 0$ still has to be true.

particular point of time, the radial velocity of the photon is 0 ($dr/dt = 0$), so we know from the defining equation of the Schwarzschild metric (6.155) that:

$$0 = \left(1 - \frac{r_s}{b}\right)c^2 dt^2 - b^2 d\varphi^2 \tag{6.174}$$

Then, (6.174) can be simplified as:

$$\left.\frac{d\varphi}{dt}\right|_{r=b} = \frac{c}{b}\sqrt{1 - \frac{r_s}{b}} \tag{6.175}$$

Substituting (6.175) back into (6.172), we can obtain:

$$\frac{L}{E} = \frac{b}{c\sqrt{1 - \frac{r_s}{b}}} \approx \frac{b}{c} \tag{6.176}$$

Note that $r_s/b \ll 1$ is true because we're only considering the scenario of weak fields here. Similarly, we'll only retrieve the lowest-order term for r_s/r in the below calculations.[45] As such, we can now substitute (6.176) back into (6.173):

$$\varphi \approx \int \frac{dr}{r^2\sqrt{\frac{1}{b^2} - \frac{1}{r^2}\left(1 - \frac{r_s}{r}\right)}} \approx \int \frac{dr}{r^2\sqrt{\frac{1}{b^2} - \frac{1}{\left(r + \frac{r_s}{2}\right)^2}}} \tag{6.177}$$

In deriving the second term in (6.177), we adopted the binomial approximation method. Then, for easier calculation, we'll suppose $(r + r_s/2)/b = x$ to simplify (6.177) as:

$$\varphi = \int \frac{x\,dx}{\left(x - \frac{r_s}{2b}\right)^2\sqrt{x^2 - 1}} \approx \int \left(\frac{1}{x^2} + \frac{r_s}{bx^3}\right)\frac{x\,dx}{\sqrt{x^2 - 1}} \tag{6.178}$$

Due to the photon's symmetrical behavior when approaching and leaving the astronomical body, the integral range should be $r = b$ to $r \to \infty$.

---

[45] From the result of (6.169), we can also tell that r_s/b is exactly the degree of light deflection. Though roughly calculated and not totally accurate, it can still serve as a reference metric. Hence, when the angle of deflection is extremely small, we can reasonably consider r_s/r as a very small value in our calculation as well.

Multiplying this result by 2, we can retrieve the range of $x \approx 1$ to $x \to \infty$.

$$\varphi = 2 \int_{1}^{\infty} \frac{dx}{x\sqrt{x^2 - 1}} + \frac{2r_s}{b} \int_{1}^{\infty} \frac{dx}{x^2\sqrt{x^2 - 1}} = \pi + \frac{2r_s}{b} \qquad (6.179)$$

Following the result of the integration above, we can substitute back the equation $r_s = 2GM/c^2$ and obtain:

$$\Delta\varphi = \frac{4GM}{bc^2} \qquad (6.180)$$

Which is the degree of light bending caused by the solar gravitational field, calculated with the theory of general relativity. Compared to the result obtained with the principle of equivalence and Newtonian approximation, the value is twice as large.

Last but not least, let's replace the parameters with actual values. For a ray of light that simply brushes over the surface of Sun, b should be replaced with Sun's radius 6.96×10^8 m, and M the solar mass of 1.99×10^{30} kg. In addition, the gravitational constant $G = 6.67 \times 10^{-11} \mathrm{m^3 kg^{-1} s^{-2}}$ should also be substituted into (6.180), which will eventually lead us to:

$$\Delta\varphi = 8.48 \times 10^{-6} \, \mathrm{rad} = 1.75 \, \mathrm{ArcSec} \qquad (6.181)$$

6.4.3 Precession of Mercury's Perihelion

Despite the title, the precession of revolving orbits is actually observed in the motion of not just Mercury but all other planets in the Solar System as well. In the case of Mercury, the closest planet to Sun, the revolving period is 88 days on Earth, while the eccentricity of its elliptic orbit is as high as 0.206, topping all the other counterparts in the system. Due to the advance of its trajectory, the day that marks the shortest distance between Mercury and Sun (i.e. when the planet moves to its perihelion) changes every year. Meanwhile, the major axis of the elliptical orbit also rotates at a certain angular velocity, so the azimuth of Mercury varies year by year as well. Such a phenomenon is called the perihelion precession, illustrated in Fig. 6.11.

So what's the reason behind the precession of Mercury's orbit? To approach this question, we can switch to a different perspective: what kind of orbit doesn't advance? The lack of precession implies that an orbit is a closed circuit

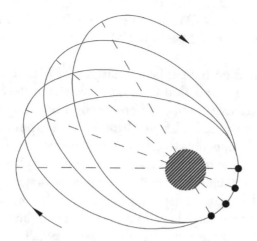

Fig. 6.11 Mercury's perihelion advance

(i.e. anything that follows this path will return to the origin after a cycle), and the most common case is elliptical orbits. Under the effect of a gravitational field with spherical symmetry, there are two possible scenarios where an orbit can be elliptic: (1) when the gravitational force is proportional to its radius, and (2) when the force is inversely proportional to the square of the radius. In the first situation, the center of the gravitational field overlaps with that of the ellipse, and in the second, sits at one of the two foci of the trajectory. For the revolution of planets around the Sun, the second scenario is true. Yet when the disturbance from external factors is present, causing changes in the inversely proportional relationship between the force of gravity and the square of the radius, the orbit will no longer be a closed circuit, thus causing orbital precession.

In 1859, Urbain Le Verrier (1811–1877) sorted through the observation data of Mercury from 1697 to 1848 in an effort to describe changes in its perihelion. Based on the data collected from over 100 years, he found out the nearest point of Mercury to Sun was moved by 574ArcSec (a circle = 360°, 1° = 60 Arcminutes = 360 Arcseconds) every 100 revolutions (i.e. every hundred years). Then, he tried to integrate the freshly-announced equation of universal gravitation into his calculation by taking into account all celestial gravity that could impact Mercury's revolution. This time, however, the result became 531ArcSec. Regarding the 43ArcSec between 531 and 574, it was later proved that the discrepancy could be explained with Einstein's field equations. In the following paragraphs, therefore, we'll calculate the perturbation exerted on Mercury's revolving pattern by other planets in the Solar System under the framework of Newtonian mechanics before justifying the deviation of 43ArcSec with the theory of general relativity.

(1) Calculating Impact of Perturbation from Other Planets in Solar System with Theories of Newtonian Mechanics

While elements causing the perihelion advance of Mercury actually include its axial precession, perturbation from other planets in the Solar System, the effect of general relativity, and the non-spherical (quadruple) shape of Sun, we'll only consider the second factor here for the purpose of simplification. Despite so, it's still impossible to calculate the impact exerted by other planets on Mercury's perihelion to perfect accuracy because the phenomenon involves multiple astronomical bodies. Therefore, we need to first simplify the scenario by considering each planet other than Mercury to be shattered pieces that's evenly distributed on a ring-shaped trajectory surrounding the Sun. In addition, we're also going to assume that all these pieces fall on the same ecliptic plane. As such, we can determine the distribution of potential energy $V(r)$ on this plane, which is point-symmetric and only changes with r, the distance to Sun.

Secondly, the mass of Sun M_0 can be deemed a non-moving point as it's way larger than that of the planets. The orbits of all planets are also postulated to be circular for simplification. If we represent the mass and orbital radius of Mercury as m and r_0, the angular velocity of its revolution will be[46]:

$$\omega_r = \sqrt{\frac{V'(r_0)}{mr_0}} \tag{6.182}$$

Taking into consideration the deviation caused to Mercury's circular orbit by the perturbation along the radial direction of the circle $(r = r_0 + \delta)$, which has the oscillating period of ω_φ, Mercury's degree of precession per revolution around Sun can be calculated to be approximately:

$$T(\omega_\varphi - \omega_r) \approx 2\pi\left(\frac{\omega_\varphi}{\omega_r} - 1\right) \tag{6.183}$$

In (6.183), $T = 2\pi/\omega_r$ is the revolving period of Mercury around the Sun. As a next step, we're going to derive the equation for expressing ω_φ. Since the distribution of potential energy is only affected by r, the force working on Mercury is a central force. Hence, we know the conservation

---

[46]The result is calculated with the centripetal force, which is obtained by multiplying the mass by the centripetal acceleration $mr_0\omega_r^2$, provided by the potential energy gradient along the radial direction $V'(r_0) = \frac{dV}{dr}\big|_{r=r_0}$. Note that $V'(r)$ denotes a one-time differentiation of $V(r)$ over r, while $V''(r)$ means twice.

of angular momentum is true, while the physical quantity is approximate:

$$L = mr_0^2 \sqrt{\frac{V'(r_0)}{mr_0}} = mr^2\dot\theta \tag{6.184}$$

Meanwhile, the radial motion of Mercury can be expressed as:

$$m\left(\ddot{r} - r\dot\theta^2\right) = -V'(r_0) \tag{6.185}$$

Then, the substitution of $\dot\theta = L/mr^2$ and $r = r_0 + \delta$ will lead us to:

$$\ddot\delta - \frac{L^2}{m^2(r_0 + \delta)^3} = -\frac{V'(r_0)}{m} \tag{6.186}$$

Since $\delta \ll r_0$, if we expand the above equation to retrieve the linear term of δ/r_0 before substituting (6.184) in, we'll be able to obtain:

$$\ddot\delta + \left(\frac{3V'(r_0)}{mr_0} + \frac{V''(r_0)}{m}\right)\delta = 0 \tag{6.187}$$

Which is the standard equation for the simple harmonic motion, with the oscillating angular frequency of:

$$\omega_\varphi = \sqrt{\frac{3V'(r_0)}{mr_0} + \frac{V''(r_0)}{m}} \tag{6.188}$$

Up to this point, we've got almost all we need to calculate the degree of precession per revolution, except for the value of $V(r)$. Therefore, let's now consider the gravitational potential energy between a ring (radius: R, total mass: M) and a point mass m whose distance from the center of the ecliptic plane is r. In addition, we'll also assume that $r \ll R$. In Fig. 6.12, the point mass m represents Mercury, while M any other planet in the Solar System. As for $V_{\text{ring}}(r)$, it's the gravitational potential energy between M and m, which can then be used to estimate the impact of the planet on the orbit of Mercury.

To calculate the gravitational potential energy between the ring in the illustration and m:

$$V_{\text{ring}}(r) = \int_0^{2\pi} \frac{GMm}{\sqrt{R^2 + r^2 - 2Rr\cos\theta}} \frac{d\theta}{2\pi}$$

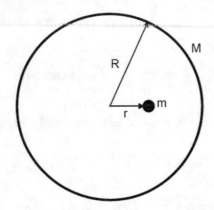

Fig. 6.12 Relative location of Mercury *m* and other planets *M*

$$\approx -\frac{GMm}{R}\left[1+\frac{1}{4}\left(\frac{r}{R}\right)^2+\frac{9}{64}\left(\frac{r}{R}\right)^4\right] \qquad (6.189)$$

If we only consider the impact of Sun and one of the planets whose mass is M while assuming that $R \gg r_0$ (orbital radius of Mercury), we know that:

$$V(r) = V_{\text{ring}}(r) - \frac{GM_0m}{r} \qquad (6.190)$$

Where the second term represents the gravitational potential energy between Mercury and the Sun. Substituting (6.190) into (6.182) and (6.188) to obtain ω_φ and ω_r, the two values can be substituted into (6.183) again to calculate Mercury's degree of precession per revolution:

$$\frac{M}{M_0}\left[\frac{3}{4}\left(\frac{r_0}{R}\right)^3+\frac{45}{32}\left(\frac{r_0}{R}\right)^5+\cdots\right] \times 360° \qquad (6.191)$$

With (6.191) established, we can consider the joint impact on Mercury's precession by Venus, Earth, and Mars. To be more specific, we're going to replace the parameters in the above equation with the actual values in the below table before applying linear superposition to all three results:

Body	Mass (/Earth's mass)	Average orbital radius (AU)
Sun	333400	–
Mercury	0.055	0.3871
Venus	0.815	0.7233
Earth	1.000	1.0000

(continued)

(continued)

Body	Mass (/Earth's mass)	Average orbital radius (AU)
Mars	0.107	1.5237

The final result shows that with every century (100 years) passed on Earth, the perihelion of Mercury advances about 480–500ArcSec (only rounding to 1-digit accuracy). The low level of accuracy results from the fact that the mass of all three planets is treated as an evenly-distributed ring through approximation, which is highly imprecise in the case of Venus. In fact, the actual ratio of the revolving periods of Venus and Mercury is 5:2, meaning the motion of the former is periodical in relation to the latter. If we did consider this periodical motion, the derivation would've led to a very different result about planetary perturbation. Therefore, the precision is one-digit only.

(2) Modification with General Relativity

In this section, we're going to discuss how to modify the result above for higher accuracy with the theory of general relativity and the Schwarzschild geometry. First, let's list all the equations that'll be used in later steps, starting with the Schwarzschild metric:

$$c^2 d\tau^2 = \left(1 - \frac{2GM}{rc^2}\right)c^2 dt^2 - \frac{dr^2}{1 - \frac{2GM}{rc^2}} - r^2 d\varphi^2 \qquad (6.192)$$

Then, we'll look back at (6.163) and (6.166), which allow us to calculate the angular momentum and energy of Mercury in the Schwarzschild geometry:

$$L = mr^2 \frac{d\varphi}{d\tau}, \quad E = m\left(1 - \frac{2GM}{rc^2}\right)c^2 \frac{dt}{d\tau} \qquad (6.193)$$

Substituting (6.193) into (6.192), we can retrieve the below equation after a bit of simplification:

$$\frac{1}{2}m\left[\left(\frac{dr}{d\tau}\right)^2 + r^2\left(\frac{d\varphi}{d\tau}\right)^2\right] - \left(\frac{GMm}{r} + \frac{GML^2}{mc^2 r^3}\right) = \frac{E^2}{2mc^2} - \frac{1}{2}mc^2$$

$$(6.194)$$

Please note that under the framework of nonrelativistic approximation, what's inside the brackets in (6.194) represents the kinetic energy of Mercury, while the expression after the equal sign is a constant. Therefore, we know the effective potential energy of Mercury is:

$$V(r) = -\frac{GMm}{r} - \frac{GML^2}{mc^2r^3} \tag{6.195}$$

Compared to the simplicity of Newtonian mechanics, the framework of general relativity is enriched with the term of gravitational potential energy, which is inversely proportional to the cube of distance, and this term is the major modification to the degree of Mercury's orbital precession. Same as in the previous section, the potential energy can be substituted into (6.182) and (6.188) to resolve ω_φ and ω_r, while the solutions can then replace the two parameters in (6.183) to help us obtain the degree of Mercury's perihelion advance per revolution:

$$2\pi\left(\frac{\omega_\varphi}{\omega_r} - 1\right) \approx \frac{6\pi GM}{c^2 r_0} \tag{6.196}$$

In (6.196), r_0 represents Mercury's orbital radius.[47] If we substitute the actual values into this equation, we know that with every hundred years passed on Earth, Mercury would experience the precession of 41ArcSec.

6.4.4   Global Positioning System

Despite the lengthy theoretical discourse we've gone through, it'd be a big mistake to think Einstein's achievement is only confined to the ivory tower of physics. In fact, the theory of general relativity is also a key driving force behind something in the real world that has changed human civilization in an unprecedented way and still carries enormous business potential—the Global Positioning System (GPS). A GPS consists of 24 satellites that complete two circles around Earth each day on their orbits about 20,000 km high. Receiving signals from cars on Earth's surface and aircrafts in the sky, these satellites use the triangulation method in calculating positions before sending the results back to the receivers on these vehicles, where the location data is displayed through internal maps. While GPS makes use of several crucial technologies, one of them is the atomic clock installed on each satellite. Based on

---

[47]The derivation is conducted under the assumption that the orbit of Mercury is roughly a circle, but in fact, it is around an ellipse that the planet orbits. If we go through a more accurate series of calculations, we'll see that for an elliptic orbit with the eccentricity of e, the precession degree per revolving period is approximately $\frac{6\pi GM}{c^2 A(1-e^2)}$, where A represents the semi-major axis of the ellipse.

the theory of special relativity, people on Earth are supposed to think these clocks tick faster because satellites move at a fairly high speed in relation to us. Meanwhile, the effect of time dilation in general relativity also causes clocks to move faster where gravitational potential energy is larger. Therefore, the time measured by the satellites needs to be modified accordingly in order to prevent desynchronization with the time on Earth, which will then lead to inaccurate positioning. In below paragraphs, we're going to calculate the actual values of the modification.

First, as the radius of Earth is already known to be $R = 6,400$ km, we know the orbital radius of satellite that revolves on the plane that's 20,000 km above Earth's surface is $r = 26,400$ km. Now, if we suppose the satellites are in a circular motion around Earth, their centripetal acceleration should equal the gravitational acceleration generated by Earth:

$$\frac{v^2}{r} = \frac{GM}{r^2} \tag{6.197}$$

In the above equation, M is the mass of Earth, while v represents the tangential velocity during the satellites' revolution and can be resolved as:

$$v = \sqrt{\frac{GM}{r}} = 3.9 \times 10^3 \text{m/s} \tag{6.198}$$

Now, let's assume that with the time period $\Delta \tau_1$ passed on Earth, observers on Earth's surface would think the time passed on the satellites is $\Delta \tau_2$. Based on the theory of special relativity, the relationship between these two parameters can be expressed as $\Delta \tau_2 = \Delta \tau_1 \sqrt{1 - v^2/c^2}$.

$$\Delta \tau_2 - \Delta \tau_1 = \Delta \tau_1 \left(\sqrt{1 - \frac{v^2}{c^2}} - 1 \right) \approx -\frac{v^2}{2c^2} \Delta \tau_1 \tag{6.199}$$

If we substitute $\Delta \tau_1 = 1$ (day) into (6.199), the result will come out to be $\Delta \tau_2 - \Delta \tau_1 = -7.3$ ms. In other words, we'd think the time on the satellites moves slower than that on Earth because the solution is a negative value.

In the meantime, the speed of clocks is also affected by gravitational potential energy, which we're now going to address with the Schwarzschild metric. Remember, the parameters t, r, θ, and φ in (6.155) all represent physical quantities from the perspective of an observer who's infinitely far from all astronomical bodies and thus not affected by any gravitational field. As for τ, it's the proper time interval measured by a free-fall observer located at

(t, r, θ, φ). When the free-fall observer is momentarily stationary in relation to the one who's infinitely far, the below equation is true:

$$d\tau = \sqrt{1 - \frac{2GM}{c^2 r}} dt \qquad (6.200)$$

When only considering the effect of general relativity, we can represent the time on Earth and the satellites with the proper time intervals measured by free-fall observers at both locations. This way, the relationship between $\Delta\tau_2$ and $\Delta\tau_1$ can be expressed approximately as[48]:

$$\frac{\Delta\tau_2}{\Delta\tau_1} = \frac{\sqrt{1 - \frac{2GM}{c^2 r}}}{\sqrt{1 - \frac{2GM}{c^2 R}}} \approx 1 + \frac{GM}{c^2}\left(\frac{1}{R} - \frac{1}{r}\right) \qquad (6.201)$$

From (6.201), the time difference caused by varying gravitational potential energy can be calculated as:

$$\Delta\tau_2 - \Delta\tau_1 = \frac{GM}{c^2}\left(\frac{1}{R} - \frac{1}{r}\right)\Delta\tau_1 \qquad (6.202)$$

Where the solution is a positive value, meaning clocks on the satellites are perceived to move faster than those on Earth. If we replace $\Delta\tau_1$ with 1(day), the result will be $\Delta\tau_2 - \Delta\tau_1 = 45.6$ ms.

Lastly, we can combine the results from both special and general relativity to conclude that for people on Earth, clocks on the satellites move faster by $45.6 - 7.3 = 38.3$ ms per day. Without this modification, the deviation of 38ms would quickly lead to inaccurate positioning for cars, ships, planes, missiles, etc. Just consider a car that's moving at 100 km/h. The mere difference of 38.3 ms translates to a 10 m deviation, which will keep accumulating with the distance traveled and eventually cause GPS dysfunction. If such deviation existed in the positioning for planes or missiles, which normally travel as fast as several hundred or even thousand kilometers per hour in the sky, then GPS would basically become worthless. With the modification from relativity, however, the deviation of automobile GPS has been reduced to less than 1m, while that for planes and missiles are controlled at no more than

---

[48]Strictly speaking, this equation is incorrect because although $\Delta\tau_2$ and $\Delta\tau_1$ in (6.201) represent the same amount of time for the infinitely-far observer, the two parameters are of different values for the free-fall observer on Earth. This said, we can still safely adopt the approximation since the scenario under discussion is a weak field.

10 m. But just imagine if engineers designing GPS weren't aware of general relativity—the system probably would've been mistaken as broken and irreparable, and this indeed shows how important the seemingly abstract discipline of physics is to the real world. In terms of the application of GPS, the US became the first to dominate the market by starting to develop related technologies since 1978, while the EU has also invested €EUR5 billion in establishing the brand new GPS ecosystem "Galileo" and expects it to go live in 2020. In the meantime, China has also joined the contest with their 30-satellite BeiDou navigation system, which is also scheduled to be rolled into general availability in 2020.

6.4.5 Geodetic Precession and Frame Dragging

In 2004, NASA launched a satellite named Gravity Probe B (or GP-B) into a polar orbit 642 km above Earth's surface to test two unverified predictions of general relativity: the geodetic effect and frame-dragging. To explain how the satellite works, we need to start with its core instrument, the gyroscopes.[49] A gyroscope consists of a high-speed rotor revolving around the spin axis. As the revolution only generates a very small amount of friction, the rotor is able to maintain its motion without any propelling force from external torque. On the other hand, due to the large angular momentum from the motor's fast spin, it requires quite a powerful torque to change the motioning direction. In other words, a gyroscope's axis is unaffected by tilting or rotation of the mounting, which makes it an optimal choice for navigation and positioning high-speed aircrafts. Now, going back to the discussion of GP-B, the satellite contained four fused quartz gyroscopes with a diameter of 38 mm, each installed with a high-precision reference telescope. For the gyroscopes to function properly, they were housed in a heavy, complicated equipment of superfluid helium, which maintained an ultra-low temperature of under 2 k (−271 °C; −456 °F). In addition, the gyroscopes themselves were also coated with an extremely thin layer of niobium so they could be held suspended with electric fields, spun up at a constant speed using a flow of helium gas.

But what really is the point of these gyroscopes? Before answering this question, let's first look at a prediction of general relativity: In a curved 4-space in the Riemann geometry, the parallel transport of a vector along a closed route doesn't guarantee that the vector would remain in its original direction when moved back to the origin. If you still remember, it was

---

[49]For details, please see https://zh.wikipedia.org/wiki/%E9%99%80%E8%9E%BA%E5%84%80.

exactly this prediction that we used when defining the Riemann curvature in Sect. 6.2. In fact, the spin of a gyroscope can also be understood as parallel-transporting a vector. Therefore, when the rotor spins an entire circle in a curved spacetime before returning to the origin of its closed route, the orientation of its spin axis might be changed as well due to the effect of axial precession.

In the GP-B experiment, the gyroscopes revolved around Earth at the height of 642 km following a polar orbit passing above both the North and South Poles. Initially, the spin directions were pointed toward HR8703 (also known as IM Pegasi), but due to the curvature of the spacetime near Earth, the spin axes of the gyroscopes deviated after they completed a circle around the planet. As for the degree of deviation, it's essentially the degree traveled by the axes of rotation due to the effect of axial precession, which can be predicted with the theory of general relativity. In other words, the experiment was devised to verify the theoretical prediction.

As a result, what we're doing next is to calculate the degree of deviation experienced by the spin axes of the gyroscopes in GP-B (see Fig. 6.13) with the equations of general relativity derived before. If we suppose the spin axis of a gyroscope lies on its plane of revolution around Earth, change in the axis direction can be divided into two components: one overlapped with,

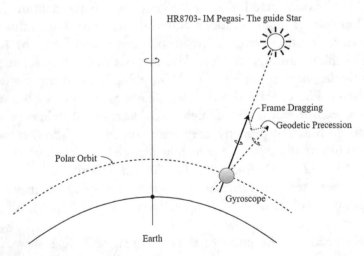

Fig. 6.13 Illustration of GP-B experiment mechanism. The radius of GP-B orbit is 642 km, the Frame Dragging is 39 mas/year, and the Geodetic Precession if 6606 mas/year, where "mas" stands for milli-arc-second. Inspired by Fig. 9.13 of Guidry [12]

the other perpendicular to that particular plane. Theoretically speaking, the former and the latter are generated by geodetic precession and frame-dragging respectively. As for the actual values, let's dive into related calculations in the following sections.[50]

(1) Geodetic Precession

In this section, we're going to disregard the effect of Earth's rotation so the spin axis of the gyroscope in question can be deemed constantly lying on the plane of revolution around the planet. If we suppose the orbital plane to be a perfect circle (radius $= R$) that's overlapped with the plane $\theta = \pi/2$[51] in a spherical coordinate system, the corresponding Schwarzschild metric can be used to derive:

$$c^2 d\tau^2 = c^2 \left(1 - \frac{2GM}{c^2 r}\right) dt^2 - \frac{dr^2}{1 - \frac{2GM}{c^2 r}} - r^2 d\varphi^2 \qquad (6.203)$$

Next, we'll need a vector to describe the direction of the spin axis (or, the angular momentum from the spinning). While this direction is a three-dimensional property, only a 4-vector can be parallel-transported and go through the transformation of coordinates in a 4-space. Therefore, we have to expand the vector to the fourth dimension, and the simplest way is to consider the spatial vector in a local inertial frame, just as we've mentioned several times before.

According to the principle of strong equivalence, any point in a four-dimensional spacetime can be found with a local inertial frame (or, a free-fall coordinate system) where the Minkowski metric can apply and all Christoffel symbols of connection are 0 (see Sect. 6.3.7). Now, suppose we have a 4-vector $s = \left(0, \vec{s}\right)$ in such a local frame with the temporal and spatial components of $s^t = 0$ and $\vec{s}$ respectively. When s is parallel-transported from x to $x + dx$ by an extremely short distance, its component on every axis will remain the same. As a next step, we'll switch to the local inertial frame at $x + dx$ and conduct another short-distance parallel transport before repeating the same steps again and again, until the vector completes an entire route and returns to the origin. Please note that the temporal component s^t is 0 in the local inertial frame of s through the entire process, and in fact, this

---

[50] See also Chap. 9 of Guidry [12].
[51] Facing the page, please rotate the book 90° clockwise to observe Fig. 6.13.

is exactly the result we want. As long as we define $\vec{s}$ as the direction of the spin axis (or, angular momentum), the 4-vector s can be used for related calculations without any unnecessary degree of freedom (details will be provided later). However, please be aware that s^t only remains 0 in local inertial frames. This said the only thing we need to consider in determining the direction of the gyroscope is the spatial component of s, nothing else.

Given all the background information above, let's now define s as a "spin 4-vector" whose components in a spherical coordinate system (from the perspective of an observer who's in a Schwarzschild metric infinitely far from Earth) can be expressed as:

$$s = (s^t, s^r, s^\theta, s^\varphi) \tag{6.204}$$

Then, we can introduce the 4-speed of the gyroscope in the spherical coordinate system:

$$u = (u^t, u^r, u^\theta, u^\varphi) = \left(c\frac{dt}{d\tau}, \frac{dr}{d\tau}, \frac{d\theta}{d\tau}, \frac{d\varphi}{d\tau} \right) \tag{6.205}$$

And of course, the 4-speed of the gyroscope in the scenario we're discussing here constantly points toward the tangential direction of the circular orbit, so we know that $u = (u^t, 0, 0, u^\varphi)$. In addition, (6.162) also tells us that:

$$u^\varphi = \frac{\Omega}{c} u^t, \quad \Omega = \sqrt{\frac{GM}{R^3}} \tag{6.206}$$

Up to this point, we've managed to describe all properties of the gyroscope's motion with the parameters s and u.

As for what we mentioned about the degrees of freedom, here comes a detailed explanation. When we describe the direction of the spin axis with three-dimensional angular momentum, the 4-vector s, which describes the angular momentum, can only have three degrees of freedom as well. Yet as s generally has four components in any coordinate system, we'll need one more equation to regulate the relationship among all of them. From the perspective of the local inertial frame of the gyroscope, the 4-speed u only contains the temporal component because its 3-speed is perceived as 0. But when it comes to its spin 4-vector s, only the spatial component is present, so the inner product of u and s must be 0 ($u \cdot s = 0$). In addition, we also know the dot product of two 4-vectors doesn't change with the transformation of coordinates, which is to say that $u \cdot s$ has to be 0 in any reference frame.

Hence, the below equation should be true in the Schwarzschild metric:

$$u \cdot s = g_{tt}s^t u^t + g_{rr}s^r u^r + g_{\varphi\varphi}s^\varphi u^\varphi = 0 \tag{6.207}$$

Please note that the component along the direction of θ doesn't need to be considered because our calculation is limited to the plane of $\theta = \pi/2$. Next, we can substitute $g_{tt} = -\left(1 - \frac{2GM}{c^2 r}\right)$, $g_{rr} = \left(1 - \frac{2GM}{c^2 r}\right)^{-1}$, $g_{\varphi\varphi} = r^2$ and (6.206) into (6.207) to obtain:

$$s^t = \frac{R^2\Omega/c}{1 - \frac{2GM}{c^2 R}}s^\varphi \tag{6.208}$$

Judging from the above equation, s^t can depend entirely on s^φ, which is why s actually has three degrees of freedom only.

In addition, we can also tell that if s^θ is set as 0 (since the spin axis lies on the plane $\theta = \pi/2$) at the beginning, it'll simply remain to be 0 due to the property of symmetry. Therefore, we can fully understand the nature of the gyroscope as long as we know how s^r and s^φ change over time. To do so, we need to revisit the definition of parallel transport. First, since $s = s^\mu \mathbf{e}_\mu$ has to be true through an extremely small-scale transport, we know that:

$$\frac{ds}{d\tau} = \frac{d(s^\mu \mathbf{e}_\mu)}{d\tau} = 0 \tag{6.209}$$

Where τ is the proper time interval for the gyroscope. Then, we can substitute the definition of the 4-speed u into (6.209) and retrieve:

$$\frac{ds^\mu}{d\tau} + \Gamma^\mu_{\kappa\nu}s^\kappa u^\nu = 0 \tag{6.210}$$

At this point, we can start observing how s^r and s^φ change with time based on (6.210). To begin with, we know s^r satisfies the below equation:

$$\frac{ds^r}{d\tau} = -\Gamma^r_{\kappa\nu}s^\kappa u^\nu \tag{6.211}$$

And since u consists only of the two non-zero components u^t and u^φ, we can rewrite the above equation with (6.206) as:

$$\frac{ds^r}{d\tau} = -\left(\Gamma^r_{\kappa t}s^\kappa + \frac{\Omega}{c}\Gamma^r_{\kappa\varphi}s^\kappa\right)u^t \tag{6.212}$$

To further simplify (6.212), let's start by observing the term $\Gamma^r_{\kappa t}$. Combining the definition of (6.40) and the fact that all non-diagonal components in the Schwarzschild metric are 0, we know that:

$$\Gamma^r_{\kappa t} = \frac{1}{2}g^{rr}\left(\frac{\partial g_{r\kappa}}{\partial t} + \frac{\partial g_{rt}}{\partial x^\kappa} - \frac{\partial g_{\kappa t}}{\partial r}\right) \tag{6.213}$$

In addition, as the differential of all metric components over t is 0 and $g_{rt} = 0$, we know that the only scenario when $\Gamma^r_{\kappa t}$ is a non-zero value is described by the below equation because $g_{\kappa t}$ is always 0 except when $\kappa = t$:

$$\Gamma^r_{tt} = \frac{1}{2}g^{rr}\left(-\frac{\partial g_{tt}}{\partial r}\right) = \frac{GM}{c^2 r^2}\left(1 - \frac{2GM}{c^2 r}\right) \tag{6.214}$$

Moving on to $\Gamma^r_{\kappa\varphi}$, we can also combine the definition of (6.40) and the fact that all non-diagonal components in the Schwarzschild metric are 0 to establish the below:

$$\Gamma^r_{\kappa\varphi} = \frac{1}{2}g^{rr}\left(\frac{\partial g_{r\kappa}}{\partial \varphi} + \frac{\partial g_{r\varphi}}{\partial x^\kappa} - \frac{\partial g_{\kappa\varphi}}{\partial r}\right) \tag{6.215}$$

And because the differential of all metric components over φ is 0 and $g_{r\varphi} = 0$, we know that the only scenario when $\Gamma^r_{\kappa\varphi}$ is a non-zero value is described by the below equation because $g_{\kappa t}$ is always 0 except when $\kappa = \varphi$:

$$\Gamma^r_{\varphi\varphi} = \frac{1}{2}g^{rr}\left(-\frac{\partial g_{\varphi\varphi}}{\partial r}\right) = -r\left(1 - \frac{2GM}{c^2 r}\right) \tag{6.216}$$

Lastly, we'll substitute (6.214) and (6.216) back into (6.212) before replacing s^t with the s^φ in (6.208) to derive:

$$\begin{aligned}
\frac{ds^r}{d\tau} &= -\left[\frac{GM}{c^2 R^2}\left(1 - \frac{2GM}{c^2 R}\right)\left(\frac{\frac{R^2\Omega}{c}}{1 - \frac{2GM}{c^2 R}}s^\varphi\right) - \frac{\Omega}{c}R\left(1 - \frac{2GM}{c^2 R}\right)s^\varphi\right]u^t \\
&= \left(1 - \frac{3GM}{c^2 R}\right)\frac{\Omega}{c}Rs^\varphi u^t
\end{aligned} \tag{6.217}$$

Again, since $u^t = cdt/d\tau$, the above equation can be rewritten as:

$$\frac{ds^r}{dt} = \left(1 - \frac{3GM}{c^2 R}\right)\Omega Rs^\varphi \tag{6.218}$$

And this is the ultimate result from the simplification of (6.212).

In the following part, we're going to discuss how s^φ changes with time. Similar to the derivation above, we know from (6.210) that:

$$\frac{ds^\varphi}{d\tau} + \Gamma^\varphi_{\kappa\nu}s^\kappa u^\nu = 0 \tag{6.219}$$

Based on (6.210) and the fact that u only has two non-zero components u^t and u^φ, (6.219) can be rewritten as:

$$\frac{ds^\varphi}{d\tau} = -\left(\Gamma^\varphi_{\kappa t}s^\kappa + \frac{\Omega}{c}\Gamma^\varphi_{\kappa\varphi}s^\kappa \right)u^t \tag{6.220}$$

To further simplify (6.220), let's first observe the term $\Gamma^\varphi_{\kappa t}$. Combining the definition of (6.40) and the fact that all non-diagonal components in the Schwarzschild metric are 0, we know that:

$$\Gamma^\varphi_{\kappa t} = \frac{1}{2}g^{\varphi\varphi}\left(\frac{\partial g_{\varphi\kappa}}{\partial t} + \frac{\partial g_{\varphi t}}{\partial x^\kappa} - \frac{\partial g_{\kappa t}}{\partial \varphi} \right) \tag{6.221}$$

In addition, as the differential of all metric components over t and φ is 0 and $g_{\varphi t} = 0$, we know that $\Gamma^\varphi_{\kappa t}$ has to be 0 constantly. Moving on to $\Gamma^\varphi_{\kappa\varphi}$, again we know the below is true based on the definition of (6.40) and the fact that all non-diagonal components in the Schwarzschild metric are 0:

$$\Gamma^\varphi_{\kappa\varphi} = \frac{1}{2}g^{\varphi\varphi}\left(\frac{\partial g_{\varphi\kappa}}{\partial \varphi} + \frac{\partial g_{\varphi\varphi}}{\partial x^\kappa} - \frac{\partial g_{\kappa\varphi}}{\partial \varphi} \right) = \frac{1}{2}g^{\varphi\varphi}\frac{\partial g_{\varphi\varphi}}{\partial x^\kappa} \tag{6.222}$$

Then, we can derive (6.223) because (6.222) only yields non-zero values when $\kappa = r$:

$$\Gamma^\varphi_{r\varphi} = \frac{1}{2}\frac{1}{r^2}(2r) = \frac{1}{r} \tag{6.223}$$

Subsequently, the substitution of (6.223) back into (6.220) will lead us to:

$$\frac{ds^\varphi}{d\tau} = -\left(\frac{\Omega}{c}\frac{1}{R}s^r \right)u^t \tag{6.224}$$

And because $u^t = cdt/d\tau$, the above equation can be rewritten as:

$$\frac{ds^\varphi}{dt} = -\frac{\Omega}{R}s^r \tag{6.225}$$

And this is the ultimate result from the simplification of (6.219).

Summarizing the entire discussion above, we just need the simultaneous solution to both (6.218) and (6.225) to determine the change of s^r and s^φ over time, which can then help us calculate the precession degree of the spin axis in the gyroscope. First, let's differentiate (6.218) over t once more to retrieve:

$$\frac{d^2 s^r}{dt^2} = \left(1 - \frac{3GM}{c^2 R}\right)\Omega R \frac{ds^\varphi}{dt} \tag{6.226}$$

Replacing the term ds^φ/dt in the above equation with the result of (6.225), we know that:

$$\frac{d^2 s^r}{dt^2} = -\left(1 - \frac{3GM}{c^2 R}\right)\Omega^2 s^r \tag{6.227}$$

Following the same logic, we can also derive from (6.225) that:

$$\frac{d^2 s^\varphi}{dt^2} = -\left(1 - \frac{3GM}{c^2 R}\right)\Omega^2 s^\varphi \tag{6.228}$$

Judging from (6.227) and (6.228), both s^r and s^φ oscillate at the angular frequency of:

$$\omega = \sqrt{1 - \frac{3GM}{c^2 R}}\,\Omega \tag{6.229}$$

With a phase difference of 90° while generating a sine wave.

Since our focus is to determine the change of degree (represented as $\Delta\varphi$) in the direction of the spin axis when the gyroscope completes a circle, we need to multiply the difference between ω and Ω with one period using (6.229):

$$\Delta\varphi = \frac{2\pi}{\Omega}\left(1 - \sqrt{1 - \frac{3GM}{c^2 R}}\right)\Omega \approx \frac{3\pi GM}{c^2 R} \tag{6.230}$$

Last but not least, after we substitute the actual values below:

(1) Gravitational constant: $G = 6.67 \times 10^{-11}\,\mathrm{m^3 kg^{-1} s^{-2}}$
(2) Earth's mass $M = 5.972 \times 10^{24}\,\mathrm{kg}$
(3) Radius of the polar orbit: $R = 6380\,\mathrm{km}\,(\text{Earth's radius}) + 642\,\mathrm{km}\,(\text{distance above Earth's surface}) = 7022\,\mathrm{km}$

$\Delta\varphi$ can be calculated to be 1.225 milli ArcSec. Then, we can also calculate the revolving period of the gyroscope around Earth with (6.228):

$$\frac{2\pi}{\Omega} = 2\pi\sqrt{\frac{R^3}{GM}} = 5858 \text{ Sec} \tag{6.231}$$

Therefore, the precession degree of the gyroscope over the course of a year is:

$$1.225 \times \frac{365.25 \times 86400 \text{ Sec}}{5858 \text{ Sec}} = 6600 \text{ milli ArcSec (mas)} \tag{6.232}$$

This result is the final prediction derived from the theory of general relativity. From the calculation above, we can reach this conclusion that when a gyroscope revolves around Earth (following a geodesic, of course), its axis of rotation will experience the phenomenon of precession on the orbital plane, which is what we call the "geodetic precession".

(2) Frame Dragging

In this section, we're going to consider the previously-omitted effect of Earth's rotation while disregarding the geodetic precession. Please note that when the rotation is taken into consideration, the Schwarzschild metric will not apply anymore because what it regulates is the spacetime geometry outside of non-rotating, spherically symmetric astronomical bodies. Based on Einstein's field equations,[52] when the angular momentum J from the rotation of a celestial body around its axis z is small enough,[53] the corresponding metric is:

$$ds^2 = ds_0^2 - \frac{4GJ}{c^3 r^2}\sin^2\theta(rd\varphi)(cdt) + O\left(J^2\right) \tag{6.233}$$

Where ds_0^2 is the Schwarzschild metric, while $O\left(J^2\right)$ is a higher-order term than J^2, which can be omitted when the value of J is small enough. Since Earth's rotation is the only effect to be discussed in this section, we can simply treat ds_0^2 as the Minkowski metric under a weak field. And for easier

---

[52]Details regarding the derivation will not be provided in this book because the result is simply a solution to the linear perturbation equation. See please Chap. 9 of Guidry [12]. As for its physical meaning, we'll be having a discussion in later paragraphs. Readers who're interested as encouraged to refer to past literature about linearized gravity.

[53]The rotation of Earth apparently satisfies this condition because its rotates at a much slower speed than light.

understanding, we'll conduct related calculations in a Cartesian coordinate system. If we suppose the revolving orbit of the gyroscope lies on the xz plane and transform (6.233) into the Cartesian form while introducing ds_0^2 as the Minkowski metric, we'll be able to obtain:

$$ds^2 = -c^2 dt^2 + dx^2 + dy^2 + dz^2 - \frac{4GJ}{c^3 r^3}(cdt)(xdy - ydx) \quad (6.234)$$

Where $r = \sqrt{x^2 + y^2 + z^2}$.

Similar to the previous section, here we'll to determine how the spin 4-vector s changes over time with (6.210). Yet to simplify the process, we'll only consider the components s^x and s^y on the xy plane while disregarding s^z and s^t.[54] In addition, u^x, u^y, and u^z will not be considered either because they're much smaller than u^t. So first, let's look at the change pattern of s^x by deriving from (6.210) that:

$$\frac{ds^x}{d\tau} + \Gamma^x_{\kappa t} s^\kappa u^t = 0 \quad (6.235)$$

As we only consider s^x and s^y, the above equation can be simplified as:

$$\frac{ds^x}{d\tau} = -\Gamma^x_{xt} s^x u^t - \Gamma^x_{yt} s^y u^t \quad (6.236)$$

For Γ^x_{xt}, we know from our previous definition that:

$$\Gamma^x_{xt} \approx \frac{1}{2} g^{xx} \left(\frac{\partial g_{xx}}{\partial t} + \frac{\partial g_{xt}}{\partial x} - \frac{\partial g_{xt}}{\partial x} \right) = 0 \quad (6.237)$$

---

[54]Readers who're interested can of course take these two components into consideration as well, but the result will prove them nearly non-influential to the derivation process. When engaging in complicated spacetime calculations, we tend to discuss one variant at a time while disregarding other factors. Otherwise, excessive complexity in the calculating process will normally make it difficult to focus on any particular element.

(Regarding the inverse matrix of the metric, only the diagonal component is considered because J is extremely small.) Following the same logic, we know the below is true for Γ^x_{yt}:

$$\Gamma^x_{yt} \approx \frac{1}{2} g^{yy} \left(\frac{\partial g_{xy}}{\partial t} + \frac{\partial g_{xt}}{\partial y} - \frac{\partial g_{yt}}{\partial x} \right) = \frac{1}{2} \left(\frac{\partial g_{xt}}{\partial y} - \frac{\partial g_{yt}}{\partial x} \right) \tag{6.238}$$

In addition, (6.234) also tells us that:

$$g_{xt} = g_{tx} = \frac{2GJ}{c^3 R^3} y, \quad g_{yt} = g_{ty} = -\frac{2GJ}{c^3 R^3} x \tag{6.239}$$

If we substitute the metric described in (6.239) back into (6.238):

$$\Gamma^x_{yt} \approx \frac{GJ}{c^3 R^3} \left(2 - 3 \frac{x^2 + y^2}{R^2} \right) \tag{6.240}$$

Then, the substitution of Γ^x_{xt} and Γ^x_{yt} back into (6.236) will lead us to:

$$\frac{ds^x}{d\tau} = -\frac{GJ}{c^3 R^3} \left(2 - 3 \frac{x^2 + y^2}{R^2} \right) s^y u^t \tag{6.241}$$

And because $u^t = cdt/d\tau$, we know that:

$$\frac{ds^x}{dt} = -\frac{GJ}{c^2 R^3} \left(2 - 3 \frac{x^2 + y^2}{R^2} \right) s^y \tag{6.242}$$

With the equation $x^2 + y^2 = R^2 \sin^2 \theta$, (6.242) can be further simplified to become:

$$\frac{ds^x}{dt} = -\frac{GJ}{c^2 R^3} \left(3 \cos^2 \theta - 1 \right) s^y \tag{6.243}$$

This is the equation describing how the component s^x of the spin 4-vector s along the x-axis over time. Similarly, we can also derive the equation that regulates the change pattern of s^y:

$$\frac{ds^y}{dt} = \frac{GJ}{c^2 R^3} \left(3 \cos^2 \theta - 1 \right) s^x \tag{6.244}$$

With (6.243) and (6.244), we know the spin axis of the gyroscope will advance on the xy plane with the below angular velocity:

$$\omega = \frac{GJ}{c^2 R^3}\left(3\cos^2\theta - 1\right) \tag{6.245}$$

As for the gyroscope's degree of precession per revolution around Earth, we need to integrate (6.245) over the orbital period by rewriting θ with $\theta = \Omega t$ and retrieve the integral value over the range of $t = 0$ to $t = 2\pi/\Omega$:

$$\Delta\varphi = \int_0^{\frac{2\pi}{\Omega}} \frac{GJ}{c^2 R^3}\left(3\cos^2\Omega t - 1\right)dt = \frac{\pi GJ}{c^2 R^3 \Omega} \tag{6.246}$$

Eventually, we can replace the parameters with the actual values—(1) angular velocity of Earth's rotation: $J = 5.825 \times 10^{33} \mathrm{kg} \cdot \mathrm{m}^3$, and (2) Ω: see (6.231). With the result of $\Delta\varphi = 7.5 \times 10^{-3}$ milli ArcSec (or mas), the precession degree over the course of a year can be calculated as:

$$7.5 \times 10^{-3} \times \frac{365.25 \times 86400\,\mathrm{Sec}}{5858\,\mathrm{Sec}} = 40\,\mathrm{milli\,ArcSec} \tag{6.247}$$

And this is the angular velocity of the precession caused by frame dragging. At 40 mas, this value is much smaller than the 6,600 mas, which is the counterpart of the geodetic precession and a lot more difficult to measure as a result. Regarding the difference between these two phenomena, geodetic precession arises from the curvature of spacetime caused by the mass of Earth, while frame dragging refers to the effect of Earth propelling nearby mass into spinning along the same direction as its own rotation.

Operating on the mechanism explained above, the detection of GP-B started on August 27th of 2004 and stopped on August 14th, 2005 after a year's data collection. About a month after that, the liquid helium carried by the gyroscopes was also depleted, showing how accurately the instruments were devised. After a multi-façade analysis of the collected data, the research team announced the following results: a geodetic drift rate of about -6602 mas/yr relative to HR8703 and a frame-dragging drift rate of -37.2 mas/yr, which were in good agreement with the general relativity predictions of -6606 mas/yr and -39.2 mas/yr respectively. Although the later value seems to be slightly more deviant, the prediction from general relativity still falls in the reasonable range as the level of uncertainty of the instruments was set at

±7.2 mas,[55] way higher than the difference between the experimental result and the actual value.

6.4.6   Gravitational Wave

In November 1915, Einstein announced his theory of general relativity and the famous field equations, combining space and time in explaining the source and working mechanism of the ubiquitous but mysterious gravity. Then four years later, when Eddington proved these results correct through the observation of the solar eclipse on the island of Príncipe, Einstein was suddenly hailed as one of the greatest scientists of all time. Ever since then, top physicists around the world have never stopped trying to prove the phenomena predicted by general relativity, hoping to leave a mark in the history of science since the association with Einstein's research has long been a guarantee of fame and wealth. For example, Joseph Taylor (1941–) and his student Russell A. Hulse (1950–) found a binary system of two neutron stars in 1974 where the orbit was slowly shrinking as it lost energy, spinning the stars into a collision. Among the two astronomical bodies, one was a binary pulsar rotating at a fast spin rate while emitting electromagnetic waves as if it were a lighthouse in the space. As the duo was actively looking for something worth researching, such a phenomenon naturally caught their attention.

Taylor and Hulse combined the electromagnetic pulse (a type of gravitational wave) data they'd collected at the Arecibo Observatory in Puerto Rico with the field equations to calculate how the orbit of the binary system was slowly shrinking as it lost energy due to the emission of gravitational radiation. As it turned out, the result was 99.7% in accordance with the values obtained through actual observations. This study didn't just verify the theory of general relativity once again but also proved the existence of gravitational waves, and both the scientists were honored with the Nobel Prize for Physics in 1993 as well. Another example of verification happened on September 14th, 2015, when the US-established Laser Interferometer Gravitational-Wave Observatory (LIGO) detected a signal of gravitational waves from two black holes (29 and 36 solar masses respectively) merging about 1.3 billion light-years from Earth, which lasted 0.5s. After that, more than five gravitational waves have been detected and put under inspection, all of them generated by mergers of black holes. Leaders of the LIGO program, including

---

[55]The research team attributed the uncertainty to external disturbance that affected the balance of the gyroscopes.

Rainer Weiss (1936–), Kip S. Thorne (1940–), and Barry Barish (1936–)[56] also won the Nobel Prize for Physics in 2017 for their contribution. The fact that the two Nobel-winning studies mentioned above are both focused on gravitational waves shows how significant this field of research is since such waves provide direct evidence of the interaction between astronomical bodies while accurate prediction and precise detection are extremely challenging.

The concept of gravitational wave was first proposed by Einstein in his 1916 paper "Approximative Integration of the Field Equations of Gravitation". According to the theory of general relativity, the spacetime curvature generated by the motion of objects with large masses would be transmitted outwards like ripples with the amplitude decreasing in the transmission. The frequency would also be reduced due to the redshift arising from the expansion of the universe/the Doppler effect, while the encounter with obstacles could cause the waves to diffract. This said, the density of mass in space is fairly low, meaning there's actually not much disturbance to such waves, so it's not impossible for observers on Earth to detect gravitational waves from the other end of the universe. Eddington was one of the scientists who took on this endeavor. After achieving overnight fame by verifying Einstein's predictions with the 1919 picture of a solar eclipse taken in Africa, he decided to continue pursuing his interest in the field of relativity by recalculating all three types of gravitational waves that Einstein mentioned in the abovementioned paper. Although he did find that two of them might be false results of incorrect use of reference frames, he also proved the third type to be unaffected by frame transformation, just as Einstein had contended. This time, however, Eddington didn't manage to take the media and public by storm because gravitational waves simply weren't that easy to detect. Without experiments to back up his claims, his research didn't arouse much enthusiasm.

The reason why these waves are hard to detect has to do with how they're generated. The gravitational waves that we know of now are all produced due to the change in gravitational force arising from the motion of asymmetrical systems, five types confirmed so far: (1) Binaries consisting of black holes, neutron stars, white dwarfs, and other large-mass astronomical bodies. In such binaries, two constituents revolve around the same axis while propagating the radiation of gravitational waves outward, which gradually leads to energy loss, increased revolving velocity, and shortened orbital radius. When

---

[56] In fact, LIGO was also co-founded by the talented Scottish experimental physicist Ronald Drever (1931–2017), who was one of the key contributors to the apparatus of the project. After the dismissal by the California Institute of Technology in 1997, however, he unfortunately fell victim to serious dementia and was often seen wandering near the university. On March 17th, seven months before the Nobel Prize was awarded to the trio leaders of LIGO, he died in a nursing center in his hometown Glasgow.

the distance between the two bodies becomes smaller than a certain value, the constituents will start to be merged, forming a black hole that will continue spinning on the energy left in the binary at the beginning but eventually come to a complete halt with its frequency[57] slowly decreasing with the energy level. (2) Neutron stars are unevenly distributed in mass and motion in a spherically asymmetric manner. When spinning independently, such a star can cause the emission of gravitational radiation if the direction of its spin axis is inconsistent with that of the gravitational field. (3) Gravitational collapse. When a supernova explodes and collapses to become a neutron star, irregular gravitational radiation will be produced therefrom unless the explosion happens in perfect spherical symmetry (i.e., unless the matter is spewed out evenly in all directions). (4) Blackhole binaries. Generally speaking, each galaxy has at its center a black hole with an infinitely large mass, which will be merged with another black hole of this kind if two galaxies collide and become one, thereby generating gravitational waves of a low frequency and high amplitude. As galaxy mergers occur more often than in binaries, the waves produced in this manner are also easier to detect. (5) The Big Bang. According to the Big Bang theory, gravitational waves were generated to form the stochastic background in the universe 10^{-24} s after the explosion of space happened.

Since our Solar System is on the verge of the Milky Way while the galaxy extends all the way to the other side of the vast universe where most gravitational waves are generated, these waves are very difficult to detect due external disturbance (e.g. due to background radiation) or small amplitudes (such as in the scenario of binary mergers). Yet because of the potential academic and commercial benefits of proving the predictions of general relativity, significant budgets have been invested in establishing precision instruments for related experiments, making the detection of gravitational waves a mainstream study in the 21st century. Below we're going to introduce a successful attempt where 4 km laser interferometers were constructed to detect the gravitational waves generated out of merging black holes. Since the amplitude of such waves can be as short as the diameter of a proton, the interferometers had to be extremely precise as well. In the following paragraphs, we'll dive into the design, mechanism, and results for the detection system.

After publishing the field equations in 1915, Einstein didn't wait much longer to propose another hypothesis the next year that drastic changes in

---

[57] If we consider the constituents in a binary to be a perfect sphere, when their merger occurs, the largest possible frequency of the gravitational waves produced therefrom will be $f = \sqrt{\rho/4\pi}$. While $\rho = M/(4/3)\pi R^3$ represents the average mass density of the sphere, M is the mean of the constituents' masses. Combining both the equations, we can obtain the expression for the wave frequency $f = (1/4\pi R)\sqrt{3M/R}$, which can then help us calculate the maximum frequency.

the mass of space would generate vibrations around the spacetime curvature caused by gravitational force, which would then be transmitted outward in the form of gravitational waves at light speed. Such an advanced prediction remained in the heads of physicists for more than half a century before the research team behind the LIGO project finally took up the challenge of examining this hypothesis with an actual experiment. In the 1970s, this team of scientists made use of the interferometer prototype, which was invented in the Michelson-Morley experiment in 1887 as an attempt to prove the existence of ether wind, to design the so-called "Laser Interferometer Gravitational-Wave Observatory", or simply LIGO.

The detection system consisted of two same-sized perpendicular vacuum tubes, both connected to a central monitoring area where two laser light transmitters of ultra-high power were deployed. When the process of detection started, the transmitters emitted into the vacuum tubes two identical rays of light, which should arrive at the reflector set up at the intersection of the tubes at the same time before returning to the laser receiver in the monitoring area. Without any impact from external factors, theoretically, the lights should enter the receiver simultaneously, but due to mutual disturbances, they could both disappear entirely, leading to absolute darkness where the receiver couldn't catch any signal at all. By contrast, if the disturbance from gravitational waves was present, the lengths of both tubes could be slightly expanded or contracted. If so, the lights would be prevented from disappearing because of entering the receiver at the exact same moment and could even intensify to a certain degree in some situations. In other words, if the rays didn't plunge the receiver into darkness or even became brighter, the experiment could thus prove the existence of gravitational waves.

In order to detect the gravitational waves from space, the instruments had to be highly sensitive because based on the predictions of the field equations, even the waves generated from the collision of two black hole-level astronomical bodies with extremely large masses can be no longer than the diameter of a proton when the vibrations are transmitted to Earth's surface. Besides, how were scientists supposed to confirm if the disturbance they detected was caused by gravitational waves from the universe or just the ubiquitous noises everywhere on Earth? So to filter all earthly factors out, the equipment was integrated with rigid insulation where the vacuum tubes were constructed in a carefully-devised damping system, while the mirrors working as the reflector and receiver were suspended with high strength steel wires above structures built with diverse shockproof functionalities. With such a precise design, the intensity of all the possible noise signals that could mislead the scientists was expected to be reduced to one millionth of the original levels.

In addition to the insulation, the most important requirement for the equipment was the ability to detect gravitational waves with an amplitude no longer than the diameter of one single proton. As the amplitude was related to the distance traveled by the laser lights, the longer the laser tubes were, the more precise the experimental equipment was. For the LIGO project, the tubes in Livingston, Louisiana, and Hanford, Washington were both as long as 4,000 m, each costing US$30 million to construct. In fact, similar instruments for detecting gravitational waves have also been established in many countries, such as the two 600m GEO600 detectors in Hannover, Germany and Glasgow, Scotland (EU€7 million) that opened in 1995, the underground project KAGRA that launched in Japan in 2010 with tubes of 3,000 m long (US$20 million), and the VIRGO detector in Pisa, Italy, where the 3,000 m tubes started working in 1996.

With such enormous investments made to detect the mysterious gravitational waves, scientists finally received good news at 5:51AM on September 14th of 2015, when the LIGO detector in Livingston sent back a weak signal from space that lasted only 0.2 s, but 6.9 ms later, the other LIGO in Hanford also detected the same signal. After deliberate calculations, researchers confirmed that the signal was the result of two black holes of 36 and 29 solar masses merging about 1.3 billion light-years from Earth. In the merging process, roughly three solar masses disappeared, triggering a series of curvature and vibration in spacetime, which took as long as 1.3 billion years to transmit to Earth at light speed in the form of gravitational waves. Although the amplitude was only the combined diameter of several protons, the gravitational force generated out of the black hole merger was approximately 50 times that of the combined gravity from all celestial bodies in the universe. By the beginning of 2018, such gravitational waves had been detected three times by the LIGO interferometers, time and again proving the validity of Einstein's theory.

Apart from the research of gravitational waves, the world has also seen many other types of space detections, such as the Viking program, which sent a pair of space probes to Mars on November 25th of 1976. Grasping the timing when the line joining Earth and Mars touched the edge of Sun in a tangential manner, scientists sent out an electromagnetic wave for the probes to receive before returning it back to Earth. Just as expected, the wave took an extra 250 ms to travel the entire path due to the bending effect of Sun's gravitational force, an experimental result 99.5% in agreement with their initial prediction. In 1997, a similar experiment was again carried out in the Cassini–Huygens space-research mission, and this time, the result agreed almost completely with the theoretical value.

Some of you might be wondering, is the theory of general relativity really worth that much money to prove? The answer is yes, and here's an example of how the theory can benefit human beings. Before the effect of relativity was taken into consideration, GPS was essentially like a broken toy due to the impractical amount of deviation. With the modification of 38 ms, though, it has now been elevated to become an indispensable instrument in cars, planes, boats, and many other types of commercial and military vehicles all around the world. Without GPS, all these fast-moving objects would act like bumper cars lacking the ability to keep their bearings. In addition, important technologies such as computed tomography, quantum computers, and mobile communications actually all work on the basis of general relativity, which again shows the research value of the experiments dedicated to verifying Einstein's theory is much higher than the monetary costs invested.

6.5   Summary and Review of General Relativity

In previous sections, we've walked through the basic knowledge of relativity theories and related applications in real life. Now that the construction of these theories has been fully reproduced, it shouldn't be too hard to tell that the framework of relativity is essentially a scientific piece of art manifesting the ultimate symmetry while explaining the most physical phenomena with the least postulates. It's where time and space, mathematics and physics, as well as experience and wisdom are perfectly integrated to yield the final field equations that regulate the relationship between the mass and energy of matter as well as the interactions among all objects in the universe.

Yet because of the great volume of theoretical description and derivation, it's not uncommon to find all the crucial points difficult to remember. Therefore, below we'll provide a comprehensive recap of how Einstein formulated special and general relativity, hoping to pull readers out of the perplexing myriad of details to look at the bigger picture from a macroscopic point of view. Readers are also encouraged to review Chaps. 5 and 6 after going through the following content, which we believe will create a resonance contributing to an even deeper understanding of relativity because this field of studies is just so full of wonders and subtleties that it's worth time-and-again revisiting.

Toward the end of the 18th century, scientists started discovering phenomena that couldn't be explained with Newtonian mechanics, the most significant ones including:

1. The transmission of electromagnetic waves, which Maxwell contended didn't require any medium and was unaffected by the choice of the reference system (i.e. the observer's perspective) in terms of speed. Such a claim challenged the absolute, fixed coordinate system in Newton's theory and how speed should be converted in the Galilean transformation. Based on Maxwell's hypothesis, electromagnetic waves were light waves transmitting in vacuum at light speed, i.e. a non-changing value.
2. The Michelson-Morley experiment, which was meant to prove but ended up negating the existence of the mysterious "ether", leading to the hypothesis that light actually didn't require any medium of transmission.
3. The different equivalent masses exhibited by electrons accelerated by an electric field when motioning parallel and perpendicular to the field direction. During the development of electrodynamics, this peculiar phenomenon was one of the most challenging puzzles in the world of physics.

As an attempt to explain the above phenomena, Einstein developed the theory of special relativity. During the research process, curiosity drove him to discard the structure of Newtonian mechanics and rethink the definition of fundamental physical quantities, such as time and space. In fact, his entire discourse was based on the following postulates:

1. The laws of physics are the same and can be stated in their simplest form in all inertial frames of reference.
2. The speed of light c is a constant, independent of the relative motion of the source.

Without any other precondition, Einstein derived the Lorentz transformations through a series of discreet thought experiments. This new set of transforming equations proved that the Galilean predecessor was still applicable to low-speed scenarios while exhibiting compatibility with Maxwell's theory under general situations. In addition, the equations even explained why the Michelson-Morley experiment turned out the way it did, so the first two questions listed above were both resolved basically.

To solve the third problem, Einstein turned to study the field of dynamics under the framework of special relativity. For energy and momentum to remain invariant through the Lorentz transformations, he redefined the expressions for both the physical quantities, which led him to find out they were inseparably related in the world of physics, just like space and time. In the end, he managed to derive an equation for momentum and used it to explain the different equivalent masses of electrons motioning along mutually perpendicular directions. In the research process, he also accidentally discovered that mass and energy exhibited similar dynamic properties, which was why he was bold enough to hypothesize that mass and energy were mutually convertible.

This hypothesis, however, was the only unproved argument in Einstein's theory of special relativity although it was discreetly made based on the concept of symmetry. Yet imagine this: if a jigsaw puzzle is missing a piece, normally we'd think it's lost during the assembling process rather than assume that piece wasn't even made at the very beginning. Similarly, when a hypothesis isn't backed by known facts or concrete proof but seems to make sense, researchers tend to presume it's correct, which is exactly how Einstein approached the mutually-convertible relationship between mass and energy. The only difference is that his exploration of the natural world was always several steps ahead of other scientists.

Due to his agile thinking, Einstein immediately sensed the incompatibilities between his theory and Newton's law of universal gravitation as well as the many limitations imposed by inertial frames of reference on gravitational field, a concept that'd always drawn his attention. Under the structure of relativity, universal gravitation is assumed to originate in fields of acceleration, which do change with the transformation of frames. By contrast, Newton considered the gravitational force to be generated out of mass, which is nevertheless an invariant in the transforming process. Faced with the apparent contradiction, Einstein claimed that to accurately describe a gravitational field, "mass" wasn't the only factor to be taken into consideration because of the mutual convertibility between mass and energy as well as the inseparability of energy and momentum. As a result, he came up with yet another pioneering hypothesis: apart from mass, energy and momentum are also sources of gravitational force.

As a next step, Einstein attempted to illustrate the working mechanism of gravitation in 4-space with the principle of weak equivalence, which he established based on a thought experiment. By way of deduction, he contended that a free-falling object wouldn't experience the downward acceleration caused by gravity because a stationary space (the object itself) in a field of acceleration (the gravitational field where the free fall took place) could be regarded as a reference frame motioning with a uniform acceleration relative to itself in weightlessness (i.e. a frame with constant proper acceleration). Therefore, the weak equivalence principle states that the outcome of any local non-gravitational experiment in a freely falling laboratory is independent of the velocity of the laboratory and its location in spacetime. Meanwhile, he also extended the application of special relativity to all the physical phenomena in reference frames with constant proper acceleration. Therefore, we know if a clock is attached to the front end of a certain object and another to the rear, the former will move faster than the latter when both are parallel with the direction of acceleration. And when the distance between the front and rear is x, the values of acceleration at both ends must satisfy $a' = a/(1 + ax/c^2)$ to prevent the disintegration of the object, which nevertheless doesn't seem possible in the gravitational field on Earth's surface. As for the reason, imagine a skyscraper that's h in height. According to the principle of weak equivalence, the gravitational forces on the top and bottom floors g' and g have to satisfy $g' = g/(1 + gh/c^2)$, which unfortunately differs from the equation of universal gravitation although Newton's law was supposed to govern all vertical changes in Earth's gravitational field. Hence, Einstein knew the combination of the weak equivalence principle and reference frames with constant proper acceleration simply wasn't enough for a seamless interpretation of the gravitational field. Just when he was stuck in the dark, he thought of the Minkowski spacetime, tailored for him by his mentor Hermann Minkowski.

When defining the concept of reference frame, Einstein compared the spacetime perceived by different observers to the spacetime regulated by different reference frames. In other words, changing observers is equivalent to transforming frames. Having decided to associate his study in physics with geometry, he then turned to his college friend Grossmann, who was then teaching at the department of mathematics at the Zürich University. Grossmann soon recommended the Riemann geometry because he believed the spacetime caused by energy and momentum that Einstein was researching could be described with the Riemann curvature tensor. Years of studying Riemann manifolds passed by. Finally, Einstein found an equation for the gravitational field that could regulate the relationship among energy, momentum, and the Riemann curvature. He also discovered that in curved spacetime, a given object would move along the route with the longest proper time interval (if a clock is attached to the object, this is the path that'd allow the most time to pass on the clock). And this route is exactly the geodesic, regulated by the geodesic equation. Up to this point, his theory of general relativity was consummate with two equations: (1) the gravitational field equation, which regulates how energy and momentum curves spacetime, and (2) the geodesic equation, which describes the motion of objects in curved spacetime. The physical notions of general relativity are in fact simpler than you could ever imagine. As for all the daunting complexities, those are mostly just mathematical details to support the major concepts.

The geodesic equation is the simpler of the two as it only describes the motion paths of objects and doesn't involve any tricky condition or phenomenon, while the field equation is a lot more complicated as it's related to physical quantities like energy and momentum. Hence, Einstein proposed that this equation had to satisfy the three basic requirements below:

1. The field equation has to be true despite any transformation of reference frames.
2. The field equation has to be compatible with Newton's law of universal gravitation in static, low-speed weak fields.
3. The field equation has to ensure the conservation of energy and momentum.

With the conditions established, Einstein then had to approach his physical research from a geometrical point of view, which he wasn't an expert at. Therefore, he spent seven years trying to figure out how to apply the Riemann geometry to his study while struggling through failure, frustration, and self-doubt. After numerous revisions, he nervously announced the final version of the gravitational field equation, which, as it turned out, was hailed as one of the greatest equations in the scientific history. In fact, Einstein's derivation process was full of guessing and conjecture, but luckily, his intuition in physics was strong and accurate enough to help him exclude asymmetrical factors and establish an equation that epitomized simplicity and harmony. In other words, it wasn't through strictly-structured reasoning or proving but intuitive thinking and hypothesizing that he obtained the final set of field equations widely known today.

Consisting of 16 simultaneous field equations, the entire set has one that's particularly famous, i.e. Newton's equation of universal gravitation. In other words, Newton's theory represents nothing but a certain scenario in the realm of general relativity. The field equations are like a lighthouse in the universe, shedding light on the mechanism of nature and opening the door to a whole new field of physics. When it comes to how Einstein developed the theories of special and general relativity, his story also serves as a reminder for us that truth is never readily discoverable but rather hidden deep in the subtleties of natural phenomena. In order to explore the wonders of the universe, we always have to resort to discreet thinking and logical reasoning. In addition, his research style is also a perfect example of keeping a sharp eye on all things happening around while smartly sifting through related details and systematically organizing information. Most important of all, he enjoyed observing nature while acquiring knowledge therefrom, which is an optimal attitude that all scientists should learn from to experience the joy of learning.

6.6   Einstein's Distraught, Lonely Late Life[58]

With the theory of relativity drastically changing the course of human civilization on all levels, it's only imaginable how much pressure must've been exerted on Einstein's shoulders due to his great achievement. The year 1919 was particularly overwhelming for him as his entire life was transformed overnight after the Times reported Eddington's discovery on November 7th. Three days after proof was published that the deviation degree of the light from the Hyades caused by solar gravity was almost in complete agreement with the prediction of general relativity, the New York Times raised him to the status of a scientific superstar with the sensational title "Lights all askew in the heavens: Einstein's theory triumphs". Commercial companies in Berlin, where people desperately needed something to believe in due to the post-war predicament, exaggeratedly advertised how general relativity could help develop the technology of wireless communication and prompted consumers to embrace changes soon to happen in their lifestyle because before they knew it, communicating devices like telephones would become just as ubiquitous and necessary as watches, books, purses, and whatnot.

In fact, the conduct of such companies was an epitome of the marketplace and society in general as people and even the government all started to benefit by associating themselves with Einstein, which inevitably led to fast-growing interest in his personal life. Although he was lavishly touted for his achievements, his privacy was also invaded to an outrageous degree, essentially making his life a Truman show. A major force behind the situation was the media, which was born at the beginning of the 20th century and began shaping people's opinions on an enormous scale. Just like a skillful manipulator, the media didn't just help Hitler and Stalin brainwash their targets but could also transforming the shy, lonely Einstein into a scientific celebrity. From then on, his fame and reputation, though a huge plus in terms of finance, started to poison his mind and spirit, sending ripples through his personal life and making it the subject of the public's scrutiny and ridicule.

Also happening to Einstein in 1919 was the divorce with his first wife Mileva Marić (1875–1948). Leaving Marić and their two sons in Switzerland, he soon married his cousin Elsa Löwenthal (1876–1936) in Berlin and settled

---

[58]Referenced from Neffe [26]

down with her two daughters. In the same year, his mother Pauline Koch (1858–1920) also moved to stay with his new family in Berlin, where he eventually had to see her die not long after. This said Berlin was a powerful stimulant to Einstein's research, just as we've mentioned before. Hanging on the wall of his office there were the portraits of Maxwell, Faraday, and Newton, on whose shoulders he stood to discover the subtleties of spacetime and the universe. Among the studies of the predecessors, Newton's absolute space is a huge contradiction to Einstein's relative framework although related data in many aspects only exhibited minor differences. When it comes to the internal theoretical structures and contexts, Newton's law is actually nothing but a small portion of Einstein's theory and only applies to low-speed motion near Earth's surface or between astronomical bodies. Once Einstein announced his results, Newton was immediately delegated although still influential in his own field. It's as if Einstein built a scientific skyscraper of relativity where Newton was allocated one room only, but of course, the room is still inhabited by passionate and competent engineers of applied mechanics propelling the progress of mankind's material civilization.

When German President Paul von Hindenburg named Hitler Chancellor of Germany in January of 1933, Einstein's life course was forever changed. Not only did the Nazis start persecuting Jews within the German territory and removing non-Aryan scholars from local universities, the disciplines of relativity and quantum physics, which Einstein himself pioneered, were also labeled as "Jewish science" and prohibited in university education due to their alleged decadent, fallacious content. Lucky enough to be visiting at the California Institute of Technology when all these happened, Einstein immediately decided not to return, which proved to be a wise choice because not long after the crackdown began, the Nazi government confiscated all his assets in Berlin, burnt all his manuscripts and documents regarding the research of relativity while offering a US$5,000 reward for his life. Publicly announcing he wouldn't be returning to Germany, he decided to retain the citizenship of Switzerland because it was where he was allowed to freely explore his thought wonderland and eventually came up with the five groundbreaking papers. Meanwhile, he also started applying for American citizenship in the hope of continuing his quest for scientific knowledge in a free, democratic country. Yet as it turned out, the process didn't go as smoothly as he'd expected, and he was forced to live there as a refugee for seven years. Though his application was eventually granted in 1940, he'd become highly disappointed by the discrepancy between America's cruel reality and its flawless façade of hope and freedom.

Completely shattering Einstein's faith in the US, though, wasn't the lengthy process of trying to acquire citizenship but John E. Hoover (1895–1972), the first Director of the Federal Bureau of Investigation (FBI). Hoover founded a dedicated program to investigate Einstein's correspondence with communist party members and support for the Anti-Fascist Alliance. In addition, he also took advantage of his power to push associated organizations under his control to attack Einstein. The Woman Patriot Corporation, for example, was compliant with Hoover's request and published a brochure to question Einstein's loyalty to America. Even after the end of WWII, the US still continued to expand the scale of nuclear research, hoping to produce even more weapons despite the country's victory. Hence, Einstein decided to establish the Emergency Committee of Atomic Scientists (ECAS) to warn people of the dangers of developing such weapons and promote the peaceful use of nuclear energy. But soon enough, the good cause made him a target of the FBI, and agents were sent out to keep a constant eye on his lab in Princeton and home in New York. Before long, the malicious treatment by intelligence services bullied him into silence and destroyed his health as well. Laughter gone and his gray, sparse hair messily strewn, Einstein started looking lost and hollow-eyed. Even when he found joy in whatever made him happy, a faint smile was all he could force out at the edge of his lips.

When the Soviet Union conducted its first successful nuclear weapon test in August of 1949, the US public descended into fear and anxiety because such destructive bombs, powerful enough to end WWII, could now be used against them. After the UK confirmed it was through undercover spies in the Manhattan Project that the Soviet Union acquired related information and technology, the FBI seized this chance to launch the most rigorous ever anti-communist program, followed by the House Un-American Activities Committee promoted by Joseph McCarthy (1908–1957), Senator from the state of Wisconsin. When McCarthyism ensued, the darkest period of America's history as a democracy broke out with never-ending suspicion, defamation, persecution, erroneous judgment, and all kinds of ridiculous events. Taking advantage of the atmosphere, Hoover conducted an investigation into the legality of Einstein's migration from Switzerland to the US in 1933, and that's when claims that he was involved in communist activities started pouring out, eventually leading to the compilation of all kinds of remotely relevant information into a report submitted to President Truman. Hastily prepared in just a month's time, the report was a mixture of false testimonies and real insanity. For example, Einstein was accused of producing a machine to read people's minds as well as a laser beam to destroy planes and tanks. A German citizen even cited a series of suspicious letters in his report to the FBI that Einstein was a former member of the German communist party, but as it turned out, he only did that to acquire American residency for himself.

Such a bleak, horrid atmosphere was what haunted Einstein's late years. Spending the last bitter winter of his life in Princeton, he gave described himself as an unorthodox researcher who integrated physics with mathematics when looking back at his legacy as a scientist. Due to his unconventional research style, scholars in neither discipline saw him as one of them. Physicists considered him a mathematician and vice versa, which gave him a growing sense of alienation. In addition, he also regarded his life as full of isolation and denial, a remark that was probably made out of the disappointment at his homeland because the country allowed the Nazis to "cleanse" the German society by persecuting Jews. Contending that Germans of all future generations would have to repent and atone for the Holocaust, he refused to be associated with official German organizations in any possible way and even turned down the fellowship offered by the Max Planck Society. Besides Germany, he was just as disappointed at Switzerland as well because the Swiss government also seized all his assets in the country during WWII at the request of the Germans, which basically included all the fortune and prizes (including the Nobel Prize in 1923) that he'd accumulated after earning himself a widespread reputation with the 1905 papers on relativity and other topics. Despite the hostile treatment, Switzerland still had a special place in his heart since it's where he conducted most of his relativity-related research, and this probably explains why he kept his Swiss citizenship till his last breath. When it comes to the US, he was left disillusioned as well because McCarthyism was not much better than the horrors of Nazism, which tore his homeland apart. In fact, he even considered the choice of seeking shelter in the US one of the most irreparable mistakes of his life. When the three countries he'd called home all turned their backs on him, it's no wonder he felt like a deserted, lonely misery.

On April 18th of 1955, Einstein's abdominal aortic aneurysm burst, leading to internal bleeding and eventually his death at 1AM in the Princeton Hospital, ending his 76-year life. While the condition was said to be a sequela of his childhood digestive dysfunction, rumor also has it that he murmured something in German to a nurse by his side, but she couldn't understand it. Several hours later, the young pathological anatomist Thomas Harvey, who arrived to start his shift, removed Einstein's brain out of the skull before dissecting the 2.5 lb organ into more than 200 pieces and preserving them in two glass jars with great care. Although he performed the procedure out of the admiration for the top talent of the century, what he did was a serious violation of the oath of ethics taken by physicians and stripped him of the doctor's license. In the 40 years that followed, Harvey moved around in an effort to hide the organ. It wasn't until he died was the brain, which unveiled the hidden wonders of the universe, returned to the University of Princeton.

When scientists finally set their hands on the carefully preserved pieces, they were just as curious as Harvey was when he performed the autopsy. Now waived of the limitation of medical ethics, they conducted a comprehensive, multi-aspect research by looking at the number of nerve cells in certain areas of the brain, the textures and patterns of sub-cranial cells of the cerebral cortex, how the brain tissues were connected and separated by fissures, etc. While all were eager to associate the brain to the creativity and amazing capability of abstract thinking of its owner, no one has been able to come up with any meaningful, convincing conclusion despite decades of research. The recent surge of genetic studies, however, again incited the hope of many scientists and even raised the prospect of reproducing Einstein's genius thoughts in cells cloned from his brain.

This said, Einstein's brain isn't just home to extraordinary competence, loving care, and pacifism but the devilish potential for inducing mass destruction as well. In fact, such a drastic contrast is also the epitome of his life. While he discovered the theory of relativity through independent thinking, this ability also posed a significant obstacle to his development of quantum mechanics. During the years of WWII, he was sharp enough to predict the imminent dangers to be caused by the Nazis and the Cold War between the US and Soviet Union, but the innocence with which he dealt with the cruel persecution by FBI was nearly ignorant that friends and colleagues all feared for his life. Compared to the sincere care that he had for fellow Jews and those injured by nuclear bombs, his bossy, overbearing attitude to his wives and kids essentially made him a tyrant at home. And speaking of marriage, his embrace of extramarital affairs as a way to satisfy his emotional needs was condemned by many as impurity, but meanwhile, he also sought to purify his mind by playing Bach's Cello Suites in leisure time. He didn't acknowledge the existence of God in the traditional sense and couldn't bear to live under any religious doctrines. In addition, he also denied any personal God concerned with fates and actions of human beings, a view he described as naïve. Nevertheless, he was indeed very faithful to a pantheistic God, who he believed controlled all things in the simple but orderly universe. To comprehensively describe this scientific giant who lived under so much contradiction he entire existence, we can only say that he's an amiable, wise man in the vast realm of truth, but a lone, isolated wanderer in the human world.

The most iconic picture of Einstein, which is widely seen as a decoration among younger generations, was taken in front of the Institute for

Advanced Study of Princeton University. Standing between the director and his wife, Einstein suddenly stuck out his tongue, which can be interpreted as an attempt to express his lonely, helpless state of mind caused by people's misunderstanding and full-swing political persecution. Meanwhile, the seemingly carefree but somehow wistful image of him also reminds us of how he conducted himself when the media put him under the limelight and plumped him as a superstar—overwhelmed, but joyous with a hint of naughtiness. In fact, this picture is no less intriguing than the Mona Lisa in the Louvre with all the possibilities behind his unique expression. What we know for sure, though, is that his playful yet perseverant attitude towards science made him the one-of-a-kind, Albert Einstein.

Appendix 6.1: Transformation between Reference Frame with Constant Proper Acceleration and Inertial Frame

For this proposition, let's start with the instantaneous inertial frame S' of an object moving with the speed of v and transform the object's acceleration to a different frame S. Under this situation, we know from the inverse transformation of (5.67) that:

$$u = \frac{u' + v}{1 + \frac{v}{c^2}u'} \tag{6.248}$$

In (6.248), $u' = dx'/dt'$ is the speed of the object in S', while $u = dx/dt$ is the same physical quantity in S. If we differentiate (6.248) over time:

$$\frac{du}{dt} = \frac{dt'}{dt} \frac{d}{dt'}\left(\frac{u' + v}{1 + \frac{v}{c^2}u'}\right)$$

$$= \frac{dt'}{\frac{dt' + vdx'/c^2}{\sqrt{1 - v^2/c^2}}} \frac{\frac{du'}{dt'}\left(1 + \frac{v}{c^2}u'\right) - (u' + v)\frac{v}{c^2}\frac{du'}{dt'}}{\left(1 + \frac{v}{c^2}u'\right)^2}$$

$$= \left(\frac{\sqrt{1 - v^2/c^2}}{1 + \frac{v}{c^2}u'}\right)^3 \frac{du'}{dt'} \tag{6.249}$$

Please note that the Lorentz transformation for t was used in establishing the second equal sign in (6.249). In this derivation above, we assumed ourselves to be inside the instantaneous inertial frame, so v is a non-changing value while u and u' change over time. As such, (6.249) is exactly the transforming equation for acceleration between S and S'. As we already know that $u' = 0$ and $du'/dt' = a$, plus the location of the object lies at the origin of S', we can also confirm that $u = v$. Then, substituting these equations into (6.249):

$$\frac{dv}{dt} = \left(1 - \frac{v^2}{c^2}\right)^{\frac{3}{2}} a \qquad (6.250)$$

Judging from (6.250), the speed v of the object is inversely proportional to the acceleration observed from S. When v approaches light speed, the acceleration will be close to 0 as well, which tells us that even after an infinite time period of acceleration, it's still impossible for the speed of any object to exceed light speed. One thing to be noted here is that in deriving (6.250), we used the below equation:

$$\frac{dt'}{dt} = \frac{dt'}{\frac{dt'+vdx'/c^2}{\sqrt{1-v^2/c^2}}} = \frac{\sqrt{1 - v^2/c^2}}{1 + \frac{v}{c^2}u'} = \sqrt{1 - v^2/c^2} \qquad (6.251)$$

Where the second equal sign was established with $u' = dx'/dt' = 0$.
To derive the relationship between v and t in S, we just need to integrate (6.250) over time:

$$at = \int_0^v \frac{dv}{\left(1 - v^2/c^2\right)^{3/2}} = \frac{v}{\sqrt{1 - v^2/c^2}} \qquad (6.252)$$

Then, (6.252) can be simplified as:

$$v(t) = \frac{at}{\sqrt{1 + (at/c)^2}} \qquad (6.253)$$

Again, the integration of (6.253) over time will lead us to the position function $x(t)$:

$$x(t) = \int_0^t \frac{at}{\sqrt{1 + (at/c)^2}} dt = \frac{c^2}{a}\left[\sqrt{1 + (at/c)^2} - 1\right] \tag{6.254}$$

As a next step, (6.254) can be simplified as:

$$\left(x + \frac{c^2}{a}\right)^2 - (ct)^2 = \left(\frac{c^2}{a}\right)^2 \tag{6.255}$$

(Note that for (6.255), we only consider $x \geq 0$.)

With (6.255) derived, we can easily draw the world line of object A (see Fig. 6.14). Judging from this above equation, the world line should be a hyperbola with the asymptote equation of $x = ct - c^2/a$. In the illustration, we've also drawn the x' axes of A's instantaneous inertial frames at P and H. As the relative acceleration of S' to S keeps increasing, the slopes of the x' axes are constantly changing as well. As for t', it's the time displayed on the clock attached to A. If we represent the time upon A's arrival at P and H as t'_P and t'_H, then for each and every point on the x'-axis of P (i.e. PQ), the time it corresponds to in S' should satisfy $t' = t'_P$. As for H, the time in S' that every single point on HR corresponds to is $t' = t'_H$. Judging from

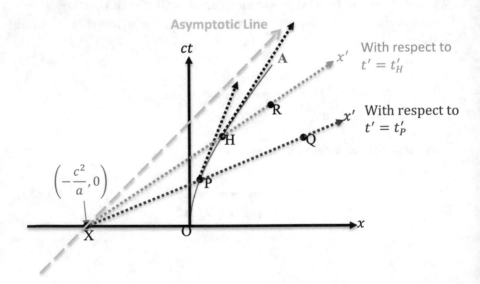

Fig. 6.14 Location change of different points in S observed from perspective of A

(6.254), the x' axes should intersect with x at the same point X^{59} with the coordinates of $(x, ct) = (-c^2/a, 0)$ despite the value of t'. Therefore, we know that each point, which can be seen as an "event" as well, located in the area between x and the asymptotic line should fall on a certain x' axis and correspond to the time of occurrence t' in S'.

With the illustration explained, let's now proceed to the equations for transforming all four coordinates. For the time axis, we can first substitute (6.253) into (6.251):

$$\frac{dt'}{dt} = \frac{1}{\sqrt{1 + (at/c)^2}} \tag{6.256}$$

Integrating the above equation, we can derive the transforming equation between t and t' (the time displayed on A's clock) at $x' = 0^{60}$ (i.e. where A is located):

$$t' = \int_0^t \frac{dt}{\sqrt{1 + (at/c)^2}} = \frac{c}{a} \sinh^{-1} \frac{at}{c} \tag{6.257}$$

Or, the result can also be converted to the below form so the equation becomes easier to handle as it doesn't contain the inverse function:

$$t|_{x'=0} = \frac{c}{a} \sinh \frac{at'}{c} \tag{6.258}$$

For (6.258), please note that it's only true when $x' = 0$. Now, the substitution of (6.258) into (6.259) can lead us to the relationship between v and t':

$$v(t') = c \tan h \frac{at'}{c} \tag{6.259}$$

Moving on to the spatial axis, please look at Fig. 6.15 as a reference. Now that the point P on the world line A is already known to correspond to the

---

[59] As the tangent slope of any point on the world line can be calculated from (6.254) and the reciprocal of a certain result is the slope of the x' axis passing through the corresponding point, we can derive the linear equation of the particular x'. Therefore, it's not hard to tell that all x' axes are supposed to intersect with x at X.

[60] Definition of a hyperbola:

$\sinh x = \frac{e^x - e^{-x}}{2}$, $\cosh x = \frac{e^x + e^{-x}}{2}$, $\tanh x = \frac{\sinh x}{\cosh x}$

$\sinh^{-1} x$ is the inverse function of $\sinh x$. Also note that $\cosh^2 x = \sinh^2 x + 1$.

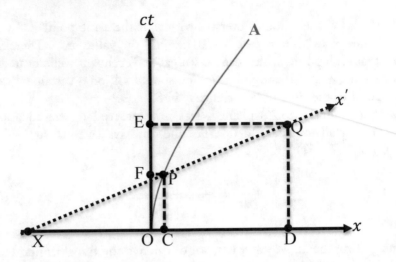

Fig. 6.15 Illustration of axis transformation

time t', we need to confirm the spacetime position of Q (spatial coordinate $= x'$) in S. Based on the discussion in Chap. 5, we know the unit length of x' is $\sqrt{(1 + \beta^2)/(1 - \beta^2)}$ times that of x, where $\beta = v/c$. Therefore, the projection of $\overline{PQ}$ on x (i.e. $\overline{CD}$) is:

$$\overline{CD} = \left(x' \sqrt{\frac{1 + \beta^2}{1 - \beta^2}} \right) \times \frac{1}{\sqrt{1 + \beta^2}} = \frac{x'}{\sqrt{1 - \beta^2}} = x' \cosh \frac{at'}{c} \quad (6.260)$$

In the above equation, results from (6.259) were used in establishing the third equal sign. As the x coordinate of Q equals $\overline{OC}$ plus $\overline{CD}$ while $\overline{OC}$ equals the x coordinate $x(t)$ of A, the substitution of (6.258) back into (6.254) lead us to:

$$\overline{OC} = x(t') = \frac{c^2}{a} \left(\cosh \frac{at'}{c} - 1 \right) \quad (6.261)$$

In (6.261), each t has been replaced with t'. Then, we can add up (6.260) and (6.261) to retrieve:

$$x = \left(x' + \frac{c^2}{a} \right) \cosh \frac{at'}{c} - \frac{c^2}{a} \quad (6.262)$$

And the result is exactly the expression for transforming the x coordinate. Similarly, we can also use the length of $\overline{OF} + \overline{FE}$ to derive the transforming equation for t. As the process isn't much different from what we've gone through, we'll leave it to readers as an exercise, but here's a heads up that the result should be:

$$ct = \left(x' + \frac{c^2}{a}\right)\sinh\frac{at'}{c} \tag{6.263}$$

At this point, all that's left to figure out is how to transform the y and z coordinates, which is in fact more than simple. If you still remember the stick-and-brush thought experiment when we were discussing special relativity (see Fig. 5.9), you'd already know that we can follow the same logic to prove the invariance of the y and z coordinates in the transformation process, i.e. $y = y'$ and $z = z'$, as shown in (5.28) and (5.29).

As a next step, let's place A at the origin $x' = 0$ of S' at $t = t' = 0$ and for it to move along with the reference frame. In addition, we also need to attach to A a clock that's been set at $t = t' = 0$ to display t' in order to determine the relationship $x(t)$ between time and A's location in S. For this purpose, we'll first have to understand how the object's acceleration transforms between S and S' by dividing the accelerating process into extremely short periods of time, each of which corresponds to an instantaneous inertial frame. When a time period is short enough, the object A can be regarded as engaging in uniform motion along a straight line in the corresponding frame.

Up to this point, we've derived the equations for transforming all four spacetime axes between S and S':

$$\begin{cases} ct = \left(x' + \frac{c^2}{a}\right)\sinh\frac{at'}{c} \\ x = \left(x' + \frac{c^2}{a}\right)\cosh\frac{at'}{c} - \frac{c^2}{a} \\ y = y' \\ z = z' \end{cases} \tag{6.264}$$

Then, let's use (6.264) to double-check some facts that we've discussed before. In Sect. 6.1.3, we mentioned that to prevent the vehicle from disintegration while it's accelerating, the proper acceleration at the point that's d away from the rear end has to be $a/(1 + ad/c^2)$ so that no part of the car is contracted, stretched, or twisted in any possible way in the proper-acceleration frame where it's located. This conclusion can now be easily confirmed with the transforming equations in (6.264), which show that for an object moving with the proper acceleration of a, the world line should satisfy:

$$x = \frac{c^2}{a} \cosh \frac{at'}{c} + \text{a certain constant} \qquad (6.265)$$

Judging from (6.262), the equation for transforming the x coordinate, the world line of an object at $x' = d$ can be expressed as:

$$x = \left(d + \frac{c^2}{a}\right) \cosh \frac{at'}{c} - \frac{c^2}{a} \qquad (6.266)$$

Then, (6.266) can be converted to the form of (6.265):

$$x = \frac{c^2}{a(1 + ad/c^2)^{-1}} \cosh \left\{ \frac{\left[a(1 + ad/c^2)^{-1}\right]\left[t'(1 + ad/c^2)\right]}{c} \right\} - \frac{c^2}{a} \qquad (6.267)$$

Last but not least, we can tell from (6.267) that the object at $x' = d$ moves with the proper acceleration of $a/(1 + ad/c^2)$ and that the clock attached to it shows the time $(1 + ad/c^2)t'$. In other words, the speed of time at $x' = d$ is $(1 + ad/c^2)$ times that at $x' = 0$, perfectly in line with our previous conclusion.

Appendix 6.2: Derivation of the Geodesic Equation

For this proposition, we'll start with the general expression for spacetime interval, $ds^2 = g_{\mu\nu}dx^\mu dx^\nu$. With $ds^2 = -c^2 d\tau^2$, the previous equation can be rewritten as:

$$d\tau^2 = -\frac{1}{c^2} g_{\mu\nu} dx^\mu dx^\nu \qquad (6.268)$$

Suppose an object moves from P_1 to P_2 in a 4-space, the time passed in the process (proper time interval) can be expressed as the integral equation below:

$$\Delta\tau = \frac{1}{c}\int_{P_1}^{P_2}\sqrt{-g_{\mu\nu}dx^\mu dx^\nu} \qquad (6.269)$$

Applying variational calculus to (6.269), we can derive the time interval for the object when its path undergoes a slight change:

$$
\begin{aligned}
\delta(\Delta\tau) &= \frac{1}{c}\int_{P_1}^{P_2}\delta\sqrt{-g_{\mu\nu}dx^\mu dx^\nu} \\[2mm]
&= \frac{1}{c}\int_{P_1}^{P_2}\frac{\delta\left(-g_{\mu\nu}dx^\mu dx^\nu\right)}{2\sqrt{-g_{\mu\nu}dx^\mu dx^\nu}} \\[2mm]
&= \frac{-1}{c^2}\int_{P_1}^{P_2}\frac{\left(\delta g_{\mu\nu}\right)dx^\mu dx^\nu + 2g_{\mu\nu}dx^\mu d(\delta x^\nu)}{2d\tau} \\[2mm]
&= \frac{-1}{c^2}\int_{P_1}^{P_2}\left(\frac{1}{2}\delta g_{\mu\nu}\frac{dx^\mu}{d\tau}dx^\nu + g_{\mu\nu}\frac{dx^\mu}{d\tau}d(\delta x^\nu)\right) \qquad (6.270)
\end{aligned}
$$

To establish the third equal sign, we applied $g_{\mu\nu}\delta(dx^\mu dx^\nu) = 2g_{\mu\nu}dx^\mu d(\delta x^\nu)$, where $\delta g_{\mu\nu}$ can also be expressed as:

$$\delta g_{\mu\nu} = \frac{\partial g_{\mu\nu}}{\partial x^\alpha}\delta x^\alpha \qquad (6.271)$$

In addition, the integral of the last equation in (6.270) can be further simplified as:

$$
\begin{aligned}
\int_{P_1}^{P_2}g_{\mu\nu}\frac{dx^\mu}{d\tau}d(\delta x^\nu) &= \int_{P_1}^{P_2}\left[d\left(g_{\mu\nu}\frac{dx^\mu}{d\tau}\delta x^\nu\right) - d\left(g_{\mu\nu}\frac{dx^\mu}{d\tau}\right)\delta x^\nu\right] \\[2mm]
&= \left(g_{\mu\nu}\frac{dx^\mu}{d\tau}\delta x^\nu\right)\Big|_{P_1}^{P_2} - \int_{P_1}^{P_2}d\left(g_{\mu\nu}\frac{dx^\mu}{d\tau}\right)\delta x^\nu
\end{aligned}
$$

$$= -\int_{P_1}^{P_2} d\left(g_{\mu\nu}\frac{dx^\mu}{d\tau}\right)\delta x^\nu \qquad (6.272)$$

Since P_1 and P_2 are both fixed, we know $\delta x^\nu = 0$ and therefore $\left(g_{\mu\nu}\frac{dx^\mu}{d\tau}\delta x^\nu\right)_{P_1}^{P_2} = 0$. Subsequently, we can substitute (6.271) and (6.272) back into (6.270) and retrieve:

$$-c^2\delta(\Delta\tau) = \int_{P_1}^{P_2}\left(\frac{1}{2}\frac{\partial g_{\mu\nu}}{\partial x^\alpha}\frac{dx^\mu}{d\tau}dx^\nu\delta x^\alpha - d\left(g_{\mu\nu}\frac{dx^\mu}{d\tau}\right)\delta x^\nu\right)$$

$$= \int_{P_1}^{P_2}\left(\frac{1}{2}\frac{\partial g_{\mu\nu}}{\partial x^\alpha}\frac{dx^\mu}{d\tau}dx^\nu\delta x^\alpha - \frac{\partial g_{\mu\nu}}{\partial x^\alpha}dx^\alpha\frac{dx^\mu}{d\tau}\delta x^\nu - g_{\mu\nu}d\left(\frac{dx^\mu}{d\tau}\right)\delta x^\nu\right)$$

$$= \int_{P_1}^{P_2}\left(\frac{1}{2}\frac{\partial g_{\mu\alpha}}{\partial x^\nu}\frac{dx^\mu}{d\tau}\frac{dx^\alpha}{d\tau} - \frac{\partial g_{\mu\nu}}{\partial x^\alpha}\frac{dx^\alpha}{d\tau}\frac{dx^\mu}{d\tau} - g_{\mu\nu}\frac{d^2x^\mu}{d\tau^2}\right)\delta x^\nu d\tau$$

$$(6.273)$$

For the last equation in (6.273), we can swap the α and ν in the first term in the parentheses as this won't change the sum. For the proper time interval of a certain path to reach the maximum value, (6.273) has to be 0, i.e.:

$$g_{\mu\nu}\frac{d^2x^\mu}{d\tau^2} + \left(\frac{\partial g_{\mu\nu}}{\partial x^\alpha} - \frac{1}{2}\frac{\partial g_{\mu\alpha}}{\partial x^\nu}\right)\frac{dx^\mu}{d\tau}\frac{dx^\alpha}{d\tau} = 0 \qquad (6.274)$$

Then, we'll further simplify (6.274) by looking at the second term, which has the below property (because μ and α are equivalent, meaning their swap won't change the final sum):

$$\frac{\partial g_{\mu\nu}}{\partial x^\alpha}\frac{dx^\mu}{d\tau}\frac{dx^\alpha}{d\tau} = \frac{\partial g_{\alpha\nu}}{\partial x^\mu}\frac{dx^\mu}{d\tau}\frac{dx^\alpha}{d\tau} \qquad (6.275)$$

As such, (6.274) can be rewritten as:

$$g_{\mu\nu}\frac{d^2x^\mu}{d\tau^2} + \frac{1}{2}\left(\frac{\partial g_{\mu\nu}}{\partial x^\alpha} + \frac{\partial g_{\alpha\nu}}{\partial x^\mu} - \frac{\partial g_{\mu\alpha}}{\partial x^\nu}\right)\frac{dx^\mu}{d\tau}\frac{dx^\alpha}{d\tau} = 0 \qquad (6.276)$$

Next, multiplying (6.276) by the inverse matrix $g^{\kappa\nu}$ of $g_{\mu\nu}$ and a bit of simplification will lead us to[61]:

$$\frac{d^2x^\kappa}{d\tau^2} + \frac{1}{2}g^{\kappa\nu}\left(\frac{\partial g_{\mu\nu}}{\partial x^\alpha} + \frac{\partial g_{\alpha\nu}}{\partial x^\mu} - \frac{\partial g_{\mu\alpha}}{\partial x^\nu}\right)\frac{dx^\mu}{d\tau}\frac{dx^\alpha}{d\tau} = 0 \qquad (6.277)$$

Which we can rewrite as the below based on the Christoffel symbols of connection in (6.40):

$$\frac{d^2x^\kappa}{d\tau^2} + \Gamma^\kappa_{\mu\alpha}\frac{dx^\mu}{d\tau}\frac{dx^\alpha}{d\tau} = 0 \qquad (6.278)$$

And (6.278), is exactly the general expression for spacetime motion, often called the "geodesic equation" as well.[62]

However, the complicated use of upper and lower indices might be quite confusing, so we're going to interpret (6.278) from a direct, geometric point of view by first rewriting (6.278) as:

$$d\left(\frac{dx^\kappa}{d\tau}\right) = -\Gamma^\kappa_{\mu\alpha}\frac{dx^\mu}{d\tau}dx^\alpha \qquad (6.279)$$

If we see $\frac{dx^\kappa}{d\tau}$ in (6.279) as the component A^κ of the vector A along the direction of $\mathbf{e}_\kappa$, this equation is basically equivalent to (6.34) and (6.35) because both regulate the parallel transport of A. Therefore, (6.278) can

---

[61] Note that $g^{\kappa\nu}g_{\mu\nu} = g^{\kappa\nu}g_{\nu\mu} = \delta^\kappa_\mu$, so $g^{\kappa\nu}g_{\mu\nu}\frac{d^2x^\mu}{d\tau^2} = \delta^\kappa_\mu\frac{d^2x^\mu}{d\tau^2} = \frac{d^2x^\kappa}{d\tau^2}$.

[62] Readers who're attentive to details might be wondering: since $\delta x^\nu(\tau)$ doesn't represent four separate functions due to the requirement that it has to satisfy (6.268), how do we know that (6.274) must be true for all values of $\delta x^\nu(\tau)$? In fact, this question can be easily answered by switching to a different perspective. If we represent the spacetime path with a more general parameter λ (i.e. $x^\mu(\lambda)$) instead of τ at the beginning, we can be sure that all four alternatives of $\delta x^\mu(\lambda)$ are independent because the relationship between λ and τ isn't defined. Similarly, we can replace the τ in (6.277) with λ by applying variational calculus and obtain:

$\frac{d^2x^\kappa}{d\lambda^2} + \frac{1}{2}g^{\kappa\nu}\left(\frac{\partial g_{\mu\nu}}{\partial x^\alpha} + \frac{\partial g_{\alpha\nu}}{\partial x^\mu} - \frac{\partial g_{\mu\alpha}}{\partial x^\nu}\right)\frac{dx^\mu}{d\lambda}\frac{dx^\alpha}{d\lambda} = 0$

Which can then be simplified as:

$0 = g_{\nu\kappa}\frac{dx^\nu}{d\lambda}\frac{d^2x^\kappa}{d\lambda^2} + \frac{1}{2}\left(\frac{\partial g_{\mu\nu}}{\partial x^\alpha} + \frac{\partial g_{\alpha\nu}}{\partial x^\mu} - \frac{\partial g_{\mu\alpha}}{\partial x^\nu}\right)\frac{dx^\nu}{d\lambda}\frac{dx^\mu}{d\lambda}\frac{dx^\alpha}{d\lambda}$

$= \left(\frac{1}{2}g_{\mu\nu}\frac{d^2x^\mu}{d\lambda^2}\frac{dx^\nu}{d\lambda} + \frac{1}{2}g_{\mu\nu}\frac{dx^\mu}{d\lambda}\frac{d^2x^\nu}{d\lambda^2}\right) + \frac{1}{2}\frac{\partial g_{\mu\nu}}{\partial x^\alpha}\frac{dx^\nu}{d\lambda}\frac{dx^\mu}{d\lambda}\frac{dx^\alpha}{d\lambda}$

$= \frac{1}{2}\frac{d}{d\lambda}\left(g_{\mu\nu}\frac{dx^\mu}{d\lambda}\frac{dx^\nu}{d\lambda}\right)$

Because the above equation is equivalent to $\frac{d}{d\lambda}\left(\frac{d\tau}{d\lambda}\right)^2 = 0$, we can safely assume that λ and τ must be related in a linear manner: $\tau = A\lambda + B$, where A and B are both constants. As such, we can also confirm that replacing all the λ in the geodesic equation with τ won't change the final result.

also be regarded as expressing the geodesic's parallel transport following the object's path at the speed of $v \equiv \frac{dx^k}{d\tau}\mathbf{e}_\kappa$. In addition, (6.278) also implies that the object's motion is always of the same direction as the previous moment, which agrees with the requirement that we set for the ant experiment for defining curved spacetime in Sect. 6.2.1. As the thought experiment turned out, a line that's formed following this requirement proves to be of maximum length.

Regarding the geodesic equation, below are two exceptions:

(1) In the Minkowski spacetime:

$$\frac{d^2t}{d\tau^2} = \frac{d^2x}{d\tau^2} = \frac{d^2y}{d\tau^2} = \frac{d^2z}{d\tau^2} = 0 \qquad (6.280)$$

Which shows that the object moves at a constant speed.

(2) In a reference frame with constant proper acceleration:

$$\frac{d}{d\tau}\left[\left(1 + \frac{gx}{c^2}\right)^2 c^2 \frac{dt}{d\tau}\right] = 0 \qquad (6.281)$$

$$\frac{d^2x}{d\tau^2} + \left(1 + \frac{gx}{c^2}\right)g\left(\frac{dt}{d\tau}\right)^2 = 0 \qquad (6.282)$$

$$\frac{d^2y}{d\tau^2} = 0 \qquad (6.283)$$

$$\frac{d^2z}{d\tau^2} = 0 \qquad (6.284)$$

With a bit of simplification, (6.281) and (6.282) will yield the below result:

$$\frac{d^2x}{dt^2} = \frac{2g}{1 + \frac{gx}{c^2}}\frac{1}{c^2}\left(\frac{dx}{dt}\right)^2 - g\left(1 + \frac{gx}{c^2}\right) \qquad (6.285)$$

And (6.285) is exactly the equation for expressing the object's motion along the x axis in the frame with constant proper acceleration.

Last but not least, to examine the validity of (6.285), we can consider two different types of scenarios. When speed is low, the term $\frac{1}{c^2}\left(\frac{dx}{dt}\right)^2$ in the equation can be disregarded, while $\frac{gx}{c^2}$ can also be omitted in weak fields. Therefore, (6.285) can be approximated to become:

$$\frac{d^2x}{dt^2} = -g \qquad (6.286)$$

Which perfectly fits the expression for gravitational field in classical mechanics.

7

The Course of Scientific Progress: Spanning Three Thousand Years of Ideological Conflict and Evolution

朝辭白帝彩雲間,千里江陵一日還.
兩岸猿聲啼不住,輕舟已過萬重山.

Leaving at dawn from the city crowned with cloud,
I have sailed a hundred miles through canyons.
While monkeys on cliffs were howling loud,
my skiff has passed over undulating mountains.
without taking a detour around.

Translated from "Leaving Baidi for Jiangling," *by Li Bai (701–762), a poet of the Tang Dynasty.*

In the last six chapters, we have tracked the progress of human scientific development, starting with the Babylonians of 1000 BC and ending 3,000 years later with Einstein's theory of relativity in 1915. This history has included dozens of scientist's stories and discoveries and used various case studies and propositions to expound on their contributions to humanity. These stories show us just how often scientists' fought ideological battles and went through philosophical evolution. Their evolution, nevertheless, moved at a snail's pace, making it almost impossible to see unless we view it on a larger time scale. But their battles had powerful undercurrents often causing dramatic, almost immeasurable changes to the scientific landscape.

These incidents of conflict or progress are dotted along the timeline of scientific history. Occasionally, they manifested independently as the birth of

F. Chen and F.-T. Hsu, *How Humankind Created Science*, https://doi.org/10.1007/978-3-030-43135-8_7

new theories and sometimes even tied up in complications like Galilei's trial at the hands of the Catholic Church. They all, however, have their own causes and effects, as outlined in detail in our last six chapters of stories, theories and case studies. As the entirety of this narrative may be many thousands of words long, we are concerned that our timeline of events might be hard to follow. In the hopes of making 3,000 years of scientific progress easier to remember, we decide to devote this chapter to looking back on and discussing the conflict and advancement it entailed. We accordingly take this chance to gain an in-depth insight into the origins and consequences of this chain of events, while identifying the human machinations or natural trends that preceded them. With this knowledge of the past, we can speculate on the direction of future scientific developments and expect to lay out a grand blueprint for a new world defined by scientific discovery.

7.1   The Evolution of Natural Philosophy: A Victory for the Geocentric Model

The collection of clay tablets discovered in the Mesopotamian river basin are the earliest known records depicting a definition of the universe—our solar system. As early as 2000 BC Babylonians from this area observing the night sky already had knowledge of the planets Venus, Mercury, Mars, Jupiter and Saturn, as well as their sluggish crawling across the horizon. These roving planets did not obediently abide by a fixed movement pattern. Instead, they occasionally shuttled back and forth, moving in retrograde and tying knots in the sky with their tracks. The early natural philosophers put forward a variety of models based on their perceptions of the universe, attempting to explain this retrograde motion. Of all the issues that this brought up, that of whether the Earth is or isn't the unmoving center of the universe incited the most disagreement. After these discussions unfolded over more than a thousand years of competition and even conflict between opposing schools of thought, it was Aristotle's (384–322 BC) geocentric model that prevailed above all others. Aristotle believed that the earth was the center of the universe, with all celestial bodies including the Sun, the Moon, the planets and the stars each attached to 55 concentric spheres with radii of varying sizes. Each sphere was said to revolve around the Earth in different directions at different speeds.

When Aristotle drew up his model of the universe, replete with over 50 celestial spheres commanding the night sky, he was still not able to explain why the planets moved in retrograde. Despite this, many natural philosophers were not concerned with retrograde motion, as they had found other methods to quench their astronomical curiosity: they used simplistic instruments to observe the stars and measure the sizes and relative distances of celestial bodies. Why were so many of this era's most intelligent people so passionate about surveying the cosmos? The answer is that knowledge of the universe had significant value—it aided ships in navigating at night, as well as caravans in traversing the desert; it was used to divine nations' future prosperity or decline, and whether their wars would end in victory or defeat; it could forecast changes in climate and predict farming seasons, among many other vital functions.

Of all these stargazers, the one with the most hallowed name was Hipparchus of Rhodes (190–120 BC). This both dexterous and sharp-minded scientific giant used his own inventions—like the armillary sphere—to demarcate the positions of over 800 fixed stars. He used his disc-shaped astrolabe to measure the latitude of the stars, his dioptra to estimate the Moon and Sun's diameters and his uncomplicated equatorial ring to observe the moment at which the equator passes the Sun's center in both spring and autumn. There was also Eratosthenes (276–194 BC) who used the shadows cast by what was known as the Babylonian cane—the gnomon—to measure the Earth's radius, and Aristarchus of Samos (310–230 BC) who used the size of the Moon's shadow during a solar eclipse to estimate sizes of the Sun and Moon, as well as the distance between them. Aristarchus's estimation is remarkably close to the actual number—a truly astonishing feat for a man of his time. But the most breathtaking discovery of this time was made by Hipparchus. After spending years monitoring how the Sun's position changes through the seasons, he found that the Earth ought to be rotating and that this rotation must have an axial precession cycle of 27,000 years—ever so close the number we know today: 25,727.

After making myriad astronomical observations and despite evidence that the earth revolves on a changing rotational axis, Aristotle's geocentric model grounded in the accumulation of human perception and experience remained unshaken. His system's unshakable status was further solidified in the second century with the great astronomer Claudius Ptolemy's (90–168) complex model of the universe. Ptolemy collated data regarding the planets' retrograde motions using Hipparchus's astronomical findings, adding the equant to Hipparchus's model of the epicycle and deferent. Within the framework of this human-invented orbit around three circles, aligning triangles and circles in his geometric analysis, Ptolemy was able to calculate data such as the planets' epicycle size, eccentric position, angular velocity and even the time it takes for some planets to complete a retrograde motion.

The Ptolemaic system of astronomy was in perfect agreement with Aristotle's theories and conformed to the geocentric teachings of the Roman Catholic Church, giving it the support it needed to become the one and only approved astronomical theory for the next thousand years. This doctrine encompassing the authority of both science and religion drew a line in the sand that most were unwilling to cross, bringing a halt to progress in astronomy for more than a millennium. During this lengthy period of stagnancy, though a scattering of new theories did appear, they were nothing more than additions of more and more epicycles to Ptolemy's model. At its height, more than 80 epicycles were crammed into our little solar system. At this point, the model had simply become too complicated.

7.2   The Astronomical Revolution of the Middle Ages: The Heliocentric System Snatches Victory from the Jaws of Defeat

It was at a point of agonizing struggles that Polish astronomer Nicolas Copernicus (1473–1543) had an awakening, as he rose up to categorically reject the geocentric system and embark on the road towards heliocentrism. He maintained that the Sun was in fact the center of the universe, with the Earth and other planets revolving around it on a circular orbit. Copernicus's heliocentric model listed six planets in terms of their distance from the sun. In order, they are Mercury, Venus, Earth, Mars, Jupiter and Saturn, while the Moon orbits Earth. This system was actually not that far off from Aristotle's geocentric model. All that is needed to make the two match is to take the moon out of Aristotle's universe and switch the Earth and Sun around. In this field in which the slightest discrepancy can cause the greatest inaccuracy, the notion of planetary retrograde that had confused astronomers for over a thousand years finally had a satisfactory explanation. Ptolemy's epicycle, deferent and equant along with all the other tedious artificial orbital constructs could be scrapped, removed from astronomical discourse. What was once overly confusing was now simple and elegant, bringing in a new dawn for astronomic research. And it was *because* of this immense contribution that Copernicus became known as the father of the astronomical revolution, a title fully deserved. However, the one thing holding the Copernican model back was that it still assumed the planets moved on circular Aristotelian orbits, which led to a clear discrepancy between Copernicus's calculations of the planets' orbits and his observational data.

Despite the relative simplicity with which heliocentrism could explain the planets' retrograde motion, Copernicus's theory still contravened the Roman Catholic Church's position that the earth was the center of the universe. His grievance towards the church's opposition meant that Copernicus's publicizing of his theory was rather low-key. Aside from this, the fact that the planetary orbits he had calculated clearly did not match up with his measurements led to numerous academics holding on to the geocentric model that had been at the core of their belief system for nearly 2,000 years. The most prominent holdout was a Danish nobleman with his own royal observatory, the renowned astronomical observer Tycho Brahe (1546–1601).

Tycho put forward a model that was a compromise between geocentrism and heliocentrism. In it, an unmoving Earth was the center of the universe with the Moon, Sun and stars revolving around it on a circular orbit, while the other five planets revolved around the Sun. As a result of Tycho's abundant resources and extensive observational data, he managed to catalog over 1,000 fixed stars, far exceeding the 700 recorded in Ptolemy's classic the *Almagest* and placing him as a leading light of the era. The wide recognition of Tycho's theory meant there was a point in time at which his system of two universal focal points, along with geocentrism and heliocentrism were all endorsed by numerous people. Even Galilei once considered Tycho's model as one of the two systems to include in his *Dialogue Concerning the Two Chief World Systems*. In the end, it was the Danish nobleman's disciple Johannes Kepler (1571–1630) who took on researching the Tychonic model.

Alluded to by his successor Newton as a scientific giant, Kepler was an astronomer with an unusual fate: a nomadic life mired in hardship, but with magnificent accomplishments. After coming into contact with Tycho's plethora of astronomic data, Kepler pushed himself to his limits in compiling it in all its enormity. But no matter how he put his calculations together on Tycho's icy Danish island, he was unable to persuade himself of the Tychonic model's accuracy.

Whether by fortune or misfortune, just at the moment when the data had Kepler at his most perplexed, complications from a bladder infection killed Tycho in the prime of his life. Kepler thus decided to put aside his research into Dane's model and focus instead on Copernicus's heliocentric system. Following this Kepler spent more than five years making 19 drafts of Mars and the Sun's relative positions. It took more than 70 attempts and nearly 1,000 pages of repeated calculation for him to finally discover that Mars moves on an elliptical orbit with the sun as its focal point. This is what is now known as the first of Kepler's laws of planetary motion: the law of orbits.

From the same set of data, Kepler found that when a planet travels along its orbit, the fan-shaped area it sweeps across covers equal areas in equal amounts of time. This became his second law of planetary motion: the law of areas. The third of his laws and perhaps the most influential—the law of periods—was also a result of Kepler's collating orbital and cyclical data on the six visible planets. He discovered that the cube of the semiaxes on the planets' elliptical orbits is proportional to the square of their orbital periods.

The discoveries of these three principles had one thing in common: they all relied on vast quantities of data. Kepler constructed concise mathematical equations in which to place the planets' orbit sizes and speeds as well as their orbital periods. This was a significant breakthrough with far-reaching implications for astronomy, if not for all physics. In terms of historical progress in astronomy, Kepler's law of orbits upended the ancient Greek model of circular orbits rooted in an aesthetic of perfect circles. His laws of areas and periods replaced elements of Greek natural philosophy steeped in theory but lacking in application, detailing the planets' movements in a quantitative equation. Kepler's three laws sounded the call to arms for an astronomic revolution that had been fairly quiet up until that point. Without this, Newton would not have been able to mount his attack on the science of old at a time when society was going through urgent philosophical changes, planting the flag for a new scientific empire on a landscape of physics previously left to rot.

On the other hand, human history often dots with pleasing coincidences. Soon after Kepler had set it in motion, the wave of quantitative scientific research he had pioneered found its way to the southern Alps of Italy, where it was found by the man Einstein later praised as the father of scientific experimentation: Galileo Galilei (1564–1642). Einstein had an extremely high estimation of Kepler because he had managed to place a model for the universe in a mathematical framework. Kepler's work included the use of simple calculations to create principles for the relationships between various celestial bodies and making geometric diagrams to describe them. A contemporary of Kepler, Galilei was a master of these methods. One of his most celebrated experiments involved allowing a copper ball to fall freely from the top of an inclined plane while measuring its velocity with a self-made water clock. He discovered from his measurements that free-falling bodies accelerate at a constant rate. Aside from this, he found that as the ball dropped from the plane's edge the parabolic movement it made was, in essence, the synthesis of a free-falling body and a constant horizontal velocity. Half a century later, both the phenomenon described by this experiment and the data provided by Galilei's measurements became the philosophical basis for Newton's law of universal gravitation while also providing him with experimental data.

In terms of his astronomical work, Galilei created his own refracting telescope that allowed him to observe the cosmos magnified by a factor of 20. He saw the universe for what it was: the moon was not a perfect sphere, but had a rough surface; while the night sky was constantly changing. The Aristotelian idea that had enjoyed primacy for a millennium was thus turned into scientific heresy. Galilei then turned his telescope to the horizon and to Jupiter where he found four satellites moving around the gas giant on different cycles and different orbits. From this point on, the universe with two centers (the earth and sun) gained a third (Jupiter). This was the last straw that broke the geocentric system's back after its eon of domination. However, because Galilei's vision of Jupiter and its satellites inferred the true structure of the solar system, he was judged a heretic by the Roman Inquisition and thrown in prison.

The more than hundred years from Copernicus to Galilei saw a series of new astronomic discoveries on continental Europe. This period, known to scientific historians as the astronomical revolution, brought on significant changes that set the wheels in motion for modern science. Prior to this what was known as science was often mixed up with religion or philosophy and lacked a specific doctrine to support its reasoning. Once these wheels began moving, it was clear that physics and astronomy would be the emissaries of science, with precise mathematics illustrating its laws.

Scientists began using these mathematical formulas to accurately calculate and predict natural phenomena, then verified and confirmed the theories they had put forward. As a whole, their reasoning formed a coherent package: a paragon of unflappable poise and confidence. Compared to the Protestant Reformation and Renaissance with which it shared a moment in time, the astronomical revolution had perhaps an even greater impact on human civilization than its contemporaries, far exceeding that of this pair of religious and cultural overhauls. And the man who was to take this astronomical revolution to its zenith was none other than Isaac Newton (1643–1727).

7.3 Ruling Physical Sciences by Law: The Law of Gravity

Just when astronomers all over continental Europe were wracking their brains over the appearance of a third center to the universe and bickering over which was the true heart of the solar system, across the English Channel a number of British academics were putting forward another question: what is the force that causes planets to revolve around a core celestial body? At the time, many academics in the public eye believed it to be magnetism. Magnetic force was first discovered all the way back in the 7th century BC in modern-day Turkey by the olive oil merchant and distinguished philosopher Thales (624–546 BC). Though soon after, due to the lack of clarity surrounding its origins, his discovery vanished from scientific relevance for 2,000 years. It was not until 1600 that it reemerged as a result of the discussion surrounding gravity when the British scientist William Gilbert (1544–1603) proposed magnetism as the force confining planets to their orbits. Later, Robert Hook (1635–1703) found that magnets would experience a reduction in their magnetic force when exposed to heat, sometimes even losing it completely. He thus believed that the scorching hot sun could not create a magnetic force capable of pulling planets. This discovery effectively removed magnetism as a candidate for the force responsible for the planets' movements.

Newton, who viewed Hook as his academic enemy, turned his nose up with contempt at the laughable idea that magnetic force had no use. But as a Cambridge professor, he could never refute other's theories without any evidence of his own. He thus used Kepler's laws of planetary motion along with Galilei's inclined plane experiment as a foundation to deduce a law determining both planetary movements in the solar system and how apples fall from trees. This, his law of universal gravitation, would later be written as[1] $F = GM_1M_2/r^2$. Newton recorded all the inferences behind his theories and all calculations relevant to his propositions in his great tome of 550 pages: the *Mathematical Principles of Natural Philosophy* (commonly referred to by its Latin abbreviation—*Principia*).

In *Principia*, generally considered one of the most influential works in scientific history, Newton uses Kepler's second law of planetary motion to prove that the sun exerts a centripetal force on the planets, and the first and third laws to define the relationship between motion and force, followed by his derivation of the law of universal gravitation. He followed this up with a series of remarkable discoveries. Newton used his gravitational law to prove the correlation between the seemingly disparate motions of the moon's orbit and an apple's drop to the ground. He then calculated tidal changes for an ocean he had never even set his eyes on, proving that the Sun, Earth and Moon's relative positions are the deciding factor in tidal fluctuations. Newton was not done yet. He went on to calculate the earth's axial precession, correcting errors made in Hipparchus's estimation and getting a result almost consistent with the actual number.

---

[1]The gravitational equation that we are familiar with today—$F = GMm/r^2$—was not included in *Principia*. This format was put together by Cambridge's Professor Henry Cavendish (1731–1810) in the 18th century.

Even though the relational equation Galilei put forward following his inclined plane experiment portrayed the laws governing how objects move in vivid clarity, it was still severely lacking. Its weaknesses were summarily remedied by Newton's law of gravity. In fact, Newton's law remedied all its issues in one fell swoop, defining the relationship between terrestrial motion and force with a definitive equation. Even planets' and comets' motions fully conformed with his functional law. The emergence of his gravitational law meant that disputes over astronomy that had been raging for over a millennium were promptly flung out from the halls of academia, replaced with a mathematical equation based on the mass of and distance between two objects. Countless scientists were astounded by the brevity and potential of the equation, for which they used the word "universal" to describe—quite fitting for the time. This miraculous theory perfectly integrated dynamics in the cosmos and dynamics on Earth: the reason an apple falls to the ground is exactly the same as the reason the Moon travels placidly along its orbit—they are both results of universal gravity.

Newton's publication of Principia was followed by a succession of wealth and acclaim, turning him smug and arrogant. He aimlessly loafed around in London's official circles and completely ignored the ample scholarly critique towards the law of universal gravitation. Fortunately for Newton, on this occasion, he had a higher power on his side, as it was as if the heavens had summoned a group of his supporters to respond to the doubters. There were scholars in mechanics like Huygens who called the law of gravity into question who asked why, since according to the law of universal gravitation there is a force of attraction between two objects with mass, universal bodies don't attract one another, gradually getting closer until they crash together. If this were the case, would the whole universe not shrink progressively and collapse eventually? They argued that this eventuality is evidently not happening, necessarily rendering Newton's theory invalid. Newton's followers came up with a response to this: as an omnipotent God has placed the planets in positions with perfectly balanced force, the gravitational force coming from all directions is perfectly offset, allowing cosmic bodies to remain steady, not budging from their original positions. From Newton's supporters' perspective, God would indeed have the opportunity to place all the celestial bodies in the universe in some sort of perfect equilibrium. This rebuttal was considered reasonable, and so this time God had rescued Newtonian mechanics.

However, there were still scholars of optics like Hook who doubted Newton's work. Since there are an infinite number of celestial bodies emitting light, they argued, then according to Newton's claim that light moves across the universe at a boundless speed, light from far and near should make its way to earth instantly. If this is the case, the Earth ought to be bathed in light at all times, regardless of time or region. But this is not so, because the Sun oversees the day and the Moon the night, while half the Earth is pitch black when facing away from the Sun during a new Moon. How could this possibly be explained?

It might have been that God's hands were tied when dealing with this challenge because the static universe and infinitely fast light that Newton championed did in fact lead to such a contradiction. Newton's mistake seemed to be that he had included these two hypotheses in the same work as the law of universal gravitation. But in reality, he did not do as his doubters had suspected, as *Principia* did not actually include such a claim. Newton thus saw their criticism as malicious attacks from nitpicking opposition. He once again dealt with the critics by paying them no attention.

However, the most severe and irrefutable criticism of Newton's work was yet to appear—that came from Germany's Gottfried Leibniz (1646–1716). Leibniz claimed that if a gravitational pull between two cosmic bodies exists, this cannot simply be proved with an equation, but requires a proper explanation regarding the origin of gravity itself. In other words, Newton ought to clearly explain why there is a gravitational pull between objects in space and what causes this force to exist. But as previously mentioned, Newton was too engaged in gallivanting around London's official circles high on his success to pay attention to criticism from those he saw as incapable mathematicians.[2]

Despite Newton's aloof disregard of Leibniz's questions, they kept doggedly chipping away at the physics world, making their presence felt in this field both strong and transparent like tempered glass, and built on the very concept of gravitation. While this chipping away could not leave a mark on the preeminent law of universal gravitation, it was still able to create a faint crack within the glass.

Sure enough, more than 200 years later, the cracks became ever-growing fissures as a result of Einstein's fierce pursuit of a satisfactory answer to Leibniz's question. In the end, the glass pane of Newton's theory was smashed to pieces with just a few corner shards retaining a semblance of its original form. During his dismantling of Newtonian mechanics, the reserved Einstein did not celebrate with much gusto; instead he chose to quietly conduct a series of thought experiments in his mind. After a gradual and understated process of solving physics' longest-running cold case, Einstein put forward his era-defining discoveries: the theories of special and general relativity. This introduced a new age for physics beginning in 1905.

---

[2] Newton had previously had a spat with Leibniz over who was first to invent infinitesimal calculus. For more details see Chap. 4, Sect. 4.10.

7.4 Perceiving the Universe Through Spacetime Relativity: A Great Correction in Our Understanding of Physics

Scientific historians dubbed 1905 as Einstein's annus mirabilis or miracle year. This was the year the great scientist published five papers, each focused on a different field. Every one of these papers introduced a new theory to its field of study, making notable contributions to the advancement of physics in the 20th century. The first discussed how to determine the molecule size. Aside from proving the existence of atoms and molecules—a highly controversial concept at the time—it was also a launchpad for the field of colloid chemistry. The second explained Brownian motion using the random movement of pollen suspended in water to create the mass diffusion equation. Jean-Baptiste Perrin (1870–1942) later conducted an experiment verifying Einstein's equation based on pollen movement, earning him the 1926 Nobel Prize in Physics.

The third and fourth papers were to do with the derivation of the special theory of relativity. Einstein put forward a brand new definition of time and space by discovering that electrons moving in an electric field display different equivalent masses when moving along the direction of the field or at a right angle to it. While deriving the mathematical expression for this, Einstein's careful calculations unintentionally led him to produce the famous equation $E = mc^2$. This lifted the lid on the enormity of energy buried deep within objects' mass.

The last of these papers, considered the most important by Einstein himself, explored the interaction between light and electrons. He discovered that light can be used to push electrons off their tracks, irrespective of amplitude, at certain light frequencies. This paper allowed Einstein to create new possibilities for 20th-century physics, as quantum mechanics was born. His work on this brought him the 1921 Nobel Prize in Physics.

After publishing these five seminal papers, Einstein was promoted from assistant examiner—level three to level two at the Swiss Patent Office where he was employed. But his achievement only served to push him toward an issue that had been lingering in his mind for so long, the very same one that Newton had scornfully avoided when it was first proposed by Leibniz: why a gravitational force exists between two objects. Why would Einstein decide to set about solving a problem that had troubled physicists for two centuries at a time like this? Precisely because he could use his principle of equivalence, which is only applicable to calculations within an inertial frame in special relativity, outside of the bounds of inertial frames. This allowed Einstein to deal with warped spacetime under gravity's force.

Following nearly ten years of scientific inquiry, he successfully applied the Ricci tensor, the Levi-Civita connection and the Christoffel symbols to the four-dimensional Riemannian manifold. What resulted was the derivation of a field equation consisting of 16 metric equations: $R_{\alpha\beta} - Rg_{\alpha\beta}/2 = \kappa T_{\alpha\beta}$. Einstein derived individual spacetime metrics from these simultaneous equations before solving each one's metric curvature. He then used this curvature in calculating the gravitational strength and acceleration at each point, allowing their velocity and then trajectory to be obtained. What's especially worth noting is the κ symbol that appears on the right side of the field equation. This can be written in full as $\kappa = 8\pi G/c^4$ with the G it contains representing the gravitational constant. It was because of this G that Newton's gravitational equation could become a prime example of an Einstein field equation. In this framework, Newtonian mechanics were still exceptionally useful when looking at low-velocity, low-gravity environments like Earth or other solar system planets. Nonetheless, once the basis of Newtonian mechanics— absolute time—had to contend with high-velocity movement or gravitational fields it gradually lost its efficacy.

7.5   Newton and Einstein: Divergent Streams of Thought Converge at the Same Destination

Despite our recognition that Newtonian mechanics share characteristics with the general theory of relativity, these two doctrines have various fundamental differences. Primarily, the universe constructed in Newton's image is static with neither time nor space being relative. Newton believed his static universe to be eternal, without a beginning or end. With absolute time as a point of departure, Newton believed that light extends to all corners of the universe at limitless speeds. Thus all matter in the universe exists concurrently while the speed at which time passes is always consistent.

But the concept of absolute time to which Newton held firm was not acknowledged by Einstein. This is because in a universe with absolute time all space, everywhere exists simultaneously—a concept really difficult to understand for us Earth-dwelling humans. Think about it, if you had to travel from Earth to another planet, your journey would take some time—perhaps a few months or even a few decades. Once you have arrived at your destination absolutely nothing will be the same. This is why Einstein's view of time and space is relative. Scales of time and space in different frames of reference differ according to the degree of movement within each frame, unlike in Newtonian mechanics in which the scale and size of time and space are not connected to relative movement.

The equations for general relativity and special relativity's mass-energy equivalence have a lot in common with Newton's gravitational equation. They are all concise and elegant to the point that they are deeply treasured, their simple clarity making them difficult to refute. These three laws, considered the most significant in the history of physics, were almost completely free of personal subjectivity or coercion from religious authorities. They did not need the intervention of thought-up mystical powers to obtain a foolproof explanation, nor did they need to lean on political powers or moralistic ideals to plug holes in their theories. Following on from these three laws, scientists' common goal became the quest for precise mathematics to explain physical phenomena within the natural world. This approach no longer allowed for uncontrollable human impulses or inviolable spiritual taboos to enter the study of physics.

Equally, Newton and Einstein's successes allow us to understand some taboos within the scientific community. Firstly, those who irrationally reject simple yet powerful theories need to be ready to face complete failure at any time. Theorists like Hook, Descartes and Leibniz who tried to refute Newton's laws were all eventually sent back with their tails between their legs. Secondly, once a theory has been proven, regardless of how many hypotheses and simplifications come up during its derivation, it is never really seen as completely incorrect. Just take the law of universal gravitation: this theory went through rigorous examination before being proven accurate, but despite being later replaced by Einstein's general relativity scientists did not consider it invalidated. They instead saw Einstein's work as a revision of Newton's law. In other words, Newton's theory was not perfect, or perhaps we could say its accuracy was based on certain conditions that were eventually defined by Einstein's theory of general relativity in 1915.

With these three treasured, almost irrefutable equations, could it be possible for civilization to allow science to develop unimpeded by wavering human emotions or religious stricture? Looking at the past 3,000 years of scientific progress it is hard to be overly optimistic. This progression has seen religious power and human manipulation both play pivotal roles in scientific exploration. Man's manipulative effect has been witnessed in the back and forth animosity between different schools of thought, leading to scientific facts being obscured from the public eye. The plethora of incidents like this, their histories complex and tangled, will be looked at more deeply in Sect. 7.7.

The intertwining of religion and science is usually marked by simple measures and obvious intentions, their tracks still evident to us after 1,000 years of scientific evolution. This is due to the close association between religion and the evolution of human thinking, from which science can obviously not hide. Next, let us revisit religion's influence on science.

7.6   A Battle Between Rationality and Sensitivity: A Love-Hate Relationship Between Science and Religion

The sixth generation Babylonian king Hammurabi (circa 1810–1750 BC) was the founding father of embroiling science with religion. In 1772 BC this arrogant ruler overestimated his own abilities in drawing up a fanciful set of natural laws, hoping to use it to define natural phenomena. In spite of its absurdity, this made-up doctrine based on selfish desire still had a lasting effect on the future of natural philosophy. From then on, a number of natural philosophers described the reasons behind changes in matter in terms of righteousness and devilry, love and hate or rationality and emotion. With matter and man combined, all changes had both a cause and a conclusion, all of which were usually under the control of the gods. Eventually, the gods' control became so complete that in the Greek era if one were to admit to being an atheist, one could be violently suppressed and perhaps even face mortal danger. One example is the Pythagorean mathematician Hippasus (circa 500 BC), who was thought to have violated the simplicity of nature because he had proved that the Pythagorean theorem produced irrational numbers. Not long after his discovery he was murdered. Another, the Socratic period philosopher Anaxagoras (500–428 BC), was driven out of Athens for claiming that the sun is not a god, but just a massive fiery rock.

Plato's narrative of the natural world was also full of religious leanings. For instance, he would often portray the planets as gods, and even wrote in his dialogue *Laws* that anyone who refuses to believe in the gods' existence has interfered with the faith of others and should be imprisoned for five years. He further stated that if the offender does not change their ways, they should receive punishments as serious as a death sentence. But there were a few exceptions to this trend. The ancient Greek author Homer's epics the *Iliad* and the *Odyssey* both mention gods, but they are appreciated as works of literature rather than religious texts. This is quite different from the God of the Jewish Torah or Islamic Quran, who holds absolute authority and does not tolerate offense.

Thought and behavior inspired by mythology were diluted considerably during the reign of Alexander the Great, but once the Roman Empire had enveloped all of the Mediterranean coast belief transformed into a tool for political governance. In order to rule over a variety of vast territories and uneducated foreign barbarians, the Roman Empire gathered gods from all regions and allowed their subjects to worship freely. There was just one condition: they had to pledge loyalty to the emperor of Rome. The result of this was that the future emperor Julius Caesar (100–44 BC) was hailed as a living god demanding the reverence of all his subjects. Refusing to kneel before the Roman emperor in worship was the reason second century Christians were found guilty of disloyalty. They were punished by being burned on the cross, turned into lampposts—warning signs to any who dared to disobey. It was not until the year 313 that, after the image of Christ appeared to Constantine the Great (272–337) in a dream inspiring him to thoroughly defeat foreign invaders in battle, the emperor issued the Edict of Milan legalizing Christianity. Then, in 380, the emperor Theodosius Augustus took this a step further, enshrining Christianity as the one and only legal religion. From this point on education based on Christian supremacy began to replace the natural philosophy passed down from the Greek period.

In the following period of more than a thousand years, the Roman Empire combined politics and religion to a previously unseen degree. With the intention of solidifying Rome's reign over vast swathes of Europe, the church adopted the Aristotelian doctrine of geocentrism, expanding it to become an inviolable core feature of their biblical teaching. Consequently, the study of astronomy that had flourished in the Greek period sunk to the depths of spuriousness for over a millennium. The extent of this dogma was evident when in the 15th century the previously mentioned pioneers of the astronomical revolution would not dare to openly declare their findings in spite of having sound evidence to support them. They were able to disprove the church-sanctioned absurdity of geocentrism but avoided openly opposing the Pope, only reticently sharing what they knew to be the truth in private circles. When any published content was found to violate biblical precepts no-one was safe, even Galilei—a personal friend of the Pope—was put behind bars.

Although many scientists had previously been afraid to speak out, after the astronomical revolution, the influence of religion over scientific progress gradually transformed from a repressive combination of faith and politics to individual subjectivity based on personal beliefs. Take Newton for example: despite his doubt over the concept of the Holy Trinity, he believed deeply in God's creation of the universe and all things living within just as it was described in scripture. It could have been his pious belief in God that led him to lay down his laws of absolute time and space on the boundless universe. He may have been marrying his reverence of the almighty with his understanding of the cosmos. His belief meant that he agreed with the biblical conception of God as an eternal and omnipotent being not bound by time or space, who created space and the stars. We could say that Newton's highly subjective and reverential definition of God is a revised version of Hammurabi's natural law. In Einstein's hands, this revision was further developed and refined.

Despite Einstein's Judaism, he believed that there were two universal gods: one who controlled the natural order of the universe and another who controlled human action and societal development. The former was a just and compassionate god who could be easily understood, while the latter was hard to comprehend, often wrathful and stern in manner. Rabbi Herbert S. Goldstein once asked Einstein to describe his idea of God in 50 words. The physicist responded in fewer words, saying he believes in a God "who reveals himself in the harmony of all that exists, not in a God who concerns himself with the fate and the doings of mankind." Einstein emphasized that the result of scientific progress was to have people see that God's creation is harmonious—a simple piece of logic. Thus, he saw the world's harmony and order as a victorious refutation over unscrupulous people who believe that any and all occurrences are totally random.

Not long before Einstein's death, he praised God for putting the laws of relativity in place when creating the world and including the necessary relationship between mass and energy within. He said with a sentimentality that when it was his time to meet his maker, he most wished to ask what God was thinking when the universe was being created. To him, everything else was merely trivial. This reflected how Einstein defined the relationship between science and religion: science without religion is crippled, while religion without science is blind. In his eyes science and religion were complementary—one could not exist without the other.

It is for this very reason that, despite scientific discoveries often running counter to or contradicting myths and legends, modern scientists won't always maintain that supernatural phenomena are purely fictitious, neither will they necessarily believe that deities are fantasies conceived of in our minds. Thus a more direct explanation of scientific research would be: the examination of natural phenomena within the scope of human understanding, done using objective methods and free from the interference of supernatural powers. This definition is necessary because if supernatural elements were acknowledged from the start, any argument could just be explained away subjectively, meaning that any and all explanations would be unverifiable.

Science is, after all, a scholarly field that demands evidence for its claims. If a theory cannot be proven by experimentation or observation, it is effectively meaningless, and could even be proven as falsehood. The above definition of science seems exceedingly impartial and accurate, but humanity's path toward scientific progress does not always advance in so straight a direction. It is frequently mangled up with human manipulation borne of individuals' selfish desires. In the end, this leads to scientific accomplishments being mutated into evils that may destroy whole civilizations. Below we will touch on a few examples of this.

7.7 Human Manipulation: Struggles for Survival and Disputes Over Personal Gain

Over the past hundred years, scientific discoveries and technological inventions have developed at the same pace, allowing people to use their scientific knowledge in producing technology for their personal gain. There are countless examples of this, their incidence rapidly increasing, while they have even become a key factor in struggles for profit between governments and civil society. Science can be used as a tool to achieve victory on the battlefield or to seek profits in the marketplace. Below we have laid out four examples elucidating the complex relationship between science and its innovators.

(A) **Science Used to Mankind's Ruin**

It was the Jewish-born German chemist Fritz Haber (1868–1934) who discovered that nitrogen and hydrogen could be separated from air and then used to synthesize ammonia for use in fertilizers. This meant that humanity no longer needed to rely on fertilizers naturally containing ammonia, massively increasing the scale of food production and relieving more than half the world's population of the threat of starvation. This brought Haber the 1918 Nobel Prize in Chemistry, however during the First World War Haber also invented the processes for creating the toxic chemicals chlorine gas and mustard gas. Their use on the battlefields opened up an unprecedented age of chemical warfare, killing millions and causing many others to go blind or lose limbs. His discovery even led his wife to suicide in a show of despair and resent.

What is more, during the Second World War although the American-invented atomic bomb forced Japan to unconditionally surrender and thus ended a conflict that had already robbed more than 50 million of their lives. This invention then sent the world into a crazed race to develop nuclear arms. Maintaining a terrifying equilibrium of mutually assured destruction, the US and the Soviet Union entered the half-century long Cold War. Their build-up of weapons sounded the death knell for human civilization and started a count down to the end of days. After the Cold War, the US proposed initiatives for the peaceful use of nuclear capabilities in an attempt to deal with their enormous nuclear facilities. They started constructing large-scale nuclear power plants that up to this day have amassed mountains of nuclear waste. Throughout this process, they have been unable to develop an effective way to dispose of this radioactive byproduct, which could leave a fatal environmental legacy for future generations.

In 1923 with the aim of improving fuel economy and increasing engine horsepower, American inventor Thomas Midgley (1889–1944) added lead to petrol to create tetraethyl lead (TEL). His invention increased engine efficiency and lifespan, making it popular with oil companies and car manufacturers who profited substantially from it. The very next year in the refinery for who's company Midgley worked 15 employees died from lead poisoning and 35 were paralyzed for life by its contamination. Others suffering from lead poisoning showed clear signs of aggression and ill-temper, while their attention spans waned. Hysterical yelling could often be heard, and violent and destructive acts would occasionally be committed. At their time of death, the victims of lead poisoning often had temperatures of up to 43 °C.

In the 1960s, at a time when the toxicity of leaded petrol had not yet been proven, the atmosphere above the US contained 625 times the amount of lead it had prior to the adoption of Midgley's invention. This caused irredeemable damage to children's development in the US and the rest of the world. Midgley, ever the one to use scientific knowledge in his pursuit of personal gain, was also the inventor of the refrigerant freon (tetrafluorochlorine). While this product was more commonly used in refrigeration machinery than its predecessors and the toxic sulphur dioxide contained within seemed safe at the time, it was later discovered that this very chemical was the chief culprit responsible for creating a hole in the ozone layer larger than the African continent. Additionally, the greenhouse effect caused by sulphur dioxide is tens of thousands times more potent than that of carbon dioxide, making it a devastating contributor to climate change. The death and destruction caused by these two of Midgley's discoveries were unprecedented, peerless in their effect on the human race.

(B) **Suffering for Science**

While the damage caused by leaded petrol was discovered just a year after Midgley's discovery, its commercial value meant that the American government and local corporations made no attempt to suppress its production. It was, in fact, marketed quite forcefully around the world. It was not until the 1960s that its poisonous nature was uncovered by the California Institute of Technology professor of geology Clair Patterson (1922–1995). A well-regarded scientist, Patterson's research into trace quantities of lead in the atmosphere was world-renowned along with his work on determining the Earth's age using half-lives of lead isotopes. Patterson discovered from ocean sediment samples that both the atmosphere and the ocean's lead content had drastically increased from 1923 onward. This was exactly the year that Midgley's TEL began being used in vast quantities.

Patterson also found that the lead content of ocean surface water was more than 20 times higher than that of ocean floor waterproof that the increase in lead came from the atmosphere. He thus deduced that atmospheric lead came from car engines emitting leaded petrol. In 1965 he collated his findings into a 45-page report to inform the world of the dangers of leaded petrol. This put him in the combined cross-hairs of oil companies and car manufacturers, who not only terminated his industry-academia research funding but also compelled the government's National Research Council to no longer hire him as an academic reviewer. They pushed the United States Public Health Service to break ties with Patterson and lobbied the Environmental Protection Agency to publicly state that they rejected his data. And these were just some of their underhanded tactics. Eventually, they even tried to compel his school Caltech into letting him go. This brazen revival of Galilei's persecution was fortunately blocked by the university president's fervent support for the professor.

After another 30 years of struggles, Patterson finally succeeded in having Congress pass a ban on leaded petrol on January 1, 1996. From then on the amount of lead allowed in gasoline has been kept within a safe threshold. A moral conscience based in science had once again defeated corporations with a vested interest. This time the vested interest was not the Roman Catholic Church, but the colossal government-industrial complex. Unfortunately, Patterson never got to see the ban pass, as just days before, on 5 December 1995 he departed from this world.

A case of a special interest group suppressing scientific discovery reared its head once more in 2002. A Nigerian immigrant to the US, doctor Bennet Omalu (1968-), when conducting an autopsy on an irregular case found that some American football players had damage to their brains caused by head trauma. These players displayed symptoms generally only found in the elderly like dementia, hallucinations, bad tempers, depression and short attention spans as middle-aged men. After studying the brains of his autopsy subjects, Omalu was certain that these early-onset neurological disorders were connected to head trauma in sport, as when the brain suffers physical trauma protein cells are created, blocking blood vessels and effecting neurological functioning.

Omalu published his findings along with two other researchers in 2005, only to be intimidated by powerful senior executives from the National Football League (NFL). As a result, the Nigerian doctor lost his job, his wife miscarried under the stress of NFL threats and his residency in the US was nearly revoked. Nevertheless, following the release of his research, ever-increasing numbers of American football players realized they were suffering from the very symptoms Omalu had described.

Finally, in 2011 around 5000 former NFL players jointly sued the NFL for misleading them about the permanent damage head trauma causes the brain, including serious cognitive impairment and Chronic Traumatic Encephalopathy (CTE), just as Omalu had reported. After two years of legal proceedings, the NFL finally buckled under their pressure and officially changed their rules in 2013. Considerations regarding the life-long injury head trauma can potentially cause the brain were included within, as players were allowed to receive medical assistance immediately after being struck to the head.

The turbulent time between the discovery of these types of wrongdoings and them being redressed had now evolved from the 30 years it took in the leaded petrol case to eight years for head trauma in the NFL. This indeed shows a considerable improvement and that the belief society places in scientific research has gradually begun to exceed the control of special interest groups. At this moment of public faith in science another question has emerged: What is science? In order to find potential answers to this question, we need to once again look back on the history of scientific advancement.

7.8   What Is Science? the Question with an Ever-Changing Answer

Before any historical records were ever made, the objects of scientific inquiry could be found all around us: fire, rain, wind, thunder, lightning, stars, light, tides, earthquakes, droughts and floods along with anything else we could observe. Ancient humans believed that these phenomena were purely natural and even had specific inherent purposes: fire is hot and can be used for cooking or destruction, wind is cold and can be used to cool things down or set machinery in motion, rain is there to nourish the crops and give the people a livelihood, the rise and fall of tides aid ships in anchoring and setting sail, and so on. This too was the objective of the natural philosophy that came out of ancient Greece—to first describe nature, then explain it from a humanistic perspective.

Looking at it from another angle, we can see scientific development as a natural part of the human thinking process. From this process of using systematic observations to come up with practical applications, we develop a desire to further understand how natural mechanisms work. But desire can only be realized once a methodology is put in place. So, what is this methodology?

From our overview of the last thousand years of scientific advancement, we can see that the sprouting of a new development will always start with asking the right question. Ancient and modern people's thought processes are vastly different. From our point of deeper understanding in this modern age, ancient humans' questions are completely off base. For example, when they saw lightning they would ask which god had been angered; today we ask where the positive and negative charges that produced this reaction came from. When they saw the tides changing they asked if the earth were quaking in a manner that caused the ocean to rise and fall; today we ask about the relative positions of the sun and the moon. Regardless of how we put forward our questions, our hope is always to understand the workings of nature. Having amassed knowledge of how science has evolved throughout history, we are able to learn how best to ask questions. By defining inquiry and exploring the answers that come up, we come to understand the laws of nature.

Let us take humanity's earliest science—the development of astronomy—as an example. Plenty of stargazers have put forward the same questions: how is the universe structured? What is the mechanism making it work? Guided by these questions, people of the Bronze age living as far back as 1000 BC made discoveries that would leave us in the modern age in awe of their achievements. In spite of this, the millennium following the ancient Greek's age of discovery saw science, with all its far-reaching allure, fall stagnant at the hands of the Roman Catholic Church's crackdown. While fragments of new astronomic data emerged during this period of censorship, the accepted model of the universe was still stuck in ancient Greece with Aristotle's geocentric system. Only once the astronomical revolution of the 16th century began did the scientific world begin to feed the wave on which contemporary science would ride.

While the scientific revolution took place in Europe, it benefited from knowledge and research from other parts of the world: Egyptian geometry, Babylonian astronomy, Arab and Indian arithmetic, as well as Chinese printing techniques and compasses. But the research principles adopted at the time were independently developed within Europe, with their most evident contribution to science being mathematics. With the advent of mathematics, the model for scientific development began moving towards its completion. The final product was the four-step process scientists have come to be familiar with: first, make a rigorous definition of a natural phenomenon; then try to understand what the principles and mechanisms driving this phenomenon are; and then, most importantly, build a theoretical model based in mathematics; and finally an experiment must be designed to prove the validity of this model.

This complete scientific model has allowed modern scientists to find the operational order that ought to exist behind natural phenomena with every new discovery they make, with the method for expressing this order being mathematics. Contemporary scientists understand that the natural world is filled with simple beauty, but it is often concealed within everyday scenes. We must thus rely on applicable mathematical functions to uncover the deep mysteries that lie within.

The integration of mathematics and experimentation is truly the best way to decode the universe's riddles and understand the workings of nature. That is why research heavily reliant on mathematics like Newtonian mechanics, Maxwell's electromagnetic radiation and Einstein's theory of relativity achieved such significant results. They all used concise mathematical symbols to create meticulous rules for the laws of nature. Guided by these rules engineers are able to build bridges, buildings and pave roads, as well as cars, planes and ocean liners; while physicists are able to predict when comets will appear and correct GPS errors. The outstanding and the ordinary of these equations contributed equally to the spread of scientific truth to all corners of the earth.

It was because of the inclusion of mathematics that the astronomical revolution became a turning point for scientific progress. Prior to the astronomical revolution explanations of natural principles were often interspersed with religious or cultural views. These subjective perceptions dictated what the theories included and how they were likely to develop. After the astronomical revolution mathematics became the primary tool for interpreting natural phenomena. Its objective laws allowed science to elegantly explain how the universe operates in accordance with God's design, before confidently taking up a position of importance ahead of both religion and the humanities.

Despite the impetus gained from a system of mathematics envisioned by man, the fruits of all scientific research are regarded as "discoveries" rather than "inventions," because natural phenomena already exist; they are not human creations. Since science is based on discovery, can it be considered an art form? The answer is perhaps not. As science is based on evidence and modeling, its exploratory process is a sequence set in stone, while its thought process is pure, cold logic. For the most part, these processes have nothing to do with culture or emotion and are even less suited to unbridled imagination that could muddy scientific facts. Inspiration might still be beneficial in galvanizing scientific theories, while aesthetics might improve the portrayal of research outcomes, and music could bring out the rhythms within the laws of science. But, precise theories will always need to be proven by careful experimentation, and not by emotionally sublimated art.

Recognizing that aesthetics and music can be eliminated as the keys to the founding of scientific theories, we can propose the following conclusion: science has never been a means to a specific end, neither could it be a form of fatalism. Science is an exploratory process filled with endless possibilities. It is evident that the mass of scientific developments made to this day are very seldom the result of scientists' original, predetermined intentions. Even if someone of miraculous talent were able to predict a certain discovery, there is no way they could determine what kind of principle would be ascertained from it or the timeline of its development. Therefore scientific advancement is without purpose or consciousness; instead it is a system of facts based on various principles, devoid of human or religious sentiment, applied broadly to the natural world.

Interestingly science is not just used in the natural world, the ingenious human race has put its discoveries to some very particular uses elsewhere. For example, nowadays countries the world overemphasize the importance of technological advancement and have thus positioned science as a tool for enhancing their economies. But this is just a transitional phenomenon, not the end goal of scientific advancement.

On the other hand, science does not bend to religious beliefs. In fact, there are multiple scientific discoveries that are in clear contradiction with religious teachings from various faiths. A considerable number of scientists agree that if our world was indeed created by a deity, scientific discoveries would prove this god's existence, as the omnipotent creator's fingerprint would be left on the natural laws that dictate the universe. However, once this godly fingerprint has passed through the hands of the human race, it will gradually become the fingerprint of humanity, going on to frequently fill our scientific world with selfishness and distraction, and slowly stripping it of the harmonious simplicity it ought to permeate. With these considerations in mind, how is the world likely to develop in the future? Next, let us once again ponder upon the answer to this question through the lens of historical progress.

7.9   A Brave New World: Facing the Anxieties of the Future

The progression of physics to its current state has given us theories of gravity and general relativity capable of explaining the movements of celestial bodies, as well as quantum mechanics capable of determining functional relationships between the particles that constitute matter. Some physicists have boldly prophesied that when we have a computer powerful enough to calculate all the movements and functions of all particles in the human body at an adequate speed, scientists will be able to attain values for our thoughts, emotions, hopes and desires. When the time comes, computers will be able to gauge the inner thoughts and outward behaviors of every person on earth, as well as predict every single event that will befall us in the future. But we can almost be certain that those hearing these kinds of prophecies will feel a sense of deep reproach towards the human race. They will likely tell themselves that if science progresses to this point, it would be more meaningful to hark back to nature and go fishing on the ocean shores, go hunting in the wilderness or foraging in the woods than engage with technology.

Be that as it may, scientists with a powerful thirst for knowledge would pay no heed to buffoons who choose fishing, hunting and gathering over pondering the future of humanity. Scientists are instead more concerned with how to construct a robot army on par with those in the Star Wars movies. Their war machines, however, would not be as carelessly thrown together as the loophole-riddled ones from the silver screen. Theirs would have heads equipped with computers capable of defeating chess grandmasters, arms with high-efficiency mechanisms capable of moving heavy loads in no time at all, legs fitted onto magnetic levitation platforms capable of leaping and gliding in all directions, and joints controlled by hundreds of variable speed brushless motors, with their power needs to be met by a nuclear fusion reactor. Of course, they would be impenetrably strong and virus-resistant, while their nimble reactions would allow them to evade any enemy firepower and their sharp instincts could swiftly deal with unpredictable situations. In other words, the world envisioned above would be one dominated by all-powerful robots. In the eyes of these intelligent robots, humans would likely be seen as nothing more than pieces of meat loaded with both emotion and reason.

After reading the satirical fantasy depicted above, you might feel the need to outright reject this kind of possibility. But when we look back on the development of the nuclear bomb throughout the 20th century, we find a situation not only just as ludicrous, but equally terrifying. The nuclear tragedy happened just this last century and played out on this very planet for nearly 50 years. Top scientists the world over combined their utmost efforts in designing a bomb tens of thousands of times more powerful than any other, and then had it hurled at Japan's malicious military powers, ending the Second World War that had already robbed 55 million of their lives. Yet while this bomb exploded, the death knell for humanity sounded and the official countdown to the end of civilization commenced. Currently, the global club of countries with nuclear weapons has enough combined firepower to eradicate all humanity tens of times over. Their precise, rapid warheads could enshroud half the world's population in a cloud of atomic tangerine radiation in just 24 h. With the history of nuclear weaponry in mind, could you really say that the absurd ramblings above won't become a reality? Perhaps all we ought to do is wait and see.

Finally, we must reemphasize that science is the establishment of a set of mathematical models made to simulate the principles of nature, followed by the design of a series of experiments to prove the validity of these models, with the ultimate goal of understanding natural phenomena and uncovering the truth. This book has collectively chronicled the stories, discoveries, exploration of cases and proposals that make up the course of scientific progress. It seems like it may even have formed into a history of science itself. While it may not be as rigorous as academic historical research, it does have the same aims: to look back on the scientific research and experiences of the past in an attempt to recognize the legitimacy of research ideas and the feasibility of research methods, and possibly inspire continued exploration of our natural world.

We believe that once you have read this book you will have discovered that humanity's development of science is the manifestation of research accumulated over multiple generations. It is the confluence of all mankind's intelligence. Yet it can also become an immensely dangerous tool, and so needs to be guided by high moral principles in order to be translated into constructive knowledge. World War II is an example of this still fresh in our minds. Its battlefields saw invaders' mechanized armies sweep across vast swathes of land and their submarines command the boundless seas. But, as World War II showed us, when the hearts of those in power become filled with hate, advanced technology becomes nothing but a malevolent tool with which to quench their maniacal ambitions. The result is evil and destruction. It has been this way since the dawn of time.

References[1]

1. D. Abulafia, *The Great Sea: The Human History of the Mediterranean* (Allen Lane, London, UK, 2011). This book serves as a reference for the social and economic background that fostered the early development of Greek natural philosophy illustrated in Chapter 1.
2. S. Chandrasekhar, *Newton's Principia for the Common Reader* (Oxford University Press, Oxford, UK, 1997). This book tried to simplify the mathematics of Newton's *the Principia*, as explained in Chapter 4.
3. F. Chen, *When Atomic Energy Is Released: The Story of Manhattan Project* (CPRD Foundation, Taipei, 2018) (In Mandarin). The mechanism and calculation regarding the chain reaction and energy release of atomic bombs in Chapter 5 are referenced from the sample calculations of this book.
4. D. Cheung, *The Journey of Electricity: Exploring How Man Harnessed the Energy* (Commonwealth Publishing Co., Ltd, Taipei, 2011) (In Mandarin). Some of

---

[1]All the entries listed here are publications where reference materials have been obtained for the composition of this book, while the ways in which these materials are made use of can also be found after each entry. We only listed the more important references instead of adopting a strictly-detailed approach to the citation as in formal research papers because, for example, the stories of many scientists that we mentioned are so well-known already that it's both unnecessary and, to be fair, quite impossible to include all the sources. If not simplified, the total number of notes would've added up to thousands, which would largely reduce the density of the entire book and lead to a tedious reading experience. This said the referenced sources of the many theories, sample calculations, as well as tables and figures explained in this book are all clearly documented. For example, some of the figures that we demonstrated are retrieved from other publications before modified to serve our purposes. How such modification is carried out can always be found in the legends of figures of this sort.

F. Chen and F.-T. Hsu, *How Humankind Created Science*,
https://doi.org/10.1007/978-3-030-43135-8

the stories regarding Faraday and Maxwell in Chapter 5 are referenced from this book.

5. M. Chown, *The Ascent of Gravity: The Quest to Understand the Force that Explains* (Pegasus Books Ltd. New York, 2017). The information about tidal movements provided in Chapter 4 is in part referenced from this book.

6. C. Chu, Archimedes' methodology of volume calculation. Dissem. Math. Knowl. **37**(Q2), 73–79 (2013) (In Mandarin). The method that Archimedes used to calculate spherical volume introduced in Chapter 1 is in part referenced from this paper.

7. "The Code," BBC Motion Gallery, Cast: Professor Marcus du Sautoy. Produced & Directed by Stephen Cooter (2011), available on Netflix 2018. Appendix 2.1 of Chapter 2 is in part referenced from this video.

8. B. Cox, J. Forshaw, *Why Does $E = mc^2$? And Why Should We Care?* (Nurnberg Publishing Ltd. 2009). This book explains the development of Einstein's relativity theories in plain language, which inspired part of the derivation in Chapter 5.

9. J. Evans, *The History and Practice of Ancient Astronomy* (Oxford University Press, Oxford, UK, 1998). This book contains interesting and practical sample calculations and theories explained, which offer good references to Chapter 3.

10. R. Feynman, *Six Not-So-Easy Pieces: Einstein's Relativity, Symmetry, and Space-Time* (California Institute of Technology, 1997). The discussion about relativity theories in Chapter 6 is in part inspired by this book.

11. David, J. Goodstein, *Feynman's Lost Lecture: The Motion of the Planets Around the Sun* (Johnathan Cape, London, 1996). Some stories about the astronomical revolution of Chapter 3 and Appendix 4.7 of Chapter 4 are in part referenced from this book.

12. M. Guidry, *Modern General Relativity—Black Holes, Gravitational Waves, and Cosmology* (Cambridge University Press, Cambridge 2019). The derivation of Section 6.2 is in part referenced from this book.

13. A.J.S. Hamilton, *General Relativity, Black Holes, and Cosmology* (2018), https://jila.colorado.edu/~ajsh/astr3740_17/grbook.pdf. The derivation of Section 6.2 is in part referenced from this book.

14. W.Y. Hsiang, H.C. Chang, From Kepler to Newton: when the long-standing puzzle was solve with Universal Gravitation. Dissem. Math. Knowl. **32**(Q2), 3–12 (2008) (In Mandarin). This paper serves as one of the references for the derivation of Kepler's first and second laws of planetary motion of Chapter 3.

15. W. Y. Hsiang, H. C. Chang, H. Yao, *A Thousand-Year Myth: Geometry, Astronomy & Physics* (Taiwan Commercial Press, Taipei, 1999) (In Mandarin). This book is one of the publications that puts equal focus on scientific theories and historical stories in introducing the development of mathematics, geometry, and physics over the course of human civilization. Content regarding the exploration of ancient mathematics, the research process of scientists like Copernicus and Kepler, and the derivation of related theories illustrated in Chapter 3 are generally referenced from this book.

16. W.Y. Hsiang, H.C. Chang, H. Yao, P.J. Chen, A simple reconstruction of Kepler's Laws of planetary motion. Sci. Educ. Mon. (329), 8–18 (2010) (In Mandarin). This paper derives Kepler's first and second laws of planetary motion from a modern perspective, which can help readers familiarize themselves with the scientist's reasoning in no more than 20 pages. The derivation of the two laws in Chapter 3 is in part referenced from this paper.

17. C.E. Hummel, *The Galileo Connection: Resolving Conflicts between Science & the Bible* (InterVarsity Press, USA, 1986). Some of the events that happened to Copernicus, Kepler, and Galileo recounted in Chapter 3 are retrieved from this book.

18. T. Jacobsen, *Planetary Systems from the Ancient Greeks to Kepler* (University of Washington, Seattle, 1999). The derivation of Kepler's laws of planetary motion in Chapter 3 is in part referenced from this book.

19. J. Kozhamthadam, *Discovery of Kepler's Law: The Interaction of Science Philosophy and Religion* (University of Notre Dame, 1995). The derivation of Kepler's laws of planetary motion in Chapter 3 is in part referenced from this book.

20. C.S. Lee, *The Universe That Rings: On Einstein's General Relativity and Gravitational Waves* (National Taiwan University Press, Taipei, 2017) (In Mandarin). The reasoning and deduction regarding the theory of general relativity in Chapter 6 are in part inspired by this book.

21. J-P. Maury, *Newton et la mécanique céleste*, (Gallimard Press, Paris, 1990). This book serves as a reference for some of Newton's life events described in Chapter 4.

22. J.C. Maxwell, *A Treatise on Electricity and Magnetism*, vol. I (Clarendon Press, Oxford, UK, 2002). Originally Maxwell's doctoral dissertation, the treatise was published by Oxford University Press as a book in modern English. The mathematical derivations of Appendix 4.6 and Maxwell's equations in Chapter 5 are referenced from this book.

23. L. Mlodinow, *The Upright Thinkers: The Human Journey from Living in Trees to Understanding the Cosmos* (Pantheon, New York, 2015). This book is one of the not-so-many works that managed to provide a comprehensive and easy-to-understand overview of the scientific development over human history, this publication greatly influenced the writing style of present authors. Many stories about the scientists in Chapters 3, 4 and 5 are referenced from this book.

24. D. Morin, *Introduction to Classical Mechanics: With Problems and Solutions* (Cambridge University Press, 2007). This book is home to the explanation of some mechanical theories and sample calculations of Chapters 3 and 4.

25. K. Moritani, *Alexander's Conquest & Myth* (Kodansha Ltd, Japan, 2016). Certain stories about the kingdom of Macedonia and Alexander the Great recounted in Chapter 1 are referenced from this book.

26. J. Neffe, *Einstein: Eine Biographie* (Rowohlt Taschenbuch Verlag GmbH, Hamburg, Germany, 2005). Some of Einstein's life stories recounted in Chapters 5 & 6 are referenced from this book.

27. C.A. Pickover, *The Physics Book: From the Big Bang to Quantum Resurrection, 250 Milestones in the History of Physics* (Sterling Publishing Co., New York, 2011). Certain information about the technological development in ancient times described in Chapter 1 is referenced from this book.

28. G. Polya, *Mathematics and Plausible Reasoning* (Princeton University Press, NJ, USA, 1954). The method that Archimedes used to calculate the volume of a sphere explained in Chapter 1 is in part referenced from this book.

29. B.C. Reed, *The History and Science of the Manhattan Project* (Springer, New York, 2007). This is a classic book providing a complete chronicle of the Manhattan Project. The sample calculations regarding nuclear design and details about how the bombs were produced in Chapter 5 are in part referenced from this book.

30. W.B. Somerville, The description of Foucault's Pendulum. R. Astr. Soc. **13**, 40–62 (1972). This paper explains the principle of Foucault's Pendulum, which helps derive the equations of Appendix 3.1 of the present book.

31. I. Stewart, *In Pursuit of the Unknown: 17 Equations That Changed the World* (Profile Books LTD, London, 2012). Certain information depicted in Chapter 5 about how Faraday and Maxwell developed their equations is referenced from this book.

32. S. Weinberg, *To Explain the World: The Discovery of Modern Science* (Harper & Perennial, New York, 2015). This book is truly a well-organized work, which possesses significant academic value and is an inspiration for how the content of this book is structured, serving as a reference for several of the sample calculations that we present to readers.

33. H. Yao, C.R. Huang, The primitive meanings of Kepler's Ellipse Law of planets. Sci. Educ. Mon. (256), 33–45 (2013) (In Mandarin). Part of the derivation process of Kepler's first law of planetary motion provided in Chapter 3 of this book is referenced from this paper.

34. H. Yao, J.H. Yu, Experimental method or theoretical framework: What was Galileo's focus? Sci. Educ. Mon. (345), 9–21 (2011) (In Mandarin). This paper serves as a reference for part of the discourse on Galileo's theories in Chapter 3 of the present book.

35. A.J. Zhao, Z.L. Jiao, A mechanical analysis of Tidal motion and similar phenomena exploration 2010 (In Mandarin), https://wenku.baidu.com/view/785645c708a1284ac8504384.html. The mathematical derivation of Appendix 4.2 is referenced from this paper.

Printed in the United States
By Bookmasters